Bioelectronics & Wearable Devices: Sensing, Signal Processing and Powering

Bioelectronics & Wearable Devices: Sensing, Signal Processing and Powering

Editors

Xiao Xiao
Gang Ge

MDPI

Basel • Beijing • Wuhan • Barcelona • Belgrade • Novi Sad • Cluj • Manchester

Editors
Xiao Xiao
Department of Electrical and Computer Engineering
National University of Singapore
Singapore

Gang Ge
Department of Electrical and Computer Engineering
National University of Singapore
Singapore

Editorial Office
MDPI
St. Alban-Anlage 66
4052 Basel, Switzerland

This is a reprint of articles from the Special Issue published online in the open access journal *Micromachines* (ISSN 2072-666X) (available at: www.mdpi.com/journal/micromachines/special_issues/Bioelectronics_Wearable_Devices_Sensing_Signal_Processing_and_Powering).

For citation purposes, cite each article independently as indicated on the article page online and as indicated below:

Lastname, A.A.; Lastname, B.B. Article Title. *Journal Name* **Year**, *Volume Number*, Page Range.

ISBN 978-3-7258-1100-7 (Hbk)
ISBN 978-3-7258-1099-4 (PDF)
doi.org/10.3390/books978-3-7258-1099-4

Contents

Preface . . . . **vii**

Jiayi Yang, Yuanyuan Chen, Shuoyan Liu, Chang Liu, Tian Ma and Zhenmin Luo et al.
Single-Line Multi-Channel Flexible Stress Sensor Arrays
Reprinted from: *Micromachines* **2023**, *14*, 1554, doi:10.3390/mi14081554 . . . . **1**

Chang Peng, Qianqian Cai, Mengyue Chen and Xiaoning Jiang
Recent Advances in Tracking Devices for Biomedical Ultrasound Imaging Applications
Reprinted from: *Micromachines* **2022**, *13*, 1855, doi:10.3390/mi13111855 . . . . **25**

Xiaoqiang Ji, Zhi Rao, Wei Zhang, Chang Liu, Zimo Wang and Shuo Zhang et al.
Airline Point-of-Care System on Seat Belt for Hybrid Physiological Signal Monitoring
Reprinted from: *Micromachines* **2022**, *13*, 1880, doi:10.3390/mi13111880 . . . . **58**

Tingting Yu, Yebo Tao, Yali Wu, Dongguang Zhang, Jiayi Yang and Gang Ge
Heterogeneous Multi-Material Flexible Piezoresistive Sensor with High Sensitivity and Wide Measurement Range
Reprinted from: *Micromachines* **2023**, *14*, 716, doi:10.3390/mi14040716 . . . . **71**

Yunfeng Hu, Bin Tang, Lexing Hu, Haibo Liang, Bin Li and Zhaohui Wu et al.
A 7.6-nW 1-kS/s 10-Bit SAR ADC for Biomedical Applications
Reprinted from: *Micromachines* **2022**, *13*, 2110, doi:10.3390/mi13122110 . . . . **83**

Guanglei Chu, Zengning Liu, Yanyan Zhang, Yemin Guo, Xia Sun and Ming Li
Simple Immunosensor Based on Carboxyl-Functionalized Multi-Walled Carbon Nanotubes @ Antimony-Doped Tin Oxide Composite Membrane for Aflatoxin B_1 Detection
Reprinted from: *Micromachines* **2023**, *14*, 996, doi:10.3390/mi14050996 . . . . **95**

Anirudh Bishoyi, Md. Anish Alam, Mohd. Rahil Hasan, Manika Khanuja, Roberto Pilloton and Jagriti Narang
Cyclic Voltammetric-Paper-Based Genosensor for Detection of the Target DNA of Zika Virus
Reprinted from: *Micromachines* **2022**, *13*, 2037, doi:10.3390/mi13122037 . . . . **106**

Gang Ma, Jie Zhang, Jing Liu, Lirong Wang and Yong Yu
A Multi-Parameter Fusion Method for Cuffless Continuous Blood Pressure Estimation Based on Electrocardiogram and Photoplethysmogram
Reprinted from: *Micromachines* **2023**, *14*, 804, doi:10.3390/mi14040804 . . . . **117**

Gang Ma, Yuhang Chen, Wenliang Zhu, Lesong Zheng, Hui Tang and Yong Yu et al.
Evaluating and Visualizing the Contribution of ECG Characteristic Waveforms for PPG-Based Blood Pressure Estimation
Reprinted from: *Micromachines* **2022**, *13*, 1438, doi:10.3390/mi13091438 . . . . **133**

Liangyu Zhang, Junxin Chen, Chenfei Ma, Xiufang Liu and Lisheng Xu
Performance Analysis of Electromyogram Signal Compression Sampling in a Wireless Body Area Network
Reprinted from: *Micromachines* **2022**, *13*, 1748, doi:10.3390/mi13101748 . . . . **145**

Yanfang Dong, Miao Zhang, Lishen Qiu, Lirong Wang and Yong Yu
An Arrhythmia Classification Model Based on Vision Transformer with Deformable Attention
Reprinted from: *Micromachines* **2023**, *14*, 1155, doi:10.3390/mi14061155 . . . . **158**

Vasileios Pentsos, Ourania Spantidi and Iraklis Anagnostopoulos
Dynamic Image Difficulty-Aware DNN Pruning
Reprinted from: *Micromachines* **2023**, *14*, 908, doi:10.3390/mi14050908 . . . . **170**

Hector Carreon-Ortiz, Fevrier Valdez, Patricia Melin and Oscar Castillo
Architecture Optimization of a Non-Linear Autoregressive Neural Networks for Mackey-Glass Time Series Prediction Using Discrete Mycorrhiza Optimization Algorithm
Reprinted from: *Micromachines* **2023**, *14*, 149, doi:10.3390/mi14010149 . . . . **179**

Junyi Liang, Jiakun Wang, Wenjuan Yao and Mianzhi Wang
Development of an Assessment Model for the Effect of the Replacement of Minimal Artificial Ossicles on Hearing in the Inner Ear
Reprinted from: *Micromachines* **2023**, *14*, 483, doi:10.3390/mi14020483 . . . . **203**

Tian Ma, Xinlei Zhou, Jiayi Yang, Boyang Meng, Jiali Qian and Jiehui Zhang et al.
Dental Lesion Segmentation Using an Improved ICNet Network with Attention
Reprinted from: *Micromachines* **2022**, *13*, 1920, doi:10.3390/mi13111920 . . . . **216**

Xin Li, Jinkang Liu, Yijing Huang, Donghao Wang and Yang Miao
Human Motion Pattern Recognition and Feature Extraction: An Approach Using Multi-Information Fusion
Reprinted from: *Micromachines* **2022**, *13*, 1205, doi:10.3390/mi13081205 . . . . **231**

Peng Su, Sikai Wang, Yuliang Lai, Qinran Zhang and Leiyu Zhang
Screw Analysis, Modeling and Experiment on the Mechanics of Tibia Orthopedic with the Ilizarov External Fixator
Reprinted from: *Micromachines* **2022**, *13*, 932, doi:10.3390/mi13060932 . . . . **250**

Leiyu Zhang, Zhenxing Jiao, Yandong He and Peng Su
Ergonomic Design and Performance Evaluation of H-Suit for Human Walking
Reprinted from: *Micromachines* **2022**, *13*, 825, doi:10.3390/mi13060825 . . . . **265**

Wenjing Xu, Meng Yang, Xuanlong Du, Hao Peng, Yue Yang and Jitao Wang et al.
Multifunctional Nanoplatform Based on Sunitinib for Synergistic Phototherapy and Molecular Targeted Therapy of Hepatocellular Carcinoma
Reprinted from: *Micromachines* **2023**, *14*, 613, doi:10.3390/mi14030613 . . . . **283**

Lulu Wang
Holographic Microwave Image Classification Using a Convolutional Neural Network
Reprinted from: *Micromachines* **2022**, *13*, 2049, doi:10.3390/mi13122049 . . . . **295**

Francesca Maria Orecchio, Vito Tommaso, Tommaso Santaniello, Sara Castiglioni, Federico Pezzotta and Andrea Monti et al.
A Novel Fluidic Platform for Semi-Automated Cell Culture into Multiwell-like Bioreactors
Reprinted from: *Micromachines* **2022**, *13*, 994, doi:10.3390/mi13070994 . . . . **314**

Haiyang Wu, Dan Dang, Xieliu Yang, Junhai Wang, Ruolong Qi and Wenguang Yang et al.
Accurate and Automatic Extraction of Cell Self-Rotation Speed in an ODEP Field Using an Area Change Algorithm
Reprinted from: *Micromachines* **2022**, *13*, 818, doi:10.3390/mi13060818 . . . . **325**

Preface

This Special Issue on "Bioelectronics and Wearable Devices: Sensing, Signal Processing, and Powering" explores an exciting and rapidly evolving field that stands at the convergence of engineering and medicine. The primary subjects of this publication are the development and application of bioelectronic devices and wearable technology in monitoring and assessing human health. The scope of this Special Issue extends from technical advancements in device engineering to innovative applications in real-world healthcare settings, aiming to push the boundaries of how these technologies can be used to enhance healthcare outcomes.

The motivation for compiling this Special Issue arises from the transformative potential of bioelectronics and wearables within the healthcare sector. As technology progresses, the integration of the Internet of Things (IoT) and advanced wireless communications like 5G is making it possible to not only collect but also share critical health data in real time. This capability is pivotal for developing personalized healthcare strategies that can respond dynamically to an individual's health needs.

This Special Issue is aimed at a diverse audience, including researchers, engineers, and clinicians, all of whom play a crucial role in advancing the field of bioelectronics. It is also intended for policymakers and health tech enthusiasts who are interested in the implications of these technologies for public health and regulatory frameworks.

The articles and reviews presented in this Special Issue have been contributed by a group of leading authors whose expertise spans a broad range of disciplines related to bioelectronics and wearable devices. Their contributions reflect the interdisciplinary nature of this field, incorporating perspectives from engineering, medicine, and data science.

We extend our sincere thanks to all the authors and peer reviewers whose diligent efforts have ensured the high quality of the published works. We are also grateful for the logistical and editorial support provided by our publishing team and the various institutions that have backed the research included in this Special Issue. Their collective assistance has been invaluable in bringing this project to fruition.

By highlighting recent developments and ongoing research in bioelectronics and wearable technologies, this Special Issue aims to foster a deeper understanding of their role in shaping the future of healthcare. We hope that it serves as a catalyst for further innovation and collaboration in this vital field.

Xiao Xiao and Gang Ge
Editors

micromachines

Review

Single-Line Multi-Channel Flexible Stress Sensor Arrays

Jiayi Yang [1,2,*,†], Yuanyuan Chen [1,†], Shuoyan Liu [3], Chang Liu [3], Tian Ma [1,2], Zhenmin Luo [2] and Gang Ge [3,*]

1 College of Computer Science and Technology, Xi'an University of Science and Technology, Xi'an 710054, China; chenyuanyuan202008@163.com (Y.C.)
2 College of Safety Science and Engineering, Xi'an University of Science and Technology, Xi'an 710054, China
3 Department of Materials Science and Engineering, National University of Singapore, Singapore 117583, Singapore
* Correspondence: jyang46@xust.edu.cn (J.Y.); ggeup@nus.edu.sg (G.G.)
† These authors contributed equally to this work.

Abstract: Flexible stress sensor arrays, comprising multiple flexible stress sensor units, enable accurate quantification and analysis of spatial stress distribution. Nevertheless, the current implementation of flexible stress sensor arrays faces the challenge of excessive signal wires, resulting in reduced deformability, stability, reliability, and increased costs. The primary obstacle lies in the electric amplitude modulation nature of the sensor unit's signal (e.g., resistance and capacitance), allowing only one signal per wire. To overcome this challenge, the single-line multi-channel signal (SLMC) measurement has been developed, enabling simultaneous detection of multiple sensor signals through one or two signal wires, which effectively reduces the number of signal wires, thereby enhancing stability, deformability, and reliability. This review offers a general knowledge of SLMC measurement beginning with flexible stress sensors and their piezoresistive, capacitive, piezoelectric, and triboelectric sensing mechanisms. A further discussion is given on different arraying methods and their corresponding advantages and disadvantages. Finally, this review categorizes existing SLMC measurement methods into RLC series resonant sensing, transmission line sensing, ionic conductor sensing, triboelectric sensing, piezoresistive sensing, and distributed fiber optic sensing based on their mechanisms, describes the mechanisms and characteristics of each method and summarizes the research status of SLMC measurement.

Keywords: pressure sensor array; tactile sensor array; resonators; array integration; soft sensor array; force sensor array; parallel signal processing

Citation: Yang, J.; Chen, Y.; Liu, S.; Liu, C.; Ma, T.; Luo, Z.; Ge, G. Single-Line Multi-Channel Flexible Stress Sensor Arrays. *Micromachines* **2023**, *14*, 1554. https://doi.org/10.3390/mi14081554

Academic Editor: Lin Zhang

Received: 19 June 2023
Revised: 1 August 2023
Accepted: 1 August 2023
Published: 3 August 2023

1. Introduction

Stress refers to the internal resistance or response that occurs within an object when it is subjected to external forces or loads. A stress sensor is an electronic device used to measure and detect mechanical strain or deformation in materials. These sensors convert the physical stress into measurable electrical signals [1–3]. Stress sensors find wide applications in engineering and materials science, where they are used to assess the integrity of structures, monitor deformations, and determine load distributions. Typically, these sensors utilize rigid materials such as metals and semiconductors. However, their limited ability to withstand significant deformations restricts their usefulness. With the advent of wearable electronic devices, flexible stress sensors have become more desirable due to their high sensitivity, good repeatability, and large stretchable range, making them ideal for wearable devices, human-computer interaction, intelligent robots, and health monitoring.

The research work of Professor Zhenan Bao and Professor John A. Rogers is of great significance in the field of flexible electronics, particularly in the development of high-performance flexible stress sensors [4–12]. Their research achievements have provided crucial technical support and guidance for stress sensors in flexible electronic applications.

For instance, in their review, professor Rogers et al. [13]. introduced unconventional methods for fabricating and patterning nanomaterials, which laid a new scientific foundation for the fabrication of small-sized and patterned structures. These methods have significant implications in specific domains and in combination with other fabrication techniques. On the other hand, professor Bao's research [14] has made significant progress in the scalable synthesis of multifunctional polyaniline hydrogels and their outstanding electrode performance, offering strong support for the preparation of flexible stress sensors. These studies provide important references for our understanding and application of flexible stress sensors.

Flexible stress sensor array, consisting of multiple flexible stress sensor units, can better quantify and analyze the magnitude and distribution of spatial stress, with advantages in flexibility, integration, and systematicity [15,16]. These sensors can be fabricated by conductive materials such as metal [17–19], carbon based [20–22], conducting polymer [23–25]. However, current implementations of flexible stress sensor arrays encounter the challenge of excessive signal wires, leading to reduced deformability, stability, and reliability, as well as increased costs [16,26,27]. While wireless signal transmission is commonly employed to reduce the number of signal wires between the sensor array and readout circuit, it does not address the issue within the flexible stress sensor array itself. The primary hindrance posed by the abundance of signal wires is the nature of the sensor unit's electric amplitude modulation signal (e.g., resistance and capacitance), which restricts the transmission of only one signal per wire.

To address this challenge, the single-line multi-channel (SLMC) signal measurement has been developed, enabling simultaneous detection of multiple sensor signals through one or two signal wires [15,16,28–40]. By utilizing SLMC technology, the flexible stress sensor array can measure the magnitude and location of the stress by one or two signal lines. This approach effectively reduces the number of signal wires, enhancing stability, deformability, and reliability, and facilitating the application of flexible stress sensor arrays in bionic robots and rehabilitation medicine. Existing research has successfully implemented various mechanisms including RLC series resonant sensing, transmission line sensing, ionic conductor sensing, triboelectric sensing, piezoresistive sensing, and distributed fiber optic sensing [15,16,28,31–34,36,41–46]. These advancements have demonstrated promising potential in the domains of bionic robots, medical rehabilitation, and other related fields, as depicted in Figure 1.

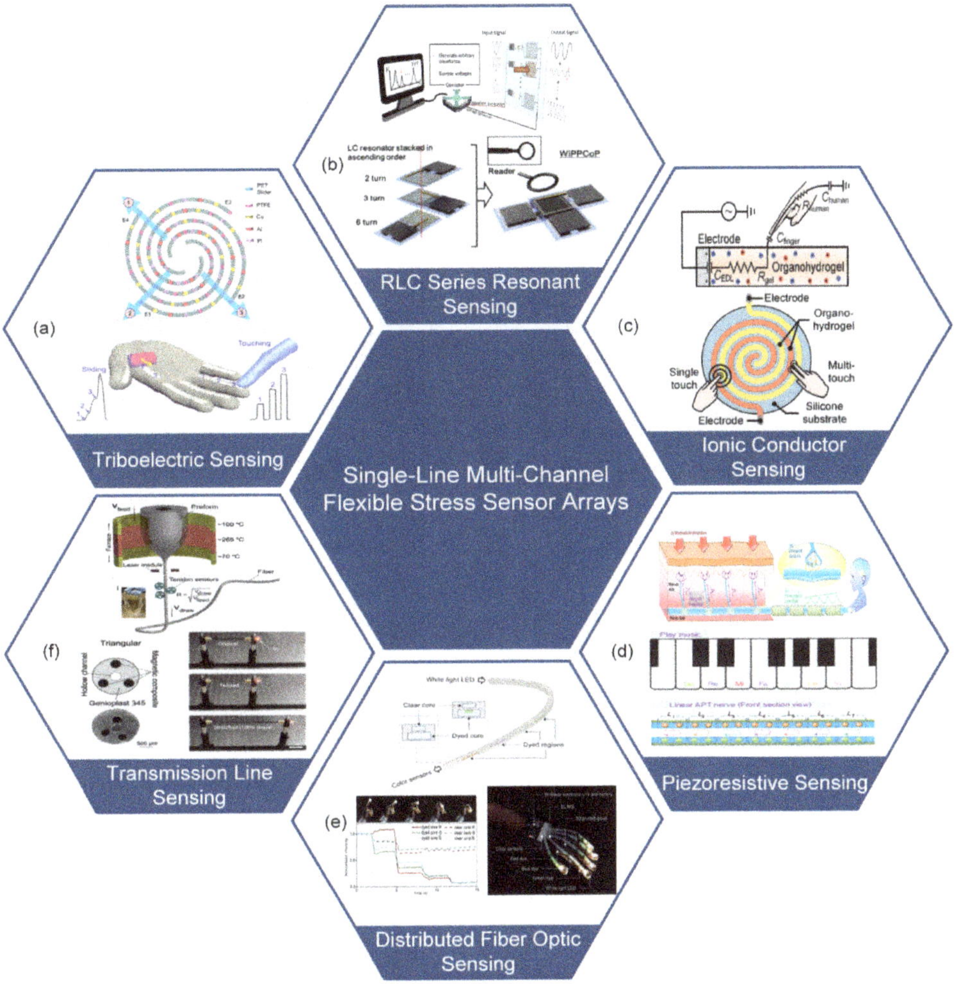

Figure 1. Implementation methods of single-line multi-channel signal measurement. (**a**) Triboelectric sensing (reprinted with permission from [42]; copyright 2022, Elsevier). (**b**) RLC series resonant sensing (reprinted with permission from [28]; copyright 2019, Wiley-VCH). (**c**) Ionic conductor sensing (reprinted with permission from [36]; copyright 2021, Elsevier). (**d**) Piezoresistive sensing (reprinted with permission from [41]; copyright 2020, Springer Nature). (**e**) Distributed fiber optic sensing (reprinted with permission from [34]; copyright 2020, AAAS). (**f**) Transmission line sensing (reprinted with permission from [47]; copyright 2023, Wiley-VCH (reprinted with permission from [48]; copyright 2022, AAAS).

2. Traditional Flexible Stress Sensor Array

Flexible stress sensors have significant potential in the fields of bionic robotics and rehabilitation medicine due to their outstanding flexibility and stretchability. Novel materials and structures are the most common approaches to enhance the performance of flexible stress sensors, which are determined by the mechanism [49]. Piezoresistive [50–52], capacitive [53–55], triboelectric [56,57], and piezoelectric [58] are the four mechanisms for flexible stress sensors, as shown in Figure 2.

2.1. Piezoresistive Flexible Stress Sensor Array Method

Piezoresistive flexible stress sensor consists of a piezoresistive material that converts mechanical stress into resistance when subjected to stress, resulting in a change in resistance by modifying the shape or relative position of the conductive filler, as shown in Figure 2a [50,59]. The equation for resistance can be expressed as:

$$R = \frac{\rho L}{wt} \tag{1}$$

where R is resistance, ρ is resistivity, L, ω and t are the length, width, and thickness of the resistor, respectively.

Several research works have been conducted to develop stress sensors using different materials and structures. Park et al. [60] developed a hyperelastic stress sensor, which is based on silicone rubber and eutectic gallium-indium (EGaIn). The sensor contains EGaIn in microchannels of the silicone rubber elastic body. The stress applied to the sensor causes deformation in the cross-sectional area of the EGaIn, which changes the resistance and enhances the sensitivity of the sensor. Figure 2a(i) is another example of EGaIn-based hyperelastic stress sensor [61]. As shown in Figure 2a(ii), Liao et al. [41] designed a sensor array that can measure stress position using conductive graphite film, paper, and an intermediate spacer. The sensor's output resistance depends on the stress position from the connecting electrode, providing high sensitivity and accurate measurement of stress position. As shown in Figure 2a(iii), Shi et al. [62] developed a transparent and flexible stress sensor that utilizes polydimethylsiloxane (PDMS) and urchin-like hollow carbon spheres (UHCS). Stress-induced changes in the relative position of UHCS in PDMS lead to changes in resistance with ultrahigh sensitivity and high transparency. As shown in Figure 2a(iv), Yun et al. [63] created a mixed material stress sensor that contains PDMS, EGaIn, and iron powder. The sensor's conductive fillers change their shape and position in response to stress, leading to changes in resistance. The conductive fillers return to their initial state when the stress is removed, and the initial resistance is restored. Figure 2a illustrates the different stress sensors developed in the works mentioned above.

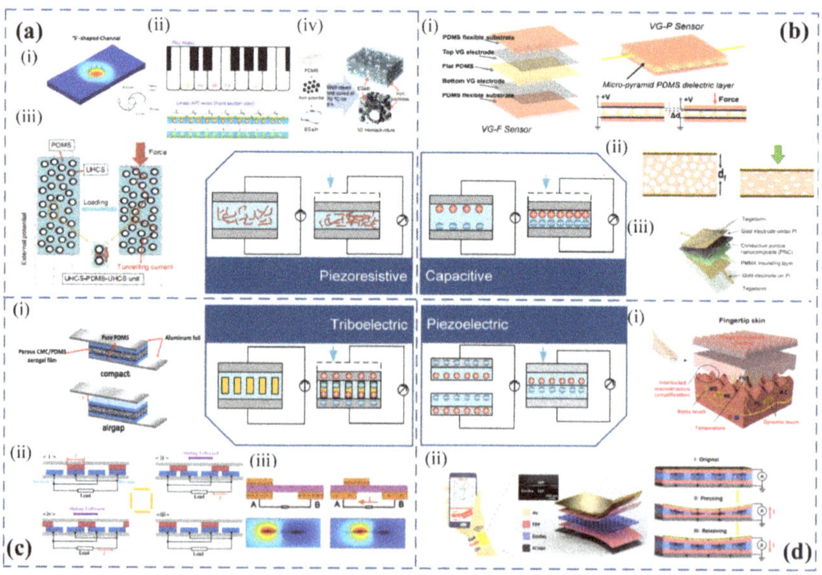

Figure 2. Sensing mechanisms and basic structures of flexible strain sensors. (**a**) Piezoresistive sensing mechanisms. (**i**) Cross-sectional area change in conductive filler eutectic gallium indium (reprinted

with permission from [61]; copyright 2016, American Chemical Society). (**ii**) Change in relative position of biomimetic sensory transmission nerve (reprinted with permission from [41]; copyright 2020, Springer Nature). (**iii**) Change in relative position of polydimethylsiloxane (PDMS) and sea urchin-shaped hollow carbon spheres (reprinted with permission from [64]; copyright 2020, Springer Nature). (**iv**) Change in shape of conductive filler (PDMS, eutectic gallium indium, Fe) (reprinted with permission from [41]; copyright 2019, Springer Nature). (**b**) Capacitive sensing mechanisms. (**i**) Variable distance type capacitive sensor (reprinted with permission from [64]; copyright 2023, MDPI). (**ii**) Variable dielectric type capacitive sensor (reprinted with from [64]; copyright 2020, Wiley-VCH). (**iii**) Hybrid capacitive sensor combining variable distance and variable dielectric types (reprinted with permission from [64]; copyright 2021, Wiley-VCH). (**c**) Triboelectric sensing mechanisms. (**i**) Contact-separation type (reprinted with permission from [64]; copyright 2017, Elsevier). (**ii**) Contact-sliding type (reprinted with permission from [64]; copyright 2014, Wiley-VCH). (**iii**) Independent type (reprinted with permission from [65]; copyright 2018, Wiley-VCH). (**d**) Piezoelectric sensing mechanisms. (**i**) Multifunctional electronic skin based on interlocking structure (reprinted with permission from [34]; copyright 2015, AAAS). (**ii**) Piezoelectric composite based on stacking structure (reprinted with permission from [64]; copyright 2020, Wiley-VCH).

There are three methods available for implementing piezoresistive flexible stress sensor arrays. These include unit-independent measurement [66], row-column multiplexing measurement [67], and anisotropic electrical impedance tomography [68,69], which are illustrated in Figure 3.

The unit-independent measurement involves using independent electrodes and signal wires for each sensing unit. While this method has high measurement accuracy, it can reduce reliability due to the large number of signal wires required. As shown in Figure 3a, Wu et al. [66] utilized this method to design a sensor array for real-time monitoring of pressure and temperature in smart insoles, using a pressure-sensitive layer based on a mixed carbon nanotube/polydimethylsiloxane (CNT/PDMS) material and a pair of independent electrodes in each sensing unit. The row–column multiplexing measurement method arranges upper and lower electrodes in a cross pattern to reduce the number of signal wires. However, this method may be susceptible to signal crosstalk between sensor units. As shown in Figure 3b, Yu et al. [67] designed a stretchable and flexible stress sensor array for electronic skin, consisting of 6×6 sensing units, using a three-layer structure in each sensing unit: top electrode, stress-sensitive medium layer, and bottom electrode. The top and bottom electrodes adopt a cross-arrangement structure. The anisotropic electrical impedance tomography calculates the multi-dimensional resistivity distribution inside a composite material to measure the stress magnitude and position, eliminating the need for a complicated flexible electrode array on the sensor surface. However, this method may also be susceptible to signal crosstalk between sensor units, and the calculation of composite material resistivity can be complex. As shown in Figure 3c, Lee et al. [68] designed a stretchable 3D stress sensor array for human–machine interfaces that used this method. Applying stress increases the resistivity, enabling measurement of stress magnitude and position. As shown in Figure 3d, Duan et al. [69] proposed using anisotropic electrical impedance tomography on piezoresistive conductive fabric to realize low-cost and large-area touch sensing for dynamic touch stress mapping of single or multiple points.

Piezoresistive flexible stress sensors are highly sensitive, offer a wide measurement range, are cost-effective, and have simple signal acquisition circuitry, making them a popular choice. These sensors can be integrated into arrays using techniques, including unit-independent measurement [66], row-column multiplexing measurement [67], and anisotropic electrical impedance tomography [68,69]. Despite their usefulness, these techniques have certain limitations, such as the requirement of numerous signal wires, intricate structures, and complicated acquisition circuits.

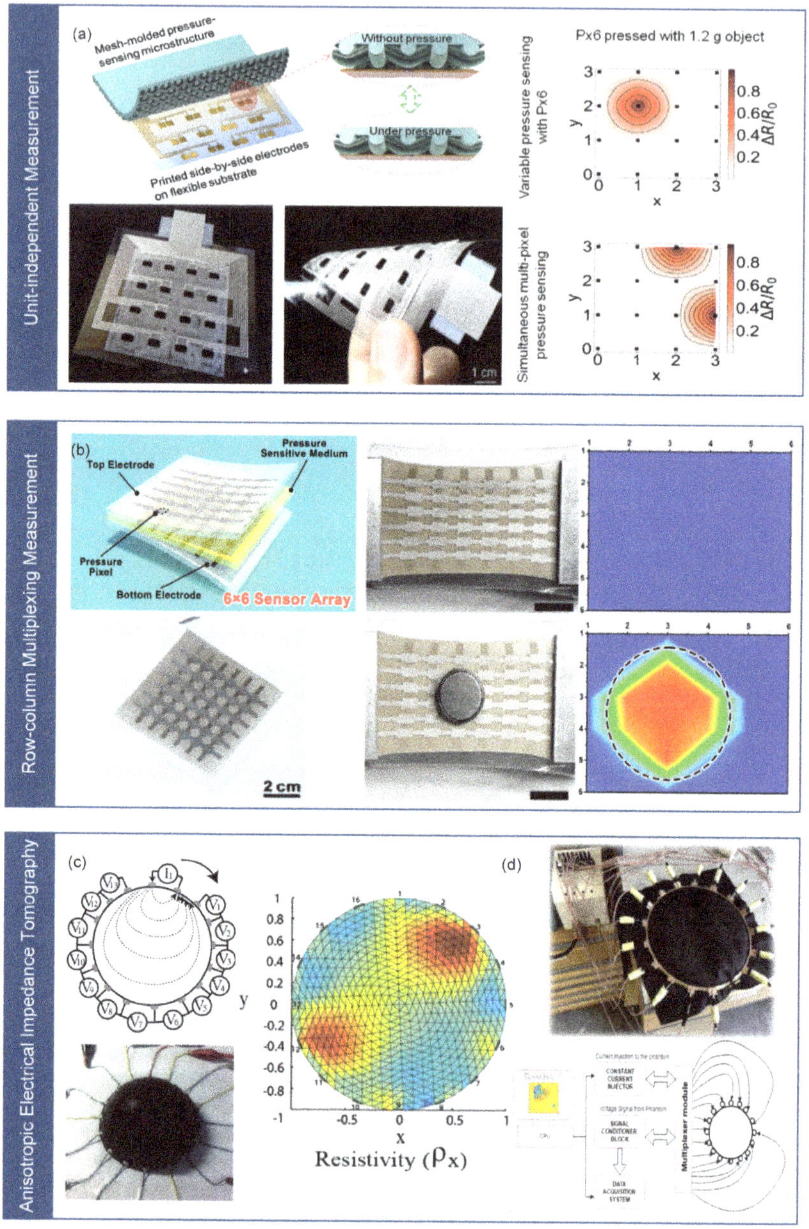

Figure 3. Array measurement methods for piezoresistive flexible stress sensors. (**a**) Unit-independent measurement (reprinted with permission from [70]; copyright 2020, Wiley-VCH). (**b**) Row-column multiplexed measurement (reprinted with permission from [70]; copyright 2020, Elsevier). (**c**) Anisotropic electrical impedance tomography (reprinted with permission from [70]; copyright 2017, Springer Nature). (**d**) Anisotropic electrical impedance tomography (reprinted with permission from [70]; copyright 2019, Springer Nature). Note: In (**c**,**d**), the same Anisotropic electrical impedance tomography is presented but from different references.

2.2. Capacitive Flexible Stress Sensor Array Method

The capacitive flexible stress sensor is composed of upper and lower electrodes separated by a compressible dielectric material [53,54,71]. Applying stress to the electrodes compresses the dielectric material, decreasing the distance between the two electrodes and converting stress into capacitance, as shown in Figure 2b [72]. The equation for the capacitive sensor can be written as:

$$C = \frac{\varepsilon_r \varepsilon_0 A}{d} \quad (2)$$

where C represents the capacitance, ε_r is the effective dielectric constant of the dielectric material, ε_0 is the vacuum dielectric constant, A is the overlap area of the electrodes, and d is the distance between the electrodes.

Applying stress changes the electrode overlap area (A) and spacing (d) of the capacitive sensor, resulting in a variation in capacitance [73,74]. Capacitive stress sensors are divided into two types: variable distance and variable dielectric sensors. The former involves one stationary electrode while the other moves relative to it. As shown in Figure 2b(i), Zhao et al. [75] proposed a micro-pyramid dielectric layer with graphene electrodes and PDMS as the dielectric layer, which increases sensitivity. In the variable dielectric type, a compressible dielectric material is placed on top of parallel finger-shaped electrodes, and stress changes the dielectric constant above the electrodes, resulting in a change in capacitance. As shown in Figure 2b(ii), Mahmoudinezhad et al. [76] designed a compressible stress sensor based on PDMS foam and finger-shaped electrodes. Our team proposed a flexible capacitive stress sensor based on liquid metal elastomer foam, which combines the variable distance and variable dielectric types. Liquid metal elastomer foam has a high dielectric constant that gradually increases during compression, providing high sensitivity and a wide measurement range. Recently, the Lu group proposed a hybrid response stress sensor that combines resistance and capacitance [77]. As shown in Figure 2b(iii), The sensor comprises porous nanocomposites (PNC) made of carbon nanotubes (CNT) and Ecoflex, placed between two parallel electrodes, and an ultra-thin insulating layer (500 nm) added between the PNC and one electrode. Applying stress transforms the PNC from a dielectric material to a conductive material, increasing the propagation distance of AC current within the PNC, and greatly reducing the effective distance between the two electrodes. This achieves high sensitivity and a large range for the capacitive stress sensor.

The measurement techniques for capacitive flexible stress sensor arrays are akin to those used for piezoresistive sensor arrays, and comprise unit-independent and row–column multiplexing measurements, as illustrated in Figure 4.

The unit-independent measurement involves independent top and bottom electrodes and signal wires for each sensor unit, providing high accuracy but reduced reliability due to the large number of signal wires. As shown in Figure 4a, Bae et al. [20] proposed a flexible sensor array with 4 × 4 units for real-time monitoring and discrimination of stress and temperature in electronic skin, where each sensing unit consists of a top part (Parylene C substrate, Ni/Ti electrodes, reduced graphene oxide thermistor, and dielectric layer) and a bottom part (Parylene C substrate and a microstructural CNT/PDMS). Stress changes the shape of the micro-protruding electrodes, reducing the electrode spacing and changing the capacitance. The proposed sensor array individually connects each sensing unit to two signal wires. In contrast, the row–column multiplexing measurement involves the cross-arrangement of the top and bottom electrodes, resulting in fewer signal wires but with the disadvantage of parasitic capacitance interference between sensing units [78]. Wang et al. [79] developed a flexible capacitive tactile sensor array for real-time contact force measurement in prosthetics, which integrates 8 × 8 sensing units, each consisting of four layers: copper electrodes arranged in a cross pattern, a PDMS film, and a PDMS bump layer for stress concentration. The sensor array is integrated into the prosthetic device, achieving real-time visualization of grip force. As shown in Figure 4b, Zhou et al.

also proposed a capacitive stress sensor array based on the row–column multiplexing measurement [80].

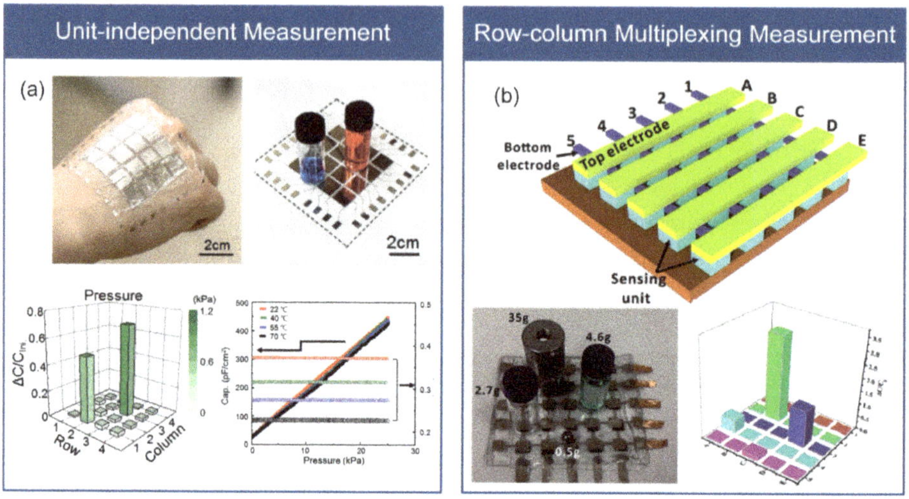

Figure 4. Array measurement methods of capacitive flexible stress sensors. (**a**) Unit-independent measurement (reprinted with permission from [70]; copyright 2018, Wiley-VCH). (**b**) Row–column multiplexed measurement (reprinted with permission from [80]; copyright 2019, The Royal Society of Chemistry).

Capacitive flexible stress sensors are advantageous due to their good temperature independence, low power consumption, high sensitivity, compact circuit layout, and simple device structure. Flexible capacitive stress sensor arrays can be fabricated using unit-independent measurement [20] and row–column multiplexing measurement [79]. Nevertheless, similar to the flexible piezoresistive stress sensor array, these methods have drawbacks such as crosstalk, multiple signal wires, and a complicated structure.

2.3. Triboelectric Flexible Stress Sensor Array Method

The triboelectric flexible stress sensor utilizes the coupling effect of frictional electrification and electrostatic induction to generate and convert mechanical energy into electrical energy for stress measurement [56]. As shown in Figure 2c, the sensor comprises two materials with different electron affinities. When the two different friction materials make contact, their varying electron affinity leads to a difference in electron attraction ability, resulting in the formation of a potential difference. The material with stronger electron affinity has a negative surface charge, while the material with weaker electron affinity has a positive surface charge. Electrons flow through an external circuit to generate current. The triboelectric nanogenerator (TENG) comprises three structures: contact-separation type, contact-sliding type, and independent type. In the contact-separation type, two materials with different electron affinity face each other, and external force causes them to contact perpendicularly, generating a potential difference that forms a current after the external force is removed. Figure 2c(i) shows a TENG based on a high-polymer aerogel film proposed by Tang et al. [81] to improve the performance of the contact-separation type TENG. The porous structure of the aerogel produces a larger potential difference at the contact surface, increasing the output of the electric signal. The contact-sliding type generates current by repeatedly sliding two seamlessly contacting materials laterally relative to each other, polarizing the electrode. Figure 2c(ii) depicts a TENG with a linear grating structure proposed by Xie et al. [82], which achieved up to 85% total conversion efficiency at low operating frequencies. The independent type differs slightly from the contact-sliding type. It comprises two electrodes placed on the same horizontal plane and separated by a certain distance. Another piece of insulating film acts as a slider that seamlessly contacts the two electrodes, generating charge on the slider surface. Zhu et al. [83] designed an independent TENG with a rotating structure, which fixes the independent friction layer above the two electrodes and rotates to produce a periodic potential difference, allowing electrons to flow between the two fixed electrodes. The sensor has an output power of up to 1.5 W, and an energy conversion efficiency of 24%. Figure 2c(iii) shows another example of a TENG with a rotating structure [65].

The triboelectric flexible stress sensor array measurement includes two approaches: the unit-independent measurement [84] and the row-column multiplexing measurement [70], as shown in Figure 5. The unit-independent measurement refers to each sensor unit having an independent upper and lower triboelectric material and signal wire. Its advantage lies in high accuracy, while the disadvantage is the poor reliability caused by the large number of signal wires. As shown in Figure 5a, Zhou et al. [84] prepared a triboelectric nanogenerator with high sensitivity and high stretchability based on thermoplastic polyurethane (TPU) and silver nanowire/reduced graphene oxide (AgNWs/rGO). The authors proposed a 5 × 5 sensor array, and each sensing unit is connected by a separate AgNWs/rGO conductive layer. Stress makes the independent polymer materials in contact, producing an output voltage signal that indicates the magnitude and position of the stress. The row–column multiplexing measurement refers to the upper and lower layers of triboelectric materials arranged in a cross pattern, which reduces the number of signal wires, but introduces signal interference between sensing units. As shown in Figure 5b, Ning et al. [70] prepared a fiber-shaped triboelectric nanogenerator with ultra-high flexibility and stretchability based on silver nanowire/carbon nanotube and PDMS. The authors proposed an 8 × 8 sensor array, and each sensing unit includes cross-arranged fiber-shaped friction layers. When touching the sensor array, the corresponding pixels of the tactile sensor array produce output voltage signals that can reflect the tactile trajectory and stress distribution, showing promising applications in human–machine interaction and health monitoring.

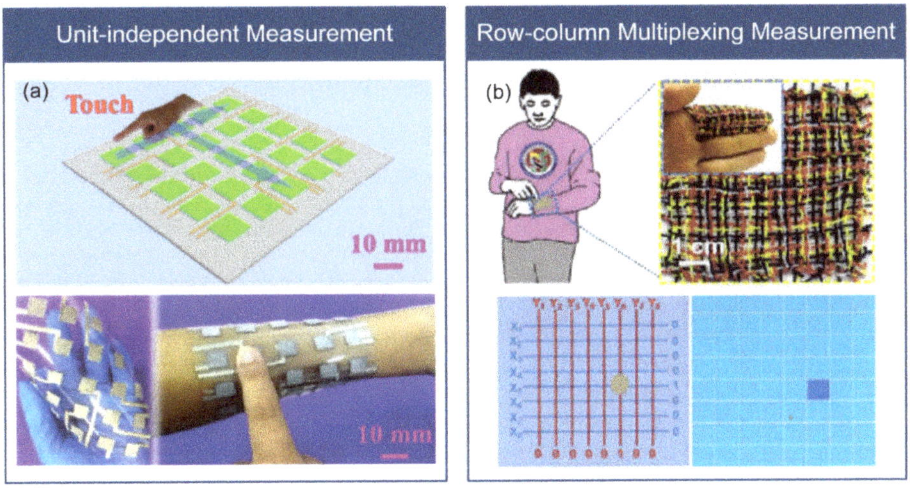

Figure 5. Triboelectric flexible stress sensor array measurement method. (**a**) Unit-independent measurement (reprinted with permission from [70]; copyright 2020, Elsevier). (**b**) Row–column multiplexed measurement (reprinted with permission from [70]; copyright 2020, Wiley-VCH).

The triboelectric flexible stress sensor array method can convert mechanical energy into electrical energy, which could potentially solve the issue of frequent recharging or battery replacement for wearable electronic devices. They offer advantages such as energy efficiency, good dynamic performance, low cost, a wide range of material options, and broad applications. However, triboelectric flexible stress sensor requires continuous friction to generate charges and are unable to detect static stress. Triboelectric flexible stress sensor arrays can be achieved through unit-independent measurements [84] and row–column multiplexing measurements [70]. Similar to capacitive sensors, these arraying methods have disadvantages such as multiple signal wires, complicated structures, and complicated acquisition circuits.

2.4. Piezoelectric Flexible Stress Sensor Array Method

Piezoelectric flexible stress sensors are made of materials that exhibit the piezoelectric effect. When stress is applied to these materials, they polarize, accumulate, or release charges, and convert stress into charge, enabling stress measurement [58]. Figure 2d shows commonly used piezoelectric materials, including single crystals, ceramics, and polymers. In addition, flexible materials such as porous polypropylene (PP) and porous polyvinylidene fluoride (PVDF) are also utilized. Park et al. [70] developed a multifunctional flexible electronic skin, shown in Figure 2d(i), based on PVDF and RGO with an interlocking microstructure and fingerprint-like patterns. This electronic skin can detect both static and dynamic tactile signals. Chu et al. [64], as shown in Figure 2d(ii), utilized piezoelectric polymers to create piezoelectric composites with a superposition structure, allowing for real-time monitoring of pulse signals at the radial artery and promising applications in wearable human health monitoring.

Piezoelectric flexible stress sensor arrays can be measured using two methods: row–column multiplexing measurement and shared electrode measurement, as illustrated in Figure 6. The row-column multiplexing measurement involves arranging the upper and lower electrodes in a cross pattern, which reduces the number of signal wires but leads to signal crosstalk between sensing units. To solve this issue, as shown in Figure 6a, Lin et al. [85] developed a flexible piezoelectric tactile sensor array with 25 sensing units arranged in a 5 × 5 matrix. They employed a double-layer comb-shaped structure for the row and column electrodes to eliminate signal crosstalk and reduce the number of connecting lines. Each sensing unit was composed of two protective layers (PDMS), two sensing layers (PVDF), and one insulation layer (PDMS). By applying stress to the protective layer, the sensing layer could convert the stress into charge, facilitating the identification of stress and position. This sensor array exhibited zero crosstalk, fast response time, high sensitivity, and durability. As shown in Figure 6b, Liu et al. [86] also developed a piezoelectric tactile sensor array, which used PVDF fibers and comprised two orthogonal arrays of 5 × 5 based on nanofiber-reinforced polyurethane films. This sensor array enabled real-time localization of the pressing position and tracking of the pressing trajectory, offering high resolution and flexibility. On the other hand, the shared electrode measurement involves independent electrodes and signal wires for each sensing unit, while sharing the same electrode and signal wire. This method provides high accuracy but still requires many signal wires. Liu et al. [87] presented a piezoelectric tactile sensor array based on PDMS and aluminum. The array consisted of four sensing units arranged in a 2 × 2 matrix, and each sensing unit was composed of a PDMS protrusion layer, a PET strip layer, a shared aluminum (Al) electrode layer, a PVDF film layer, an Al electrode layer, and a PET strip layer. The upper electrode layer, PVDF film layer, and lower electrode layer formed the piezoelectric sensing unit, with the four units sharing the same upper electrode. By applying three-dimensional stress to the top of the protrusion, different charge changes were generated in the four sensing units, enabling the measurement of normal and tangential stress and identification of roughness at variable speeds. As shown in Figure 6c, Yu et al. present a piezoelectric sensor array based on shared electrode method [88].

Piezoelectric flexible stress sensors offer advantages such as high sensitivity, good dynamic performance, and a simple structure. However, they are susceptible to temperature changes, leading to poor temperature stability. Similarly to flexible triboelectric stress sensors, piezoelectric flexible stress sensors are unable to measure static stress. The piezoelectric stress sensors array also present challenges, including a high number of signal wires, complicated structures, and complicated acquisition circuits.

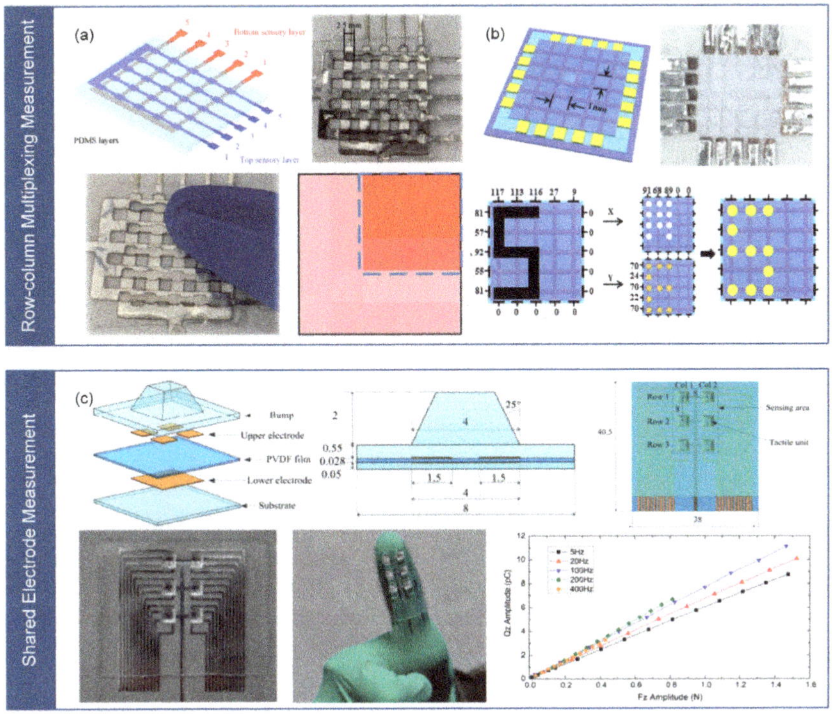

Figure 6. Arrayed measurement methods for piezoelectric flexible stress sensors. (**a**) Row-column multiplexing measurement (reprinted with permission from [85]; copyright 2021, Wiley-VCH). (**b**) Row-column multiplexing measurement (reprinted with permission from [86]; copyright 2021, American Chemical Society). (**c**) Shared electrode measurement (reprinted with permission from [88]; copyright 2016, MDPI). Note: In (**a**,**b**), the same Row-column multiplexing measurement is presented but from different references.

In summary, piezoresistive, capacitive, triboelectric, and piezoelectric flexible stress sensors have all been employed in sensor arrays. While these methods fulfill the requirements for spatial stress distribution and high spatial resolution, increasing the integration levels in existing methods results in a significant rise in the number of signal wires, which decreases in flexibility and stretchability of the sensor array, along with increased complexity and cost. To address these issues, researchers have explored a novel sensor array measurement method known as the single-line multi-channel (SLMC) measurement. This approach utilizes one or two signal wires to simultaneously measure signals from multiple sensors, thereby reducing the number of signal wires required for the sensor array. The SLMC measurement has the potential to enhance the deformability, reliability, and stability of flexible stress sensor arrays, facilitating their practical application.

3. Single-Line Multi-Channel Measurement

SLMC measurement can simultaneously measure multiple sensor signals using one or two signal wires. Depending on the mechanism, SLMC measurement can be classified into various types, including RLC resonant sensing [15,28,29,32], transmission line sensing [16], ionic conductor sensing [33,36], triboelectric sensing [42,43], piezoresistive sensing [41], and distributed fiber optic sensing [34].

3.1. RLC Resonant Sensing

The RLC resonant sensing utilizes the interaction between inductance and capacitance. By adjusting the resonant frequency of the RLC resonant circuit, it is possible to obtain multiple sensor signals using only two signal lines. Implementing this technology requires selecting appropriate inductance and capacitance components, as well as designing and fabricating the circuit. An RLC resonator consists of resistors, inductors, and capacitors connected in series, forming a resonant circuit with a specific frequency and frequency-selective capability. By parallel connecting multiple RLC resonant circuits and adjusting their resonant frequencies, a multi-frequency resonant signal can be generated. Each resonant frequency corresponds to the output characteristics of a sensor unit, which can modify the resonant frequency or amplitude of the multi-frequency signal by converting stress into changes in resistance, inductance, or capacitance, as shown in Figure 7.

As shown in Figure 7a, Wen et al. [15] developed a flexible wireless passive LC stress sensor using a flexible composite protein film as the substrate. By adding regenerated silk fibroin film, which makes better air permeability and water permeability. The sensor employs a commercial printed circuit board and magnetron sputtering methods to create the inductor and capacitor in the LC circuit. The study combines radio frequency identification technology and the LC resonant circuit to design a wireless insole sensor for measuring sole stress distribution. The RLC series resonant sensing simplifies sensor wiring and reduces costs. As shown in Figure 7b, inspired by the somatosensory system, Lee et al. [28] proposed a stress sensor system using a wireless system to differentiate parallel signals. The sensor system comprises multiple stress sensors with polypyrrole-coated microstructured PDMS placed on top of electrodes, which operates as a capacitive sensor. By altering the number of turns of the coil, the system assigns specific resonant frequencies to each sensor. A stacked coil structure is adopted to conserve space and minimize crosstalk between coils. The sensor can simultaneously receive resonant signals from multiple stress sensors and utilize a convolutional neural network to train data for simultaneous measurement of stress magnitude and location. As shown in Figure 7c, Kim et al. [32] introduced a flexible sensor array composed of multiple bandpass filters. The array consists of parallel bandpass filters (RLC resonant circuits) and two wires. Each sensing unit corresponds to a bandpass filter, which can be assigned different filtering frequencies by adjusting the values of inductance and capacitance. The sensor array consists of an inductor, a capacitor, and a variable resistor made of a microfluidic channel filled with EGaIn. This enables separate measurements while sharing the same signal wire. The article also demonstrates the application of the flexible sensor array in tactile sensing and foot pressure measurement.

The RLC resonant circuit allows for the selection of a specific frequency, facilitating the acquisition of multiple sensor signals using only two signal wires. By connecting RLC resonant circuits in parallel and series while adjusting their resonant frequencies, multiple resonant frequencies can be generated. Each resonant frequency in the multi-frequency resonant signal corresponds to the output characteristic of a sensor unit. The flexible sensor array based on RLC resonant sensing utilizes only two external wires to scan the sensor array, regardless of the number of sensors. The RLC sensing has the advantages of two signal lines, tunable sensitivity, and multi-model sensing. However, due to the RLC resonant circuit, the sensor array has a complicated circuit design, and has a crosstalk between different RLC resonators. RLC resonant sensing has wide applications in areas such as wireless communication, biomedical, and environmental monitoring. In wireless sensor networks, it enables wireless monitoring and data transmission of various physical parameters [31] (temperature, humidity, etc.). In bio-sensors, it can be used to monitor physiological parameters [5] and disease indicators within the human body.

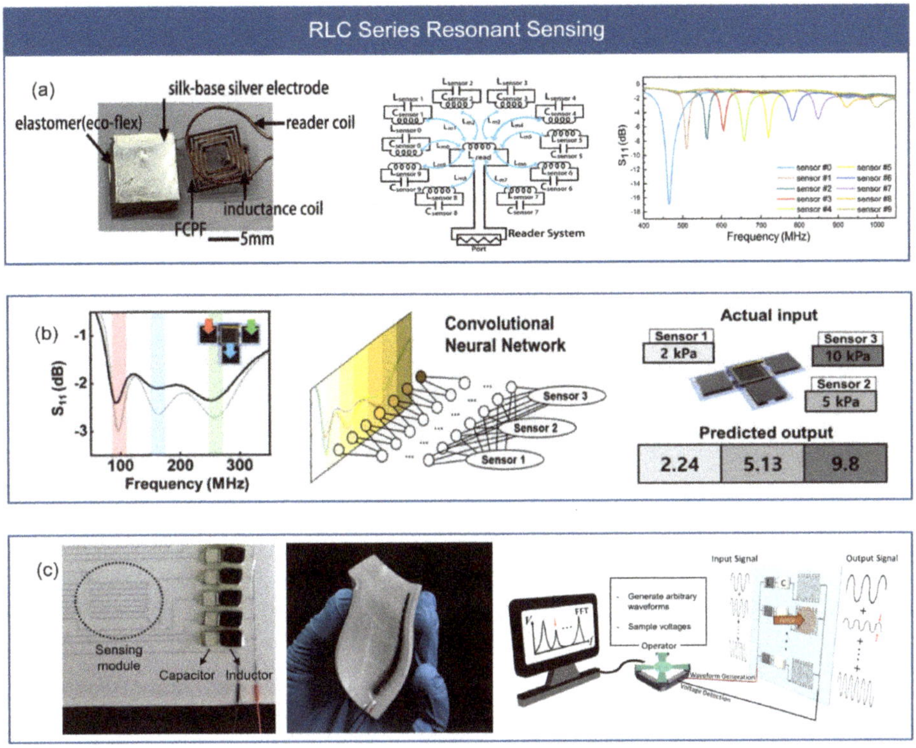

Figure 7. RLC series resonant sensing. (**a**) Wireless passive LC stress sensor (reprinted with permission from [15]; copyright 2021, Wiley-VCH). (**b**) Stress sensor system based on wireless system to differentiate parallel signals (reprinted with permission from [28]; copyright 2019, Wiley-VCH). (**c**) Flexible sensor array based on multiple band-pass filters (reprinted with permission from [32]; copyright 2022, Springer Nature).

3.2. Transmission Line Sensing

The transmission line sensing measures the voltage or current on the transmission line to obtain information about the target physical quantity. When the target physical quantity changes, it affects the electromagnetic field distribution on the transmission line, thereby altering the characteristics of the transmission line, such as resistance, inductance, and capacitance. Implementing this technology requires selecting appropriate transmission line types and parameters, as well as designing and manufacturing the sensors. The transmission line, composed of multi-core cables, coaxial cables, optical fibers, or wireless devices, can transmit electromagnetic energy. The transmission performance is influenced by several factors, such as transmission rate, signal bandwidth, data transmission quality, and anti-interference capability. As the working frequency increases and the wavelength decreases, the voltage and current on the transmission line vary with spatial position, resulting in wave-like fluctuations. These fluctuations of the electromagnetic signal on the transmission line can indicate the magnitude of stress at different positions, enabling the achievement of a SLMC measurement.

Leber et al. [16] proposed a stretchable transmission line for measuring stress at various positions using a single line. This transmission line is created by enclosing a liquid metal (65% Ga, 22% In, and 10% Sn) in a thermoplastic elastomer dielectric (pol(styrene-b-(ethylene-co-butylene)-b-styrene), SEBS) via thermal stretching and uses time-domain reflectometry to send high-frequency pulses. Stress magnitude and location are determined by analyzing the reflection resulting from applied stress. The stretchable transmission line exhibits high sensitivity, and repeatability, making it suitable for measuring stress distributions in electronic textiles, wearable electronic devices, and electronic skins. This approach offers several advantages, including complexity reduction and high productivity.

In transmission line sensing, the reflected electromagnetic pulse signal conveys information about the magnitude of stress at different locations by exploiting the fluctuations of voltage and current along the spatial position of the transmission line. The transmission line sensing only has one signal line, which has the advantage of high integration. While it has drawbacks, including high cost, complicated fabrication, and complexity of information decoding.

3.3. Ionic Conductor Sensing

The ionic conductor sensing utilizes the contact resistance and capacitance between the human hand and the ionic conductor under AC current. Ionic conductors can be made from ion-containing hydrogels or organic hydrogels. When exposed to high-frequency AC current, human touch on the ionic conductor causes changes in the contact resistance and capacitance within the electric double layer structure of the ionic conductor. The contact resistance is measured to determine the position of touch, while the newly formed contact capacitance after human touch is used to measure the magnitude of stress, as shown in Figure 8. The ionic conductor sensing enables the determination of stress magnitude and location information through two single-signal lines, providing a low-cost, easily operated for measuring stress distribution in electronic skins, wearable electronic devices. In addition, implementing this technology requires the selection of suitable ion conductor materials and the design of corresponding circuits.

Gao et al. [33] developed a pressure-sensitive touchpad with self-healing properties using an ionic conductor, as shown in Figure 8a. The ionic conductor is a polyzwitterion-clay nanocomposite hydrogel, which has flexibility, stretchability, transparency, and pressure-sensitivity. When touched, the hydrogel exhibits changes in the resistance and capacitance of the electric double layer structure, enabling the measurement of touch position and magnitude. The touch position is determined by the resistance, whereas the capacitance is dependent on the distance between the touch point and the electrode. This study provides an innovative solution for human–machine interfaces with self-healing properties of polymer nanocomposite hydrogels, along with the recognition and differentiation of multi-touch signals. As shown in Figure 8b, Wu et al. [36] proposed a multifunctional, wearable, and transparent human–machine interface based on ionic hydrogels. An anti-drying sodium alginate/polyacrylamide/ionic liquid double-network organohydrogel was synthesized using a wet spinning method. The interface utilizes a parallel mutual detection structure that allows for both single-point touch and multi-point touch using only two electrodes. The interface is composed of a pair of horizontal or spiral ionic hydrogel strips embedded in a flexible substrate. The ionic hydrogel exhibits excellent stretchability, transparency, and durability. When a finger touches the interface, the ionic hydrogel between the electrode and the touch point acts as a resistor that is connected to the circuit and is in series with the electric double-layer capacitor. This configuration allows for the determination of touch position and magnitude through variations in current and capacitance. This study employs only two electrodes for touch sensing, which simplifies maintenance and operation. Furthermore, the detection of multi-touch enhances the diversity of human–machine interaction and advances the development of the Internet of Things.

The ionic conductor exhibits varying resistances and capacitances based on the magnitude and location of the touch when it contacts with the human body. The resistance of the sensor indicates different positions on the sensor, whereas the capacitance corresponds to the intensity of the touch force. Two-dimensional planar touch sensing can be accomplished by utilizing four wires, offering the benefits of a streamlined circuit layout and a straightforward device structure. The ionic conductor sensing technology has the advantages of two signal lines and multi model measurement. However, it still needs to address challenges such as low temperature stability and complicated signal decoding. The ionic conductor sensing finds application potentials in fields such as human–machine interfaces and biosensing.

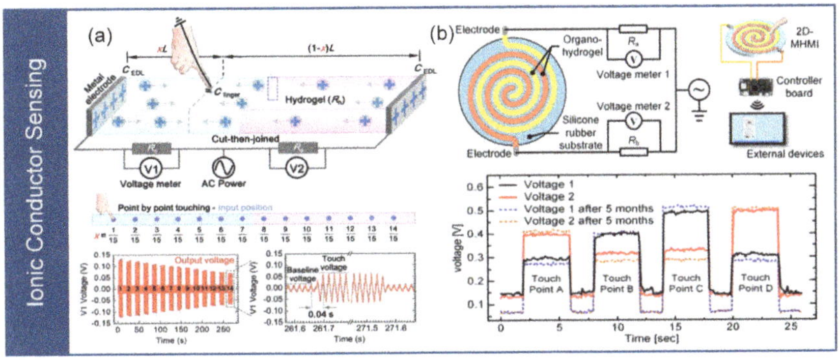

Figure 8. Ionic conductor sensing. (**a**) Self-healing composite hydrogel sensors (reprinted with permission from [33]; copyright 2020, Wiley-VCH). (**b**) Parallel interdigitated electrode structure-based hydrogel sensors (reprinted with permission from [36]; copyright 2021, Elsevier).

3.4. Triboelectric Sensing

Triboelectric sensing is a sensing array composed of triboelectric materials with unique physical and chemical properties (electron affinity, mobility, adhesion energy, and surface potential), which measures the change in the applied stress through the charge variation generated by the friction between the triboelectric materials. These materials are arranged in a way that generates triboelectric signals exhibiting diverse voltage or charge density. When subjected to triboelectric force, the different positions of the materials produce distinct signal responses, enabling the simultaneous detection of multiple frictional positions, as shown in Figure 9.

To reduce the number of connection wires and achieve high spatial resolution, Jang et al. [42] proposed a wearable single-ended triboelectric tactile sensor array that can detect multi-point friction, as shown in Figure 9a. This array consists of different stretchable block copolymer elastomeric dielectrics that are patterned with UV light. By adjusting the UV exposure time of the thermoplastic block copolymer film, triboelectric patterns with varying relative charge densities can be generated. With just one electrode, the tactile sensor array can discern the touch position. Guo et al. [43] developed a self-powered electronic skin based on an Archimedean spiral structure, as shown in Figure 9b. The electronic skin comprises a triboelectric layer (such as polymer, metal), patterned AgNW electrodes, and a PDMS substrate, capable of determining the frictional position without the need for an external power supply. These materials are distributed on the electrodes through a multilayer alignment transfer process. By designing the electrode with an Archimedean spiral structure, the triboelectric signal is distinguished into digital "0" or "1" based on the intrinsic electron affinity of the materials. Only four electrodes are needed to identify 280 frictional positions. The electronic skin easily integrable into portable electronic devices, such as laptops and healthcare devices, showing application potential in the fields of human–machine interface and artificial intelligence.

Applying friction to triboelectric sensors generates voltage or charge density based on the different friction positions and the triboelectric materials with different properties, therefore enabling SLMC measurement. Triboelectric sensing technology has the advantages of two signal lines, low energy consumption, high dynamic performance, but it still needs to solve the challenges of poor static measurement and low signal stability. Triboelectric sensing has the application potential in the fields of friction and wear monitoring and tactile sensing.

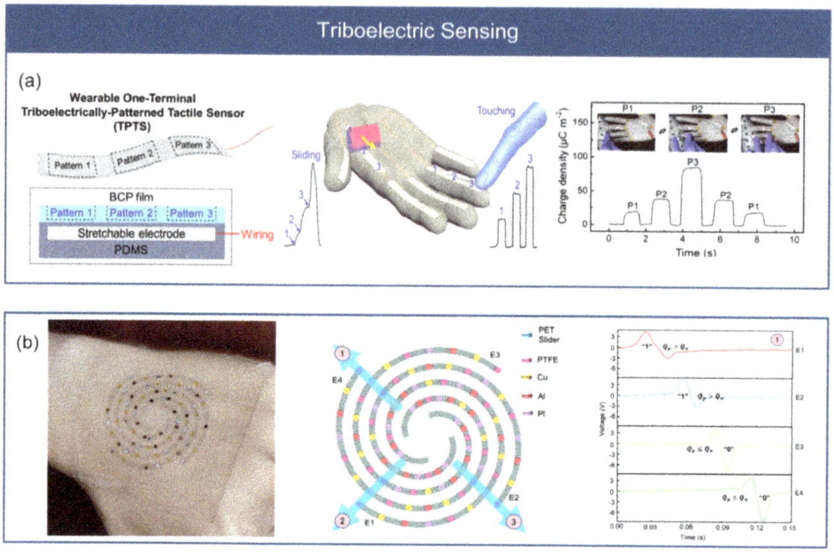

Figure 9. Triboelectric sensing. (**a**) Wearable single-ended frictional tactile sensor array (reprinted with permission from [42]; copyright 2022, Elsevier). (**b**) Self-powered electronic skin based on Archimedean spiral structure (reprinted with permission from [43]; copyright 2021, Wiley-VCH).

3.5. *Piezoresistive Sensing*

The piezoresistive sensing consists of materials with piezoresistivity that transforms the stress position into a corresponding resistance value. Applying stress at different positions leads to different resistance variation, which enables the realization of a SLMC measurement, as shown in Figure 10.

As shown in Figure 10a, Liao et al. [41] introduced a biomimetic sensory transmission neural sensor array capable of accurately determining the position of stress. The sensor array consists of a conductive graphite film, a grid paper, and a sheet in between. The proposed sensor array, known as the artificial perception and transmission neural sensor array, employs an independent two-layer electrode structure with progressively decreasing resistance in each electrode. Applying stress contacts the two electrodes, and the resulting resistance determines the stress position. The sensor array exhibits remarkable attributes such as rapid response, robustness, durability, and flexibility, enabling diverse applications such as multi-functional touch interaction. Furthermore, by integrating the spatiotemporal resolution function of the sensor array with artificial intelligence algorithms, this study demonstrated its applicability in neural prostheses and robots. The sensor array designed with a distinct double-layer structure, effectively resolves challenges related to complicated structures, intricate interconnections, and signal transmission interference.

By manipulating the resistance at different positions within the piezoresistive material, piezoresistive sensing-based sensor array can determine the stress location by analyzing the resulting resistance. Piezoresistive sensing has several advantages, such as one signal line, simple structure, low cost, and simple signal acquisition circuit. However, for some application scenarios, it is necessary to consider its stress magnitude and multiple-channel unmeasurably. Piezoresistive sensing technology is widely used in the field of structural monitoring. The sensors can be used to monitor stress changes in multiple key locations such as buildings or bridges to assess the health and safety of the structure. In the medical field, piezoresistive sensing technology can be applied to the measurement of stress distribution on human body, such as on mattresses, to improve treatment.

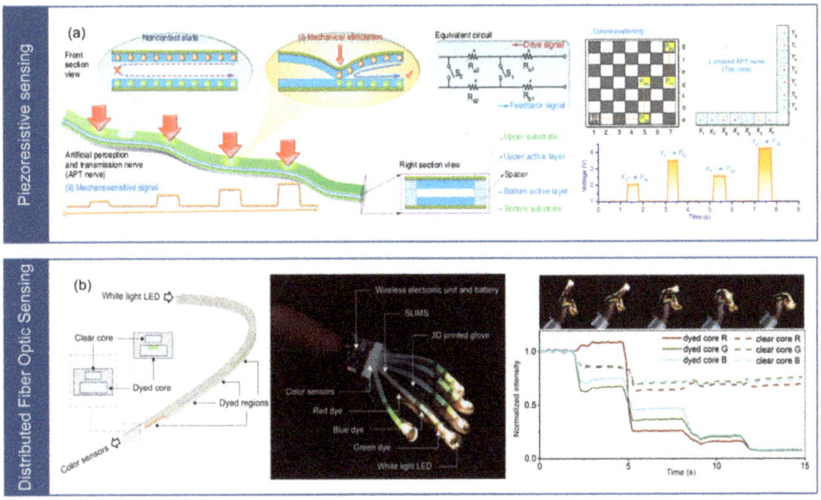

Figure 10. Piezoresistive sensing and distributed fiber optic sensing. (**a**) Biomimetic sensory nerve sensor (reprinted with permission from [41]; copyright 2020, Springer Nature). (**b**) Stretchable distributed fiber optic sensor (reprinted with permission from [34]; copyright 2020, AAAS).

3.6. Distributed Fiber Optic Sensing

Distributed fiber optic sensing utilizes the continuous one-dimensional characteristic of optical fibers. The distributed fiber optic sensing relies on optical time-domain reflectometry, which is based on the principle of backscattering of light emitted by a source as it propagates through the fiber. The intensity of the backscattered light decreases in proportion to the distance and time traveled. By detecting changes in the time and intensity of the backscattered light, an optical detector can determine the magnitude of stress at different positions along the fiber, as shown in Figure 10.

As shown in Figure 10b, Bai et al. [34] developed a stretchable fiber optic sensor based on silicon dioxide. The sensor consists of a dual-core lightguide with one dyed core doped with absorbing dyes at discrete locations and one clear core, separated by silicone cladding. The sensor can measure the location, magnitude, and mode (stretching, bending, or twisting) of deformations using an elastic optical waveguide containing continuous or discrete colored patterns. Applying stress changes the chromaticity and intensity of the waveguide, allowing the position and magnitude of the stress to be determined through total internal reflection and absorption phenomena. The researchers successfully integrated this stretchable fiber optic sensor into a glove and demonstrated its effectiveness in monitoring various hand movements.

Applying stress to the distributed fiber optic sensor modifies the reflection and refraction of light sources within the fiber, resulting in variations in the detection time and intensity of the light source. These variations can indicate the magnitude and position of stress on the sensing array, making it possible to implement SLMC measurement. Distributed fiber optic sensing provides significant advantages, including one signal line, high spatial resolution, and multi-model detection. However, it also faces challenges such as fiber brittleness, signal parsing and data processing complexity.

Fiber Bragg Grating Sensors (FBGs) are used to realize fiber sensing by introducing a periodic index modulation structure into the fiber. Specifically, FBGs are composed of refractive index periodic modulation regions in optical fibers that cause specific wavelengths of light to be reflected or transmitted. Applying stress or temperature to FBGs changes their periodic refractive index structure, resulting in a shift in the reflected or transmitted wavelength, and then the stress or temperature is measured. FBGs has the advantages of high precision, fast response and anti-electromagnetic interference, and has been widely used in the fields of structure monitoring, pipeline inspection, aerospace and medical diagnosis. However, higher costs and flexibility limitations are some of the drawbacks of FBGs. Li et al. [89] proposed a sensor manufacturing method for joint motion monitoring in stroke rehabilitation by embedding silicon fiber grating in silicone tubes. The materials used include silicon fiber FBG and silicon tubes. The sensor has demonstrated excellent performance in detecting and differentiating joint movements, making it a promising medical tool for rehabilitation of stroke patients and potential detection of Parkinson's tremor in stroke survivors.

Intensity modulation polymer optical fiber (POF) sensor is a fiber based sensing technology, its working principle is based on the intensity modulation of the optical signal. The sensor uses a sensitive material or structure that causes the intensity of the light signal to change when affected by the measured parameters. This intensity change can be measured and analyzed by the photodetector, so that the measurement parameters can be monitored and measured. Intensity modulated POF sensors have the advantages of simplicity, economy and high sensitivity, and are suitable for applications that measure changes in the intensity of optical signals, but are limited by changes in light intensity and the influence of ambient light [90].

FBGs uses the grating structure in fiber to measure optical signals, which is suitable for point measurement and has high spatial resolution and complicated reading system. The distributed optical fiber sensor is suitable for continuous measurement, with low spatial resolution and simplified reading principle. Intensity modulated POF sensors measure by adjusting the intensity of the optical signal, which is different from FBGs and distributed optical fiber sensors in working principle, spatial resolution and application field.

We summarized the advantages and disadvantages of the existing single-wire multi-channel measurement technology, as shown in Table 1.

Table 1. The advantages and disadvantages of the existing sensor array measurement method.

Method	Advantages	Limitations
RLC Resonant Sensing	• Two signal lines • Tunable sensitivity • Multi-model sensing	• Complicated circuit design • Crosstalk between frequencies
Transmission Line Sensing	• One signal line • High integration	• High cost • Complicated fabrication • Complicated signal decoding
Ionic Conductor Sensing	• Two signal lines • Multi model measurement	• Low temperature stability • Complicated signal decoding
Triboelectric Sensing	• Two signal lines • Low energy consumption • High dynamic performance	• Poor static measurement • Low signal stability
Piezoresistive Sensing	• One signal line • Simple signal acquisition • Low cost	• Stress magnitude unmeasurably • Multiple-channel unmeasurably
Distributed Fiber Optic Sensing	• One signal line • High spatial resolution • Multi-model detection	• Fiber brittleness • Complicated signal decoding
Unit-Independent Measurement	• High precision	• Tedious signal line • Complicated design
Row–Column Multiplexed Measurement	• High precision • Reduced signal line	• Signal crosstalk • Tedious signal line
Anisotropic Electrical Impedance Tomography	• Reduce design complexity	• Signal crosstalk • Low resolution • Complicated signal processing • Tedious signal line
Shared Electrode Measurement	• Reduced signal line	• Complicated design • Signal crosstalk • Tedious signal line

In summary, the SLMC measurement reduces the need for multiple signal wires, enhancing the flexibility, stability, and reliability of flexible stress sensor arrays while also reducing costs. Additionally, the SLMC measurement introduces new concepts to the fundamental theory of measuring flexible stress sensor arrays, overcomes technical challenges, and promotes widespread adoption of soft stress sensor arrays.

4. Summary and Prospect

Flexible stress sensor arrays have gained popularity due to their high sensitivity, functionality, deformability, and stretchability. However, the sensor signal is an electric amplitude modulation signal (such as resistance and capacitance), which cannot simultaneously distinguish multiple signals through a single-signal wire. Thus, a large number of signal wires are needed to meet the measurement requirements over a large measurement area, which greatly increases costs, decreases equipment reliability and operability, and hinders their widespread application in wearable electronic devices such as electronic skin and health monitoring.

In this review, we introduce the stress sensing mechanisms based on piezoresistive, capacitive, triboelectric, and piezoelectric principles, as well as their array structures and performance. We focus on SLMC measurement that can transmit multiple signals through one or two signal wires. According to the mechanism of SLMC measurement, we classify the methods into RLC resonant sensing, transmission line sensing, ionic conductor sensing, triboelectric sensing, piezoresistive sensing, and distributed fiber optic sensing. We describe the mechanisms and characteristics of each method and summarize the research status of SLMC measurement.

Several techniques mentioned in this article have successfully verified the feasibility and reliability of the SLMC measurement, which reduces the number of signal wires, reduces costs, and improves the flexibility, stability, and reliability of flexible sensor arrays. Nevertheless, the scalability of each technology from sensors to large-scale commercial applications requires improvement. We can promote the application of SLMC signal transmission methods on flexible sensor arrays by optimizing structural design, material selection, and signal processing.

In previous studies, multiple sensing units were concatenated to achieve multimodal measurement, resulting in poor operability, reliability, and efficiency. To improve operability and efficiency, simple structural designs such as vertical stacking structures can be used to achieve higher-dimensional measurement [91–93]. Wearable electronic devices, such as electronic skin, are vulnerable to damage due to their flexible material, which negatively impacts usability and reduces efficiency. The use of self-healing polymer materials, such as composite hydrogels with self-repairing properties, can provide quick self-repair during use, which improves the stability and durability of wearable electronic devices [94–96]. Previous signal processing was based on simple threshold judgments, which is inadequate for signal decoupling that is affected by noise, signal damage, or multiple signals. The use of machine learning algorithms can improve the accuracy and efficiency of sensing. In conclusion, SLMC measurement is a new breakthrough in signal processing for flexible stress sensor arrays [27,97,98]. By optimizing structural design, material selection, and signal processing, they can promote implementation in health monitoring, human–machine interaction.

The SLMC measurement technologies reduce the number of signal lines, enhancing the deformability, stability, and reliability of the flexible stress sensor array. These advantages make the SLMC technology well suited for a wide range of applications, including wearable devices, biomedical sensing, structural health monitoring, and robotics, where accurate and real-time measurement of spatial stress distribution is crucial for performance optimization and safety enhancement. However, it is important to acknowledge that challenges may arise in certain aspects of the technology.

The distributed fiber sensing has the property of brittle fracture. However, distributed fiber sensing provides continuous, real-time monitoring along the entire fiber length, enabling the measurement of strain, temperature, and other physical parameters at high spatial resolution. This makes the SLMC suitable for applications that require distributed sensing over large areas or long distances. Additionally, future advances in optical fiber materials such as the development of bendable fiber will greatly improve the durability and flexibility of the fiber. RLC resonant sensing often requires complicated circuit design, due to resistors, inductors, and capacitors. Future research can combine RLC resonant sensing with other sensing materials to achieve simultaneous measurement of multiple parameters (humidity, temperature, etc.), and simplify circuit design to achieve higher integration. Transmission line sensing still presents challenges in terms of the complexity of fabrication and cost, and future research could simplify fabrication processes by applying screen printing, gravure printing or spin coating and use less expensive materials. Ionic conductor sensing and triboelectric sensing are easily disturbed by environmental factors, and future research can develop new ionic/triboelectric conductor materials to expand its application field and improve its stability.

This review introduces the mechanisms of flexible stress sensors, as well as discusses the advantages and disadvantages of the design of flexible stress sensor arrays. To address the issue of excessive signal wires in current flexible stress sensor arrays, solutions for flexible stress sensor arrays are summarized, and the implementation principle and pros and cons of existing SLMC measurement are analyzed. In summary, flexible stress sensor arrays allow for quantification and analysis of the magnitude and distribution of spatial stress, while also offering advantages such as flexibility, integration, and systematization. However, increasing the number of signal wires may negatively impact the stability and reliability of the sensor array. Thus, it is essential to develop more efficient, reliable, and cost-effective methods for arraying flexible stress sensors to promote their application in various fields.

Author Contributions: Conceptualization, J.Y. and Y.C.; methodology, J.Y., Y.C. and G.G.; software, Y.C.; validation, J.Y. and Y.C.; formal analysis, Y.C.; investigation, J.Y. and Y.C.; resources, J.Y., G.G., T.M. and Z.L.; data curation, Y.C., S.L. and C.L.; writing-original draft preparation, J.Y. and Y.C.; writing-review and editing, J.Y., Y.C. and G.G.; visualization, Y.C.; supervision, J.Y.; project administration, J.Y.; funding acquisition, J.Y. and G.G. All authors have read and agreed to the published version of the manuscript.

Funding: This research was funded by the National Natural Science Foundation of China, grant number 62101432, and the China Postdoctoral Science Foundation, grant number 2022M722558.

Data Availability Statement: Data available on request from the authors.

Conflicts of Interest: The authors declare no conflict of interest.

References

1. Hammock, M.L.; Chortos, A.; Tee, B.C.; Tok, J.B.; Bao, Z. 25th anniversary article: The evolution of electronic skin (e-skin): A brief history, design considerations, and recent progress. *Adv. Mater.* **2013**, *25*, 5997–6038. [PubMed]
2. Yang, R.; Zhang, W.; Tiwari, N.; Yan, H.; Li, T.; Cheng, H. Multimodal Sensors with Decoupled Sensing Mechanisms. *Adv. Sci.* **2022**, *9*, e2202470. [CrossRef]
3. Trung, T.Q.; Lee, N.E. Flexible and Stretchable Physical Sensor Integrated Platforms for Wearable Human-Activity Monitoringand Personal Healthcare. *Adv. Mater.* **2016**, *28*, 4338–4372. [CrossRef] [PubMed]
4. Chen, G.; Rastak, R.; Wang, Y.; Yan, H.; Feig, V.; Liu, Y.; Jiang, Y.; Chen, S.; Lian, F.; Molina-Lopez, F.; et al. Strain- and Strain-Rate-Invariant Conductance in a Stretchable and Compressible 3D Conducting Polymer Foam. *Matter* **2019**, *1*, 205–218. [CrossRef]
5. Chen, L.Y.; Tee, B.C.K.; Chortos, A.L.; Schwartz, G.; Tse, V.; Lipomi, D.J.; Wong, H.S.P.; McConnell, M.V.; Bao, Z.A. Continuous wireless pressure monitoring and mapping with ultra-small passive sensors for health monitoring and critical care. *Nat. Commun.* **2014**, *5*, 5028. [CrossRef]
6. O'Neill, S.J.K.; Gong, H.; Matsuhisa, N.; Chen, S.; Moon, H.; Wu, H.C.; Chen, X.; Chen, X.; Bao, Z. A Carbon Flower Based Flexible Pressure Sensor Made from Large-Area Coating. *Adv. Mater. Interfaces* **2020**, *7*, 2000875. [CrossRef]
7. Mannsfeld, S.C.; Tee, B.C.; Stoltenberg, R.M.; Chen, C.V.; Barman, S.; Muir, B.V.; Sokolov, A.N.; Reese, C.; Bao, Z. Highly sensitive flexible pressure sensors with microstructured rubber dielectric layers. *Nat. Mater.* **2010**, *9*, 859–864. [CrossRef]
8. Fan, J.A.; Yeo, W.H.; Su, Y.; Hattori, Y.; Lee, W.; Jung, S.Y.; Zhang, Y.; Liu, Z.; Cheng, H.; Falgout, L.; et al. Fractal design concepts for stretchable electronics. *Nat. Commun.* **2014**, *5*, 3266. [CrossRef]

9. Won, S.M.; Wang, H.; Kim, B.H.; Lee, K.; Jang, H.; Kwon, K.; Han, M.; Crawford, K.E.; Li, H.; Lee, Y.; et al. Multimodal Sensing with a Three-Dimensional Piezoresistive Structure. *ACS Nano* **2019**, *13*, 10972–10979. [CrossRef]
10. Yu, X.; Xie, Z.; Yu, Y.; Lee, J.; Vazquez-Guardado, A.; Luan, H.; Ruban, J.; Ning, X.; Akhtar, A.; Li, D.; et al. Skin-integrated wireless haptic interfaces for virtual and augmented reality. *Nature* **2019**, *575*, 473–479. [CrossRef]
11. Oh, Y.S.; Kim, J.H.; Xie, Z.; Cho, S.; Han, H.; Jeon, S.W.; Park, M.; Namkoong, M.; Avila, R.; Song, Z.; et al. Battery-free, wireless soft sensors for continuous multi-site measurements of pressure and temperature from patients at risk for pressure injuries. *Nat. Commun.* **2021**, *12*, 5008. [CrossRef] [PubMed]
12. Xu, S.; Zhang, Y.; Jia, L.; Mathewson, K.E.; Jang, K.I.; Kim, J.; Fu, H.; Huang, X.; Chava, P.; Wang, R.; et al. Soft microfluidic assemblies of sensors, circuits, and radios for the skin. *Science* **2014**, *344*, 70–74. [CrossRef] [PubMed]
13. Xia, Y.; Rogers, J.A.; Paul, K.E.; Whitesides, G.M. Unconventional Methods for Fabricating and Patterning Nanostructures. *Chem. Rev.* **1999**, *99*, 1823–1848. [CrossRef] [PubMed]
14. Pan, L.; Yu, G.; Zhai, D.; Lee, H.R.; Zhao, W.; Liu, N.; Wang, H.; Tee, B.C.; Shi, Y.; Cui, Y.; et al. Hierarchical nanostructured conducting polymer hydrogel with high electrochemical activity. *Proc. Natl. Acad. Sci. USA* **2012**, *109*, 9287–9292. [CrossRef]
15. Wen, H.; Chen, C.; Li, S.; Shi, Y.; Wang, H.; Guo, W.; Liu, X. Array Integration and Far-Field Detection of Biocompatible Wireless LC Pressure Sensors. *Small Methods* **2021**, *5*, e2001055. [CrossRef]
16. Leber, A.; Dong, C.Q.; Chandran, R.; Das Gupta, T.; Bartolomei, N.; Sorin, F. Soft and stretchable liquid metal transmission lines as distributed probes of multimodal deformations. *Nat. Electron.* **2020**, *3*, 316–326. [CrossRef]
17. Aksoy, B.; Hao, Y.; Grasso, G.; Digumarti, K.M.; Cacucciolo, V.; Shea, H. Shielded soft force sensors. *Nat. Commun.* **2022**, *13*, 4649. [CrossRef]
18. Cooper, C.B.; Arutselvan, K.; Liu, Y.; Armstrong, D.; Lin, Y.; Khan, M.R.; Genzer, J.; Dickey, M.D. Stretchable Capacitive Sensors of Torsion, Strain, and Touch Using Double Helix Liquid Metal Fibers. *Adv. Funct. Mater.* **2017**, *27*, 1605630. [CrossRef]
19. Ge, G.; Zhang, Y.Z.; Zhang, W.; Yuan, W.; El-Demellawi, J.K.; Zhang, P.; Di Fabrizio, E.; Dong, X.; Alshareef, H.N. Ti(3)C(2)T(x) MXene-Activated Fast Gelation of Stretchable and Self-Healing Hydrogels: A Molecular Approach. *ACS Nano* **2021**, *15*, 2698–2706. [CrossRef]
20. Bae, G.Y.; Han, J.T.; Lee, G.; Lee, S.; Kim, S.W.; Park, S.; Kwon, J.; Jung, S.; Cho, K. Pressure/Temperature Sensing Bimodal Electronic Skin with Stimulus Discriminability and Linear Sensitivity. *Adv. Mater.* **2018**, *30*, e1803388. [CrossRef]
21. Zhao, X.F.; Hang, C.Z.; Wen, X.H.; Liu, M.Y.; Zhang, H.; Yang, F.; Ma, R.G.; Wang, J.C.; Zhang, D.W.; Lu, H.L. Ultrahigh-Sensitive Finlike Double-Sided E-Skin for Force Direction Detection. *ACS Appl. Mater. Interfaces* **2020**, *12*, 14136–14144. [CrossRef] [PubMed]
22. Zhang, Z.M.; Zhang, Y.X.; Jiang, X.; Bukhari, H.; Zhang, Z.X.; Han, W.H.; Xie, E.Q. Simple and efficient pressure sensor based on PDMS wrapped CNT arrays. *Carbon* **2019**, *155*, 71–76. [CrossRef]
23. Ge, G.; Wang, Q.; Zhang, Y.Z.; Alshareef, H.N.; Dong, X.C. 3D Printing of Hydrogels for Stretchable Ionotronic Devices. *Adv. Funct. Mater.* **2021**, *31*, 2107437. [CrossRef]
24. Ge, G.; Mandal, K.; Haghniaz, R.; Li, M.; Xiao, X.; Carlson, L.; Jucaud, V.; Dokmeci, M.R.; Ho, G.W.; Khademhosseini, A. Deep Eutectic Solvents-based Ionogels with Ultrafast Gelation and High Adhesion in Harsh Environments. *Adv. Funct. Mater.* **2023**, *33*, 2207388. [CrossRef] [PubMed]
25. Ge, G.; Lu, Y.; Qu, X.; Zhao, W.; Ren, Y.; Wang, W.; Wang, Q.; Huang, W.; Dong, X. Muscle-Inspired Self-Healing Hydrogels for Strain and Temperature Sensor. *ACS Nano* **2020**, *14*, 218–228. [CrossRef] [PubMed]
26. Wang, M.; Luo, Y.; Wang, T.; Wan, C.; Pan, L.; Pan, S.; He, K.; Neo, A.; Chen, X. Artificial Skin Perception. *Adv. Mater.* **2021**, *33*, e2003014. [CrossRef]
27. Kim, T.; Lee, S.; Hong, T.; Shin, G.; Kim, T.; Park, Y.L. Heterogeneous sensing in a multifunctional soft sensor for human-robot interfaces. *Sci. Robot.* **2020**, *5*, eabc6878. [CrossRef]
28. Lee, G.H.; Park, J.K.; Byun, J.; Yang, J.C.; Kwon, S.Y.; Kim, C.; Jang, C.; Sim, J.Y.; Yook, J.G.; Park, S. Parallel Signal Processing of a Wireless Pressure-Sensing Platform Combined with Machine-Learning-Based Cognition, Inspired by the Human Somatosensory System. *Adv. Mater.* **2020**, *32*, e1906269. [CrossRef]
29. Xu, B.C.; Chen, D.; Wang, Y.; Tang, R.L.; Yang, L.A.; Feng, H.; Liu, Y.J.; Wang, Z.P.; Wang, F.; Zhang, T. Wireless and Flexible Tactile Sensing Array Based on an Adjustable Resonator with Machine-Learning Perception. *Adv. Electron. Mater.* **2023**, *9*, 2201334. [CrossRef]
30. Wu, X.; Ahmed, M.; Khan, Y.; Payne, M.E.; Zhu, J.; Lu, C.; Evans, J.W.; Arias, A.C. A potentiometric mechanotransduction mechanism for novel electronic skins. *Sci. Adv.* **2020**, *6*, eaba1062. [CrossRef]
31. Wu, C.G.; Han, L.; Dong, Y.P.; Guo, M.M.; Wang, R.; Si, J.W. Wireless Battery-Free Flexible Sensing System for Continuous Wearable Health Monitoring. *Adv. Mater. Technol.* **2023**, *8*, 2201662. [CrossRef]
32. Kim, J.; Kim, S.; Park, Y.L. Single-input single-output multi-touch soft sensor systems using band-pass filters. *Npj Flex. Electron.* **2022**, *6*, 65. [CrossRef]
33. Gao, G.; Yang, F.; Zhou, F.; He, J.; Lu, W.; Xiao, P.; Yan, H.; Pan, C.; Chen, T.; Wang, Z.L. Bioinspired Self-Healing Human-Machine Interactive Touch Pad with Pressure-Sensitive Adhesiveness on Targeted Substrates. *Adv. Mater.* **2020**, *32*, e2004290. [CrossRef]
34. Bai, H.; Li, S.; Barreiros, J.; Tu, Y.; Pollock, C.R.; Shepherd, R.F. Stretchable distributed fiber-optic sensors. *Science* **2020**, *370*, 848–852. [CrossRef]
35. Yang, Y.N.; Li, X.; Shen, Y.J. Electrode Array-Free Tactile Sensor for Addressable Force Sensing Assisted by a Neural Network. *ACS Appl. Polym. Mater.* **2022**, *4*, 4551–4557. [CrossRef]

36. Wu, W.; Ren, Y.; Jiang, T.; Hou, L.; Zhou, J.; Jiang, H. Anti-drying, transparent, ion-conducting, and tough organohydrogels for wearable multifunctional human–machine interfaces. *Chem. Eng. J.* **2022**, *430*, 132635. [CrossRef]
37. Wang, P.F.; Liu, P.; Feng, H.; Li, Y.W.; Zhang, Q.; Hu, R.H.; Liu, C.X.; Xing, K.; Song, A.G.; Yang, X.M.; et al. Flexible and Wireless Normal-Tangential Force Sensor Based on Resonant Mechanism for Robotic Gripping Applications. *Adv. Mater. Technol.* **2022**, *7*, 2101385. [CrossRef]
38. Nie, B.; Huang, R.; Yao, T.; Zhang, Y.; Miao, Y.; Liu, C.; Liu, J.; Chen, X. Textile-Based Wireless Pressure Sensor Array for Human-Interactive Sensing. *Adv. Funct. Mater.* **2019**, *29*, 1808786. [CrossRef]
39. Lee, G.H.; Lee, G.S.; Byun, J.; Yang, J.C.; Jang, C.; Kim, S.; Kim, H.; Park, J.K.; Lee, H.J.; Yook, J.G.; et al. Deep-Learning-Based Deconvolution of Mechanical Stimuli with $Ti_3C_2T_x$ MXene Electromagnetic Shield Architecture via Dual-Mode Wireless Signal Variation Mechanism. *ACS Nano* **2020**, *14*, 11962–11972. [CrossRef]
40. Ho, D.H.; Roe, D.G.; Choi, Y.Y.; Kim, S.; Choi, Y.J.; Kim, D.H.; Jo, S.B.; Cho, J.H. Non-von Neumann multi-input spike signal processing enabled by an artificial synaptic multiplexer. *Sci. Adv.* **2022**, *8*, eabn1838. [CrossRef]
41. Liao, X.; Song, W.; Zhang, X.; Yan, C.; Li, T.; Ren, H.; Liu, C.; Wang, Y.; Zheng, Y. A bioinspired analogous nerve towards artificial intelligence. *Nat. Commun.* **2020**, *11*, 268. [CrossRef]
42. Jang, J.; Kim, D.W.; Lee, J.H.; Choi, C.; Go, M.; Kim, J.K.; Jeong, U. Triboelectric UV patterning for wearable one-terminal tactile sensor array to perceive dynamic contact motions. *Nano Energy* **2022**, *98*, 107320. [CrossRef]
43. Guo, H.; Wang, H.B.; Xiang, Z.H.; Wu, H.X.; Wan, J.; Xu, C.; Chen, H.T.; Han, M.D.; Zhang, H.X. Soft Human-Machine Interface with Triboelectric Patterns and Archimedes Spiral Electrodes for Enhanced Motion Detection. *Adv. Funct. Mater.* **2021**, *31*, 2103075. [CrossRef]
44. Cheng, G.Y.; Xu, H.T.; Gao, N.F.; Zhang, M.Q.; Gao, H.L.; Sun, B.H.; Gu, M.X.; Yu, L.Y.; Lin, Y.C.; Liu, X.Q.; et al. Carbon nanotubes field-effect transistor pressure sensor based on three-dimensional conformal force-sensitive gate modulation. *Carbon* **2023**, *204*, 456–464. [CrossRef]
45. Lin, Y.C.; Cai, M.X.; Yang, Y.J. A wireless passive pressure sensor using microstructured ferromagnetic films with tunable effective permeability. *J. Micromech. Microeng.* **2021**, *31*, 045017. [CrossRef]
46. Deng, W.-J.; Wang, L.-F.; Dong, L.; Huang, Q.-A. LC Wireless Sensitive Pressure Sensors With Microstructured PDMS Dielectric Layers for Wound Monitoring. *IEEE Sens. J.* **2018**, *18*, 4886–4892. [CrossRef]
47. Banerjee, H.; Leber, A.; Laperrousaz, S.; La Polla, R.; Dong, C.; Mansour, S.; Wan, X.; Sorin, F. Soft Multimaterial Magnetic Fibers and Textiles. *Adv. Mater.* **2023**, e2212202. [CrossRef]
48. Dong, C.; Leber, A.; Yan, D.; Banerjee, H.; Laperrousaz, S.; Das Gupta, T.; Shadman, S.; Reis, P.M.; Sorin, F. 3D stretchable and self-encapsulated multimaterial triboelectric fibers. *Sci. Adv.* **2022**, *8*, eabo0869. [CrossRef]
49. Zhu, M.L.; He, T.Y.Y.; Lee, C.K. Technologies toward next generation human machine interfaces: From machine learning enhanced tactile sensing to neuromorphic sensory systems. *Appl. Phys. Rev.* **2020**, *7*, 031305. [CrossRef]
50. Duan, L.Y.; D'hooge, D.R.; Cardon, L. Recent progress on flexible and stretchable piezoresistive strain sensors: From design to application. *Prog. Mater. Sci.* **2020**, *114*, 100617. [CrossRef]
51. Xu, J.; Wang, Z.M.; Wang, X.X.; Wu, Y.L.; Xing, R.Z.; Yu, T.T.; Li, Y.; Ao, J.P.; Tao, Y.B.; Bai, B.W.; et al. Breathable Encapsulated Liquid Metal Foam-Based Soft Stress Sensor. *Adv. Mater. Technol.* **2023**, *8*, 2201193. [CrossRef]
52. Yu, T.T.; Zhang, D.G.; Wu, Y.L.; Guo, S.Z.; Lei, F.; Li, Y.; Yang, J.Y. Graphene foam pressure sensor based on fractal electrode with high sensitivity and wide linear range. *Carbon* **2021**, *182*, 497–505. [CrossRef]
53. Yang, J.Y.; Tang, D.; Ao, J.P.; Ghosh, T.; Neumann, T.V.; Zhang, D.G.; Piskarev, Y.; Yu, T.T.; Truong, V.K.; Xie, K.; et al. Ultrasoft Liquid Metal Elastomer Foams with Positive and Negative Piezopermittivity for Tactile Sensing. *Adv. Funct. Mater.* **2020**, *30*, 2002611. [CrossRef]
54. Liang, C.Y.; Wu, Y.L.; Zhang, D.G.; Tao, Y.B.; Yu, T.T.; Xing, R.Z.; Yang, J.Y. Biomimetic Liquid Metal Elastomer Foam With Stress Sensing. *IEEE Sens. J.* **2023**, *23*, 5841–5848. [CrossRef]
55. Xue, X.T.; Zhang, D.G.; Wu, Y.L.; Xing, R.Z.; Li, H.; Yu, T.T.; Bai, B.W.; Tao, Y.B.; Dickey, M.D.; Yang, J.Y. Segregated and Non-Settling Liquid Metal Elastomer via Jamming of Elastomeric Particles. *Adv. Funct. Mater.* **2023**, *33*, 2210553. [CrossRef]
56. Zhou, Z.H.; Chen, K.; Li, X.S.; Zhang, S.L.; Wu, Y.F.; Zhou, Y.H.; Meng, K.Y.; Sun, C.C.; He, Q.; Fan, W.J.; et al. Sign-to-speech translation using machine-learning-assisted stretchable sensor arrays. *Nat. Electron.* **2020**, *3*, 571–578. [CrossRef]
57. Lai, Y.C.; Lu, H.W.; Wu, H.M.; Zhang, D.G.; Yang, J.Y.; Ma, J.; Shamsi, M.; Vallem, V.; Dickey, M.D. Elastic Multifunctional Liquid-Metal Fibers for Harvesting Mechanical and Electromagnetic Energy and as Self-Powered Sensors. *Adv. Energy Mater.* **2021**, *11*, 2100411. [CrossRef]
58. Zhong, J.; Ma, Y.; Song, Y.; Zhong, Q.; Chu, Y.; Karakurt, I.; Bogy, D.B.; Lin, L. A Flexible Piezoelectret Actuator/Sensor Patch for Mechanical Human-Machine Interfaces. *ACS Nano* **2019**, *13*, 7107–7116. [CrossRef]
59. Wu, Y.; Li, S.; Zhao, Z.; Zhang, D.; Xiong, X.; Yu, T.; Dickey, M.D.; Yang, J. Convex Microarrays-Based Liquid Metal Soft Piezoresistive Stress Sensor With High Sensitivity and Large Measurement Range. *IEEE Sens. J.* **2023**, *23*, 9176–9182. [CrossRef]
60. Park, Y.L.; Majidi, C.; Kramer, R.; Berard, P.; Wood, R.J. Hyperelastic pressure sensing with a liquid-embedded elastomer. *J. Micromech. Microeng.* **2010**, *20*, 125029. [CrossRef]
61. Yeo, J.C.; Kenry; Yu, J.; Loh, K.P.; Wang, Z.; Lim, C.T. Triple-State Liquid-Based Microfluidic Tactile Sensor with High Flexibility, Durability, and Sensitivity. *ACS Sens.* **2016**, *1*, 543–551. [CrossRef]

62. Shi, L.; Li, Z.; Chen, M.; Qin, Y.; Jiang, Y.; Wu, L. Quantum effect-based flexible and transparent pressure sensors with ultrahigh sensitivity and sensing density. *Nat. Commun.* **2020**, *11*, 3529. [CrossRef]
63. Yun, G.; Tang, S.Y.; Sun, S.; Yuan, D.; Zhao, Q.; Deng, L.; Yan, S.; Du, H.; Dickey, M.D.; Li, W. Liquid metal-filled magnetorheological elastomer with positive piezoconductivity. *Nat. Commun.* **2019**, *10*, 1300. [CrossRef] [PubMed]
64. Chu, Y.; Zhong, J.W.; Liu, H.L.; Ma, Y.; Liu, N.; Song, Y.; Liang, J.M.; Shao, Z.C.; Sun, Y.; Dong, Y.; et al. Human Pulse Diagnosis for Medical Assessments Using a Wearable Piezoelectret Sensing System. *Adv. Funct. Mater.* **2018**, *28*, 1803413. [CrossRef]
65. Li, W.; Liu, G.; Jiang, D.; Wang, C.; Li, W.; Guo, T.; Zhao, J.; Xi, F.; Liu, W.; Zhang, C. Interdigitated Electrode-Based Triboelectric Sliding Sensor for Security Monitoring. *Adv. Mater. Technol.* **2018**, *3*, 1800189. [CrossRef]
66. Wu, X.D.; Khan, Y.; Ting, J.; Zhu, J.; Ono, S.; Zhang, X.X.; Du, S.X.; Evans, J.W.; Lu, C.H.; Arias, A.C. Large-Area Fabrication of High-Performance Flexible and Wearable Pressure Sensors. *Adv. Electron. Mater.* **2020**, *6*, 1901310. [CrossRef]
67. Yu, Z.; Ying, W.B.; Pravarthana, D.; Li, Y.Y.; Mao, G.Y.; Liu, Y.W.; Hu, C.; Zhang, W.X.; He, P.X.; Zhong, Z.C.; et al. Stretchable tactile sensor with high sensitivity and dynamic stability based on vertically aligned urchin-shaped nanoparticles. *Mater. Today Phys.* **2020**, *14*, 100219. [CrossRef]
68. Lee, H.; Kwon, D.; Cho, H.; Park, I.; Kim, J. Soft Nanocomposite Based Multi-point, Multi-directional Strain Mapping Sensor Using Anisotropic Electrical Impedance Tomography. *Sci. Rep.* **2017**, *7*, 39837. [CrossRef]
69. Duan, X.; Taurand, S.; Soleimani, M. Artificial skin through super-sensing method and electrical impedance data from conductive fabric with aid of deep learning. *Sci. Rep.* **2019**, *9*, 8831. [CrossRef]
70. Ning, C.; Dong, K.; Cheng, R.W.; Yi, J.; Ye, C.Y.; Peng, X.; Sheng, F.F.; Jiang, Y.; Wang, Z.L. Flexible and Stretchable Fiber-Shaped Triboelectric Nanogenerators for Biomechanical Monitoring and Human-Interactive Sensing. *Adv. Funct. Mater.* **2021**, *31*, 2006679. [CrossRef]
71. Zhang, D.G.; Zhong, Y.Q.; Wu, Y.L.; Zhang, X.F.; Dickey, M.D.; Yang, J.Y. Liquid metal elastomer with flytrap-inspired pillar structure for stress sensing. *Compos. Sci. Technol.* **2021**, *216*, 109066. [CrossRef]
72. Zhang, D.G.; Zhang, J.; Wu, Y.L.; Xiong, X.Y.; Yang, J.Y.; Dickey, M.D. Liquid Metal Interdigitated Capacitive Strain Sensor with Normal Stress Insensitivity. *Adv. Intell. Syst.* **2022**, *4*, 2100201. [CrossRef]
73. Zhang, D.G.; Wang, X.X.; Wu, Y.L.; Song, H.L.; Ma, Z.; Zhang, X.Y.; Yang, X.F.; Xing, R.Z.; Li, Y.; Yang, J.Y. Passive Particle Jamming Variable Stiffness Material-Based Flexible Capacitive Stress Sensor with High Sensitivity and Large Measurement Limit. *Adv. Mater. Technol.* **2021**, *6*, 2100106. [CrossRef]
74. Yang, J.Y.; Kwon, K.Y.; Kanetkar, S.; Xing, R.Z.; Nithyanandam, P.; Li, Y.; Jung, W.; Gong, W.; Tuman, M.; Shen, Q.C.; et al. Skin-Inspired Capacitive Stress Sensor with Large Dynamic Range via Bilayer Liquid Metal Elastomers. *Adv. Mater. Technol.* **2022**, *7*, 2101074. [CrossRef]
75. Zhao, K.; Han, J.; Ma, Y.; Tong, Z.; Suhr, J.; Wang, M.; Xiao, L.; Jia, S.; Chen, X. Highly Sensitive and Flexible Capacitive Pressure Sensors Based on Vertical Graphene and Micro-Pyramidal Dielectric Layer. *Nanomaterials* **2023**, *13*, 701. [CrossRef] [PubMed]
76. Hesam Mahmoudinezhad, M.; Anderson, I.; Rosset, S. Interdigitated Sensor Based on a Silicone Foam for Subtle Robotic Manipulation. *Macromol. Rapid Commun.* **2021**, *42*, e2000560. [CrossRef]
77. Ha, K.H.; Zhang, W.; Jang, H.; Kang, S.; Wang, L.; Tan, P.; Hwang, H.; Lu, N. Highly Sensitive Capacitive Pressure Sensors over a Wide Pressure Range Enabled by the Hybrid Responses of a Highly Porous Nanocomposite. *Adv. Mater.* **2021**, *33*, e2103320. [CrossRef]
78. Kim, S.; Jung, Y.; Oh, S.; Moon, H.; Lim, H. Parasitic Capacitance-Free Flexible Tactile Sensor with a Real-Contact Trigger. *Soft Robot.* **2022**, *9*, 119–127. [CrossRef]
79. Wang, Y.C.; Xi, K.L.; Liang, G.H.; Mei, M.G.; Chen, Z.C. A Flexible Capacitive Tactile Sensor Array for Prosthetic Hand Real-Time Contact Force Measurement. In Proceedings of the 2014 IEEE International Conference on Information and Automation (ICIA), Hailar, China, 28–30 July 2014; pp. 937–942.
80. Zhou, Q.; Ji, B.; Wei, Y.Z.; Hu, B.; Gao, Y.B.; Xu, Q.S.; Zhou, J.; Zhou, B.P. A bio-inspired cilia array as the dielectric layer for flexible capacitive pressure sensors with high sensitivity and a broad detection range. *J. Mater. Chem. A* **2019**, *7*, 27334–27346. [CrossRef]
81. Tang, Y.F.; Zheng, Q.F.; Chen, B.; Ma, Z.Q.; Gong, S.Q. A new class of flexible nanogenerators consisting of porous aerogel films driven by mechanoradicals. *Nano Energy* **2017**, *38*, 401–411. [CrossRef]
82. Xie, Y.N.; Wang, S.H.; Niu, S.M.; Lin, L.; Jing, Q.S.; Yang, J.; Wu, Z.Y.; Wang, Z.L. Grating-Structured Freestanding Triboelectric-Layer Nanogenerator for Harvesting Mechanical Energy at 85% Total Conversion Efficiency. *Adv. Mater.* **2014**, *26*, 6599–6607. [CrossRef] [PubMed]
83. Zhu, G.; Chen, J.; Zhang, T.; Jing, Q.; Wang, Z.L. Radial-arrayed rotary electrification for high performance triboelectric generator. *Nat. Commun.* **2014**, *5*, 3426. [CrossRef] [PubMed]
84. Zhou, K.K.; Zhao, Y.; Sun, X.P.; Yuan, Z.Q.; Zheng, G.Q.; Dai, K.; Mi, L.W.; Pan, C.F.; Liu, C.T.; Shen, C.Y. Ultra-stretchable triboelectric nanogenerator as high-sensitive and self-powered electronic skins for energy harvesting and tactile sensing. *Nano Energy* **2020**, *70*, 104546. [CrossRef]
85. Lin, W.; Wang, B.; Peng, G.; Shan, Y.; Hu, H.; Yang, Z. Skin-Inspired Piezoelectric Tactile Sensor Array with Crosstalk-Free Row+Column Electrodes for Spatiotemporally Distinguishing Diverse Stimuli. *Adv. Sci.* **2021**, *8*, 2002817. [CrossRef]
86. Liu, Q.; Jin, L.; Zhang, P.; Zhang, B.; Li, Y.; Xie, S.; Li, X. Nanofibrous Grids Assembled Orthogonally from Direct-Written Piezoelectric Fibers as Self-Powered Tactile Sensors. *ACS Appl. Mater. Interfaces* **2021**, *13*, 10623–10631. [CrossRef] [PubMed]

87. Liu, W.T.; Yu, P.; Gu, C.X.; Cheng, X.Y.; Fu, X. Fingertip Piezoelectric Tactile Sensor Array for Roughness Encoding Under Varying Scanning Velocity. *IEEE Sens. J.* **2017**, *17*, 6867–6879. [CrossRef]
88. Yu, P.; Liu, W.; Gu, C.; Cheng, X.; Fu, X. Flexible Piezoelectric Tactile Sensor Array for Dynamic Three-Axis Force Measurement. *Sensors* **2016**, *16*, 819. [CrossRef]
89. Li, L.Q.; He, R.J.; Soares, M.S.; Savovic, S.; Hu, X.H.; Marques, C.; Min, R.; Li, X.L. Embedded FBG-Based Sensor for Joint Movement Monitoring. *IEEE Sens. J.* **2021**, *21*, 26793–26798. [CrossRef]
90. Min, R.; Hu, X.; Pereira, L.; Simone Soares, M.; Silva, L.C.B.; Wang, G.; Martins, L.; Qu, H.; Antunes, P.; Marques, C.; et al. Polymer optical fiber for monitoring human physiological and body function: A comprehensive review on mechanisms, materials, and applications. *Opt. Laser Technol.* **2022**, *147*, 107626. [CrossRef]
91. Liu, S.; Shah, D.S.; Kramer-Bottiglio, R. Highly stretchable multilayer electronic circuits using biphasic gallium-indium. *Nat. Mater.* **2021**, *20*, 851–858. [CrossRef]
92. Wang, M.; Wang, K.; Ma, C.; Uzabakiriho, P.C.; Chen, X.; Zhao, G. Mechanical Gradients Enable Highly Stretchable Electronics Based on Nanofiber Substrates. *ACS Appl. Mater. Interfaces* **2022**, *14*, 35997–36006. [CrossRef] [PubMed]
93. Tang, L.; Shang, J.; Jiang, X. Multilayered electronic transfer tattoo that can enable the crease amplification effect. *Sci. Adv.* **2021**, *7*, eabe3778. [CrossRef] [PubMed]
94. Yang, J.Y.; Nithyanandam, P.; Kanetkar, S.; Kwon, K.Y.; Ma, J.W.; Im, S.; Oh, J.H.; Shamsi, M.; Wilkins, M.; Daniele, M.; et al. Liquid Metal Coated Textiles with Autonomous Electrical Healing and Antibacterial Properties. *Adv. Mater. Technol.* **2023**, *8*, 2202183. [CrossRef]
95. Song, J.; Tan, Y.; Chu, Z.; Xiao, M.; Li, G.; Jiang, Z.; Wang, J.; Hu, T. Hierarchical Reduced Graphene Oxide Ridges for Stretchable, Wearable, and Washable Strain Sensors. *ACS Appl. Mater. Interfaces* **2019**, *11*, 1283–1293. [CrossRef] [PubMed]
96. Suh, D.; Faseela, K.P.; Kim, W.; Park, C.; Lim, J.G.; Seo, S.; Kim, M.K.; Moon, H.; Baik, S. Electron tunneling of hierarchically structured silver nanosatellite particles for highly conductive healable nanocomposites. *Nat. Commun.* **2020**, *11*, 2252. [CrossRef]
97. Luo, Y.; Li, Y.; Sharma, P.; Shou, W.; Wu, K.; Foshey, M.; Li, B.; Palacios, T.; Torralba, A.; Matusik, W. Learning human–environment interactions using conformal tactile textiles. *Nat. Electron.* **2021**, *4*, 193–201. [CrossRef]
98. Liu, C.; Sun, Y.; Liu, P.; Ma, F.; Wu, S.; Li, J.; Li, S.; Hu, R.; Wang, Z.; Wang, Y.; et al. Fabrication and characterization of highly sensitive flexible strain sensor based on biodegradable gelatin nanocomposites and double strain layered structures with crack for gesture recognition. *Int. J. Biol. Macromol.* **2023**, *231*, 123568. [CrossRef]

micromachines

Review

Recent Advances in Tracking Devices for Biomedical Ultrasound Imaging Applications

Chang Peng [1,*], Qianqian Cai [2], Mengyue Chen [2] and Xiaoning Jiang [2,*]

1 School of Biomedical Engineering, ShanghaiTech University, Shanghai 201210, China
2 Department of Mechanical and Aerospace Engineering, North Carolina State University, Raleigh, NC 27695, USA
* Correspondence: pengchang@shanghaitech.edu.cn (C.P.); xjiang5@ncsu.edu (X.J.)

Abstract: With the rapid advancement of tracking technologies, the applications of tracking systems in ultrasound imaging have expanded across a wide range of fields. In this review article, we discuss the basic tracking principles, system components, performance analyses, as well as the main sources of error for popular tracking technologies that are utilized in ultrasound imaging. In light of the growing demand for object tracking, this article explores both the potential and challenges associated with different tracking technologies applied to various ultrasound imaging applications, including freehand 3D ultrasound imaging, ultrasound image fusion, ultrasound-guided intervention and treatment. Recent development in tracking technology has led to increased accuracy and intuitiveness of ultrasound imaging and navigation with less reliance on operator skills, thereby benefiting the medical diagnosis and treatment. Although commercially available tracking systems are capable of achieving sub-millimeter resolution for positional tracking and sub-degree resolution for orientational tracking, such systems are subject to a number of disadvantages, including high costs and time-consuming calibration procedures. While some emerging tracking technologies are still in the research stage, their potentials have been demonstrated in terms of the compactness, light weight, and easy integration with existing standard or portable ultrasound machines.

Keywords: tracking; ultrasound imaging; optical tracking; electromagnetic tracking; 3D ultrasound imaging; ultrasound–guided interventions; ultrasound image fusion

Citation: Peng, C.; Cai, Q.; Chen, M.; Jiang, X. Recent Advances in Tracking Devices for Biomedical Ultrasound Imaging Applications. *Micromachines* **2022**, *13*, 1855. https://doi.org/10.3390/mi13111855

Academic Editors: Xiao Xiao and Gang Ge

Received: 8 October 2022
Accepted: 27 October 2022
Published: 29 October 2022

1. Introduction

An object tracking system locates a moving object (or multiple objects) through time and space [1]. The main aim of a tracking system is to identify an object regarding the position and orientation in space recorded in an extension of time, characterized by precision, accuracy, working range as well as degree-of-freedom (DOF), depending on the systems and applications [2]. With the rapid development of computational and sensing technologies, nowadays tracking systems have been widely utilized in various fields, including robotics [3,4], military [5,6], medicine [7,8] and sports [9,10]. In the medical field, tracking of rotation and translation of medical instruments or patients plays a substantial role in many important applications, such as diagnostic imaging [11], image-guided navigation systems for intervention and therapy [12,13], as well as rehabilitation medicine [14].

Ultrasound (US) imaging is a well-established imaging modality that has been widely utilized in clinical practice for diagnosing diseases or guiding decision-making in therapy [15]. Compared with other medical imaging modalities, such as computed tomography (CT) and magnetic resonance imaging (MRI), US shows the major advantages of real-time imaging, non-radiation exposure, low-cost, and ease to apply [16]. Despite its many advantages, ultrasonography is considered to be highly operator-dependent [17]. Manually guiding of the US probe to obtain reproducible image acquisition is challenging. Moreover, in order to correctly interpret the information acquired by the scanning, rich clinical experience is required for sonographers. Besides the operator dependency that brings the high risk of interpretive error influencing the diagnosis and therapy results, the restricted field of view (FOV) of US probe poses challenges for image visualization and feature localization, thus limiting diagnosis or therapy accuracy. The integration of object tracking system with US imaging can resolve the above-mentioned limitations. By integrating tracking devices with US probes, an extended FOV of US probe can be obtained, resulting in a less operator-dependent scanning procedure and more accurate results. Over the past decade, there has been a significant growth of studies on integration of various tracking systems with US imaging systems for biomedical and healthcare applications. The applying of emerging tracking systems for biomedical US imaging applications has resulted in improved accuracy and intuitiveness of US imaging and navigation with less reliance on operator skills, thereby benefiting the medical diagnosis and therapy.

The purpose of this article is to provide a literature review on the various tracking systems for biomedical US imaging applications, as illustrated in Figure 1. The rest of the article is organized as follows: in Section 2, the principles of different tracking techniques, including optical tracking, electromagnetic tracking, mechanical tracking, acoustic tracking and inertial tracking, are summarized. The typical tracking systems and their technical performances, such as accuracy and latency are provided in Section 3. Section 4 details the advancement of different tracking systems for US imaging applications, including freehand 3D US imaging, US image fusion, US-guided diagnosis, and US-guided therapy. Finally, a summary and concluding remarks are presented in Section 5.

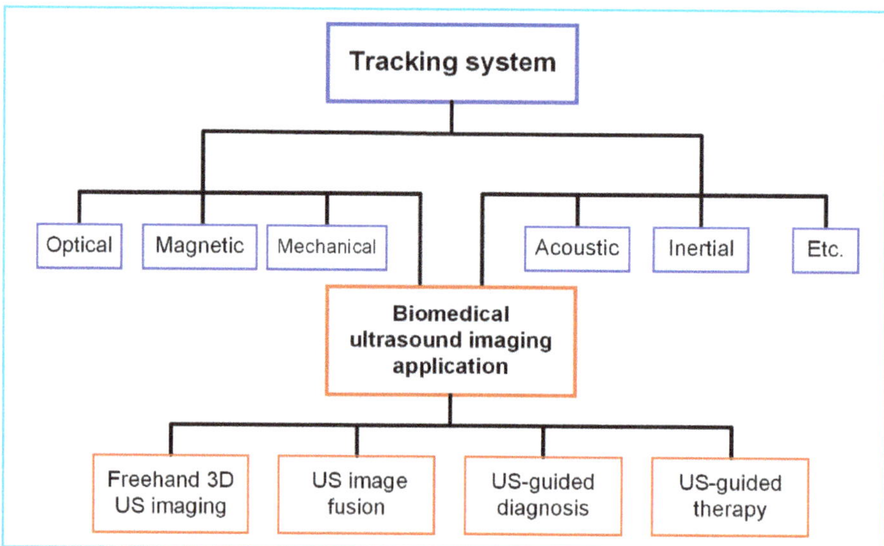

Figure 1. Various tracking systems for biomedical US imaging applications.

2. Physical Principles of Tracking Technologies

The latest advancements in tracking technologies have enabled conventional medical devices to be equipped with more advanced functions. In biomedical US imaging, object tracking technologies are key to locate US probes and other medical tools for precise operation and intuitive visualization. The underlying physical principles behind the most common tracking technologies will be reviewed in this section.

2.1. Optical Tracking

An optical tracking system is among the most precise tracking technologies with 6 DOF that achieves a sub-millimeter accuracy level. Multiple spatially synchronized cameras track the markers attached to the target in the designed space. There are two types of markers: active and passive [18]. Infrared light emitting diodes (LEDs) are used in active markers for the purpose of emitting invisible light that can be detected by cameras. A passive marker is covered with a retro-reflective surface that can reflect incoming infrared light back to the camera. There are usually three or more unsymmetrical markers in a target object. The 6 DOF position and orientation of the object are determined by triangulation [19].

Similar to the human vision system, optical tracking requires at least two cameras that are fixed at a known distance from each other. Adding additional cameras will improve tracking accuracy and robustness. In Figure 2, a trinocular vision system is illustrated as an example of triangulation [20]. P is observed simultaneously by three lenses. Additionally, it generates three projection points on the focal plane. A right-handed, Y-up coordinate is assigned to each lens, with coordinate origins at O_1, O_2, and O_3. The reference coordinate $X_cY_cZ_c$ coincides with the coordinate $X_2Y_2Z_2$ of the middle lens. A baseline l_{ij} is defined to be the distance between any two lenses. Three parallel principal axes are present in the three fixed lenses. There is a perpendicular relationship between the principal axes and the baseline.

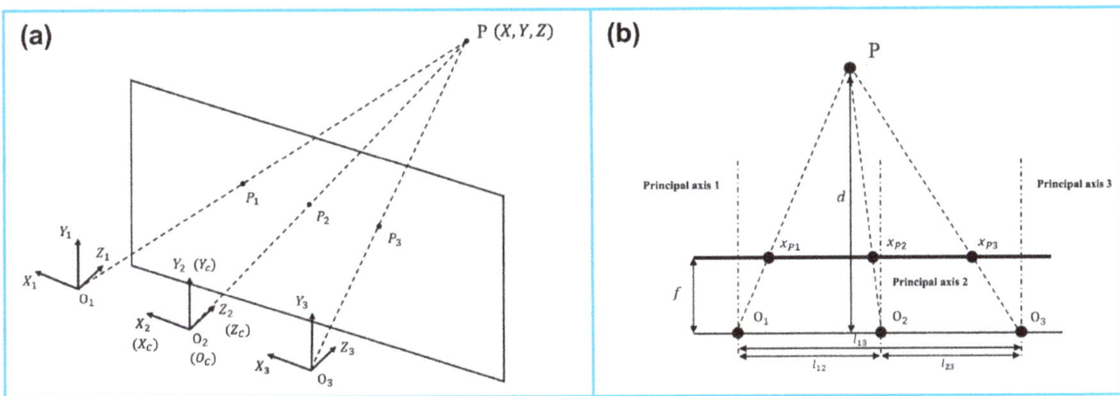

Figure 2. Principle of the camera-based optical tracking technique. (**a**) The three coordinates represent the three lenses integrated in the camera bar. The camera bar is looking at point $P(x, y, z)$, where P_1, P_2, and P_3 are the intersections on the image plane. (**b**) The top view demonstrates the similar triangles used to calculate the position information. The depth d of point P can be determined via triangulation. Reprinted from [20] with permission.

To perceive the depth with two lenses i and j, where $i, j \in \{1, 2, 3\}$, and $i \neq j$. With the known focal length f, and the disparity $x_{pi} - x_{pj}$ representing the offset between the two projections in the XOZ plane, the depth Z can be derived as

$$Z = \frac{f l_{ij}}{x_{pi} - x_{pj}} \tag{1}$$

Furthermore, the other two coordinates of $P(X, Y, Z)$ can thus be calculated as

$$\begin{aligned} X &= \frac{x_{pi} Z}{f} \\ Y &= \frac{y_{pj} Z}{f} \end{aligned} \tag{2}$$

Given all the markers' positions, the orientation of the marker set is determined. With the known positions of all markers, the orientation of the target is also determined. The 6 DOF poser information is delivered in the form of a transformation matrix T_M^C, with the subscript and superscript representing the marker set coordinate (M) to the camera coordinate (C), respectively.

$$T_M^C = \begin{bmatrix} \mathbf{R} & \mathbf{p} \\ \mathbf{0} & 1 \end{bmatrix} \quad (3)$$

where $p = (X_M, Y_M, Z_M)^T$ is the offset between the two origins of coordinates M and C. Additionally, R is a 3×3 rotational matrix in the form of

$$R = \begin{bmatrix} cos\varphi cos\theta & cos\varphi sin\theta sin\psi - sin\varphi cos\psi & cos\varphi sin\theta cos\psi + sin\varphi sin\psi \\ sin\varphi cos\theta & sin\varphi sin\theta sin\psi + cos\varphi cos\psi & sin\varphi sin\theta cos\psi - cos\varphi sin\psi \\ -sin\theta & cos\theta sin\psi & cos\theta cos\psi \end{bmatrix} \quad (4)$$

From the above matrix, the orientations φ (yaw), θ (pitch), and ψ (roll) can thus be solved as follows:

$$\begin{cases} \varphi = arctan\frac{R_{12}}{R_{11}} \\ \theta = -arcsinR_{31} \\ \psi = arctan\frac{R_{32}}{R_{33}} \end{cases} \quad (5)$$

2.2. Electromagnetic Tracking

Tracking systems using electromagnetic signals can also provide sub-millimeter accuracy in dynamic and real-time 6 DOF tracking. Its advantages include being lightweight and free of line-of-sight. In biomedical engineering, it is commonly used for the navigation of medical tools.

Electromagnetic tracking systems consist of four modules: transmission circuits, receiving circuits, digital signal processing units, and microcontrollers. Based on Faraday's law, electromagnetic tracking systems use transmitted voltages to estimate the position and orientation of objects in alternating magnetic fields when the object is coupled to a receiver sensor [21,22].

Figure 3 illustrates the involved coordinate systems. The reference coordinate system is denoted as $X_0Y_0Z_0$, which is fixed at the emission coil. $X_sY_sZ_s$ is the coordinate system fixed at the receiver sensor. The location of its origin $O_1(X, Y, Z)$ with respect to $X_0Y_0Z_0$, can also be denoted as $O_1(R, \alpha, \beta)$ in spherical coordinate. Additionally, the orientation is represented as Euler angles φ, θ, and ψ.

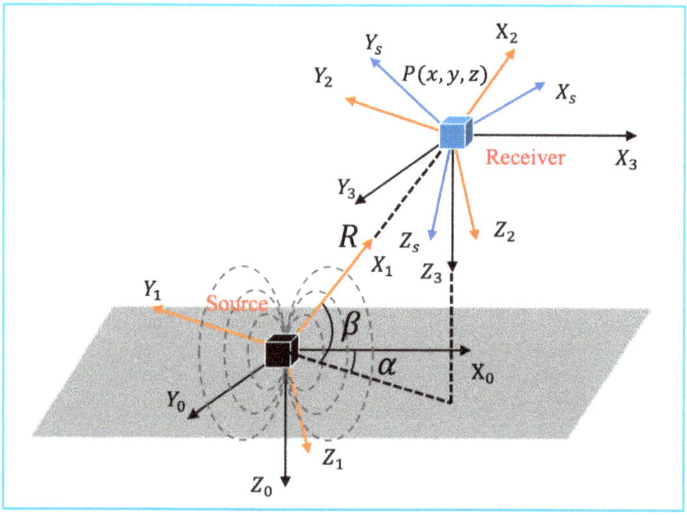

Figure 3. Coordinate systems involved in the 3D electromagnetic tracking (adapted from [21]).

Assuming the excitation current $\mathrm{i(t) = I_i \sin(\omega t + \phi)}$, the transmitter parameter along each direction defined as $\mathrm{C_i} = \frac{\mu N_i I_i S_i}{4\pi}$, where $i = x,\ y,\ z$, N_i is the number of turns in the coil, and S_i is the area of the coil, the excitation signal can be defined as

$$f_0 = C = \begin{bmatrix} C_x & 0 & 0 \\ 0 & C_y & 0 \\ 0 & 0 & C_z \end{bmatrix} \tag{6}$$

Accordingly, the receiver parameter can be written as

$$K = \begin{bmatrix} K_x & 0 & 0 \\ 0 & K_y & 0 \\ 0 & 0 & K_z \end{bmatrix} \tag{7}$$

where $\mathrm{k_i} = \omega \mathrm{n_i g_i s_i}$, with ω representing the radian frequency of the source excitation signal, $\mathrm{n_i}$ and $\mathrm{s_i}$ denoting the number of turns in the coil and the area of the coil, and g_i indicating the system gain. According to Faraday's law of induction, the amplitude of the voltage from the receiver coil is expressed as

$$S_{ij} = \omega K_j C_i h(R, \alpha, \beta, \varphi, \theta, \psi) = \omega K_j B_{ij} \tag{8}$$

With the position and orientation of the receiver fixed, the value of $\mathrm{h}(\mathrm{R}, \alpha, \beta, \varphi, \theta, \psi)$ is determined. $\mathrm{B_{ij}}$ is the amplitude of the magnetic field produced at that location.

From Equation (8), by defining the final sensor output to be

$$\mathrm{f_s} = \begin{bmatrix} \mathrm{s_{xx}} & \mathrm{s_{yx}} & \mathrm{s_{zx}} \\ \mathrm{s_{xy}} & \mathrm{s_{yy}} & \mathrm{s_{zy}} \\ \mathrm{s_{xz}} & \mathrm{s_{yz}} & \mathrm{s_{zz}} \end{bmatrix} \tag{9}$$

the magnetic field expressed in $\mathrm{X_s Y_s Z_s}$ is

$$B = K^{-1} f_s = \begin{bmatrix} \frac{\mathrm{s_{xx}}}{\mathrm{k_x}} & \frac{\mathrm{s_{xy}}}{\mathrm{k_y}} & \frac{\mathrm{s_{xz}}}{\mathrm{k_x}} \\ \frac{\mathrm{s_{yx}}}{\mathrm{k_y}} & \frac{\mathrm{s_{yy}}}{\mathrm{k_y}} & \frac{\mathrm{s_{yz}}}{\mathrm{k_y}} \\ \frac{\mathrm{s_{zx}}}{\mathrm{k_z}} & \frac{\mathrm{s_{zy}}}{\mathrm{k_z}} & \frac{\mathrm{s_{zz}}}{\mathrm{k_z}} \end{bmatrix} \tag{10}$$

When the equivalent transmitter coil along each direction is excited, the square amplitude of the magnetic field P can be expressed as

$$P = \begin{bmatrix} \frac{\mathrm{s_{xx}}}{\mathrm{k_x}}^2 + \frac{\mathrm{s_{yx}}}{\mathrm{k_x}}^2 + \frac{\mathrm{s_{zx}}}{\mathrm{k_x}}^2 \\ \frac{\mathrm{s_{xy}}}{\mathrm{k_y}}^2 + \frac{\mathrm{s_{yy}}}{\mathrm{k_y}}^2 + \frac{\mathrm{s_{zy}}}{\mathrm{k_y}}^2 \\ \frac{\mathrm{s_{xz}}}{\mathrm{k_z}}^2 + \frac{\mathrm{s_{yz}}}{\mathrm{k_z}}^2 + \frac{\mathrm{s_{zz}}}{\mathrm{k_z}}^2 \end{bmatrix} = \begin{bmatrix} \frac{4C_x^2}{r^6}\left(x^2 + \frac{1}{4}y^2 + \frac{1}{4}z^2\right) \\ \frac{4C_y^2}{r^6}\left(\frac{1}{4}x^2 + y^2 + \frac{1}{4}z^2\right) \\ \frac{4C_z^2}{r^6}\left(\frac{1}{4}x^2 + \frac{1}{4}y^2 + z^2\right) \end{bmatrix} \tag{11}$$

Canceling out the unknown position $(x,\ y,\ z)$ by summing up all three entries, the only unknown parameter r can be deduced.

With two rotational matrices $T(\alpha)$ and $T(\beta)$ as

$$\begin{gathered} T(\alpha) = \begin{bmatrix} cos\alpha & sin\alpha & 0 \\ -sin\alpha & cos\alpha & 0 \\ 0 & 0 & 1 \end{bmatrix} \\ T(\beta) = \begin{bmatrix} cos\beta & 0 & -sin\beta \\ 0 & 1 & 0 \\ sin\beta & 0 & cos\beta \end{bmatrix} \end{gathered} \tag{12}$$

A key matrix F can be defined as

$$
\begin{aligned}
F &= r^6 \left(f_0^T\right)^{-1} f_s^T K^{-1} K^{-1} f_s f_0^{-1} \\
&= \begin{bmatrix} 1+3\cos^2\alpha\cos^2\beta & 3\sin\alpha\cos\alpha\cos^2\beta & -3\cos\alpha\sin\beta\cos\beta \\ 3\sin\alpha\cos\alpha\cos^2\beta & 1+3\sin^2\alpha\cos^2\alpha & -3\sin\alpha\sin\beta\cos\beta \\ -3\cos\alpha\sin\beta\cos\beta & -3\sin\alpha\sin\beta\cos\beta & 1+3\sin^2\beta \end{bmatrix}
\end{aligned} \tag{13}
$$

As s_{ij}, ω, C_i, K_j as known parameters, the unknown α and β can be solved as

$$
\begin{cases} \alpha = arctan\frac{F_{23}}{F_{13}} \\ \beta = arcsin\sqrt{\frac{F_{33}-1}{3}} \end{cases} \tag{14}
$$

From the spherical coordinates, the position of the target P can be written as

$$
\begin{cases} x = rcos\beta cos\alpha \\ y = rcos\beta sin\alpha \\ z = rsin\beta \end{cases} \tag{15}
$$

Substituting

$$
\begin{aligned}
T(\varphi) &= \begin{bmatrix} 1 & 0 & 0 \\ 0 & cos\varphi & sin\varphi \\ 0 & -sin\varphi & cos\varphi \end{bmatrix} \\
T(\theta) &= \begin{bmatrix} cos\theta & 0 & -sin\theta \\ 0 & 1 & 0 \\ sin\theta & 0 & cos\theta \end{bmatrix} \\
T(\psi) &= \begin{bmatrix} cos\psi & sin\psi & 0 \\ -sin\psi & cos\psi & 0 \\ 0 & 0 & 1 \end{bmatrix}
\end{aligned} \tag{16}
$$

into

$$
f_s = KB = \frac{2}{r^3} KT(\varphi)T(\theta)T(\psi)T(-\alpha)T(-\beta)ST(\beta)T(\alpha)f_0 \tag{17}
$$

matrix T is defined as

$$
\begin{aligned}
T &= T(\varphi)T(\theta)T(\psi) \\
&= \begin{bmatrix} \cos\varphi\cos\psi & \cos\varphi\sin\psi & -\sin\varphi \\ \sin\theta\sin\varphi\cos\psi - \cos\theta\sin\psi & \sin\varphi\sin\theta\sin\psi + \cos\theta\cos\psi & \sin\theta\cos\varphi \\ \cos\theta\sin\varphi\cos\psi + \sin\theta\sin\psi & \cos\theta\sin\varphi\sin\psi + \sin\theta\cos\psi & \cos\theta\cos\varphi \end{bmatrix}
\end{aligned} \tag{18}
$$

Therefore, the target's orientation can be solved as

$$
\begin{cases} \varphi = arctan\left(-\frac{T_{13}}{\left(T_{23}^2+T_{33}^2\right)^{\frac{1}{2}}}\right) \\ \theta = arctan\left(\frac{T_{23}}{T_{33}}\right) \\ \psi = arctan\left(\frac{T_{12}}{T_{11}}\right) \end{cases} \tag{19}
$$

2.3. Mechanical Tracking

Robotic tracking systems use articulated robotic arms to manipulate the target attached to the end effector. Typically, industrial robots are composed of a number of joints and links. Joint movement is continuously detected by potentiometers and encoders in-stalled on each joint. The real-time position and orientation of the effector can be deter-mined by calculating homogeneous transformations from the collected robotic dynamics. In clinical practice, the operator can either control the movement of the robot to a certain location with the desired orientation, or specify the destination, and the robot solves the path using inverse dynamics based on the spatial information of the destination and the architecture of the robot. Following the Denavit and Hartenberg notation, the forward dynamics will be applied to illustrate how the 6 DOF pose information is transformed be-tween adjacent joints and links, as illustrated in Figure 4 [23].

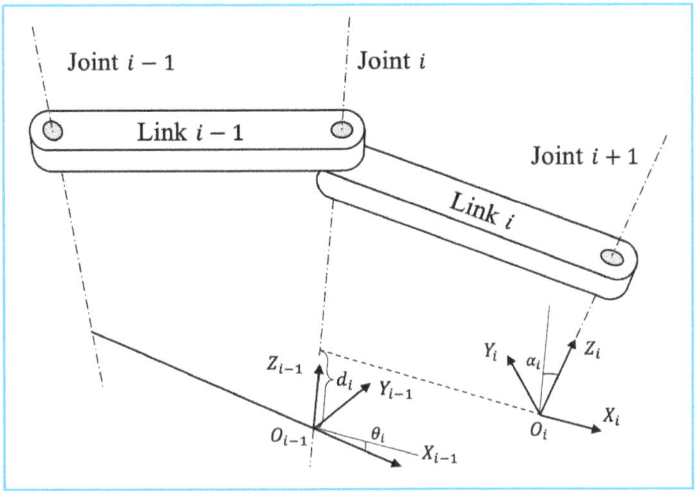

Figure 4. Two adjacent links with a revolute joint with the relative position denoted using Denavit-Hartenberg convention (adapted from [23]).

The 4×4 homogeneous transformation matrix for each step is shown below.

$$A_1 = \begin{bmatrix} cos\theta_i & -sin\theta_i & 0 & 0 \\ sin\theta_i & cos\theta_i & 0 & 0 \\ 0 & 0 & 1 & 0 \\ 0 & 0 & 0 & 1 \end{bmatrix}, A_2 = \begin{bmatrix} 1 & 0 & 0 & a_i \\ 0 & 1 & 0 & 0 \\ 0 & 0 & 1 & 0 \\ 0 & 0 & 0 & 1 \end{bmatrix}$$
$$A_3 = \begin{bmatrix} 1 & 0 & 0 & 0 \\ 0 & 1 & 0 & 0 \\ 0 & 0 & 1 & d_i \\ 0 & 0 & 0 & 1 \end{bmatrix}, A_1 = \begin{bmatrix} 1 & 0 & 0 & 0 \\ 0 & cos\alpha_i & -sin\alpha_i & 0 \\ 0 & sin\alpha_i & cos\alpha_i & 0 \\ 0 & 0 & 0 & 1 \end{bmatrix} \quad (20)$$

$$\begin{bmatrix} x_i \\ y_i \\ z_i \\ 1 \end{bmatrix} = \underbrace{A_1 A_2 A_3 A_4}_{T} \begin{bmatrix} x_i \\ y_i \\ z_i \\ 1 \end{bmatrix} = \begin{bmatrix} cos\theta_i & -sin\theta_i cos\alpha_i & sin\theta_i sin\alpha_i & a_i cos\theta_i \\ sin\theta_i & cos\theta_i cos\alpha_i & -cos\theta_i sin\alpha_i & a_i sin\theta_i \\ 0 & sin\alpha_i & cos\alpha_i & d_i \\ 0 & 0 & 0 & 1 \end{bmatrix} \begin{bmatrix} x_i \\ y_i \\ z_i \\ 1 \end{bmatrix} \quad (21)$$

The transformation matrix T can also be represented in terms of position (p_x, p_y, p_z) in the reference coordinate and orientation (φ, θ, ψ) in yaw-pitch-roll representation.

$$T = \begin{bmatrix} cos\varphi cos\theta & cos\varphi sin\theta sin\psi - sin\varphi cos\psi & cos\varphi sin\theta cos\psi + sin\varphi sin\psi & p_x \\ sin\varphi cos\theta & sin\varphi sin\theta sin\psi + cos\varphi cos\psi & sin\varphi sin\theta cos\psi - cos\varphi sin\psi & p_y \\ -sin\theta & cos\theta sin\psi & cos\theta cos\psi & p_z \\ 0 & 0 & 0 & 1 \end{bmatrix} \quad (22)$$

Based on Equations (21) and (22), the 6 DOF pose information can be solved as

$$\begin{cases} p_x = a_i cos\theta_i \\ p_y = a_i sin\theta_i \\ p_z = d_i \end{cases} \quad (23)$$

$$\begin{cases} \theta = arcsin(-T_{31}) \\ \varphi = arccos\left(\frac{T_{33}}{cos\theta}\right) \\ \psi = arccos\left(\frac{T_{11}}{cos\theta}\right) \end{cases} \tag{24}$$

2.4. Acoustic Tracking

An acoustic tracking system is one of the three DOF positional tracking systems. To determine the spatial location of the target object, an ultrasonic transmitter transmits a carrier signal that is received by multiple receivers operating at the same frequency. Specifically, by estimating the actual travel/arrival times (TOF/TOA) or the time difference between travel/arrival (TDOF/TDOA), the 3D coordinates of the object (x, y, z) can be determined to centimeter accuracy levels with receivers fixed at known locations, as shown in Figure 5. A TDOF/TDOA algorithm is more practical and accurate than a TOF/TOA algorithm since it circumvents the synchronization issue between the transmitter and receiver. A limitation of acoustic tracking is that the accuracy of the tracking is affected by the temperature and air turbulence in the environment [24]. This problem can be addressed by including the speed of sound (c) as an unknown parameter in the calculation [25].

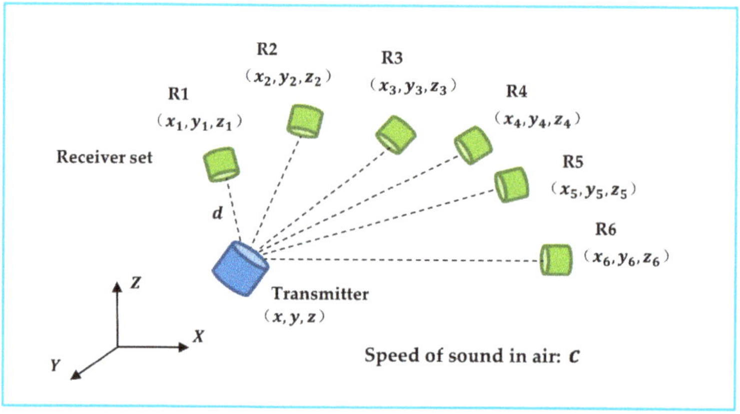

Figure 5. The transmitter sends out carrier signals to the receiver set. The unknow transmitter's position and the speed of sound can be solved based of the time difference of arrival.

The predetermined geometry of the receivers was notated as (x_i, y_i, z_i), where i represents the ith receiver, where i {1, 2, 3, 4, 5, 6}. The reference distance between the transmitter and receiver R1 is denoted as d. ΔT_{1j} indicates the TDOF between receiver $R1$ and Rj, where $j \in \{2, 3, 4, 5, 6\}$.

$$\begin{bmatrix} 2x_1-2x_2 & 2y_1-2y_2 & 2z_1-2z_2 & -2\Delta T_{12} & -2\Delta T_{12}^2 \\ 2x_1-2x_3 & 2y_1-2y_3 & 2z_1-2z_3 & -2\Delta T_{13} & -2\Delta T_{13}^2 \\ 2x_1-2x_4 & 2y_1-2y_4 & 2z_1-2z_4 & -2\Delta T_{14} & -2\Delta T_{14}^2 \\ 2x_1-2x_5 & 2y_1-2y_5 & 2z_1-2z_5 & -2\Delta T_{15} & -2\Delta T_{15}^2 \\ 2x_1-2x_6 & 2y_1-2y_6 & 2z_1-2z_6 & -2\Delta T_{16} & -2\Delta T_{16}^2 \end{bmatrix} \begin{bmatrix} x \\ y \\ z \\ cd \\ c^2 \end{bmatrix} = \begin{bmatrix} x_1^2+y_1^2+z_1^2-x_2^2-y_2^2-z_2^2 \\ x_1^2+y_1^2+z_1^2-x_3^2-y_3^2-z_3^2 \\ x_1^2+y_1^2+z_1^2-x_4^2-y_4^2-z_4^2 \\ x_1^2+y_1^2+z_1^2-x_5^2-y_5^2-z_5^2 \\ x_1^2+y_1^2+z_1^2-x_6^2-y_6^2-z_6^2 \end{bmatrix} \tag{25}$$

Occlusion can also affect the accuracy of an acoustic tracking system. A receiver configuration should be taken into serious consideration when implementing such a system for biomedical US imaging applications [26]. In addition to reducing occlusion, an optimal configuration also contributes to improved tracking performance. As an acoustic tracking system is not able to identify a target's object, other tracking systems, such as inertial tracking, are frequently required [27].

2.5. Inertial Tracking

Inertial tracking systems are based on an inertial measurement unit (IMU), which is a small, lightweight, cost-effective sensor enabled by microelectromechanical systems (MEMS) (Figure 6) [28]. An IMU sensor with 9 axes that integrates accelerometers, gyroscopes, and magnetometers is commonly used for 6 DOF object tracking. Accelerometers measure the target's acceleration. The angular velocity of a target is measured by a gyroscope. Additionally, a magnetometer detects the magnetic field strength at the target's location. With sensor fusion of the raw measurements, an IMU sensor is able to obtain more accurate readings [29]. As a result, after calibration and compensation for drifts and errors, the position and orientation of the target can be determined [27].

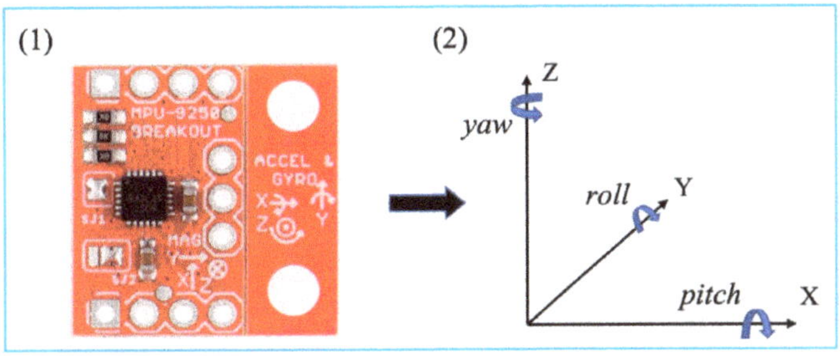

Figure 6. MPU 9250 and its coordinate system with Euler angles. Reprinted from [27] with permission.

The tri-axial measurements of the accelerometer are accelerations of each axis, where $a = [a_x, a_y, a_z]^T = \left[\frac{d^2x}{dt^2}, \frac{d^2y}{dt^2}, \frac{d^2z}{dt^2}\right]^T$. Readings from the gyroscope indicate the angular rates of the sensor when rotated, where $\omega = [\omega_x, \omega_y, \omega_z]^T = \left[\frac{d^2\varphi}{dt^2}, \frac{d^2\theta}{dt^2}, \frac{d^2\psi}{dt^2}\right]^T$. Additionally, $[pitch, roll, yaw]^T$ is denoted as $\mathbf{\Phi} = [\varphi, \theta, \psi]^T$.

By taking integration of the angular velocity from time t_{k-1} to t_k,

$$\mathbf{\Phi} = \int_{t_{k-1}}^{t_k} m\omega(\tau)\, d \tag{26}$$

where τ is the discrete time. The solution of the orientation, under the assumption, can be written as

$$\mathbf{\Phi} = \frac{\omega_{k-1} + \omega_k}{2}(t_k - t_{k-1}) \tag{27}$$

For simplicity, three rotation matrices were defined as follows.

$$R_{pitch} = \begin{bmatrix} cos\theta & 0 & -sin\theta \\ 0 & 1 & 0 \\ sin\theta & 0 & cos\theta \end{bmatrix}$$

$$R_{roll} = \begin{bmatrix} 1 & 0 & 0 \\ 0 & cos\psi & sin\psi \\ 0 & -sin\psi & cos\psi \end{bmatrix}$$

$$R_{yaw} = \begin{bmatrix} cos\varphi & -sin\varphi & 0 \\ sin\varphi & cos\varphi & 0 \\ 0 & 0 & 1 \end{bmatrix}$$

The rotational matrix R is expressed as

$$\boldsymbol{R} = R_{pitch} \cdot R_{roll} \cdot R_{yaw} \tag{28}$$

Due to the effect of the Earth's gravity,

$$\dot{\upsilon} = \boldsymbol{R}\boldsymbol{a} - \boldsymbol{g} \tag{29}$$

where υ is the velocity of the object and $\boldsymbol{g} = [0, 0, 9.8]^T$.

According to the midpoint method, the velocity

$$\upsilon_k = \upsilon_{k-1} + \left(\frac{\boldsymbol{R}_{k-1}\boldsymbol{a}_{k-1} + \boldsymbol{R}_k\boldsymbol{a}_k}{2} - \boldsymbol{g} \right)(t_k - t_{k-1}) \tag{30}$$

Thus, the position $\boldsymbol{p} = [x, y, z]^T$ at time k is

$$\boldsymbol{p}_k = \boldsymbol{p}_{k-1} + \upsilon_{k-1}(t_k - t_{k-1}) + \frac{1}{2}\left(\frac{\boldsymbol{R}_{k-1}\boldsymbol{a}_{k-1} + \boldsymbol{R}_k\boldsymbol{a}_k}{2} - \boldsymbol{g} \right)(t_k - t_{k-1})^2 \tag{31}$$

3. Tracking Systems

Different types of tracking systems have been developed and marketed over the past decade. In this section, the main tracking systems in the market are reviewed in terms of their technical specifications.

3.1. Optical Tracking Systems

A large number of manufacturers have developed various kinds of optical tracking systems for biomedical applications, as summarized in Table 1. The main technical specifications that relate to the tracking performances are measurement volume (or FOV), resolution, volumetric accuracy, average latency, and measurement rates. Due to angle of view, the shape of measurement volume is usually a pyramid, which can be represented by radius × width × height. Some manufacturers prefer to use the term "field of view", i.e., horizontal degree × vertical degree, to show the dimension of work volume. In addition, the advances of cameras promote the resolution of the image, and further increase the volumetric accuracy. To date, some advanced optical tracking system can obtain 26 megapixels (MP) resolution with 0.03 mm volumetric accuracy [30]. However, the high resolution of the captured images will burden the processor for data analysis, causing the increase of average latency and reduce of measurement rates.

Table 1. Summary of commercially available optical tracking systems.

Manufacturer	Model	Measurement Volume (Radius × Width × Height) or FOV	Resolution	Volumetric Accuracy (RMS)	Average Latency	Measurement Rate
Northern Digital Inc., Waterloo, ON, Canada	Polaris Vega ST [31]	2400 × 1566 × 1312 mm^3 (Pyramid Volume:) 3000 × 1856 × 1470 mm^3 (Extended Pyramid)	N/A	0.12 mm (Pyramid Volume) 0.15 mm (Extended Pyramid)	<16 ms	60 Hz
	Polaris Vega VT [32]	2400 × 1566 × 1312 mm^3 (Pyramid Volume) 3000 × 1856 × 1470 mm^3 (Extended Pyramid)	N/A	0.12 mm (Pyramid Volume) 0.15 mm (Extended Pyramid)	<16 ms	60 Hz
	Polaris Vega XT [33]	2400 × 1566 × 1312 mm^3 (Pyramid Volume) 3000 × 1856 × 1470 mm^3 (Extended Pyramid)	N/A	0.12 mm (Pyramid Volume) 0.15 mm (Extended Pyramid)	<3 ms	400 Hz
	Polaris Vicra [34]	1336 × 938 × 887 mm^3	N/A	0.25 mm	N/A	20 Hz

Table 1. *Cont.*

Manufacturer	Model	Measurement Volume (Radius × Width × Height) or FOV	Resolution	Volumetric Accuracy (RMS)	Average Latency	Measurement Rate
ClaroNav Inc., Toronto, ON, Canada	H3-60 [35]	2400 × 2000 × 1600 mm^3	1280 × 960	0.20 mm	~60 ms	16 Hz
	SX60 [35]	1150 × 700 × 550 mm^3	640 × 480	0.25 mm	~20 ms	48 Hz
	HX40 [35]	1200 × 1200 × 900 mm^3	1024 × 768	0.20 mm	~50 ms	20 Hz
	HX60 [35]	2000 × 1300 × 1000 mm^3	1024 × 768	0.35 mm	~50 ms	20 Hz
BTS Bioengineering Corp., Quincy, MA, USA	SMART DX 100 [36]	2000 × 2000 × 2000 mm^3	0.3 MP	<0.20 mm	N/A	280 FPS
	SMART DX 400 [36]	4000 × 3000 × 3000 mm^3	1.0 MP	<0.30 mm	N/A	300 FPS
	SMART DX 700 [36]	4000 × 3000 × 3000 mm^3	1.5 MP	<0.10 mm	N/A	1000 FPS
	SMART DX 6000 [36]	4000 × 3000 × 3000 mm^3	2.2 MP	<0.10 mm	N/A	2000 FPS
	SMART DX 7000 [36]	6000 × 3000 × 3000 mm^3	4.0 MP	<0.10 mm	N/A	2000 FPS
NaturalPoint, Inc., Corvallis, OR, USA	OptiTrack $Prime^X$ 41 [37]	FOV 51° × 51°	4.1 MP	0.10 mm	5.5 ms	250+ FPS
	OptiTrack $Prime^X$ 22 [38]	FOV 79° × 47°	2.2 MP	0.15 mm	2.8 ms	500+ FPS
	OptiTrack $Prime^X$ 13 [39]	FOV 56° × 46°	1.3 MP	0.20 mm	4.2 ms	1000 FPS
	OptiTrack $Prime^X$ 13W [40]	FOV 82° × 70°	1.3 MP	0.30 mm	4.2 ms	1000 FPS
	OptiTrack $Slim^X$ 13 [41]	FOV 82° × 70°	1.3 MP	0.30 mm	4.2 ms	1000 FPS
	OptiTrack V120: Trio [42]	FOV 47° × 43°	640 × 480	N/A	8.33 ms	120 FPS
	OptiTrack V120: Duo [43]	FOV 47° × 43°	640 × 480	N/A	8.33 ms	120 FPS
	OptiTrack Flex 13 [44]	FOV 56° × 46°	1.3 MP	N/A	8.3 ms	120 FPS
	OptiTrack Flex 3 [45]	FOV 58° × 45°	640 × 480	N/A	10 ms	100 FPS
	OptiTrack Slim 3U [46]	FOV 58° × 45°	640 × 480	N/A	8.33 ms	120 FPS
	TrackIR 4 [47]	46° (Horizontal)	355 × 288	N/A	N/A	120 FPS
	TrackIR 5 [47]	51.7° (Horizontal)	640 × 480	N/A	N/A	120 FPS
Qualisys Inc., Gothenburg, Sweden	Arqus A5 [30]	FOV 77° × 62°	5 MP (normal) 1MP (high-speed)	0.06 mm	N/A	700 FPS (normal) 1400 FPS (high-speed)
	Arqus A9 [30]	FOV 82° × 48°	9 MP (normal) 2.5 MP (high-speed)	0.05 mm	N/A	300 FPS (normal) 590 FPS (high-speed)
	Arqus A12 [30]	FOV 70° × 56°	12 MP (normal) 3 MP (high-speed)	0.04 mm	N/A	300 FPS (normal) 1040 FPS (high-speed)
	Arqus A26 [30]	FOV 77° × 77°	26 MP (normal) 6.5 MP (high-speed)	0.03 mm	N/A	150 FPS (normal) 290 FPS (high-speed)
	Miqus M1 [48]	FOV 58° × 40°	1 MP	0.14 mm	N/A	250 FPS
	Miqus M3 [48]	FOV 80° × 53°	2 MP (normal) 0.5 MP (high-speed)	0.11 mm	N/A	340 FPS (normal) 650 FPS (high-speed)

Table 1. *Cont.*

Manufacturer	Model	Measurement Volume (Radius × Width × Height) or FOV	Resolution	Volumetric Accuracy (RMS)	Average Latency	Measurement Rate
Qualisys Inc., Gothenburg, Sweden	Miqus M5 [48]	FOV 49° × 49°	4 MP (normal) 1 MP (high-speed)	0.07 mm	N/A	180 FPS (normal) 360 FPS (high-speed)
	Miqus Hybrid [49]	FOV 62° × 37°	2 MP	N/A	N/A	340 FPS
	3+ [50]	N/A	1.3 MP (normal) 0.3 MP (high-speed)	N/A	N/A	500 FPS (normal) 1750 FPS (high-speed)
	5+ [50]	49° (Horizontal)	4 MP (normal) 1 MP (high-speed)	N/A	N/A	180 FPS (normal) 360 FPS (high-speed)
	6+ [50]	56° (Horizontal)	6 MP (normal) 1.5 MP (high-speed)	N/A	N/A	450 FPS (normal) 1660 FPS (high-speed)
	7+ [50]	54° (Horizontal)	12 MP (normal) 3 MP (high-speed)	N/A	N/A	300 FPS (normal) 1100 FPS (high-speed)
Vicon Industries Inc., Hauppauge, NY, USA	Valkyrie VK26 [51]	FOV 72° × 72°	26.2 MP	N/A	N/A	150 FPS
	Valkyrie VK16 [51]	FOV 72° × 56°	16.1 MP	N/A	N/A	300 FPS
	Valkyrie VK8 [51]	FOV 72° × 42°	8.0 MP	N/A	N/A	500 FPS
	Valkyrie VKX [51]	FOV 66° × 66°	7.2 MP	N/A	N/A	380 FPS
	Vantage+ V16 [52]	FOV 76.4° × 76.4°	16 MP (normal) 4.2 MP (high-speed)	N/A	8.3 ms	120 FPS (normal) 500 FPS (high-speed)
	Vantage+ V8 [52]	FOV 61.7° × 47°	8 MP (normal) 2.2 MP (high-speed)	N/A	5.5 ms	260 FPS (normal) 910 FPS (high-speed)
	Vantage+ V5 [52]	FOV 63.5° × 55.1°	5 MP (normal) 1.8 MP (high-speed)	N/A	4.7 ms	420 FPS (normal) 1070 FPS (high-speed)
	Vero v2.2 [53]	FOV 98.1° × 50.1°	2.2 MP	N/A	3.6 ms	330 FPS
	Vero v1.3 [53]	FOV 55.2° × 43.9°	1.3 MP	N/A	3.4 ms	250 FPS
	Vero v1.3 X [53]	FOV 79.0° × 67.6°	1.3 MP	N/A	3.4 ms	250 FPS
	Vero Vertex [53]	FOV 100.6° × 81.1°	1.3 MP	N/A	3.4 ms	120 FPS
	Vue [54]	FOV 82.7° × 52.7°	2.1 MP	N/A	N/A	60 FPS
	Viper [55]	FOV 81.8° × 49.4°	2.2 MP	N/A	3.2 ms	240 FPS
	ViperX [56]	FOV 50.2° × 50.2°	6.3 MP	N/A	3.2 ms	240 FPS
Atracsys LLC., Puidoux, Switzerland	fusionTrack 500 [57]	2000 × 1327 × 976 mm^3	2.2 MP	0.09 mm	~ 4 ms	335 Hz
	fusionTrack 250 [58]	1400 × 1152 × 900 mm^3	2.2 MP	0.09 mm	~ 4 ms	120 Hz
	spryTrack 180 [59]	1400 × 1189 × 1080 mm^3	1.2 MP	0.19 mm	<25 ms	54 Hz
	spryTrack 300 [60]	1400 × 805 × 671 mm^3	1.2 MP	0.14 mm	<25 ms	54 Hz

Table 1. *Cont.*

Manufacturer	Model	Measurement Volume (Radius × Width × Height) or FOV	Resolution	Volumetric Accuracy (RMS)	Average Latency	Measurement Rate
Motion Analysis Corp., Rohnert Park, CA, USA	Kestrel 4200 [61]	N/A	4.2 MP	N/A	N/A	200 FPS
	Kestrel 2200 [62]	N/A	2.2 MP	N/A	N/A	332 FPS
	Kestrel 1300 [63]	N/A	1.3 MP	N/A	N/A	204 FPS
	Kestrel 300 [64]	N/A	0.3 MP	N/A	N/A	810 FPS
STT Systems, Donostia-San Sebastian, Spain	EDDO Biomechanics [65]	N/A	N/A	1 mm	N/A	120 FPS
Advanced Realtime Tracking GmbH & Co. KG, Oberbayern, Germany	ARTTRACK6/M [66]	FOV 135° × 102°	1280 × 1024	N/A	N/A	180 Hz
	ARTTRACK5 [67]	FOV 98° × 77°	1280 × 1024	N/A	10 ms	150 Hz
	SMARTTRACK3 [68]	FOV 135° × 102°	1280 × 1024	N/A	9 ms	150 Hz

3.2. Electromagnetic Tracking Systems

Compared to optical tracking systems, electromagnetic tracking system can cover a larger volume of measurement space, but normally has lower position accuracy. Table 2 summarizes the specification of some representative, commercially available electromagnetic tracking systems. Since it does not require the transmission of light, electromagnetic tracking systems are promising in intracorporeal biomedical applications. For example, Polhemus Inc. (Colchester, VT, USA) developed a miniatured electromagnetic motion tracking sensors with outer diameter of 1.8 mm. It can be inserted into human vessel with a catheter for both position and orientation tracking [69].

Table 2. Summary of commercially available electromagnetic tracking systems.

Manufacturer	Model	Tracking Distance	Position Accuracy (RMS)	Orientation Accuracy (RMS)	Average Latency	Measurement Rate
Northern Digital Inc., Waterloo, ON, Canada	Aurora-Cube Volume-5DOF [70]	N/A	0.70 mm	0.2°	N/A	40 Hz
	Aurora-Cube Volume-6DOF [70]	N/A	0.48 mm	0.3°	N/A	40 Hz
	Aurora-Dome Volume-5DOF [70]	660 mm	1.10 mm	0.2°	N/A	40 Hz
	Aurora-Dome Volume-6DOF [70]	660 mm	0.70 mm	0.3°	N/A	40 Hz
	3D Guidance trakSTAR-6DOF [71]	660 mm	1.40 mm	0.5°	N/A	80 Hz
	3D Guidance driveBAY-6DOF [71]	660 mm	1.40 mm	0.5°	N/A	80 Hz
Polhemus Inc., Colchester, VT, USA	Viper [72]	N/A	0.38 mm	0.10°	1 ms	960 Hz
	Fastrak [73]	N/A	0.76 mm	0.15°	4 ms	120 Hz
	Patriot [74]	N/A	1.52 mm	0.40°	18.5 ms	60 Hz
	Patriot Wireless [75]	N/A	7.62 mm	1.00°	20 ms	50 Hz
	Liberty [76]	N/A	0.76 mm	0.15°	3.5 ms	240 Hz
	Liberty Latus [77]	N/A	2.54 mm	0.50°	5 ms	188 Hz
	G4 [78]	N/A	2.00 mm	0.50°	<10 ms	120 Hz

3.3. Mechanical Tracking Systems

Unlike other tracking systems, the development of mechanical tracking systems, especially for biomedical applications, is limited. This might be due to the fact that mechanical tracking systems are usually bulky and heavy. Meta motion. Inc. presented a mechanical tracking system, named Gypsy 7, decades ago, which had a position accuracy of 0.125° [79]. However, this kind of exoskeleton system consists of 14 joint sensors and the total weight is 4 kg.

3.4. Acoustic Tracking Systems

Most commercially available acoustic tracking system is related to marine positioning, the use for in-door positioning is still in its infancy. Sonitor Technologies, Inc. developed a Forkbeard system, which applied 40 kHz US for echo location [80]. Although it can cover a floor, the volumetric accuracy is 1–2 feet, while the latency is 1–2 s. The nature of low accuracy and high latency of acoustic tracking hamper its applications in biomedical field. However, considering that US has the benefit of non-ionizing radiation, it might be promising for some specific biomedical applications.

3.5. Inertial Tracking Systems

Inertial tracking systems are also commercially available for many years. Some products that can be purchased on the market are summarized in Table 3. Due to the fact that it does not require both transmitters and receivers as optical tracking, electromagnetic tracking or acoustic tracking, the size of the inertial tracking device can be very compact, such as a dot [81]. This feature contributes to a friendly and comfortable condition for tracking objects, which hardly affect the normal motion of objects. However, the inertial tracking system can only provide a position information relatively, and it always needs the assistance from other kinds of tracking systems.

Table 3. Summary of commercially available inertial tracking systems.

Manufacturer	Model	Static Accuracy (Roll/Pitch)	Static Accuracy (Heading)	Dynamic Accuracy (Roll/Pitch)	Dynamic Accuracy (Heading)	Average Latency	Update Rate
Xsens Technologies B.V., Enschede, The Netherlands	MTw Awinda [82]	0.5°	1.0°	0.75°	1.5°	30 ms	120 Hz
	Xsens DOT [81]	0.5°	1.0°	1.0°	2.0°	30 ms	60 Hz
	MTi-1 [83]	0.5°	N/A	N/A	N/A	N/A	100 Hz
	MTi-2 [83]	0.5°	N/A	0.8°	N/A	N/A	100 Hz
	MTi-3 [83]	0.5°	N/A	0.8°	2.0°	N/A	100 Hz
	MTi-7 [83]	0.5°	N/A	0.5°	1.5°	N/A	100 Hz
	MTi-8 [83]	0.5°	N/A	0.5°	1.0°	N/A	100 Hz
	MTi-20 [84]	0.2°	N/A	0.5°	N/A	N/A	N/A
	MTi-30 [84]	0.2°	N/A	0.5°	1.0°	N/A	N/A
	MTi-200 [84]	0.2°	N/A	0.3°	N/A	<10 ms	N/A
	MTi-300 [84]	0.2°	N/A	0.3°	1.0°	<10 ms	N/A
	MTi-710 [84]	0.2°	N/A	0.3°	0.8°	<10 ms	400 Hz
	MTi-610 [85]	N/A	N/A	N/A	N/A	N/A	400 Hz
	MTi-620 [85]	0.2°	N/A	0.25°	N/A	N/A	400 Hz
	MTi-630 [85]	0.2°	N/A	0.25°	1.0°	N/A	400 Hz
	MTi-670 [85]	0.2°	N/A	0.25°	0.8°	N/A	400 Hz
	MTi-680 [85]	0.2°	N/A	0.25°	0.5°	N/A	400 Hz
STT Systems, Donostia-San Sebastian, Spain	iSen system [86]	N/A	N/A	<0.5°	<2.0°	N/A	400 Hz
VectorNav Technologies, Dallas, TX, USA	VN-100 [87]	0.5°	N/A	1.0°	2.0°	N/A	800 Hz
	VN-110 [88]	0.05°	N/A	N/A	2.0°	N/A	800 Hz
	VN-200 [89]	0.5°	2.0°	0.2°, 1σ	0.03°, 1σ	N/A	800 Hz
	VN-210 [90]	0.05°	2.0°	0.015°, 1σ	0.05–0.1°, 1σ	N/A	800 Hz
	VN-300 [91]	0.5°	2.0°	0.03°, 1σ	0.2°, 1σ	N/A	400 Hz
	VN-310 [92]	0.05°	2.0°	0.015°, 1σ	0.05–0.1°, 1σ	N/A	800 Hz

Table 3. *Cont.*

Manufacturer	Model	Static Accuracy (Roll/Pitch)	Static Accuracy (Heading)	Dynamic Accuracy (Roll/Pitch)	Dynamic Accuracy (Heading)	Average Latency	Update Rate
Advanced Navigation, Sydney, Australia	Motus [93]	0.05°	0.8°	N/A	N/A	N/A	1000 Hz
	Orientus [94]	0.2°	0.8°	0.6°	1.0°	0.3 ms	1000 Hz
	Boreas D90 [95]	0.005° *	0.006° *	N/A	N/A	N/A	1000 Hz
	Spatial FOG Dual [96]	0.005° *	0.007°	N/A	N/A	N/A	1000 Hz
	Certus Evo [97]	0.01° *	0.01°	N/A	N/A	N/A	1000 Hz
	Certus [98]	0.03° *	0.06°	N/A	N/A	N/A	1000 Hz
	Spatial [99]	0.04° *	0.08°	N/A	N/A	0.4 ms	1000 Hz
	GNSS Compass [100]	0.4°	0.4°	N/A	N/A	N/A	200 Hz
Inertial Labs, Paeonian Springs, VA, USA	KERNEL-100 [101]	0.05°	0.08°	N/A	N/A	<1 ms	2000 Hz
	KERNEL-110 [102]	0.05°	0.08°	N/A	N/A	<1 ms	2000 Hz
	KERNEL-120 [102]	0.05°	0.08°	N/A	N/A	<1 ms	2000 Hz
	KERNEL-210 [103]	0.05°	0.08°	N/A	N/A	<1 ms	2000 Hz
	KERNEL-220 [103]	0.05°	0.08°	N/A	N/A	<1 ms	2000 Hz
	IMU-P [104]	0.05°	0.08°	N/A	N/A	<1 ms	2000 Hz

4. Biomedical Ultrasound Imaging Applications

Over the past decade, a range of commercial and research tracking systems have been developed for biomedical US imaging-related applications. Following sections categorically review the applications of different tracking systems reported till date.

4.1. Freehand 3D Ultrasound Imaging

Over the past few decades, US imaging has become a valuable tool in clinical diagnostic and therapeutic procedures across a broad range of fields, ranging from routine screening, early cancer detection, diagnosis of cardiovascular disease to real-time monitoring [105]. Compared with CT, MRI, and PET, US demonstrates the advantages of safe for patients (no risk of ionizing radiation or high magnetic fields), real-time imaging, portability, and low-cost [11,106,107]. In clinical practice, a handheld US probe typically composed of a 1D linear US transducer array is routinely used to generate 2D US images in real-time, displaying cross-sectional images of the human anatomy. While 2D US imaging offers several advantages for medical applications, it can only acquire selectively-sampled, cross-sectional slice images of a 3D anatomic structure, and the orientation of each image plane depends on how the operator positions the handheld probe (i.e., operator dependency) [108]. If clinicians need to view 3D anatomic structures, they have to imagine the 3D volume with the planar 2D images mentally, thus limiting the diagnostic accuracy.

In order to overcome the limitations of 2D US, volumetric 3D US imaging has been developed, allowing direct visualization of the arbitrary plane of 3D volume and helping obtain a more accurate view of the shape, size, and location of the organ and lesion [16]. Up to now, three different types of methods have been utilized for the construction of 3D US volumes: employing a 2D phased array transducer, mechanical 3D US scanning, and freehand 3D US scanning [109,110]. Instead of using 1D array ultrasonic transducer for conventional 2D US systems, 3D US volume can also be generated by using a 2D phased array ultrasonic transducer with its elements spreading on a 2D aperture, which can deflect and focus the ultrasonic beam in a volumetric space [111,112]. Since the US beams are steered and focused on the region of interest by electronic scanning, the 2D array remains stationary during the procedure. Although this approach can acquire 3D volume straightforwardly and in real-time, manufacturing process of a 2D phased array transducer is complex and manufacturing cost is high due to a large number of array elements and the electrical connection of each element [113]. Another approach to obtain a 3D US volume is via mechanical 3D scanning using conventional linear array transducer. In this method, a mechanical motor is used to control the transducer rotation, tilt, or translation with designed scanning trajectory [109]. The 3D US volumes can then be reconstructed by using the acquired 2D US images with their predefined positions and orientations. While the 3D US imaging systems based on this kind of method can be operated conveniently by controlling mechanical motor, the whole system is bulky due to a mechanical motor integrated and the system flexibility is low due to the controlled movement limitation.

In addition to the above-mentioned approaches for 3D US imaging, freehand 3D US has become the most rapidly advancing technique over the years due to the advantages of scanning flexibility, convenience to operate and low cost. Freehand 3D US images are acquired by rigidly attaching a 6-DOF position sensor to a handheld US probe that generates a sequence of B-mode US images [114]. The position sensor records the positions and orientations of the probe during the scanning procedure, and then the 3D volumes are constructed by combining the sequence of the 2D US images along with the corresponding position information (Figure 7a). It is noted that for reconstructing a 3D US volume, the position and orientation data of each 2D US image is required. Various techniques have been reported for obtaining the position and orientation data of the US probe during freehand US scanning. The most commonly used position sensors during freehand 3D US imaging are optical tracking sensor and electromagnetic tracking sensor.

In a typical optical tracking system, either light-reflective makers (passive markers) or light-emitting markers (active markers) are attached to the US probe and the markers are monitored by two or more cameras fixed in a position (Figure 7b). Passive markers are usually matt spheres coated with retroreflective material and reflect light back to the cameras. Three or more markers are usually arranged asymmetrically, allowing the cameras to infer the orientation in space. Contrary to passive markers that reflect light generated by the external sources to the cameras, active markers are made of infrared LEDs, powered by themselves to emit infrared light. In a typical electromagnetic tracking system, a time-varying 3D magnetic field is transmitted through the volume in which the US scanning is to be conducted. Three sensor coils are attached to an US probe and utilized to obtain the field in the 3D Cartesian coordinates (x, y, z) (Figure 7c). This information enables the position and orientation of the sensor coils to be acquired [109].

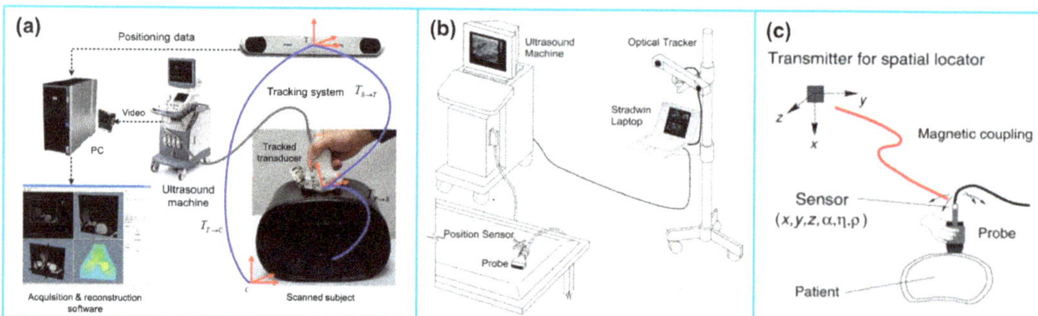

Figure 7. (**a**) The typical configuration of a freehand 3D US imaging system. Reprinted from [115] with permission. (**b**) An optical tracker based freehand 3D US imaging system. Reprinted from [109] with permission. (**c**) An electromagnetic sensor based freehand 3D US imaging system. Reprinted from [109] with permission.

Due to the advantages of flexible operation and simultaneous visualization, freehand 3D US imaging is increasingly gaining popularity in medical applications. For example, Chung et al. [116] reported an imaging system based on optical motion tracking technique with the objective of developing a carotid artery contour detection procedure for carotid atherosclerosis diagnosis (Figure 8a). The 3D motion tracking system consisted of 8 Eagle digital CCD cameras for motion detection in 3D space and 4 passive fluorescent markers attached to an US probe, showing spatial and temporal resolutions of 10 μm and 0.01 s, respectively. Daoud et al. [117] developed a freehand 3D US imaging system using a 3D electromagnetic position tracking system (trakSTAR, NDI, ON, Canada). The position and orientation of the US probe in 3D space were tracked by one of the electromagnetic sensors attached to the probe (Figure 8b). Herickhoff et al. [17] invented a volumetric 3D US imaging system at a very low cost (under USD 250) by using a single IMU sensor for orientation acquisition and a light-weight fixture customized to the US probe (Figure 8c). The preliminary results demonstrated the capability of the low-cost method for reconstructing a 3D US image volume, providing a solution for solving the problem of operator dependence. In another study, Chen and Huang [118] reported a freehand 3D US imaging system that could obtain volume reconstruction and visualization during data acquisition at real-time level. The real-time freehand 3D US system mainly consisted of a linear probe, an electromagnetic sensing system, and a computer with a GPU for image data reconstruction and visualization of the 3D volume image. A summary of the various reported freehand 3D US imaging system during the past decade is provided in Table 4.

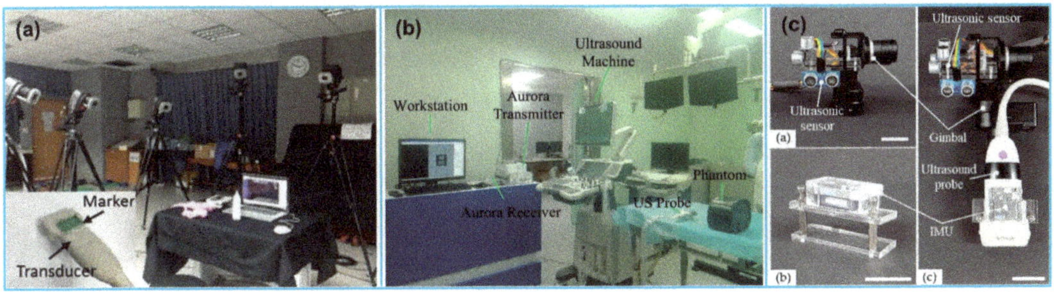

Figure 8. (**a**) A freehand 3D US imaging system with optical motion tracking system settings. Reprinted from [116] with permission. (**b**) The setup of a freehand 3D US imaging system with electromagnetic tracking. Reprinted from [122] with permission. (**c**) A low-cost 3D US image acquisition method. Reprinted from [126] with permission.

Table 4. A summary of freehand 3D US imaging study.

Reference	Tracking Principle	Tracking System	Accuracy	Application
Chung et al. [116]	Optical tracking	Motion Analysis, Santa Rosa, CA, USA	Spatial: 10 μm Temporal: 0.01 s	Carotid atherosclerotic stenosis detection
Pelz et al. [119]	Electromagnetic tracking	Curefab CS system (Curefab Technologies GmbH, Munich, Germany)	None	Internal carotid artery stenosis diagnosis
Miller et al. [120]	Optical tracking	VectorVision2 navigation system (BrainLAB, Munich, Germany)	None	Image-guided surgery
Mercier et al. [121]	Optical tracking	Polaris (Northern Digital, Waterloo, ON, Canada)	Spatial: 0.49–0.74 mm Temporal: 82 ms	Neuronavigation
Chen et al. [122]	Electromagnetic tracking	Aurora (NDI, ON, Canada)	None	Image-guided surgery
Wen et al. [115]	Optical tracking	Polaris (Northern Digital, Waterloo, ON, Canada)	None	Image-guided intervention
Sun et al. [123]	Optical tracking	OptiTrack V120:Trio (NaturalPoint Inc., Corvallis, OR, USA)	Spatial: <1 mm	Image-guided intervention

Table 4. *Cont.*

Reference	Tracking Principle	Tracking System	Accuracy	Application
Worobey et al. [124]	Optical tracking	Vicon Motion Systems; Centennial, Colorado	None	Scapular position
Passmore et al. [125]	Optical tracking	Vicon Motion Systems, Oxford, UK	None	Femoral torsion measurement
Daoud et al. [117]	Electromagnetic tracking	trakSTAR, NDI, ON, Canada	None	3D US imaging
Chen and Huang [118]	Electromagnetic tracking	MiniBird, Ascension Technology Corp., Burlington, VT, USA	None	Real-time 3D imaging
Cai et al. [20]	Optical tracking	OptiTrack V120: Trio (NaturalPoint Inc., Corvallis, OR, USA)	Positional: 0.08–0.69 mm Rotational: 0.33–0.62°	3D US imaging
Herickhoff et al. [17]	Inertial tracking	IMU sensor (iNEMO-M1; STMicroelectronics, Geneva, Switzerland)	None	Low-cost 3D imaging platform
Kim et al. [126]	Inertial tracking	Ultrasonic sensor + IMU sensor (HC-SR04, Shenzhen AV, Shenzhen, China)	Spatial: 0.79–1.25 mm	Low-cost 3D imaging platform
Lai et al. [127]	Optical tracking	T265, Intel, Santa Clara, CA, USA	Spatial: 2.9 ± 1.8°	Scoliosis assessment
Jiang et al. [128]	Electromagnetic tracking	Ascension Technology, Burlington, VT, USA	None	Scoliosis assessment

4.2. Ultrasound Image Fusion in Multimodality Imaging

Medical image fusion refers to the co-display of registered images from the same or different imaging modalities, such as US, CT, MRI, and PET [129]. Since the fused image contains all the important features from each input image, it can offer a more comprehensive, more reliable and better description of lesions, so as to assist the preclinical research and clinical diagnosis as well as therapy, such as routine staging, surgical navigation, radiotherapy planning, etc. [130] Percutaneous interventional procedures, particularly percutaneous biopsy and percutaneous tumor ablation, play an important role in caring for patients with cancer. To guide percutaneous interventional procedures, US imaging is the most widely used imaging modality owing to its real-time capability, no radiation exposure, and easy accessibility [131]. However, compared with CT and MRI, US imaging shows a narrower field of view and lower contrast resolution. In addition, the imaging performance is reduced by the presence of gas and fat in human body [132]. To localize and characterize lesions more precisely, applying US fusion imaging allows exploitation of the strengths of different imaging modalities simultaneously, eliminating or minimizing the weakness of every single modality [133]. The procedures of fusing CT/MRI images and US images are detailed in reference [134], the interested readers can refer to it. After the image fusion procedure, the CT/MRI images will be displayed on the monitor side-by-side with the real-time US images in a synchronous manner and updated simultaneously according to the change in position and imaging plane of US probe. A process of US and MRI fusion is illustrated in Figure 9.

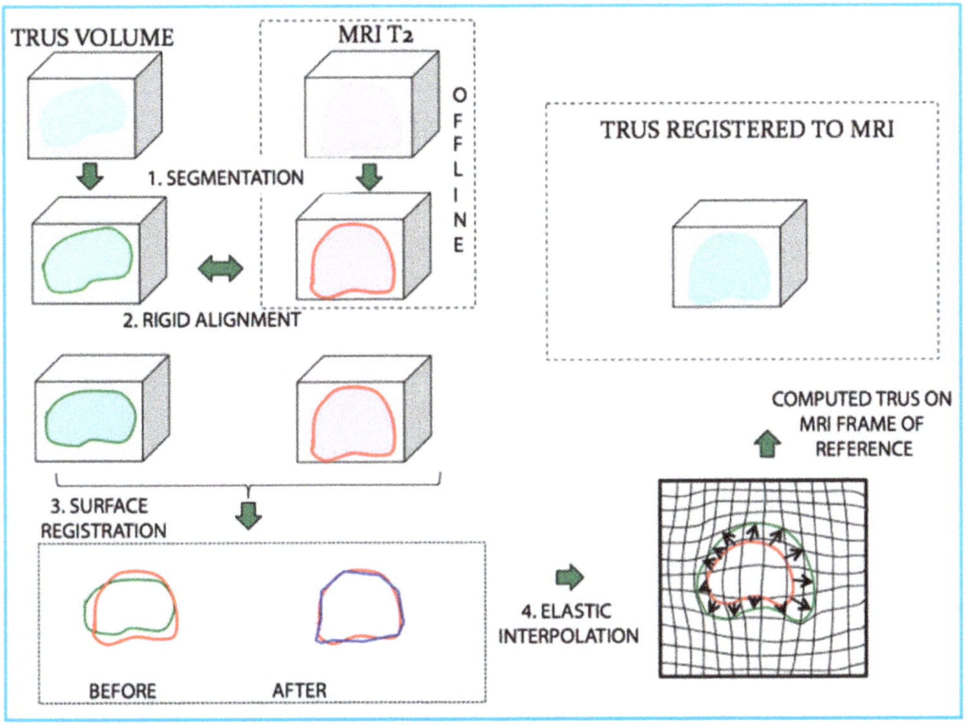

Figure 9. Schematic illustration of the process of US/MRI fusion. MR and transrectal US (TRUS) images were (1) segmented and then (2) rigidly aligned. Fusion then proceeded, involving (3) a surface registration, and (4) elastic (non-rigid) interpolation. Reprinted from [135] with permission.

As we have discussed in Section 3, to track an US probe in 3D space, there are 5 available tracking techniques. However, for US image fusion applications in percutaneous interventional procedures, the electromagnetic tracking system is the one mostly implemented [129], as shown in Figure 10. For instance, Krucker et al. [136] developed an Aurora (Northern Digital Inc, Waterloo, ON, Canada) electromagnetic tracking system to fuse real-time US with CT, providing real-time visualization of tracked interventional needles within preprocedural CT scans. Appelbaum et al. [137] compared conventional CT-guided biopsy to biopsy employing a U.S. Food and Drug Administration–approved electromagnetic biopsy navigation system (Veran IG4, Veran Medical Technologies). Phantom model study results showed that by using electromagnetic tracking system, needle placement accuracy had been improved and radiation exposure had been reduced compared with conventional CT techniques. Venkatesan et al. [138] fused US image to CT and ^{18}F-FDG-PET/CT with an electromagnetic tracking system (Northern Digital Inc, Waterloo, ON, Canada) for biopsy of technically challenging FDG-avid targets. By using conventional US imaging, a total number of 36 lesion samples could not be well seen or were completely inapparent during the biopsy procedures. However, by using the combined electromagnetic tracking and US/CT/^{18}F-FDG-PET fusion, 31 out of 36 biopsies were diagnostic.

In recent decade, US image fusion has developed significantly and can now perform crucial roles in diagnosis and clinical management across various anatomical regions [129,139–143]. One of the most widely applied examples in clinics is US fused with MRI images for percutaneous image-guided prostate biopsy [144,145]. Although US is the commonest modality utilized for real-time guidance during biopsy, it is limited in its ability to visualize deep targets. In addition, the biopsy procedure is performed targeting only the different anatomic locations of the prostate, thus the underdetection rate of transrectal US-guided biopsy is high. Fusing US images with MRI images allows the information from MRI to be used to direct biopsy needles under US guidance. It combines the superior diagnostic accuracy of MRI for detecting suspicious lesions in the prostate with the practicality and familiarity of US [145,146]. Several U.S. FDA approved systems for fusion imaging of real-time US with MRI are commercially available (summarized in Table 5), as shown in Figure 11.

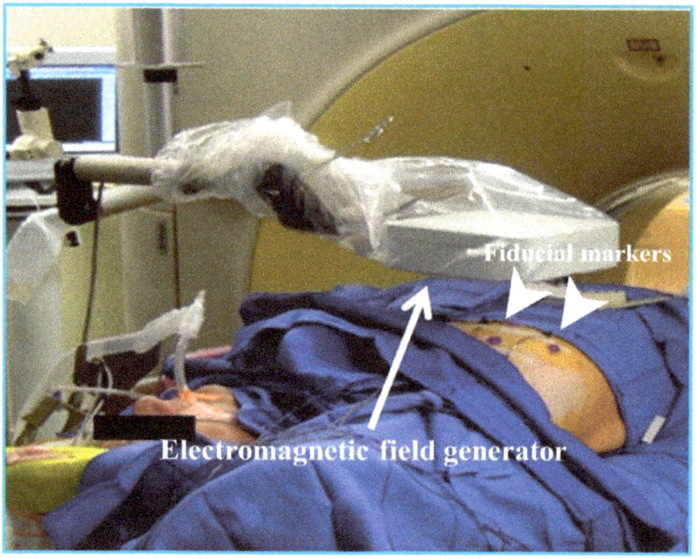

Figure 10. Equipment setup for electromagnetic tracking during interventional procedures. Reprinted from [138] with permission.

In addition to US/MRI fusion-guided prostate biopsy, US image fusion has been investigated for clinical applications in various anatomical regions including liver, kidney, pancreas, and musculoskeletal system. A summary of US image fusion for applications in different anatomical regions is illustrated in Table 6.

Table 5. U.S. Food and Drug Administration approved US/MRI fusion system [147].

System Type	Manufacturer	Year of FDA Approval	US Image Acquisition	Tracking Principle	Biopsy Route
UroNav	Philips	2005	Manual sweep	Electromagnetic tracking	Transrectal
Artemis	Eigen	2008	Manual rotation	Mechanical arm	Transrectal
Urostation	Koelis	2010	Automatic US probe rotation	Real-time registration	Transrectal
HI-RVS	Hitachi	2010	Real-time biplanar transrectal US	Electromagnetic tracking	Transrectal or transperineal
GeoScan	BioJet	2012	Manual sweep	Mechanical arm	Transrectal or transperineal

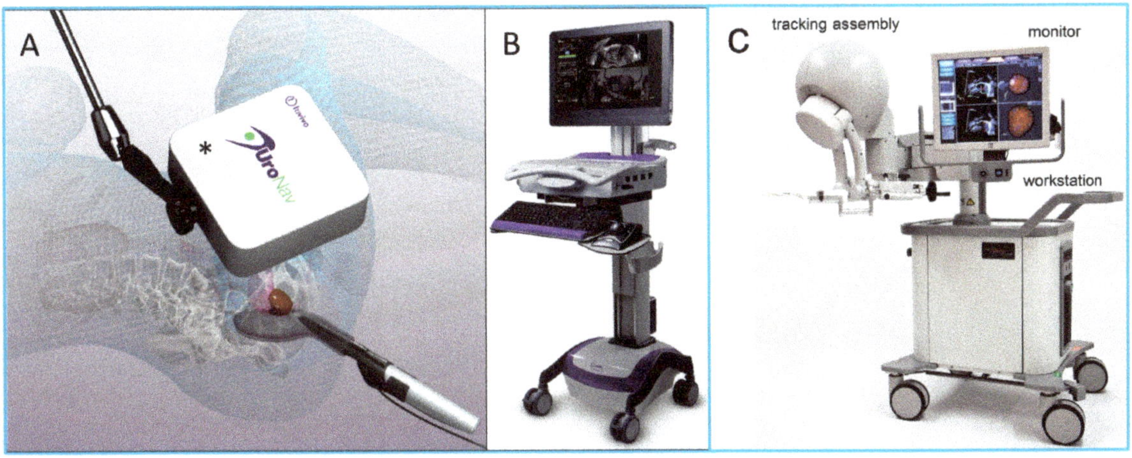

Figure 11. (**A**,**B**) Photo images of the UroNav US/MRI fusion system. (**A**) An electromagnetic field generator enables tracking of the transrectal US probe; (**B**) US/MRI fusion device. (**C**) Artemis US/MRI fusion system. Reprinted from [144] with permission.

Table 6. A summary of US image fusion applications.

Reference	Modality for Fusion	Tracking Principle	Application
Park et al. [148]	Liver CT or MRI	Electromagnetic tracking	Biopsy of focal hepatic lesions with poor conspicuity on conventional B-mode US image
Lee et al. [149]	Liver CT or MRI	Electromagnetic tracking	Lesion detection of small hepatocellular carcinomas (HCCs)
Song et al. [150]	Liver CT or MRI	Plane registration and point registration	Improve sonographic conspicuity of HCC and feasibility of percutaneous radiofrequency ablation for HCCs not visible on conventional US images
Helck et al. [151]	Renal CT or MRI	Electromagnetic tracking	Identifiability and assessment of the dignity of renal lesions
Andersson et al. [152]	Renal CT	Electromagnetic tracking	Image-guided percutaneous radiofrequency ablation of small renal masses
Zhang et al. [153]	Pancreatic CT	Real-time registration	Image-guided percutaneous catheter drainage in treatment of acute pancreatitis
Klauser et al. [143]	Musculoskeletal CT	Internal landmarks	Image-guided sacroiliac joint injection
Lee et al. [141]	Thigh MRI	Real-time registration	Selecting the appropriate biopsy site in patients with suspected myopathies
Rubenthaler et al. [154]	Renal MRI/contrast enhanced US	Electromagnetic tracking	Classification of unclear and difficult renal lesions

4.3. *Ultrasound-Guided Diagnosis*

Percutaneous needle biopsy plays an important role in the diagnosis, staging, and treatment planning for various tumors [155,156]. The success of needle insertion procedures mainly depends on accurate needle placement to minimize complications and to avoid damage to neighboring tissues [156]. In many applications, US guidance has been shown to increase the safety and success rate of the procedure due to its real-time imaging capability, easy operation, portability, etc. [157–159]. During the procedure, the physician manually manipulates the needle and the US probe simultaneously while mentally relating US images acquired to locations inside a patient's body [160]. Practically, it is very challenging for the physician to visualize the needle trajectory inside the patient tissue just by checking the US image [156]. In order to let the needle tip follow the desired trajectory and hit the target location in the image plane, it is beneficial and necessary to track the pose of the needle with respect to the coordinate system of the US image.

Three different types of tracking systems have been applied for US-guided needle insertion: electromagnetic, optical and mechanical trackers. For electromagnetic trackers, Franz et al. [161] assessed the precision and accuracy of a compact electromagnetic field generator (Aurora, Northern Digital Inc., Waterloo, ON, Canada) attached to 6 different US probes with various operating frequencies. Based on the assessment results, the error of the field generator was <0.2 mm; the positional accuracy was <1.0 mm. Xu et al. [162] evaluated the effectiveness of magnetic navigation in US-guided interventional procedures (Figure 12). A commercially available magnetic navigation system (GE Healthcare, Milwaukee, WI, USA) was applied. They found that compared with conventional US guidance, magnetic navigation in US-guided interventional procedure was especially useful for some complicated clinical situations, such as liver tumor ablation. In addition, Hakime et al. [163] evaluated the accuracy and safety of electromagnetic needle tracking for US-guided liver biopsy. An electromagnetic transmitter was placed near the scanning area and a pair of electromagnetic receiving sensors were attached to the US probe. The clinical results demonstrated that the overall diagnostic success rate of liver lesion was 91%. März et al. [164] proposed an interventional imaging system based on a mobile electromagnetic field generator (Aurora, Northern Digital Inc., Waterloo, ON, Canada) attached to an US probe. The tracking and calibration accuracy of the system was assessed in a clinical setting. The tracking accuracy was tested to be <1 mm and the calibration error was 1–2 mm.

For optical trackers, Wang et al. [165] utilized a low-cost Kinect sensor (a stereo camera) for interventional needle tracking. The accuracy of needle tracking was measured, ranging from 2.6 ± 1.7 to 6.9 ± 5.1 mm. Stolka et al. [166] developed a camera-based tracking system for US-guided interventions, consisting of an optical sensing head mounted on an US probe. The head could be mounted to support both in- or out-of-plane interventions. The phantom test results showed that the mean accuracy of the system was 3.27 ± 2.28 mm. Najafi et al. [167] proposed a single camera-based tracking system for US-guided needle insertion (Figure 13). The camera was directly mounted on the US probe and the needle location was tracked by using the needle markers. A needle tracking accuracy of 0.94 ± 0.46 mm was achieved, which was higher than that of the existing solutions. Daoud et al. [168] also reported a camera-based tracking system for US-guided needle interventions. An USB web camera (IceCam2, Macally Peripherals, Ontario, CA, USA) was attached to a 3D curvilinear US probe using a plastic housing. Dynamic needle tracking in a sequence of 3D US volumes was achieved. Based on the ex vivo animal experiments, the maximum error rate of 1.2 mm for the needle tip was measured in individual US volumes.

In addition to the magnetic and optical tracking devices, Ho et al. [169] invented an US-guided robotic system for transperineal prostate intervention, consisting of a gantry, a gun-holder, and an US probe holder (Figure 14). The system was constructed based on the dual-cone concept, ensuring that any part of the prostate can be accessed with minimal skin puncture. The egg phantom experimental results illustrated the system accuracy was <1 mm. Orhan et al. [170] reported design and modeling of a 5-DOF parallel robot for autonomous US-guided biopsy. The robot was composed of 5-DOF and 3 main stages; front stage, back stage, syringe mechanism. The biopsy needle connected to the syringe mechanism passed through the gimbal in the front stage. Poquet et al. [171] designed a 6-DOF, serial robotic co-manipulator system for assisting endorectal prostate biopsy. The robotic system consisted of three brakes and three motors. The system could provide freedom to the urologist to position the probe with respect to the prostate in the free mode while leaving him/her to focus on insertion only during locked mode.

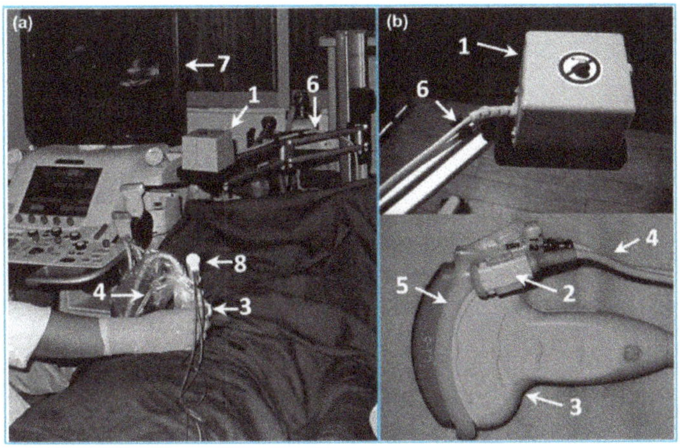

Figure 12. A magnetic navigation system for liver cancer ablation procedures. (**a**) The setup of the magnetic navigation system; (**b**) Magnetic field generator and magnetic receivers attached to an US probe. Reprinted from [162] with permission.

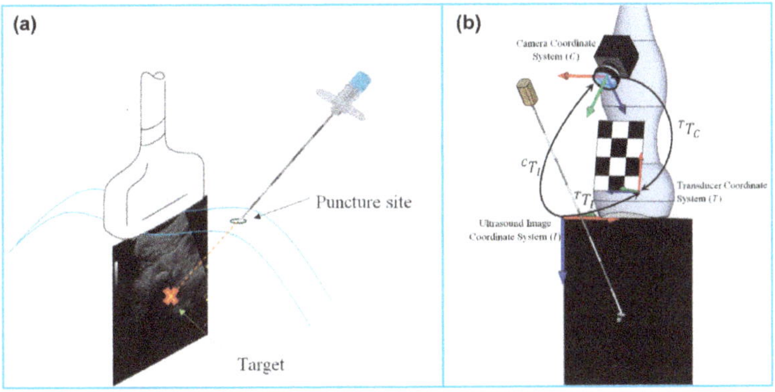

Figure 13. (**a**) Out-of-plane needle trajectory planning. (**b**) Schematic of the coordinate systems and needle movement. Reprinted from [167] with permission.

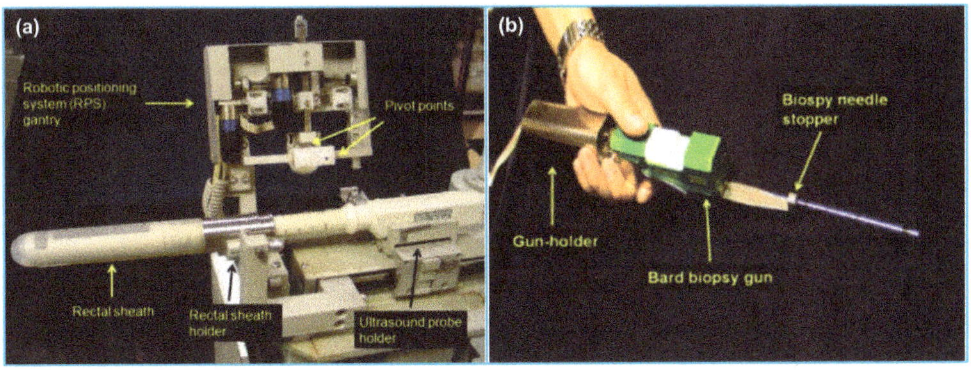

Figure 14. A robotic US-guided prostate intervention system. (**a**) Gantry and US probe holder. (**b**) Gun-holder and biopsy gun. Reprinted from [169] with permission.

4.4. Ultrasound-Guided Therapy

US-guided surgery is an area of minimally invasive surgery where surgical procedures are performed with the aid of US imaging throughout the operation. Contrary to traditional surgical access, US-guided surgery uses computer-based systems to provide real-time US images to help the physician precisely visualize and target the surgical site by updating the intraoperative information [12]. While other imaging modalities, such as CT and MRI, have also been applied for surgery navigation, US-guided surgery shows several advantages, including real-time imaging, equipment portability, low cost and reduced hospital stays [13]. Figure 15 shows a basic process of 3D US-guided surgery navigation [172]. In order to utilize US to guide surgical procedures, the US probe must be tracked. Although several tracking technologies are commercially available today, which are review in the last section, the most widely used solutions are optical and electromagnetic systems.

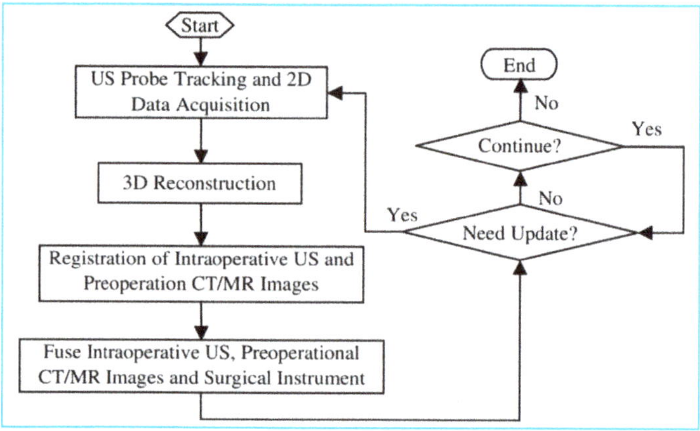

Figure 15. A basic process of 3D US image-guided surgery navigation. Reprinted from [172] with permission.

For instance, Stoll et al. [173] presented a novel approach for tracking surgical instruments in 3D US imaging by using a series of passive echogenic markers. The markers were attached near the distal end of the surgical instrument, and the marker position and orientation could be simply determined in a 3D US volume using image processing. Since the markers were completely passive, they can be easily implemented without prior integration with the imaging system. Moreover, the error of registering the tracking coordinate frame to the image frame can be eliminated. In another study, Li et al. [174] systematically compared the real-time US-guided percutaneous nephrolithotomy (PCNL) using SonixGPS navigation system with conventional US-guided PCNL using an US machine for the treatment of complex kidney stones. Based on their clinical results, the SonixGPS system was superior to the conventional method in terms of stone clearance rate and puncture accuracy. Hamamoto et al. [175] investigated the efficacy of applying real-time virtual sonography (RVS) guidance for renal puncture for endoscopic combined intrarenal surgery (ECIRS) treatment of large renal calculi (Figure 16). The RVS system synchronized real-time US images with CT images via a magnetic navigation system to provide volume and position data side by side. Compared with US-guided puncture, RVS-guided renal puncture illustrated lower incidence of bleeding-related complications. In addition, Gomes-Fonseca et al. [176] assessed the performance of electromagnetic tracking system guidance for percutaneous renal access in the operating room environment. Their experimental results demonstrated that ureterorenoscopes and 2D US probe did not affect the precision and accuracy of the electromagnetic tracking systems, suggesting that these instruments may be used for a safe percutaneous renal access.

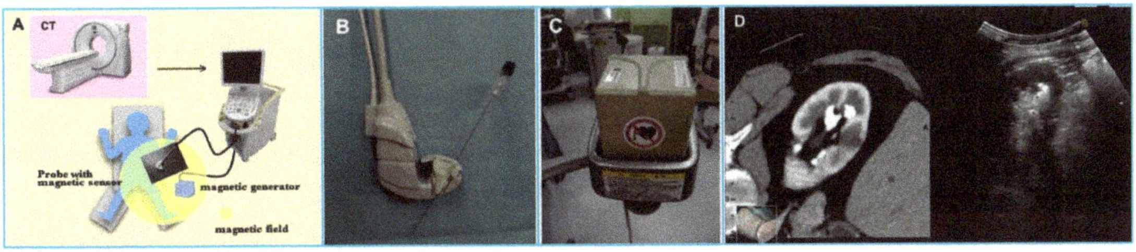

Figure 16. A navigation system for percutaneous renal puncture. (**A**) The main components of the system; (**B**) The magnetic sensor attached to an US probe; (**C**) The magnetic field generator; (**D**) US image and CT volume data displaying side by side on the same monitor. Reprinted from [175] with permission.

Bharat et al. [177] measured the accuracy of the electromagnetic tracking system for identification of the position and shape of the treatment catheters in high-dose-rate (HDR) prostate brachytherapy (Figure 17). The tracking experiments were performed in both a controlled laboratory environment and a typical brachytherapy operating room. The robotic validation of the electromagnetic system found that the mean accuracy of the system was <0.5 mm, illustrating the potential value of using electromagnetic tracking for catheter mapping in HDR brachytherapy. Schwaab et al. [178] developed an US based motion tracking method for real-time motion correction in ion beam therapy. It was found that by using US tracking, it can yield nearly real-time position information at high frame rate of moving targets. Yu et al. [179] also evaluated the accuracy and precision of a transperineal US image-guided system (Clarity Autoscan US system (Elekta, Stockholm, Sweden)) for prostate radiotherapy. Based on a male pelvic phantom experimental result, the accuracy of US tracking performance in the lateral direction was better than that in the axial direction; the precision of US tracking performance in the axial (superior-inferior) direction was better than that in the lateral (left-right) direction.

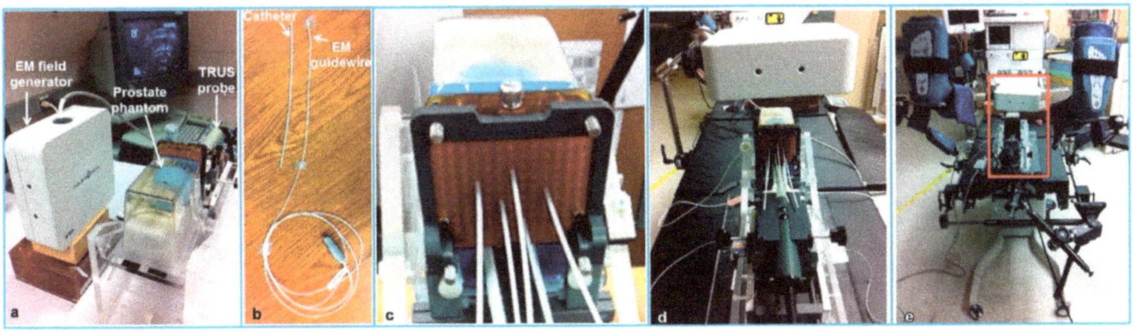

Figure 17. The experimental setup for catheter tracking in a controlled laboratory environment (**a**–**c**) and in a brachytherapy operating room (**d**,**e**). (**a**) The experimental phantom setup; (**b**) The flexible electromagnetic-tracked guidewire and catheter; (**c**) Catheters inserted into the prostate model through the grid; (**d**) The experimental setup positioned on the treatment table in the operating room; (**e**) Mimicking a typical brachytherapy setup. Reprinted from [177] with permission.

In addition to US-guided surgical navigation and radiotherapy, US-guided catheterization has also attracted the attention of many researchers. Jakola et al. [180] reported a method to guide the placement of ventricular catheters using 3D US navigation system. The US-based navigation system (Sonowand Invite, Sonowand AS, Trondheim, Norway) consisted of an US probe integrated with an optical tracking system. Based on the patient studies, this 3D US navigation system was promising for accurate placement of catheters.

Brattain et al. [181] designed a probe-mounted US guidance system for US-guided procedures. The system consisted of a lockable, articulating needle guide that attached to an US probe and a user-interface that provided real-time visualization of the predicted needle trajectory overlaid on the US image. The system illustrated the potential to increase efficiency, safety, quality, and reduce costs for US-guided procedures. Kobayashi et al. [182] invented an US-guided needle insertion manipulator for central venous catheterization (Figure 18). The performance of the manipulator was evaluated in vivo in a porcine model. The animal study results found that a venous placement rate of 80% could be obtained with opened skin, and this system was especially effective for jugular venous puncture of opened skin.

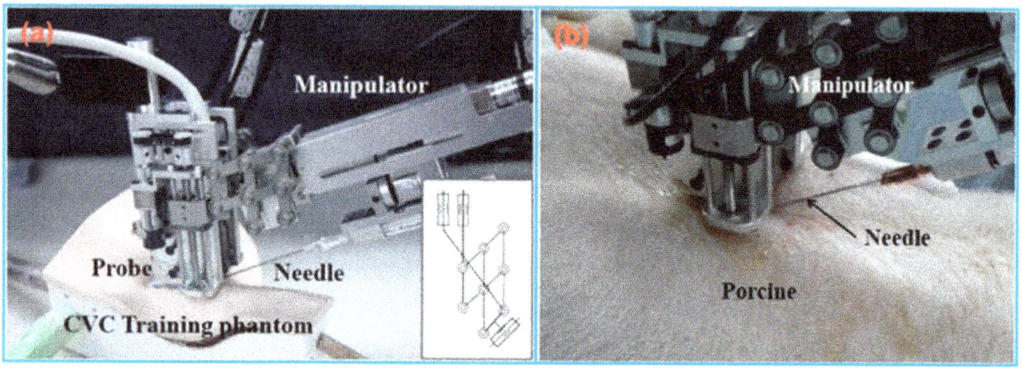

Figure 18. (**a**) Needle insertion manipulator for central venous catheterization. (**b**) Overview of venous puncture experiment in a porcine model. Reprinted from [182] with permission.

5. Conclusions

In this paper, we categorized and reviewed different types of tracking devices for biomedical US imaging applications based on the different tracking principles. The applications of various tracking systems reported in the literature in the past decade were categorized into four types: freehand 3D US imaging, US image fusion, US-guided diagnosis as well as US-guided therapy. In this review article, the working principles of different tracking technologies were analyzed in terms of their advantages and disadvantages for biomedical applications. A comprehensive overview of the state-of-the-art tracking devices on the market is provided in terms of their technical specifications, including accuracy, update rate and latency. With the rapid advancement of various tracking devices over the past decade, the usefulness of different tracking systems has been illustrated by a diverse range of biomedical applications, as reviewed in this paper.

6. Future Perspectives

Although the utilization of tracking device is becoming more and more essential for providing better information and navigation for biomedical applications, there is still much room for improvement. Nowadays, many different types of commercial tracking devices have been introduced and no significant specification differences have been found among them. For the biomedical applications, such as image-guided surgery, perhaps the existing tracking technologies do not fully meet the requirements, and the best choice of tracking device is highly application dependent. The future research of tracking systems may be focused on further improving accuracy and reducing the registration error of these technologies for medical applications. While freehand 3D US has already demonstrated its benefits for obstetrics, cardiology, and image-guided intervention applications, more preclinical studies are required to allow physicians to integrate 3D US imaging effectively and safely into US-guided interventional procedures. In addition, while real-time US image fusion has demonstrated its usefulness in different anatomical regions, such as prostate, liver, and kidney, future studies need to explore its effectiveness in imaging other anatomical regions or during surgery. Although the advancement of different tracking devices has accelerated the development of US image-guided systems, most of these systems are still in the prototype stage, and so far, only limited clinical trials have been carried out. As surgery continues to move toward minimally invasive interventions, US image-guided systems will increasingly be used to improve the precision and quality of medical procedures. More studies from the fields of biomedical engineering, medical physics as well as clinical research are necessary to move this technology from laboratory to hospital to improve patient care.

Author Contributions: Conceptualization, X.J.; writing—original draft preparation, C.P., Q.C. and M.C.; writing—review and editing, C.P., Q.C., M.C. and X.J.; supervision, project administration, and funding acquisition, X.J. All authors have read and agreed to the published version of the manuscript.

Funding: We would like to acknowledge the financial support from the Bill and Melinda Gates Foundation under Award #OPP1191684.

Acknowledgments: We appreciate the anonymous reviewers for their careful reading of our manuscript and their many insightful comments and suggestions to help improve and clarify this manuscript.

Conflicts of Interest: The authors declare no conflict of interest.

Abbreviations

The following abbreviations are used in this manuscript:

2D	Two dimensions
3D	Three dimensions
CCD	Charge-coupled device
Corp.	Corporation
CT	Computed tomography
DOF	Degree-of-freedom
ECIRS	Endoscopic combined intrarenal surgery
FDA	Food and drug administration
FOV	Field of view
FPS	Frames per second
GPU	Graphics processing unit
HCC	Hepatocellular carcinoma
HDR	High-dose-rate
IMU	Inertial measurement unit
Inc.	Incorporated
LED	Light emitting diode
MEMS	Microelectromechanical system
MP	Megapixel
MRI	Magnetic resonance imaging
NDI	Northern Digital Inc.
PCNL	Percutaneous nephrolithotomy
PET	Positron emission tomography
RMS	Root mean square
RVS	Real-time virtual sonography
TDOA	Time difference of arrival
TDOF	Time difference of flight
TOA	Time of arrival
TOF	Time of flight
TRUS	Transrectal ultrasound
US	Ultrasound
USB	Universal serial bus

References

1. Kothiya, S.V.; Mistree, K.B. A Review on Real Time Object Tracking in Video Sequences. In Proceedings of the Electrical, Electronics, Signals, Communication and Optimization (EESCO), 2015 International Conference on, Visakhapatnam, India, 24–25 January 2015; pp. 1–4.
2. Octorina Dewi, D.E.; Supriyanto, E.; Lai, K.W. Position Tracking Systems for Ultrasound Imaging: A Survey. In *Medical Imaging Technology*; Springer: Berlin/Heidelberg, Germany, 2015; pp. 57–89.
3. Karayiannidis, Y.; Rovithakis, G.; Doulgeri, Z. Force/Position Tracking for a Robotic Manipulator in Compliant Contact with a Surface Using Neuro-Adaptive Control. *Automatica* **2007**, *43*, 1281–1288. [CrossRef]
4. Chang, Y.-C.; Yen, H.-M. Design of a Robust Position Feedback Tracking Controller for Flexible-Joint Robots. *IET Control Theory Appl.* **2011**, *5*, 351–363. [CrossRef]
5. Liu, M.; Yu, J.; Yang, L.; Yao, L.; Zhang, Y. Consecutive Tracking for Ballistic Missile Based on Bearings-Only during Boost Phase. *J. Syst. Eng. Electron.* **2012**, *23*, 700–707. [CrossRef]
6. Kendoul, F. Survey of Advances in Guidance, Navigation, and Control of Unmanned Rotorcraft Systems. *J. Field Robot.* **2012**, *29*, 315–378. [CrossRef]
7. Ren, H.; Rank, D.; Merdes, M.; Stallkamp, J.; Kazanzides, P. Multisensor Data Fusion in an Integrated Tracking System for Endoscopic Surgery. *IEEE Trans. Inf. Technol. Biomed.* **2011**, *16*, 106–111. [CrossRef] [PubMed]
8. Huang, Q.-H.; Yang, Z.; Hu, W.; Jin, L.-W.; Wei, G.; Li, X. Linear Tracking for 3-D Medical Ultrasound Imaging. *IEEE Trans. Cybern.* **2013**, *43*, 1747–1754. [CrossRef] [PubMed]
9. Leser, R.; Baca, A.; Ogris, G. Local Positioning Systems in (Game) Sports. *Sensors* **2011**, *11*, 9778–9797. [CrossRef]
10. Hedley, M.; Zhang, J. Accurate Wireless Localization in Sports. *Computer* **2012**, *45*, 64–70. [CrossRef]
11. Mozaffari, M.H.; Lee, W.-S. Freehand 3-D Ultrasound Imaging: A Systematic Review. *Ultrasound Med. Biol.* **2017**, *43*, 2099–2124. [CrossRef]
12. Cleary, K.; Peters, T.M. Image-Guided Interventions: Technology Review and Clinical Applications. *Annu. Rev. Biomed. Eng.* **2010**, *12*, 119–142. [CrossRef]
13. Lindseth, F.; Langø, T.; Selbekk, T.; Hansen, R.; Reinertsen, I.; Askeland, C.; Solheim, O.; Unsgård, G.; Mårvik, R.; Hernes, T.A.N. Ultrasound-Based Guidance and Therapy. In *Advancements and Breakthroughs in Ultrasound Imaging*; IntechOpen: London, UK, 2013.
14. Zhou, H.; Hu, H. Human Motion Tracking for Rehabilitation—A Survey. *Biomed. Signal Process. Control* **2008**, *3*, 1–18. [CrossRef]
15. Moran, C.M.; Thomson, A.J.W. Preclinical Ultrasound Imaging—A Review of Techniques and Imaging Applications. *Front. Phys.* **2020**, *8*, 124. [CrossRef]
16. Fenster, A.; Downey, D.B.; Cardinal, H.N. Three-Dimensional Ultrasound Imaging. *Phys. Med. Biol.* **2001**, *46*, R67. [CrossRef]
17. Herickhoff, C.D.; Morgan, M.R.; Broder, J.S.; Dahl, J.J. Low-Cost Volumetric Ultrasound by Augmentation of 2D Systems: Design and Prototype. *Ultrason. Imaging* **2018**, *40*, 35–48. [CrossRef] [PubMed]
18. Schlegel, M. *Predicting the Accuracy of Optical Tracking Systems*; Technical University of Munich: München, Germany, 2006.
19. Abdelhamid, M. Extracting Depth Information from Stereo Vision System: Using a Correlation and a Feature Based Methods. Master's Thesis, Clemson University, Clemson, SC, USA, 2011.
20. Cai, Q.; Peng, C.; Lu, J.; Prieto, J.C.; Rosenbaum, A.J.; Stringer, J.S.A.; Jiang, X. Performance Enhanced Ultrasound Probe Tracking with a Hemispherical Marker Rigid Body. *IEEE Trans. Ultrason. Ferroelectr. Freq. Control* **2021**, *68*, 2155–2163. [CrossRef] [PubMed]
21. Zhigang, Y.; Kui, Y. An Improved 6DOF Electromagnetic Tracking Algorithm with Anisotropic System Parameters. In *International Conference on Technologies for E-Learning and Digital Entertainment*; Springer: Berlin/Heidelberg, Germany, 2006; pp. 1141–1150.
22. Zhang, Z.; Liu, G. The Design and Analysis of Electromagnetic Tracking System. *J. Electromagn. Anal. Appl.* **2013**, *5*, 85–89. [CrossRef]
23. Craig, J.J. *Introduction to Robotics: Mechanics and Control*, 4th ed.; Pearson Education: New York, NY, USA, 2018.
24. Gueuning, F.; Varlan, M.; Eugene, C.; Dupuis, P. Accurate Distance Measurement by an Autonomous Ultrasonic System Combining Time-of-Flight and Phase-Shift Methods. In Proceedings of the Quality Measurement: The Indispensable Bridge between Theory and Reality, Brussels, Belgium, 4–6 June 1996; Volume 1, pp. 399–404.
25. Mahajan, A.; Walworth, M. 3D Position Sensing Using the Differences in the Time-of-Flights from a Wave Source to Various Receivers. *IEEE Trans. Robot. Autom.* **2001**, *17*, 91–94. [CrossRef]
26. Ray, P.K.; Mahajan, A. A Genetic Algorithm-Based Approach to Calculate the Optimal Configuration of Ultrasonic Sensors in a 3D Position Estimation System. *Rob. Auton. Syst.* **2002**, *41*, 165–177. [CrossRef]
27. Cai, Q.; Hu, J.; Chen, M.; Prieto, J.; Rosenbaum, A.J.; Stringer, J.S.A.; Jiang, X. Inertial Measurement Unit Assisted Ultrasonic Tracking System for Ultrasound Probe Localization. *IEEE Trans. Ultrason. Ferroelectr. Freq. Control* **2022**. [CrossRef]
28. Filippeschi, A.; Schmitz, N.; Miezal, M.; Bleser, G.; Ruffaldi, E.; Stricker, D. Survey of Motion Tracking Methods Based on Inertial Sensors: A Focus on Upper Limb Human Motion. *Sensors* **2017**, *17*, 1257. [CrossRef]
29. Patonis, P.; Patias, P.; Tziavos, I.N.; Rossikopoulos, D.; Margaritis, K.G. A Fusion Method for Combining Low-Cost IMU/Magnetometer Outputs for Use in Applications on Mobile Devices. *Sensors* **2018**, *18*, 2616. [CrossRef] [PubMed]
30. Arqus. Available online: https://www.qualisys.com/cameras/arqus/#!%23tech-specs (accessed on 27 September 2022).
31. Polaris Vega® ST. Available online: https://www.ndigital.com/optical-measurement-technology/polaris-vega/polaris-vega-st/ (accessed on 27 September 2022).

32. Polaris Vega® VT. Available online: https://www.ndigital.com/optical-measurement-technology/polaris-vega/polaris-vega-vt/ (accessed on 27 September 2022).
33. Polaris Vega® XT. Available online: https://www.ndigital.com/optical-measurement-technology/polaris-vega/polaris-vega-xt/ (accessed on 27 September 2022).
34. Polaris Vicra®. Available online: https://www.ndigital.com/optical-measurement-technology/polaris-vicra/ (accessed on 27 September 2022).
35. ClaroNav MicronTracker Specification. Available online: https://www.claronav.com/microntracker/microntracker-specifications/ (accessed on 27 September 2022).
36. Smart DX. Available online: https://www.btsbioengineering.com/products/smart-dx-motion-capture/ (accessed on 27 September 2022).
37. PrimeX 41. Available online: https://optitrack.com/cameras/primex-41/specs.html (accessed on 27 September 2022).
38. PrimeX 22. Available online: https://optitrack.com/cameras/primex-22/specs.html (accessed on 27 September 2022).
39. PrimeX 13. Available online: https://optitrack.com/cameras/primex-13/specs.html (accessed on 27 September 2022).
40. PrimeX 13W. Available online: https://optitrack.com/cameras/primex-13w/specs.html (accessed on 27 September 2022).
41. SlimX 13. Available online: https://optitrack.com/cameras/slimx-13/specs.html (accessed on 27 September 2022).
42. V120:Trio. Available online: https://optitrack.com/cameras/v120-trio/specs.html (accessed on 27 September 2022).
43. V120:Duo. Available online: https://optitrack.com/cameras/v120-duo/specs.html (accessed on 27 September 2022).
44. Flex 13. Available online: https://optitrack.com/cameras/flex-13/specs.html (accessed on 27 September 2022).
45. Flex 3. Available online: https://optitrack.com/cameras/flex-3/specs.html (accessed on 27 September 2022).
46. Slim 3U. Available online: https://optitrack.com/cameras/slim-3u/specs.html (accessed on 27 September 2022).
47. TrackIR 4. vs. TrackIR 5. Available online: https://www.trackir.com/trackir5/ (accessed on 27 September 2022).
48. Miqus. Available online: https://www.qualisys.com/cameras/miqus/#tech-specs (accessed on 27 September 2022).
49. Miqus Hybrid. Available online: https://www.qualisys.com/cameras/miqus-hybrid/#tech-specs (accessed on 27 September 2022).
50. 5+, 6+ and 7+ Series. Available online: https://www.qualisys.com/cameras/5-6-7/#tech-specs (accessed on 27 September 2022).
51. Valkyrie. Available online: https://www.vicon.com/hardware/cameras/valkyrie/ (accessed on 27 September 2022).
52. Vantage+. Available online: https://www.vicon.com/hardware/cameras/vantage/ (accessed on 27 September 2022).
53. Vero. Available online: https://www.vicon.com/hardware/cameras/vero/ (accessed on 27 September 2022).
54. Vue. Available online: https://www.vicon.com/hardware/cameras/vue/ (accessed on 27 September 2022).
55. Viper. Available online: https://www.vicon.com/hardware/cameras/viper/ (accessed on 27 September 2022).
56. ViperX. Available online: https://www.vicon.com/hardware/cameras/viper-x/ (accessed on 27 September 2022).
57. FusionTrack 500. Available online: https://www.atracsys-measurement.com/fusiontrack-500/ (accessed on 27 September 2022).
58. FusionTrack 250. Available online: https://www.atracsys-measurement.com/fusiontrack-250/ (accessed on 27 September 2022).
59. SpryTrack 180. Available online: https://www.atracsys-measurement.com/sprytrack-180/ (accessed on 27 September 2022).
60. SpryTrack 300. Available online: https://www.atracsys-measurement.com/sprytrack-300/ (accessed on 27 September 2022).
61. Kestrel 4200. Available online: https://motionanalysis.com/blog/cameras/kestrel-4200/ (accessed on 27 September 2022).
62. Kestrel 2200. Available online: https://motionanalysis.com/blog/cameras/kestrel-2200/ (accessed on 27 September 2022).
63. Kestrel 1300. Available online: https://motionanalysis.com/blog/cameras/kestrel-130/ (accessed on 27 September 2022).
64. Kestrel 300. Available online: https://motionanalysis.com/blog/cameras/kestrel-300/ (accessed on 27 September 2022).
65. EDDO Biomechanic. Available online: https://www.stt-systems.com/motion-analysis/3d-optical-motion-capture/eddo/ (accessed on 27 September 2022).
66. ARTTRACK6/M. Available online: https://ar-tracking.com/en/product-program/arttrack6m (accessed on 27 September 2022).
67. ARTTRACK5. Available online: https://ar-tracking.com/en/product-program/arttrack5 (accessed on 27 September 2022).
68. SMARTTRACK3 & SMARTTRACK3/M. Available online: https://ar-tracking.com/en/product-program/smarttrack3 (accessed on 27 September 2022).
69. Micro Sensor 1.8. Available online: https://polhemus.com/micro-sensors/ (accessed on 27 September 2022).
70. Aurora. Available online: https://www.ndigital.com/electromagnetic-tracking-technology/aurora/ (accessed on 27 September 2022).
71. 3D Guidance. Available online: https://www.ndigital.com/electromagnetic-tracking-technology/3d-guidance/ (accessed on 27 September 2022).
72. Viper. Available online: https://polhemus.com/viper (accessed on 27 September 2022).
73. Fastrak. Available online: https://polhemus.com/motion-tracking/all-trackers/fastrak (accessed on 27 September 2022).
74. Patriot. Available online: https://polhemus.com/motion-tracking/all-trackers/patriot (accessed on 27 September 2022).
75. Patriot Wireless. Available online: https://polhemus.com/motion-tracking/all-trackers/patriot-wireless (accessed on 27 September 2022).
76. Liberty. Available online: https://polhemus.com/motion-tracking/all-trackers/liberty (accessed on 27 September 2022).
77. Liberty Latus. Available online: https://polhemus.com/motion-tracking/all-trackers/liberty-latus (accessed on 27 September 2022).
78. G4. Available online: https://polhemus.com/motion-tracking/all-trackers/g4 (accessed on 27 September 2022).
79. Gypsy 7. Available online: https://metamotion.com/gypsy/gypsy-motion-capture-system.htm (accessed on 27 September 2022).
80. Forkbeard. Available online: https://www.sonitor.com/forkbeard (accessed on 27 September 2022).
81. Xsens DOT. Available online: https://www.xsens.com/xsens-dot (accessed on 27 September 2022).

82. MTw Awinda. Available online: https://www.xsens.com/products/mtw-awinda (accessed on 27 September 2022).
83. MTi 1-Series. Available online: https://mtidocs.xsens.com/sensor-specifications$mti-1-series-performance-specifications (accessed on 27 September 2022).
84. MTi 10/100-Series. Available online: https://mtidocs.xsens.com/output-specifications$orientation-performance-specification (accessed on 27 September 2022).
85. MTi 600-Series. Available online: https://mtidocs.xsens.com/sensor-specifications-2$mti-600-series-performance-specifications-nbsp (accessed on 27 September 2022).
86. Inertial Motion Capture. Available online: https://www.stt-systems.com/motion-analysis/inertial-motion-capture/ (accessed on 27 September 2022).
87. VN-100. Available online: https://www.vectornav.com/products/detail/vn-100 (accessed on 27 September 2022).
88. VN-110. Available online: https://www.vectornav.com/products/detail/vn-110 (accessed on 27 September 2022).
89. VN-200. Available online: https://www.vectornav.com/products/detail/vn-200 (accessed on 27 September 2022).
90. VN-210. Available online: https://www.vectornav.com/products/detail/vn-210 (accessed on 27 September 2022).
91. VN-300. Available online: https://www.vectornav.com/products/detail/vn-300 (accessed on 27 September 2022).
92. VN-310. Available online: https://www.vectornav.com/products/detail/vn-310 (accessed on 27 September 2022).
93. Motus. Available online: https://www.advancednavigation.com/imu-ahrs/mems-imu/motus/ (accessed on 27 September 2022).
94. Orientus. Available online: https://www.advancednavigation.com/imu-ahrs/mems-imu/orientus/ (accessed on 27 September 2022).
95. BOREAS D90. Available online: https://www.advancednavigation.com/inertial-navigation-systems/fog-gnss-ins/boreas/ (accessed on 27 September 2022).
96. Spatial FOG Dual. Available online: https://www.advancednavigation.com/inertial-navigation-systems/fog-gnss-ins/spatial-fog-dual/ (accessed on 27 September 2022).
97. Certus Evo. Available online: https://www.advancednavigation.com/inertial-navigation-systems/mems-gnss-ins/certus-evo/ (accessed on 27 September 2022).
98. Certus. Available online: https://www.advancednavigation.com/inertial-navigation-systems/mems-gnss-ins/certus/ (accessed on 27 September 2022).
99. Spatial. Available online: https://www.advancednavigation.com/inertial-navigation-systems/mems-gnss-ins/spatial/ (accessed on 27 September 2022).
100. GNSS Compass. Available online: https://www.advancednavigation.com/inertial-navigation-systems/satellite-compass/gnss-compass/ (accessed on 27 September 2022).
101. Kernel-100. Available online: https://inertiallabs.com/wp-content/uploads/2021/12/IMU-Kernel_Datasheet.rev_.2.9_December_2021.pdf (accessed on 27 September 2022).
102. Kernel-110, 120. Available online: https://inertiallabs.com/wp-content/uploads/2022/09/IMU-Kernel-110-120_Datasheet.rev1_.7_September20_2022.pdf (accessed on 27 September 2022).
103. Kernel-210, 220. Available online: https://inertiallabs.com/wp-content/uploads/2022/09/IMU-Kernel-210-220_Datasheet.rev1_.6_Sept20_2022.pdf (accessed on 27 September 2022).
104. IMU-P. Available online: https://inertiallabs.com/wp-content/uploads/2022/09/IMU-P_Datasheet.rev4_.1_Sept20_2022.pdf (accessed on 27 September 2022).
105. Peng, C.; Chen, M.; Spicer, J.B.; Jiang, X. Acoustics at the Nanoscale (Nanoacoustics): A Comprehensive Literature Review. Part II: Nanoacoustics for Biomedical Imaging and Therapy. *Sens. Actuators A Phys.* **2021**, *332*, 112925. [CrossRef] [PubMed]
106. Rajaraman, P.; Simpson, J.; Neta, G.; de Gonzalez, A.B.; Ansell, P.; Linet, M.S.; Ron, E.; Roman, E. Early Life Exposure to Diagnostic Radiation and Ultrasound Scans and Risk of Childhood Cancer: Case-Control Study. *BMJ* **2011**, *342*, d472. [CrossRef] [PubMed]
107. Huang, Q.; Zeng, Z. A Review on Real-Time 3D Ultrasound Imaging Technology. *Biomed. Res. Int.* **2017**, *2017*, 6027029. [CrossRef]
108. Morgan, M.R.; Broder, J.S.; Dahl, J.J.; Herickhoff, C.D. Versatile Low-Cost Volumetric 3-D Ultrasound Platform for Existing Clinical 2-D Systems. *IEEE Trans. Med. Imaging* **2018**, *37*, 2248–2256. [CrossRef]
109. Prager, R.W.; Ijaz, U.Z.; Gee, A.H.; Treece, G.M. Three-Dimensional Ultrasound Imaging. *Proc. Inst. Mech. Eng. Part H J. Eng. Med.* **2010**, *224*, 193–223. [CrossRef]
110. Fenster, A.; Parraga, G.; Bax, J. Three-Dimensional Ultrasound Scanning. *Interface Focus* **2011**, *1*, 503–519. [CrossRef]
111. Yen, J.T.; Steinberg, J.P.; Smith, S.W. Sparse 2-D Array Design for Real Time Rectilinear Volumetric Imaging. *IEEE Trans. Ultrason. Ferroelectr. Freq. Control* **2000**, *47*, 93–110. [CrossRef]
112. Yen, J.T.; Smith, S.W. Real-Time Rectilinear 3-D Ultrasound Using Receive Mode Multiplexing. *ieee Trans. Ultrason. Ferroelectr. Freq. Control* **2004**, *51*, 216–226. [CrossRef]
113. Turnbull, D.H.; Foster, F.S. Fabrication and Characterization of Transducer Elements in Two-Dimensional Arrays for Medical Ultrasound Imaging. *IEEE Trans. Ultrason. Ferroelectr. Freq. Control* **1992**, *39*, 464–475. [CrossRef]
114. Gee, A.; Prager, R.; Treece, G.; Berman, L. Engineering a Freehand 3D Ultrasound System. *Pattern Recognit. Lett.* **2003**, *24*, 757–777. [CrossRef]
115. Wen, T.; Yang, F.; Gu, J.; Wang, L. A Novel Bayesian-Based Nonlocal Reconstruction Method for Freehand 3D Ultrasound Imaging. *Neurocomputing* **2015**, *168*, 104–118. [CrossRef]
116. Chung, S.-W.; Shih, C.-C.; Huang, C.-C. Freehand Three-Dimensional Ultrasound Imaging of Carotid Artery Using Motion Tracking Technology. *Ultrasonics* **2017**, *74*, 11–20. [CrossRef] [PubMed]

117. Daoud, M.I.; Alshalalfah, A.-L.; Awwad, F.; Al-Najar, M. Freehand 3D Ultrasound Imaging System Using Electromagnetic Tracking. In Proceedings of the 2015 International Conference on Open Source Software Computing (OSSCOM), Amman, Jordan, 10–13 September 2015; pp. 1–5.
118. Chen, Z.; Huang, Q. Real-Time Freehand 3D Ultrasound Imaging. *Comput. Methods Biomech. Biomed. Eng. Imaging Vis.* **2018**, *6*, 74–83. [CrossRef]
119. Pelz, J.O.; Weinreich, A.; Karlas, T.; Saur, D. Evaluation of Freehand B-Mode and Power-Mode 3D Ultrasound for Visualisation and Grading of Internal Carotid Artery Stenosis. *PLoS ONE* **2017**, *12*, e0167500. [CrossRef] [PubMed]
120. Miller, D.; Lippert, C.; Vollmer, F.; Bozinov, O.; Benes, L.; Schulte, D.M.; Sure, U. Comparison of Different Reconstruction Algorithms for Three-dimensional Ultrasound Imaging in a Neurosurgical Setting. *Int. J. Med. Robot. Comput. Assist. Surg.* **2012**, *8*, 348–359. [CrossRef]
121. Mercier, L.; Del Maestro, R.F.; Petrecca, K.; Kochanowska, A.; Drouin, S.; Yan, C.X.B.; Janke, A.L.; Chen, S.J.-S.; Collins, D.L. New Prototype Neuronavigation System Based on Preoperative Imaging and Intraoperative Freehand Ultrasound: System Description and Validation. *Int. J. Comput. Assist. Radiol. Surg.* **2011**, *6*, 507–522. [CrossRef]
122. Chen, X.; Wen, T.; Li, X.; Qin, W.; Lan, D.; Pan, W.; Gu, J. Reconstruction of Freehand 3D Ultrasound Based on Kernel Regression. *Biomed. Eng. Online* **2014**, *13*, 1–15. [CrossRef]
123. Sun, S.-Y.; Gilbertson, M.; Anthony, B.W. Probe Localization for Freehand 3D Ultrasound by Tracking Skin Features. In *Medical Image Computing and Computer-Assisted Intervention—MICCAI 2014*; Springer: Berlin/Heidelberg, Germany, 2014; pp. 365–372.
124. Worobey, L.A.; Udofa, I.A.; Lin, Y.-S.; Koontz, A.M.; Farrokhi, S.S.; Boninger, M.L. Reliability of Freehand Three-Dimensional Ultrasound to Measure Scapular Rotations. *J. Rehabil. Res. Dev.* **2014**, *51*, 985–994. [CrossRef]
125. Passmore, E.; Pandy, M.G.; Graham, H.K.; Sangeux, M. Measuring Femoral Torsion in Vivo Using Freehand 3-D Ultrasound Imaging. *Ultrasound Med. Biol.* **2016**, *42*, 619–623. [CrossRef]
126. Kim, T.; Kang, D.-H.; Shim, S.; Im, M.; Seo, B.K.; Kim, H.; Lee, B.C. Versatile Low-Cost Volumetric 3D Ultrasound Imaging Using Gimbal-Assisted Distance Sensors and an Inertial Measurement Unit. *Sensors* **2020**, *20*, 6613. [CrossRef] [PubMed]
127. Lai, K.K.-L.; Lee, T.T.-Y.; Lee, M.K.-S.; Hui, J.C.-H.; Zheng, Y.-P. Validation of Scolioscan Air-Portable Radiation-Free Three-Dimensional Ultrasound Imaging Assessment System for Scoliosis. *Sensors* **2021**, *21*, 2858. [CrossRef] [PubMed]
128. Jiang, W.; Chen, X.; Yu, C. A Real-time Freehand 3D Ultrasound Imaging Method for Scoliosis Assessment. *J. Appl. Clin. Med. Phys.* **2022**, *23*, e13709. [CrossRef] [PubMed]
129. Ewertsen, C.; Săftoiu, A.; Gruionu, L.G.; Karstrup, S.; Nielsen, M.B. Real-Time Image Fusion Involving Diagnostic Ultrasound. *Am. J. Roentgenol.* **2013**, *200*, W249–W255. [CrossRef] [PubMed]
130. Li, X.; Zhou, F.; Tan, H.; Zhang, W.; Zhao, C. Multimodal Medical Image Fusion Based on Joint Bilateral Filter and Local Gradient Energy. *Inf. Sci.* **2021**, *569*, 302–325. [CrossRef]
131. Klibanov, A.L.; Hossack, J.A. Ultrasound in Radiology: From Anatomic, Functional, Molecular Imaging to Drug Delivery and Image-Guided Therapy. *Investig. Radiol.* **2015**, *50*, 657. [CrossRef]
132. Baad, M.; Lu, Z.F.; Reiser, I.; Paushter, D. Clinical Significance of US Artifacts. *Radiographics* **2017**, *37*, 1408–1423. [CrossRef]
133. European Society of Radiology (ESR) communications@ myesr. org D'Onofrio Mirko Beleù Alessandro Gaitini Diana Corréas Jean-Michel Brady Adrian Clevert Dirk. Abdominal Applications of Ultrasound Fusion Imaging Technique: Liver, Kidney, and Pancreas. *Insights Imaging* **2019**, *10*, 6. [CrossRef]
134. Chien, C.P.Y.; Lee, K.H.; Lau, V. Real-Time Ultrasound Fusion Imaging–Guided Interventions: A Review. *Hong Kong J. Radiol.* **2021**, *24*, 116. [CrossRef]
135. Natarajan, S.; Marks, L.S.; Margolis, D.J.A.; Huang, J.; Macairan, M.L.; Lieu, P.; Fenster, A. Clinical Application of a 3D Ultrasound-Guided Prostate Biopsy System. In *Urologic Oncology: Seminars and Original Investigations*; Elsevier: Amsterdam, The Netherlands, 2011; Volume 29, pp. 334–342.
136. Krücker, J.; Xu, S.; Venkatesan, A.; Locklin, J.K.; Amalou, H.; Glossop, N.; Wood, B.J. Clinical Utility of Real-Time Fusion Guidance for Biopsy and Ablation. *J. Vasc. Interv. Radiol.* **2011**, *22*, 515–524. [CrossRef]
137. Appelbaum, L.; Sosna, J.; Nissenbaum, Y.; Benshtein, A.; Goldberg, S.N. Electromagnetic Navigation System for CT-Guided Biopsy of Small Lesions. *Am. J. Roentgenol.* **2011**, *196*, 1194–1200. [CrossRef] [PubMed]
138. Venkatesan, A.M.; Kadoury, S.; Abi-Jaoudeh, N.; Levy, E.B.; Maass-Moreno, R.; Krücker, J.; Dalal, S.; Xu, S.; Glossop, N.; Wood, B.J. Real-Time FDG PET Guidance during Biopsies and Radiofrequency Ablation Using Multimodality Fusion with Electromagnetic Navigation. *Radiology* **2011**, *260*, 848–856. [CrossRef] [PubMed]
139. Lee, M.W. Fusion Imaging of Real-Time Ultrasonography with CT or MRI for Hepatic Intervention. *Ultrasonography* **2014**, *33*, 227. [CrossRef] [PubMed]
140. Sumi, H.; Itoh, A.; Kawashima, H.; Ohno, E.; Itoh, Y.; Nakamura, Y.; Hiramatsu, T.; Sugimoto, H.; Hayashi, D.; Kuwahara, T. Preliminary Study on Evaluation of the Pancreatic Tail Observable Limit of Transabdominal Ultrasonography Using a Position Sensor and CT-Fusion Image. *Eur. J. Radiol.* **2014**, *83*, 1324–1331. [CrossRef] [PubMed]
141. Lee, K.-H.; Lau, V.; Gao, Y.; Li, Y.-L.; Fang, B.X.; Lee, R.; Lam, W.W.-M. Ultrasound-MRI Fusion for Targeted Biopsy of Myopathies. *AJR Am. J. Roentgenol.* **2019**, *212*, 1126–1128. [CrossRef] [PubMed]
142. Burke, C.J.; Bencardino, J.; Adler, R. The Potential Use of Ultrasound-Magnetic Resonance Imaging Fusion Applications in Musculoskeletal Intervention. *J. Ultrasound Med.* **2017**, *36*, 217–224. [CrossRef]

143. Klauser, A.S.; De Zordo, T.; Feuchtner, G.M.; Djedovic, G.; Weiler, R.B.; Faschingbauer, R.; Schirmer, M.; Moriggl, B. Fusion of Real-Time US with CT Images to Guide Sacroiliac Joint Injection in Vitro and in Vivo. *Radiology* **2010**, *256*, 547–553. [CrossRef]
144. Sonn, G.A.; Margolis, D.J.; Marks, L.S. Target Detection: Magnetic Resonance Imaging-Ultrasound Fusion–Guided Prostate Biopsy. In *Urologic Oncology: Seminars and Original Investigations*; Elsevier: Amsterdam, The Netherlands, 2014; Volume 32, pp. 903–911.
145. Costa, D.N.; Pedrosa, I.; Donato, F., Jr.; Roehrborn, C.G.; Rofsky, N.M. MR Imaging–Transrectal US Fusion for Targeted Prostate Biopsies: Implications for Diagnosis and Clinical Management. *Radiographics* **2015**, *35*, 696–708. [CrossRef]
146. Appelbaum, L.; Mahgerefteh, S.Y.; Sosna, J.; Goldberg, S.N. Image-Guided Fusion and Navigation: Applications in Tumor Ablation. *Tech. Vasc. Interv. Radiol.* **2013**, *16*, 287–295. [CrossRef]
147. Marks, L.; Young, S.; Natarajan, S. MRI–Ultrasound Fusion for Guidance of Targeted Prostate Biopsy. *Curr. Opin. Urol.* **2013**, *23*, 43. [CrossRef]
148. Park, H.J.; Lee, M.W.; Lee, M.H.; Hwang, J.; Kang, T.W.; Lim, S.; Rhim, H.; Lim, H.K. Fusion Imaging–Guided Percutaneous Biopsy of Focal Hepatic Lesions with Poor Conspicuity on Conventional Sonography. *J. Ultrasound Med.* **2013**, *32*, 1557–1564. [CrossRef] [PubMed]
149. Lee, M.W.; Rhim, H.; Cha, D.I.; Kim, Y.J.; Lim, H.K. Planning US for Percutaneous Radiofrequency Ablation of Small Hepatocellular Carcinomas (1–3 Cm): Value of Fusion Imaging with Conventional US and CT/MR Images. *J. Vasc. Interv. Radiol.* **2013**, *24*, 958–965. [CrossRef] [PubMed]
150. Song, K.D.; Lee, M.W.; Rhim, H.; Cha, D.I.; Chong, Y.; Lim, H.K. Fusion Imaging–Guided Radiofrequency Ablation for Hepatocellular Carcinomas Not Visible on Conventional Ultrasound. *Am. J. Roentgenol.* **2013**, *201*, 1141–1147. [CrossRef]
151. Helck, A.; D'Anastasi, M.; Notohamiprodjo, M.; Thieme, S.; Sommer, W.; Reiser, M.; Clevert, D.A. Multimodality Imaging Using Ultrasound Image Fusion in Renal Lesions. *Clin. Hemorheol. Microcirc.* **2012**, *50*, 79–89. [CrossRef] [PubMed]
152. Andersson, M.; Hashimi, F.; Lyrdal, D.; Lundstam, S.; Hellström, M. Improved Outcome with Combined US/CT Guidance as Compared to US Guidance in Percutaneous Radiofrequency Ablation of Small Renal Masses. *Acta Radiol.* **2015**, *56*, 1519–1526. [CrossRef] [PubMed]
153. Zhang, H.; Chen, G.; Xiao, L.; Ma, X.; Shi, L.; Wang, T.; Yan, H.; Zou, H.; Chen, Q.; Tang, L. Ultrasonic/CT Image Fusion Guidance Facilitating Percutaneous Catheter Drainage in Treatment of Acute Pancreatitis Complicated with Infected Walled-off Necrosis. *Pancreatology* **2018**, *18*, 635–641. [CrossRef]
154. Rübenthaler, J.; Paprottka, K.J.; Marcon, J.; Reiser, M.; Clevert, D.A. MRI and Contrast Enhanced Ultrasound (CEUS) Image Fusion of Renal Lesions. *Clin. Hemorheol. Microcirc.* **2016**, *64*, 457–466. [CrossRef]
155. Guo, Z.; Shi, H.; Li, W.; Lin, D.; Wang, C.; Liu, C.; Yuan, M.; Wu, X.; Xiong, B.; He, X. Chinese Multidisciplinary Expert Consensus: Guidelines on Percutaneous Transthoracic Needle Biopsy. *Thorac. Cancer* **2018**, *9*, 1530–1543. [CrossRef]
156. Beigi, P.; Salcudean, S.E.; Ng, G.C.; Rohling, R. Enhancement of Needle Visualization and Localization in Ultrasound. *Int. J. Comput. Assist. Radiol. Surg.* **2021**, *16*, 169–178. [CrossRef]
157. Holm, H.H.; Skjoldbye, B. Interventional Ultrasound. *Ultrasound Med. Biol.* **1996**, *22*, 773–789.
158. Stone, J.; Beigi, P.; Rohling, R.; Lessoway, V.; Dube, A.; Gunka, V. Novel 3D Ultrasound System for Midline Single-Operator Epidurals: A Feasibility Study on a Porcine Model. *Int. J. Obstet. Anesth.* **2017**, *31*, 51–56. [CrossRef] [PubMed]
159. Scholten, H.J.; Pourtaherian, A.; Mihajlovic, N.; Korsten, H.H.M.; Bouwman, R.A. Improving Needle Tip Identification during Ultrasound-guided Procedures in Anaesthetic Practice. *Anaesthesia* **2017**, *72*, 889–904. [CrossRef] [PubMed]
160. Boctor, E.M.; Choti, M.A.; Burdette, E.C.; Webster Iii, R.J. Three-dimensional Ultrasound-guided Robotic Needle Placement: An Experimental Evaluation. *Int. J. Med. Robot. Comput. Assist. Surg.* **2008**, *4*, 180–191. [CrossRef] [PubMed]
161. Franz, A.M.; März, K.; Hummel, J.; Birkfellner, W.; Bendl, R.; Delorme, S.; Schlemmer, H.-P.; Meinzer, H.-P.; Maier-Hein, L. Electromagnetic Tracking for US-Guided Interventions: Standardized Assessment of a New Compact Field Generator. *Int. J. Comput. Assist. Radiol. Surg.* **2012**, *7*, 813–818. [CrossRef] [PubMed]
162. Xu, H.-X.; Lu, M.-D.; Liu, L.-N.; Guo, L.-H. Magnetic Navigation in Ultrasound-Guided Interventional Radiology Procedures. *Clin. Radiol.* **2012**, *67*, 447–454. [CrossRef]
163. Hakime, A.; Deschamps, F.; De Carvalho, E.G.M.; Barah, A.; Auperin, A.; De Baere, T. Electromagnetic-Tracked Biopsy under Ultrasound Guidance: Preliminary Results. *Cardiovasc. Intervent. Radiol.* **2012**, *35*, 898–905. [CrossRef]
164. März, K.; Franz, A.M.; Seitel, A.; Winterstein, A.; Hafezi, M.; Saffari, A.; Bendl, R.; Stieltjes, B.; Meinzer, H.-P.; Mehrabi, A. Interventional Real-Time Ultrasound Imaging with an Integrated Electromagnetic Field Generator. *Int. J. Comput. Assist. Radiol. Surg.* **2014**, *9*, 759–768. [CrossRef]
165. Wang, X.L.; Stolka, P.J.; Boctor, E.; Hager, G.; Choti, M. The Kinect as an Interventional Tracking System. In *Medical Imaging 2012: Image-Guided Procedures, Robotic Interventions, and Modeling*; SPIE: Washington, DC, USA, 2012; Volume 8316, pp. 276–281.
166. Stolka, P.J.; Foroughi, P.; Rendina, M.; Weiss, C.R.; Hager, G.D.; Boctor, E.M. Needle Guidance Using Handheld Stereo Vision and Projection for Ultrasound-Based Interventions. In *Medical Image Computing and Computer-Assisted Intervention—MICCAI 2014*; Springer: Cham, Switzerland, 2014; pp. 684–691.
167. Najafi, M.; Abolmaesumi, P.; Rohling, R. Single-Camera Closed-Form Real-Time Needle Tracking for Ultrasound-Guided Needle Insertion. *Ultrasound Med. Biol.* **2015**, *41*, 2663–2676. [CrossRef]
168. Daoud, M.I.; Alshalalfah, A.-L.; Mohamed, O.A.; Alazrai, R. A Hybrid Camera-and Ultrasound-Based Approach for Needle Localization and Tracking Using a 3D Motorized Curvilinear Ultrasound Probe. *Med. Image Anal.* **2018**, *50*, 145–166. [CrossRef]

169. Ho, H.S.S.; Mohan, P.; Lim, E.D.; Li, D.L.; Yuen, J.S.P.; Ng, W.S.; Lau, W.K.O.; Cheng, C.W.S. Robotic Ultrasound-guided Prostate Intervention Device: System Description and Results from Phantom Studies. *Int. J. Med. Robot. Comput. Assist. Surg.* **2009**, *5*, 51–58. [CrossRef]
170. Orhan, S.O.; Yildirim, M.C.; Bebek, O. Design and Modeling of a Parallel Robot for Ultrasound Guided Percutaneous Needle Interventions. In Proceedings of the IECON 2015—41st Annual Conference of the IEEE Industrial Electronics Society, Yokohama, Japan, 9–12 November 2015; pp. 5002–5007.
171. Poquet, C.; Mozer, P.; Vitrani, M.-A.; Morel, G. An Endorectal Ultrasound Probe Comanipulator with Hybrid Actuation Combining Brakes and Motors. *IEEE/ASME Trans. Mechatron.* **2014**, *20*, 186–196. [CrossRef]
172. Chen, X.; Bao, N.; Li, J.; Kang, Y. A Review of Surgery Navigation System Based on Ultrasound Guidance. In Proceedings of the 2012 IEEE International Conference on Information and Automation, Shenyang, China, 6–8 June 2012; pp. 882–886.
173. Stoll, J.; Ren, H.; Dupont, P.E. Passive Markers for Tracking Surgical Instruments in Real-Time 3-D Ultrasound Imaging. *IEEE Trans. Med. Imaging* **2011**, *31*, 563–575. [CrossRef] [PubMed]
174. Li, X.; Long, Q.; Chen, X.; He, D.; He, H. Assessment of the SonixGPS System for Its Application in Real-Time Ultrasonography Navigation-Guided Percutaneous Nephrolithotomy for the Treatment of Complex Kidney Stones. *Urolithiasis* **2017**, *45*, 221–227. [CrossRef] [PubMed]
175. Hamamoto, S.; Unno, R.; Taguchi, K.; Ando, R.; Hamakawa, T.; Naiki, T.; Okada, S.; Inoue, T.; Okada, A.; Kohri, K. A New Navigation System of Renal Puncture for Endoscopic Combined Intrarenal Surgery: Real-Time Virtual Sonography-Guided Renal Access. *Urology* **2017**, *109*, 44–50. [CrossRef]
176. Gomes-Fonseca, J.; Veloso, F.; Queirós, S.; Morais, P.; Pinho, A.C.M.; Fonseca, J.C.; Correia-Pinto, J.; Lima, E.; Vilaça, J.L. Assessment of Electromagnetic Tracking Systems in a Surgical Environment Using Ultrasonography and Ureteroscopy Instruments for Percutaneous Renal Access. *Med. Phys.* **2020**, *47*, 19–26. [CrossRef] [PubMed]
177. Bharat, S.; Kung, C.; Dehghan, E.; Ravi, A.; Venugopal, N.; Bonillas, A.; Stanton, D.; Kruecker, J. Electromagnetic Tracking for Catheter Reconstruction in Ultrasound-Guided High-Dose-Rate Brachytherapy of the Prostate. *Brachytherapy* **2014**, *13*, 640–650. [CrossRef] [PubMed]
178. Schwaab, J.; Prall, M.; Sarti, C.; Kaderka, R.; Bert, C.; Kurz, C.; Parodi, K.; Günther, M.; Jenne, J. Ultrasound Tracking for Intra-Fractional Motion Compensation in Radiation Therapy. *Phys. Med.* **2014**, *30*, 578–582. [CrossRef]
179. Yu, A.S.; Najafi, M.; Hristov, D.H.; Phillips, T. Intrafractional Tracking Accuracy of a Transperineal Ultrasound Image Guidance System for Prostate Radiotherapy. *Technol. Cancer Res. Treat.* **2017**, *16*, 1067–1078. [CrossRef]
180. Jakola, A.S.; Reinertsen, I.; Selbekk, T.; Solheim, O.; Lindseth, F.; Gulati, S.; Unsgård, G. Three-Dimensional Ultrasound–Guided Placement of Ventricular Catheters. *World Neurosurg.* **2014**, *82*, 536.e5–536.e9. [CrossRef]
181. Brattain, L.J.; Floryan, C.; Hauser, O.P.; Nguyen, M.; Yong, R.J.; Kesner, S.B.; Corn, S.B.; Walsh, C.J. Simple and Effective Ultrasound Needle Guidance System. In Proceedings of the 2011 Annual International Conference of the IEEE Engineering in Medicine and Biology Society, Boston, MA, USA, 30 August–3 September 2011; pp. 8090–8093.
182. Kobayashi, Y.; Hamano, R.; Watanabe, H.; Koike, T.; Hong, J.; Toyoda, K.; Uemura, M.; Ieiri, S.; Tomikawa, M.; Ohdaira, T. Preliminary in Vivo Evaluation of a Needle Insertion Manipulator for Central Venous Catheterization. *ROBOMECH J.* **2014**, *1*, 1–7. [CrossRef]

micromachines

MDPI

Article

Airline Point-of-Care System on Seat Belt for Hybrid Physiological Signal Monitoring

Xiaoqiang Ji [1,†], Zhi Rao [1,†], Wei Zhang [2,†], Chang Liu [3], Zimo Wang [3], Shuo Zhang [1], Butian Zhang [4], Menglei Hu [5,*], Peyman Servati [5] and Xiao Xiao [6,*]

1 School of Life Science and Technology, Changchun University of Science and Technology, Changchun 130022, China
2 Department of Biomedical Engineering, College of Design and Engineering, National University of Singapore, Singapore 117583, Singapore
3 Department of Materials Science and Engineering, College of Design and Engineering, National University of Singapore, Singapore 117583, Singapore
4 Department of Imaging, China-Japan Union Hospital of Jilin University, Changchun 130033, China
5 Department of Electrical and Computer Engineering, University of British Columbia, Vancouver, BC V6T 1Z4, Canada
6 Department of Electrical and Computer Engineering, College of Design and Engineering, National University of Singapore, Singapore 117583, Singapore
* Correspondence: mengleih@ece.ubc.ca (M.H.); xiao.xiao@u.nus.edu (X.X.)
† These authors contributed equally to this work.

check for updates

Citation: Ji, X.; Rao, Z.; Zhang, W.; Liu, C.; Wang, Z.; Zhang, S.; Zhang, B.; Hu, M.; Servati, P.; Xiao, X. Airline Point-of-Care System on Seat Belt for Hybrid Physiological Signal Monitoring. *Micromachines* **2022**, *13*, 1880. https://doi.org/10.3390/mi13111880

Academic Editors: Nam-Trung Nguyen and Angeliki Tserepi

Received: 13 October 2022
Accepted: 29 October 2022
Published: 1 November 2022

Abstract: With a focus on disease prevention and health promotion, a reactive and disease-centric healthcare system is revolutionized to a point-of-care model by the application of wearable devices. The convenience and low cost made it possible for long-term monitoring of health problems in long-distance traveling such as flights. While most of the existing health monitoring systems on aircrafts are limited for pilots, point-of-care systems provide choices for passengers to enjoy healthcare at the same level. Here in this paper, an airline point-of-care system containing hybrid electrocardiogram (ECG), breathing, and motion signals detection is proposed. At the same time, we propose the diagnosis of sleep apnea-hypopnea syndrome (SAHS) on flights as an application of this system to satisfy the inevitable demands for sleeping on long-haul flights. The hardware design includes ECG electrodes, flexible piezoelectric belts, and a control box, which enables the system to detect the original data of ECG, breathing, and motion signals. By processing these data with interval extraction-based feature selection method, the signals would be characterized and then provided for the long short-term memory recurrent neural network (LSTM-RNN) to classify the SAHS. Compared with other machine learning methods, our model shows high accuracy up to 84–85% with the lowest overfit problem, which proves its potential application in other related fields.

Keywords: point-of-care system; wearable biosensors; machine-learning; airline industry security

1. Introduction

Point-of-care systems have received great attention in recent years since it aims to provide diagnosis and therapeutic services with imperceptible devices [1–4]. Traditional healthcare systems are always lagging and can often only be applied after disease onset or diagnosis [5,6]. Moreover, some chronic diseases require continuous monitoring and long-term nursing, which causes a huge burden on public healthcare resources [7–9]. To this end, point-of-care shows its advantages in serving patients with economic and convenient healthcare monitoring, which enable individual diagnosis ahead of time [10–12]. Combined with variable wearable devices, such as seatbelts and bracelets [13–15], the new-type point-of-care system can be comfortably used by patients in their daily life. As a result, the systems supply sufficient and instant medical services for patients with long-term monitoring, which decrease the medical burden for patients and hospitals, and provide an attractive alternative for customers to enjoy health care in flight.

For the airline industry, there is nothing more important than the safety of passengers [16–18]. However, during air travel, passengers are inevitably exposed to unusual situations such as overweight, weightlessness, low atmospheric pressure, and the fluctuations [19,20]. The occurrence of in-flight medical emergencies is rare while associated costs derived from en route diversion might significantly influence airlines' budgetary equilibrium [21]. Published data suggest that the incidence of inflight deaths from medical conditions is between 0.3 and 1 per million passengers [22]. For a person who recently underwent chest surgery, it has a 25% expansion of any residual trapped gas with compromise to both circulation and respiration [23]. It is more reasonable for passengers to choose airlines that provide onboard medical services. Considering that associated costs derived from en route diversion might significantly influence airlines' budgetary equilibrium and the occurrence of inflight medical emergency is rare [21], which has a death incidence between 0.3 and 1 per million passengers [22], the healthcare system can act as an inexpensive way for aviation corporations to reduce the risks of diversion. For some special customers, such as the persons who recently underwent chest surgery recently and therefore have a 25% expansion of any residual trapped gas with compromise to both circulation and respiration [23], it is more reasonable for them to choose airlines that provide onboard medical services. The growing population of older people suffers more medical emergency under these extreme conditions because the elders are often accompanied by various cardiovascular disease [24,25]. Multiple assessment methods based on ECG signals have been applied for the pilot to observe the physiological characteristics during flight operation [26]. Current wearable monitoring equipment on the market for pilots mainly uses belt type and cap type wearable physiological parameters monitoring equipment; belt type equipment can measure and record parameters such as ECG, respiration, and three-phase acceleration in real time, while cap type equipment can measure and record parameters such as head pulse signal and blood oxygen saturation in real time. These devices are usually only given to pilots and captains of dedicated aircrafts. Therefore, retrofitting an existing long-distance transport seat into a novel and modern standardized point-of-care system for general passengers is necessary and offers significant cost savings [21,27].

Sleeping on long trips is almost inevitable, so it makes sense to monitor sleep-related disorders during the flight. Among sleep-related disorders, SAHS is highly prevalent in the general population and is associated with an increased risk of hypertension, diabetes, coronary disease, stroke, and myocardial infarction [28]. However, 75–85% of patients with SAHS are undiagnosed and untreated, which causes irreparable consequences [29,30]. Since the risk of death is associated with SAHS, potential patients are suffering a shortage of being diagnosed. The severity of SAHS is assessed by the apnea-hypopnea index (AHI), defined as the number of apnea and insufficient breathing per hour during sleep [31,32]. In addition, the standard diagnostic test of SAHS is done by polysomnography (PSG) [33]. In clinical settings, patients go to a professional sleep laboratory to complete the acquisition of several signals under the guidance of professional staff, mainly ECG, electroencephalography, electromyography, respiratory airflow signal, blood oxygen saturation, nasal pressure, snoring sounds, etc. Abnormalities of airflow signals or ECG signals in the sensors are used as a basis for clinical judgment. This method quantitatively analyzes the sleep quality of patients and then accurately diagnoses the incidence of SAHS.

Herein, we suppose a point-of-care system with wearable chest belts for detection of ECG, breath, and motion signals, combined with a designed algorithm to classify SAHS for airlines. This work proposes a system to detect SAHS through a wearable device based on flexible dry ECG electrodes and a machine learning-based classifier algorithm. The signal acquisition terminal is a smart chest band consisting of flexible ECG electrodes and a control box that integrates various signal processing circuits and Bluetooth modules. The wearable ECG electrodes are used to detect ECG signal, while the control box with control circuits is used to detect tri-axis acceleration, and flexible piezoelectric material breathing coil belt is used to detect the breath signal. The detected signals are dealt with by these circuits and then transferred by the Bluetooth module to the Microcontroller Unit (MCU). These signals can then be dealt with by our designed algorithm, which includes the ECG signal denoising, RR interval data extraction, and extraction and selection of HRV features. Then, the characterized signals are analyzed through our designed LSTM to diagnose whether the user is a potential SAHS patient. The results show that the average recognition accuracy of the ECG-SAHS discriminant model is up to 84–85%. Besides that, the chest band design and the material of the electrode made longtime SAHS detection more comfortable, which is especially suitable for transoceanic flights.

Creativity:

1. Originality: Retrofitting existing safety belts, first extending of the point-of-care system on the aircraft to passengers besides pilots.
2. Feasibility: The cost of the equipment is lower than nearly 90% of commercial products since our design is a cost-saving approach by retrofitting of existing safety belts.
3. Accuracy: With the LSTM-RNN based algorithm, the average recognition accuracy of the model is up to 84–85%.
4. Extensibility: Data of hybrid ECG, breathing, and motion signals can also be used for monitoring other diseases.

2. Design and Methods

At the hardware design aspect, the system contains a controller box, printed circuit board (PCB), textile electrode, and electrode button on the surface of human skin [34]. The connection of these layer structures is displayed in Figure 1a, while the well-fabricated device and wear model is shown in Figure 1b. Figure 1c indicates that the cause of sleep apnea is the obstruction of throat airflow, which can be further explained as the different aspirations between normal breath. Therefore, we designed the wearing position on the wearer's chest for better detection of breathing-related signals as Figure 1d shows.

The design of software includes both the ECG signal and the breathing signal. Meanwhile, the motion capture is added into the system as the final classification criterion considering the movement of the user. After A/D conversion, the signal is processed to MCU and then transferred by Bluetooth module to the user's client on smartphones. The users could access their sleep assessment through the mobile terminal, which is presented in Figure 1e. A learning-based model will classify the user's aspiration model in Figure 1f. As stated before, the design of our system aims at the application of long-distance traveling. The ECG electrodes are connected to a chest band which could be integrated into safety belts to make the detection comfortable, as Figure 1g shows.

2.1. Hardware Design and Signals Acquisition

The flexible dry ECG electrodes with good biocompatibility are selected for the wearable terminal, which is also soft, washable, comfortable, durable, reusable, and easy to integrate with clothing [35–37]. Meanwhile, the sensor close to the skin could collect complete and reliable signals. The wearable terminal consists of single lead and double electrodes, which aim to simulate the standard 1-lead to realize the acquisition of ECG signal. As Figure 2a shows, the elastic chest strap is used to fix the flexible electrode to human skin. The fabrication embeds polymers such as polyvinylidene difluoride (PVDF) and other material to form piezoelectric film into the elastic chest strap of the wearable terminal to form a breathing coil, and the front-end collection of human breathing signals is realized through the breathing coil [38–40]. The piezoelectric film is subjected to force by human respiration and periodic respiratory rhythm deformation [41]. The surficial charge generated by the piezoelectric film also changes. Human respiration stresses the piezoelectric film and produces periodic deformation with the respiratory rhythm, and the amount of charge generated by the piezoelectric film also changes. By extracting the surface charge of the piezoelectric film and converting it into voltage amount, and then after signal conditioning, the respiratory signal can be obtained. Therefore, the acquisition method used in this paper is the piezoelectric breath tracing method.

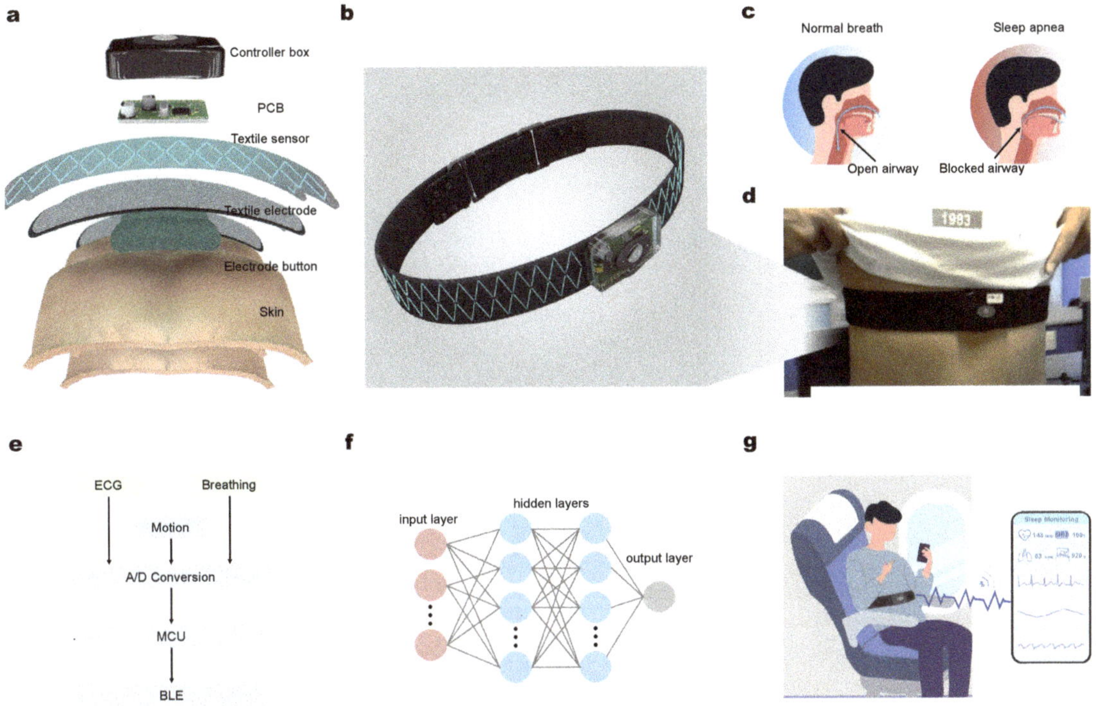

Figure 1. Airline point-of-care system on seat belt. (**a**) Illustration of the layered hardware structure of our designed chest band and (**b**) the fabricated chest band device. (**c**) Aspiration difference between normal breath and sleep apnea caused by airflow. (**d**) Physical photos of the chest belt on a human subject. (**e**) Working flowchart of the monitoring system with hardware and software. (**f**) Neural network that applied in the system as a classifier. The red nodes represent input layer, the blue nodes represent hidden layers, and the gray node represent the output layer in the network. (**g**) Schematic of airline point-of-care monitoring system in the user's client with Bluetooth-based data transmission to the mobile terminal for hybrid physiological signal detection.

Compared with the patch electrode, the flexible electrode can fit closely with the skin surface and change its shape to maintain continuous contact with the skin, which is useful for wearable ECG signal monitoring equipment [42]. The electrodes commonly used in clinical practice are traditional patch electrodes, which can cause skin irritation to the user over a long period of time. In addition, the impedance between the electrode interface and the skin will become larger as the detection time increases, resulting in a decrease in signal sensitivity and signal-to-noise ratio (SNR). Applying flexible fabric electrodes is to attach conductive materials to the fabric through various coating processes and braiding the lead wire on the clothing to prevent the lead wire from winding [43]. The inner side of the elastic chest strap facing the skin is the base layer and the conductive sensing layer. The base layer is made of polyester fiber. The base-layer fabric is wear-resistant, elastic, and easily washable. The conductive sensing layer is made of copper-plated conductive fabric and then blended with acrylic fiber. As a thinner electrode results in a larger area and better ECG signal quality [44], the experimental results show that the flexible electrode has a higher signal-to-noise ratio and higher stability than the conventional electrode [45]. Figure 2d shows the flexible ECG electrodes. From the inner side of the elastic chest strap facing the skin in the direction of the base layer, the elastic support layer, and the conductive fabric sensing layer, in turn, greatly reduces the thickness of the electrodes and improves the wearing comfort and is suitable for dynamic and long-time ECG monitoring. The system sews the two flexible fabric electrodes closely to the left and right inner side of the elastic chest strap near the controller box so that they can fit close to the skin, and the ECG signals collected using the flexible fabric electrodes need to be processed before the next step.

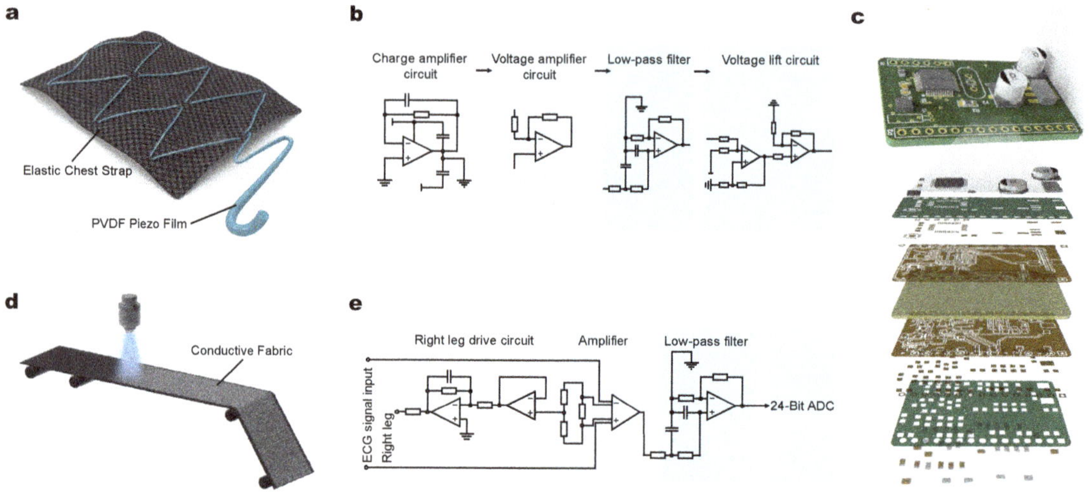

Figure 2. Architecture of the components of airline point-of-care system. (**a**) Fabrication process of breathing coil in elastic chest strap of the wearable terminal, the green line represents PVDF. (**b**) Breathing signal conditioning circuit pattern, including charge amplifier, voltage amplifier, low-pass filter, and voltage lift circuit. (**c**) PCB board displayed in layers. (**d**) Fabrication process of flexible ECG electrodes: from the inner side are the basal layer, the elastic support layer and the conductive fabric, respectively. (**e**) ECG signal conditioning module including right leg drive circuit, amplifier, and low-pass filter, finally 24 Bit ADC.

The circuit pattern of the wearable signal acquisition terminal is shown in Figure 2b. The signal collected through the breathing coil is relatively weak, with a frequency of 0.1–1 Hz, easily disturbed by various noises. Therefore, the breathing signal needs to be conditioned, including signal charge amplification, voltage method, low pass filtering, and voltage boosting, and the system designs a conditioning circuit for the breathing signal based on the TLV2464 chip, which is used to extract the effective breathing signal [46]. The charge amplification circuit of the respiratory signal and the voltage amplification circuit are designed through the first op-amp and the second op-amp of the TLV2464 chip, respectively. The amplification of the amplifier circuit is 80 times. A second-order active low-pass filter circuit is designed to filter out high-frequency noise with a cutoff frequency of 88.46 Hz. A voltage boost circuit is designed based on the fourth op-amp of the TLV2464 chip to boost the signal to a standard that the microprocessor can recognize, and the signal is transmitted to the microprocessor through the TLV_OUT pin. Use the A/D module in the microprocessor for analog-to-digital conversion processing.

The signal acquisition terminal consists of a controller box, a biomedical sensor, and an elastic chest strap, which enables 8 h of continuously monitoring. The extended monitoring time of the flexible electrode shows almost no influence on the signal sensitivity and the SNR. The collected ECG signal is processed by the circuit and sent to the microprocessor, and then transmitted to the computer through a Bluetooth module for further analysis and processing. The removable controller box contains a power supply, a communication module, a signal processing module, the main control chip, and the memory, which is integrated on a fast-prototyping PCB board that patterned using a UV laser system (LPKF; Protolaser U4), as shown in Figure 2c. Four electrode buttons are mounted on the elastic chest strap. The biomedical sensor is a flexible ECG electrode stitched tightly to the elastic chest strap to ensure continuous contact with the skin surface. The control box of the belt comes with a three-axis acceleration sensor MPU9150 to collect motion signals, MPU9150 sensor role is to measure the X, Y, Z-axis acceleration signal into electrical signals, through the analog-to-digital conversion and then convert the measured analog signal into a digital signal, so as to complete the extraction of human body position information of the subject.

In collecting ECG signals, certain noise will be introduced, requiring the circuit to have a high common-mode rejection ratio (CMRR) and high-precision analog-to-digital conversion (ADC). Therefore, the high-precision medical chip ADS1292 is used to design the ECG signal conditioning circuit, which has a built-in amplifier circuit and a high-resolution ADC and can reduce common mode interference [47]. The pin configuration and the peripheral circuit of ADS1292 are shown in Figure 2e. The pin configuration and peripheral circuit diagram of ADS1292 can be found in the supplementary Figure S1. After the ECG signal is input through the CH2 channel, the ADS1292 will amplify the ECG signal and perform A/D conversion through the internal modulation circuit, where the gain is set to 1000. The system directly feeds back the output of the right leg drive to the two detecting electrodes to realize the function of the right leg drive circuit. The SPI communication interface transmits the ECG signals after ADC to the microprocessor (EFM32), in which the ADS1292 is the slave device for communication. EFM32 is responsible for the overall control of the system, including initializing, configuring Bluetooth, ADC, and reading and storing ECG data. The photo of the industrial produced PCB is shown in Figure S2.

After we built the signal acquisition terminal, 18 adult volunteers aged between 20 and 39, including 9 males and 9 females, were selected for experimental verification. They had good sleep quality and did not suffer from respiratory and heart diseases. The subjects knew the content of the experiment in advance and agreed with it. The subjects were required to wear the terminal for a 3 h sleep experiment in a quiet room with both normal breathing and apnea. During the experiment, each subject was asked to discontinue breathing randomly, and the time and duration of artificial apnea were recorded. The details of the 18 adult volunteers can be found in the Supplementary Material Table S1.

The ECG signal collected by the terminal was transmitted to the computer by a Bluetooth module. The collected data then dealt with denoising and classification in the next steps. The details of collection methods will be discussed in the next section.

We computationally extracted three shallow feature signals of ECG: the RR interval signal (RRI), the R peak amplitude sequence signal ($RAMP$), and the ECG-derived respiratory (EDR), and used them as features for the detection of SAHS. It is important to note that the EDR signal is obtained starting from the R peaks detected by the Pan–Tompkins algorithm. It is just a pseudo-breathing signal. The calculated EDR signal has a certain amplitude error compared with the real respiratory signal, which cannot completely replace the respiratory signal. However, the EDR signal is an effective feature for detecting SAHS. Although SAHS have many subcase categories due to causes, since they have similar performance, we did not distinguish these subcases here.

These three signals are calculated using only the R-peak position index (RP_idx) and the denoised ECG ($ECGF$, F is just a marker to distinguish from denoised ECG and ECG, no special meaning) signal. The RRI signal refers to the time interval between adjacent R-peak positions. The RRI signal refers to the time interval between adjacent R-peak positions as Equation (1) shows.

$$RRI = RP\_idx_{i+1} - RP\_idx_i \tag{1}$$

The $RAMP$ signal refers to the amplitude sequence of the R-peak position in the denoised ECG signal $ECGF$ in Equation (2), where L is the length of the current R-peak position index.

$$RAMP_i = ECGF[RP\_idx_i], 0 <= i <= L \tag{2}$$

The EDR signal can show the current respiratory airflow and is often used in the detection of SAHS. In this paper, the kurtosis of the ECG signal between the adjacent R peaks and the ECG signal between the adjacent R peaks is calculated to obtain the sampling of the EDR signal at the current R peak, and then the complete EDR signal is obtained by sample bar interpolation. First, the kurtosis of the ECG signal between the adjacent R peaks is calculated as Equation (3) [48]

$$K_i = \frac{(1/n)\sum_{t=0}^{J_i-1}\left(ECGFS_{it} - \overline{ECGFS_{it}}\right)^4}{(1/n)\sum_{t=0}^{J_i-1}\left\{\left(ECGFS_{it} - \overline{ECGFS_i}\right)^2\right\}^2} - 3,\ 0 <= i <= L-1 \tag{3}$$

$$J_i = RP\_idx_{i+1} - RP\_idx_i + 1 \tag{4}$$

where $ECGFS_i$ is the segment of the signal $ECGF$ from RP_idx_i to RP_idx_{i+1}, $\overline{ECGFS_i}$ is the arithmetic mean of $ECGFS_i$, and J_i is the length of the $ECGFS_i$ signal segment, J_i can be calculated as Equation (4). After obtaining the kurtosis of the ECG signal between adjacent R peaks, the peak EDR signal sampled on that segment of the ECG signal can be obtained as Equation (5),

$$EDR_i = \left\{\frac{\left(\sigma^4 + (\gamma^4 - 3\sigma^4)K_i\right)^{1/2} - \sigma^2}{\gamma^4 - 3\sigma^4}\right\}^{1/2},\ 0 <= i <= L-1 \tag{5}$$

$$\gamma^4 = \exp\left\{\left(ECGFS_i - \exp(ECGFS_i)\right)^4\right\} \tag{6}$$

where σ is the standard deviation of the signal $ECGFS_i$, γ^4 is the fourth-order central moment of the signal $ECGFS_i$ that can be calculated as Equation (6), exp is the expectation, and K is the kurtosis of the signal $ECGFS_i$. At last, the EDR signal can be obtained by using spline interpolation for the discrete EDR signal.

2.2. ECG and Breathing Signal Collecting and Processing

The ECG signals were collected at a normal sleep state by asking the subjects to close their eyes and breathe normally. The data were filtered by cutting the beginning and denoising to avoid subject psychological factors or other interference, finally the comparison of raw and denoising data under the normal sleep model can be observed as a waveform in Figure 3a. Although there are some small disturbances caused by body movement and electrode sliding, such as baseline drifts and interference, the ECG characteristics can still be clearly identified through the waveform in raw data. Through the noise reduction process, the ECG characteristics of the signal become more obvious and easier for the subsequent classification process. Then, the subjects were asked to simulate sleep apnea by changing the breath model. After a period of normal breathing, when the subjects' breathing was stable, their breath situations were measured and then randomly held their breath for 10 s or more. When the airflow through the mouth and nose stopped and the chest and abdomen breathing disappeared, the apnea could be confirmed. The ECG signal waveforms of the subjects under sleep apnea are shown in Figure 3b, which also contain the comparison between raw ECG data and denoising ECG. Then, the denoised ECG signals of normal and abnormal waveforms were dealt with time between the successive R peaks to obtain the R-R interval in Figure 3c. Then, the HRV time-domain and frequency-domain analyses from RRI data were processed, which could extract and select features for further classification.

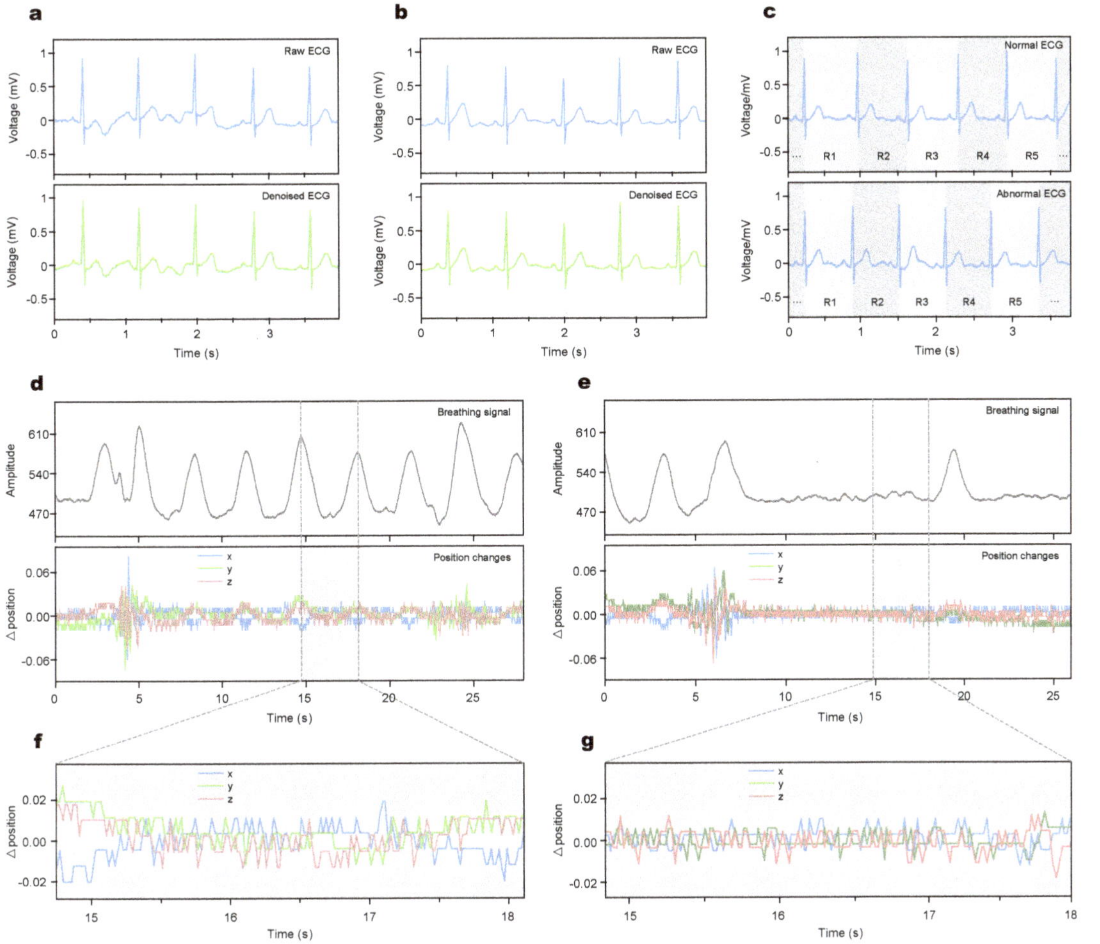

Figure 3. The display of hybrid physiological signal. (**a**) ECG signal before and after denoising at normal breath model. (**b**) ECG signal before and after denoising without apnea occurs. (**c**) ECG signals in normal and apneic RR classification. (**d**) Breathing signal and position signal in normal condition. (**e**) Breathing signal and position signal when apnea occurs. (**f**,**g**) Details of local fragments of position signals.

For the breathing signal, the breathing and position changes in a three-dimensional direction were collected and analyzed. The breathing data could help analyze the breathing signal to verify the situation of the user because when SAHS occurs, the breathing and breathing signal will also vary to a lower stable waveform [49]. Our system is designed to detect the breathing and position signals on two scales. For the large-scale time, in Figure 3d,e, the waveforms show the difference between normal conditions and when SAHS occurs. In addition, the R-peak in the ECG signal can characterize the respiratory activity, with inspiration and expiration corresponding to a decrease and an increase in the R-wave peak, respectively. During the onset of OSA, the corresponding R-wave peak will be lower than normal due to the lungs inhaling harder than normal. The R-peak interval of the ECG signal in the normal segment is mainly between 0.8 and 1.0 s, and the R-peak amplitude fluctuates in the range of 0.6 and 0.8 mV. In contrast, the R-peak interval of the ECG signal in the OSA fragment was mainly between 1.1 and 1.2 s, which was significantly larger than the R-peak interval of the normal ECG signal. In addition, the R-peak amplitude fluctuated slightly around 0.6 mV, which was significantly lower than that of the normal ECG signal. Generally, when the abnormal waveforms last for 8 s, it means that the SAHS occurs. The large-scale waveform that directly reflects the user's situation suffers obstacles from the abnormal stable waveform's appearance in Figure 3e, while the system also designed a small-scale detection for accurate analysis. In Figure 3f,g, the change of position is recorded in terms of three-dimensional displacement, which has an accuracy of less than 0.01. Our design of hardware of flexible electrodes enables detection accuracy. This accuracy ensures the waveforms on a large scale with credibility and operability. Although these two kinds of signals contribute a little to the system for the function of verification of SAHS, they will be helpful for other diseases' point-of-care detection.

2.3. Machine Learning-Enabled SAHS Recognition

To distinguish the ECG signal, the classification algorithm can be used to deal with the data efficiently. The above signals are not unstable for a typical man, and by coupling with modern computer science technology [49], multiple methods including Support vector machine (SVM), Random Forest Classifier (RFC) [50], Decision Trees (DT), K-Nearest Neighbour (KNN) [51], Adaptive Boosting Algorithm (ADABoosta), Gaussian naïve Bayes Classifier (GNB), Quadratic Discriminant Analysis (QDA), and BP are used to classify these data. Statistical methods mainly use mathematical and statistical analysis methods to establish models such as autoregressive models, binary tree classification models, and classify and analyze signals through them. However, due to the characteristics of ECG signals, it is difficult to establish a reasonable classification model and ensure the accuracy of it. The deep convolutional neural network (CNN) model seems to be a feasible choice as the sample data in this paper are small, but the severe overfitting problem of it is yet to be solved.

Here we choose (LSTM-RNN) to avoid the overfitting problem. LSTM has the function of memory effect, which is more efficient for dealing with time series data. As an extension model of RNN, it also solves the Vanishing Gradient problem of gradient back-propagation process due to gradual reduction. The working mechanism of LSTM-RNN is presented in Figure 4a. Three manual features were first extracted from the ECG signal as RR interval signal RRI, R peak amplitude sequence signal $RAMP$, and respiratory signal EDR derived from the ECG signal (not the respiratory signal acquired directly by the instrument), then fed into the model. Because traditional methods only use the RRI signal as the input signal of the neural network, only the variation of heart rate can be reflected, and the variation of respiratory airflow and the variation of chest and abdomen during breathing. The model consists of three LSTM loop layers and four fully connected layers with nodes 128, 64, 32, and 1. The output is either 0 or 1 to determine the presence or absence of apnea in the one-minute ECG signal segment. The training results of 50 epochs were recorded in supplementary material Table S2, and these data are presented in the form of a graph in Figure 4b. Despite suffering from overfitting starting at around 10 epochs, the model achieves an accuracy of over 80 percent in Apnea-ECG database [52]. The detail of Apnea-ECG database could be found in the SI Apnea-ECG database section.

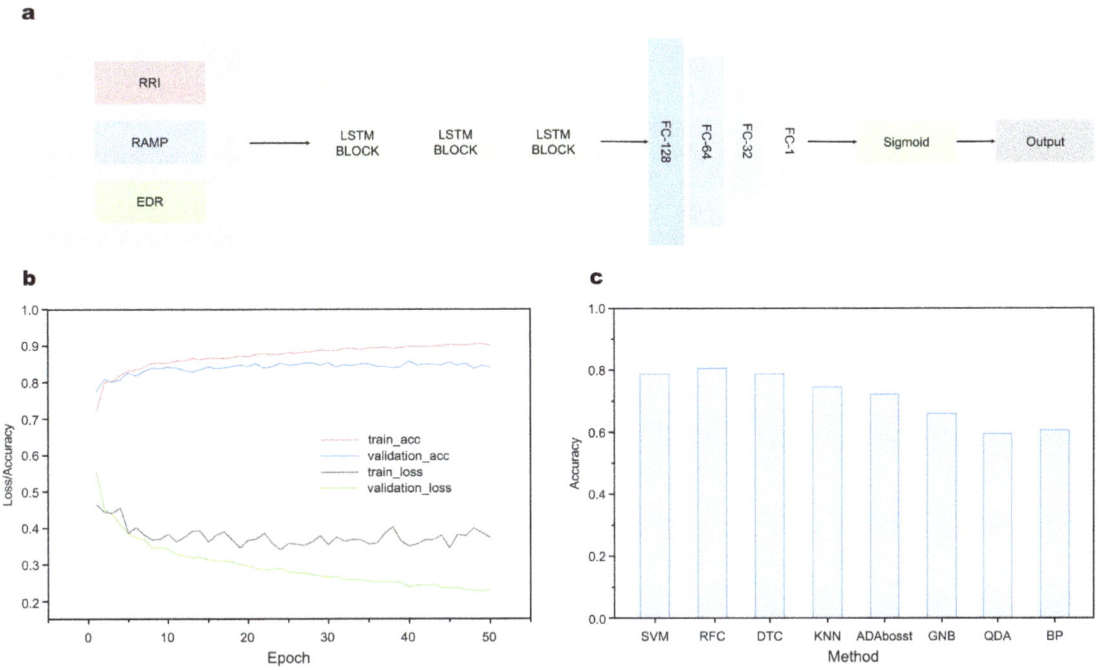

Figure 4. Structure and performance of SAHS detection system. (**a**) Structure of LSTM-RNN. (**b**) Loss/accuracy versus epoch for train and validation sets. (**c**) Comparison of different methods in terms of accuracy.

Binary cross-entropy as loss function was given as follows,

$$L(Y,P) = -\sum_x Y_x lg(P_x) \quad (7)$$

where P_x represents the prediction of the LSTM-RNN network for ECG fragment x, i.e., the Y of the output in the LSTM-RNN network, represents the true label of ECG fragment x. The final trained loss and accuracy are approaching 0.4 and 0.9, whereas the validation loss and accuracy are approaching to 0.8 and 0.2. Compared with other networks stated before, SVM, RFC, DTC, KNN, ADAboosta, GNB, QDA, and BP, our model gets the highest accuracy, which is shown in Figure 4c. The details of comparison of different methods in terms of accuracy could be found in supplementary material Table S3. For a fair comparison, the same manual features and training strategies were used.

To evaluate the classifier, Se, Sp, +PV, Acc, and F-score were applied as judgment categories, where Se represents the correct recognition rate of SAHS, Sp represents the detection rate of normal samples, +PV represents the actual response rate determined as SAHS, Acc represents the detection accuracy of any sample, and F-score is a comprehensive evaluation of the classification performance of our model. The detail of each evaluation criteria could be found in the supplementary material Table S4. Finally, we use the model of this paper to test its performance on the collected 18 subjects. The result shows high accuracy up to 84–85%.

3. Conclusions

In this paper, we build a point-of-care system model based on ECG electrode, flexible piezoelectric materials, and a control box to achieve ECG, breathing, and position signal detection. The diagnosis of SAHS for flight traveling is taken as a long-term application of this paper. Results show that the cost of the equipment is lower than nearly 90% of commercial products, while the average recognition accuracy of the model is up to 84–85%. The system originally retrofits existing safety belts to our system, firstly expanding point-of-care system from pilots to regular passengers in the post-epidemic era. By using flexible materials in the hardware design, the device is highly fitted to the human body, thus guaranteeing the accuracy of the breathing signal obtained by the piezoelectric sensor. At the same time, the data features extracted from the mixed acquisition of ECG, breath, and motion signals can reduce the diagnosis error rate since these signals reflect the variations of airflow and chest. At last, LSTM-RNN is applied to avoid the overfitting problem while ensuring high classification accuracy. The designed system also has important research significance and broad application prospects since the ECG, breathing, and motion signals could be applied to diagnose other diseases. It is worth believing that the system could have potential applications in future human aeronautics and astronautical space exploration.

Supplementary Materials: The following supporting information can be downloaded at https://www.mdpi.com/article/10.3390/mi13111880/s1, Table S1: The Screening Model Test Results; Table S2: Loss/accuracy versus epoch for train and validation sets; Table S3: Comparison of different methods in terms of accuracy; Table S4: Classifier evaluations index; Table S5: Cost of each component for the point-of-care system in detail.; Figure S1: Pin configuration and peripheral circuit diagram of ADS1292; Figure S2: Photo of fabricated PCB.

Author Contributions: X.X. and M.H. are the corresponding authors, supervised the whole project, presented the whole idea, and revised the paper. X.J., Z.R., W.Z., S.Z., M.H. and X.X. designed and conducted the research. B.Z. analyzed the subjects' data. Z.R. and S.Z. designed the program. W.Z., M.H. and S.Z. wrote the first draft of the manuscript and all authors contributed to writing the report. C.L. and P.S. organized and check the manuscript. All authors have read and agreed to the published version of the manuscript.

Funding: This work was supported by the Natural Science Foundation of Jilin Province (20190304130YY), National key R & D plan key special project (SQ2020YFF0421726), National Research Foundation, Singapore, and A*STAR under its RIE2020 Industry Alignment Fund—Industry Collaboration Projects (IAF-ICP) grant call (Grant No. I2001E0059).

Data Availability Statement: All data generated or analyzed during this study are included in this published article and its supplementary information files.

Acknowledgments: Xiao Xiao acknowledges support from the SIA-NUS Digital Aviation Corporate Laboratory, the N.1 Institute for Health, and the Institute for Health Innovation and Technology, NUS.

Conflicts of Interest: The authors have filed a provisional patent application (CN202010936602.3) based on this work. Xiao Xiao, the corresponding author on this paper, is the Guest Editor of Micromachines, but he had no involvement in the peer review process used to assess this work submitted to Micromachines. This paper was assessed and the corresponding peer review managed by another editor.

References

1. Gubala, V.; Harris, L.F.; Ricco, A.J.; Tan, M.X.; Williams, D.E. Point of Care Diagnostics: Status and Future. *Anal. Chem.* **2012**, *84*, 487–515. [CrossRef] [PubMed]
2. Shrivastava, S.; Trung, T.Q.; Lee, N.-E. Recent Progress, Challenges, and Prospects of Fully Integrated Mobile and Wearable Point-of-Care Testing Systems for Self-Testing. *Chem. Soc. Rev.* **2020**, *49*, 1812–1866. [CrossRef] [PubMed]
3. Elledge, S.K.; Zhou, X.X.; Byrnes, J.R.; Martinko, A.J.; Lui, I.; Pance, K.; Lim, S.A.; Glasgow, J.E.; Glasgow, A.A.; Turcios, K. Engineering Luminescent Biosensors for Point-of-Care SARS-CoV-2 Antibody Detection. *Nat. Biotechnol.* **2021**, *39*, 928–935. [CrossRef] [PubMed]
4. Bianchi, D.W. From Prenatal Genomic Diagnosis to Fetal Personalized Medicine: Progress and Challenges. *Nat. Med.* **2012**, *18*, 1041–1051. [CrossRef] [PubMed]
5. Davenport, M.; Mach, K.E.; Shortliffe, L.M.D.; Banaei, N.; Wang, T.-H.; Liao, J.C. New and Developing Diagnostic Technologies for Urinary Tract Infections. *Nat. Rev. Urol.* **2017**, *14*, 296–310. [CrossRef]
6. Chen, G.; Xiao, X.; Zhao, X.; Tat, T.; Bick, M.; Chen, J. Electronic Textiles for Wearable Point-of-Care Systems. *Chem. Rev.* **2022**, *122*, 3259–3291. [CrossRef]
7. Gurwitz, J.H.; Pearson, S.D. Novel Therapies for an Aging Population: Grappling with Price, Value, and Affordability. *JAMA* **2019**, *321*, 1567–1568. [CrossRef]

8. Osier, F.; Ting, J.P.Y.; Fraser, J.; Lambrecht, B.N.; Romano, M.; Gazzinelli, R.T.; Bortoluci, K.R.; Zamboni, D.S.; Akbar, A.N.; Evans, J. The Global Response to the COVID-19 Pandemic: How Have Immunology Societies Contributed? *Nat. Rev. Immunol.* **2020**, *20*, 594–602. [CrossRef]
9. Yip, W.; Fu, H.; Chen, A.T.; Zhai, T.; Jian, W.; Xu, R.; Pan, J.; Hu, M.; Zhou, Z.; Chen, Q. 10 Years of Health-Care Reform in China: Progress and Gaps in Universal Health Coverage. *Lancet* **2019**, *394*, 1192–1204. [CrossRef]
10. Hamburg, M.A.; Collins, F.S. The Path to Personalized Medicine. *N. Engl. J. Med.* **2010**, *363*, 301–304. [CrossRef]
11. Schork, N.J. Personalized Medicine: Time for One-Person Trials. *Nature* **2015**, *520*, 609–611. [CrossRef] [PubMed]
12. Hood, L.; Friend, S.H. Predictive, Personalized, Preventive, Participatory (P4) Cancer Medicine. *Nat. Rev. Clin. Oncol.* **2011**, *8*, 184–187. [CrossRef] [PubMed]
13. Quesada-González, D.; Merkoçi, A. Nanomaterial-Based Devices for Point-of-Care Diagnostic Applications. *Chem. Soc. Rev.* **2018**, *47*, 4697–4709. [CrossRef] [PubMed]
14. Yang, M.; Liu, Y.; Jiang, X. Barcoded Point-of-Care Bioassays. *Chem. Soc. Rev.* **2019**, *48*, 850–884. [CrossRef]
15. Khan, Y.; Ostfeld, A.E.; Lochner, C.M.; Pierre, A.; Arias, A.C. Monitoring of Vital Signs with Flexible and Wearable Medical Devices. *Adv. Mater.* **2016**, *28*, 4373–4395. [CrossRef]
16. Tahanisaz, S.; Shokuhyar, S. Evaluation of Passenger Satisfaction with Service Quality: A Consecutive Method Applied to the Airline Industry. *J. Air Transp. Manag.* **2020**, *83*, 101764. [CrossRef]
17. Merkert, R.; Swidan, H. Flying with(out) a Safety Net: Financial Hedging in the Airline Industry. *Transp. Res. E Logist. Transp. Rev.* **2019**, *127*, 206–219. [CrossRef]
18. Gerstle, C.R. Parallels in Safety between Aviation and Healthcare. *J. Pediatr. Surg.* **2018**, *53*, 875–878. [CrossRef]
19. Wang, H. Big Data Visualization and Analysis of Various Factors Contributing to Airline Delay in the United States. In Proceedings of the 2022 International Conference on Big Data, Information and Computer Network (BDICN), Sanya, China, 20–22 January 2022; pp. 177–181.
20. Boyd, D.D.; Scharf, M.; Cross, D. A Comparison of General Aviation Accidents Involving Airline Pilots and Instrument-Rated Private Pilots. *J. Saf. Res.* **2021**, *76*, 127–134. [CrossRef]
21. Borges do Nascimento, I.J.; Jerončić, A.; Arantes, A.J.R.; Brady, W.J.; Guimarães, N.S.; Antunes, N.S.; Carim Junior, G.; Marcolino, M.S. The Global Incidence of In-Flight Medical Emergencies: A Systematic Review and Meta-Analysis of Approximately 1.5 Billion Airline Passengers. *Am. J. Emerg. Med.* **2021**, *48*, 156–164. [CrossRef]
22. Jagoda, A.; Pietrzak, M. MEDICAL EMERGENCIES IN COMMERCIAL AIR TRAVEL. *Emerg. Med. Clin. N. Am.* **1997**, *15*, 251–260. [CrossRef]
23. DeHart, R.L. Health Issues of Air Travel. *Annu. Rev. Public Health* **2003**, *24*, 133–151. [CrossRef]
24. Jou, R.-C.; Kuo, C.-W.; Chiu, Y.-C. Bidding Behaviors for International Airline Seats in Short/Long Distance Flights. *Transp. Res. Part A Policy Pr.* **2022**, *163*, 55–79. [CrossRef]
25. Pedroso Fabrin, B.H.; Ferrari, D. Investigation of Airborne Exposure Risk to Infectious Diseases during Aircraft Boarding Process Using Agent-Based Modeling. In Proceedings of the AIAA AVIATION 2022 Forum, Chicago, IL, USA, 27 June–1 July 2022; p. 3616.
26. Shao, S.; Zhou, Q.; Liu, Z. A New Assessment Method of the Pilot Stress Using ECG Signals during Complex Special Flight Operation. *IEEE Access* **2019**, *7*, 185360–185368. [CrossRef]
27. Shen, S.; Xiao, X.; Xiao, X.; Chen, J. Wearable Triboelectric Nanogenerators for Heart Rate Monitoring. *Chem. Commun.* **2021**, *57*, 5871–5879. [CrossRef]
28. Smith, D.F.; Amin, R.S. OSA and Cardiovascular Risk in Pediatrics. *Chest* **2019**, *156*, 402–413. [CrossRef] [PubMed]
29. Bratton, D.J.; Gaisl, T.; Wons, A.M.; Kohler, M. CPAP vs. Mandibular Advancement Devices and Blood Pressure in Patients with Obstructive Sleep Apnea: A Systematic Review and Meta-Analysis. *JAMA* **2015**, *314*, 2280–2293. [CrossRef]
30. Jean, R.E.; Duttuluri, M.; Gibson, C.D.; Mir, S.; Fuhrmann, K.; Eden, E.; Supariwala, A. Improvement in Physical Activity in Persons with Obstructive Sleep Apnea Treated with Continuous Positive Airway Pressure. *J. Phys. Act. Health* **2017**, *14*, 176–182. [CrossRef]
31. Hwang, S.H.; Lee, Y.J.; Jeong, D.-U.; Park, K.S. Apnea–Hypopnea Index Prediction Using Electrocardiogram Acquired during the Sleep-Onset Period. *IEEE Trans. Biomed. Eng.* **2016**, *64*, 295–301.
32. Kapur, V.K.; Auckley, D.H.; Chowdhuri, S.; Kuhlmann, D.C.; Mehra, R.; Ramar, K.; Harrod, C.G. Clinical Practice Guideline for Diagnostic Testing for Adult Obstructive Sleep Apnea: An American Academy of Sleep Medicine Clinical Practice Guideline. *J. Clin. Sleep Med.* **2017**, *13*, 479–504. [CrossRef]
33. Gottlieb, D.J.; Punjabi, N.M. Diagnosis and Management of Obstructive Sleep Apnea: A Review. *JAMA* **2020**, *323*, 1389–1400. [CrossRef] [PubMed]
34. Fujiwara, K.; Miyajima, M.; Yamakawa, T.; Abe, E.; Suzuki, Y.; Sawada, Y.; Kano, M.; Maehara, T.; Ohta, K.; Sasai-Sakuma, T. Epileptic Seizure Prediction Based on Multivariate Statistical Process Control of Heart Rate Variability Features. *IEEE Trans. Biomed. Eng.* **2015**, *63*, 1321–1332. [PubMed]
35. Yao, G.; Xu, L.; Cheng, X.; Li, Y.; Huang, X.; Guo, W.; Liu, S.; Wang, Z.L.; Wu, H. Bioinspired Triboelectric Nanogenerators as Self-Powered Electronic Skin for Robotic Tactile Sensing. *Adv. Funct. Mater.* **2020**, *30*, 1907312. [CrossRef]
36. Luo, Y.; Xiao, X.; Chen, J.; Li, Q.; Fu, H. Machine-Learning-Assisted Recognition on Bioinspired Soft Sensor Arrays. *ACS Nano* **2022**, *16*, 6734–6743. [CrossRef]

37. Gong, D.; Celi, N.; Zhang, D.; Cai, J. Magnetic Biohybrid Microrobot Multimers Based on Chlorella Cells for Enhanced Targeted Drug Delivery. *ACS Appl. Mater. Interfaces* **2022**, *14*, 6320–6330. [CrossRef] [PubMed]
38. Lin, W.; Wang, B.; Peng, G.; Shan, Y.; Hu, H.; Yang, Z. Skin-Inspired Piezoelectric Tactile Sensor Array with Crosstalk-Free Row+Column Electrodes for Spatiotemporally Distinguishing Diverse Stimuli. *Adv. Sci.* **2021**, *8*, 2002817. [CrossRef]
39. Zhang, Y.; Ju, F.; Wei, X.; Wang, D.; Wang, Y. A Piezoelectric Tactile Sensor for Tissue Stiffness Detection with Arbitrary Contact Angle. *Sensors* **2020**, *20*, 6607. [CrossRef]
40. Qu, H.; Xiao, X.; Han, Z.; Hu, M.; Shen, S.; Yang, L.; Jia, F.; Wang, T.; Ye, Z.; Sun, W. Graphene Oxide Nanofiltration Membrane Based on Three-Dimensional Size-Controllable Metal–Organic Frameworks for Water Treatment. *ACS Appl. Nano Mater.* **2022**, *5*, 5196–5207. [CrossRef]
41. Xiao, X.; Chen, G.; Libanori, A.; Chen, J. Wearable Triboelectric Nanogenerators for Therapeutics. *Trends Chem.* **2021**, *3*, 279–290. [CrossRef]
42. Xiao, X.; Xiao, X.; Zhou, Y.; Zhao, X.; Chen, G.; Liu, Z.; Wang, Z.; Lu, C.; Hu, M.; Nashalian, A.; et al. An Ultrathin Rechargeable Solid-State Zinc Ion Fiber Battery for Electronic Textiles. *Sci. Adv.* **2021**, *7*, eabl3742. [CrossRef]
43. Lv, J.; Yin, J.; Qin, Y.; Dai, Y.; Cheng, Z.; Luo, L.; Liu, X. Post-Construction of Weaving Structure in Aramid Fiber towards Improvements of Its Transverse Properties. *Compos. Sci. Technol.* **2021**, *208*, 108780. [CrossRef]
44. Yokus, M.A.; Jur, J.S. Fabric-Based Wearable Dry Electrodes for Body Surface Biopotential Recording. *IEEE Trans. Biomed. Eng.* **2015**, *63*, 423–430. [CrossRef] [PubMed]
45. Pani, D.; Dessì, A.; Saenz-Cogollo, J.F.; Barabino, G.; Fraboni, B.; Bonfiglio, A. Fully Textile, PEDOT: PSS Based Electrodes for Wearable ECG Monitoring Systems. *IEEE Trans. Biomed. Eng.* **2015**, *63*, 540–549. [CrossRef]
46. Claude, A.; Robin, O.; Gehin, C.; Massot, B. Design and Evaluation of a Novel Technology for Ambulatory Monitoring of Bruxism Events. *Sens. Actuators A Phys.* **2019**, *295*, 532–540. [CrossRef]
47. Zhao, D.; Wang, L.; Cheng, S. Adaptive Deep Brain Stimulation System Based on ADS1292. In Proceedings of the 2019 IEEE 7th International Conference on Bioinformatics and Computational Biology (ICBCB), Hangzhou, China, 21–23 March 2019; pp. 83–87.
48. Ding, S.; Zhu, X.; Chen, W.; Wei, D. Derivation of Respiratory Signal from Single- Channel ECGs Based on Source Statistics. *Int. J. Bioelectromagn.* **2004**, *6*, 41–48.
49. Murray, L.P.; Mace, C.R. Usability as a Guiding Principle for the Design of Paper-Based, Point-of-Care Devices—A Review. *Anal. Chim. Acta* **2020**, *1140*, 236–249. [CrossRef]
50. Elola, A.; Aramendi, E.; Irusta, U.; del Ser, J.; Alonso, E.; Daya, M. ECG-Based Pulse Detection during Cardiac Arrest Using Random Forest Classifier. *Med. Biol. Eng. Comput.* **2019**, *57*, 453–462. [CrossRef]
51. Venkatesan, C.; Karthigaikumar, P.; Varatharajan, R. A Novel LMS Algorithm for ECG Signal Preprocessing and KNN Classifier Based Abnormality Detection. *Multimed. Tools Appl.* **2018**, *77*, 10365–10374. [CrossRef]
52. Penzel, T.; Moody, G.B.; Mark, R.G.; Goldberger, A.L.; Peter, J.H. The Apnea-ECG Database. In Proceedings of the Computers in Cardiology 2000, Cambridge, MA, USA, 24–27 September 2000; (Cat. 00CH37163). Volume 27, pp. 255–258.

micromachines

MDPI

Article

Heterogeneous Multi-Material Flexible Piezoresistive Sensor with High Sensitivity and Wide Measurement Range

Tingting Yu [1], Yebo Tao [2], Yali Wu [3], Dongguang Zhang [3,*], Jiayi Yang [4,*] and Gang Ge [5,*]

1 School of Aerospace Science and Technology, Xidian University, Xi'an 710071, China
2 Intelligent Manufacturing College, Jiaxing Vocational & Technical College, Jiaxing 314036, China
3 College of Mechanical and Vehicle Engineering, Taiyuan University of Technology, Taiyuan 030024, China
4 College of Computer Science and Technology, Xi'an University of Science and Technology, Xi'an 710054, China
5 Department of Electrical and Computer Engineering, National University of Singapore, Singapore 117583, Singapore
* Correspondence: zhangdongguang@tyut.edu.cn (D.Z.); jyang46@xust.edu.cn (J.Y.); ggeup@nus.edu.sg (G.G.)

Abstract: Flexible piezoresistive sensors (FPSs) have the advantages of compact structure, convenient signal acquisition and fast dynamic response; they are widely used in motion detection, wearable electronic devices and electronic skins. FPSs accomplish the measurement of stresses through piezoresistive material (PM). However, FPSs based on a single PM cannot achieve high sensitivity and wide measurement range simultaneously. To solve this problem, a heterogeneous multi-material flexible piezoresistive sensor (HMFPS) with high sensitivity and a wide measurement range is proposed. The HMFPS consists of a graphene foam (GF), a PDMS layer and an interdigital electrode. Among them, the GF serves as a sensing layer, providing high sensitivity, and the PDMS serves as a supporting layer, providing a large measurement range. The influence and principle of the heterogeneous multi-material (HM) on the piezoresistivity were investigated by comparing the three HMFPS with different sizes. The HM proved to be an effective way to produce flexible sensors with high sensitivity and a wide measurement range. The HMFPS-10 has a sensitivity of 0.695 kPa^{-1}, a measurement range of 0–14,122 kPa, fast response/recovery (83 ms and 166 ms) and excellent stability (2000 cycles). In addition, the potential application of the HMFPS-10 in human motion monitoring was demonstrated.

Keywords: soft sensors; pressure sensors; stress sensors; high dynamics range; graphene

Citation: Yu, T.; Tao, Y.; Wu, Y.; Zhang, D.; Yang, J.; Ge, G. Heterogeneous Multi-Material Flexible Piezoresistive Sensor with High Sensitivity and Wide Measurement Range. *Micromachines* **2023**, *14*, 716. https://doi.org/10.3390/mi14040716

Academic Editor: Lin Zhang

Received: 1 March 2023
Revised: 17 March 2023
Accepted: 21 March 2023
Published: 23 March 2023

1. Introduction

Flexible stress sensors are flexible electronics that conform to curved surfaces and convert stress into electrical signals [1–4]; they are widely used in motion detection [5–7], wearable electronics [8,9], human–computer interaction [10–12], and electronic skin [13–15]. Flexible stress sensors can be classified into piezoresistive [16], capacitive [17,18], piezoelectric [19] and triboelectric [20] sensors according to different sensing mechanisms. Among them, flexible piezoresistive sensors (FPSs) can convert stress into electrical signals through changes in the conductive paths, which have the advantages of simple structure, convenient signal acquisition, fast dynamic response and cost-efficiency [21,22]. Two important parameters of FPSs are sensitivity and measurement range [18,23]. High sensitivity enables the sensor to accurately identify small stress [24], while a wide measurement range is the key to ensuring the applicability of the sensor in various stress scenarios. Therefore, it is of great significance to achieve an FPS with high sensitivity and a wide measurement range [25].

FPSs usually consist of electrodes and the piezoresistive materials (PM), which determine the sensitivity and measurement range of the sensor. There are three typical piezoresistive materials: conductive films (CFs) [26,27], conductive composites (CCs) [28,29] and conductive porous foam (CPFs) [30,31]. CFs are composed of conductive films on flexible substrates [26,27]. Cracks or breaks appear on the surface's conductive layer after it is stressed, which changes the conductive path and causes a variation in resistance. CFs have the advantage of high sensitivity, while the measurement range of CFs is limited due to the low deformability [32]. CCs are conductive ingredient-doped polymers [33,34]. Appling stress changes the relative position of the conductive filler inside the CCs, resulting the establishment or break of the conductive paths. The change in the conductive paths requires a decent strain, resulting in a large measurement range and a low sensitivity. CPFs are conductive material-coated porous polymers [31,35]. Applying stress deforms the three-dimensional conductive skeleton of the porous polymer, which establishes conductive paths through contact with conductive skeletons. CPFs has better flexibility and piezoresistive properties. Compared with CFs, the three-dimensional porous skeleton can provide a larger measurement range. Comparing with CCs, the three-dimensional porous skeleton has a lower modulus and can provide higher sensitivity.

To improve the sensitivity and measurement range of CPF-based sensors, surface microstructures (such as micropillars [36], domes [37,38], micro-pyramids [39], microgrooves [40], and folds [41]) or gradient porous structures [18,42] are proposed. Although these works improve the sensitivity or measurement range, achieving high sensitivity and wide measurement range simultaneously is still a severe bottleneck that restricts the practical applications of these sensors.

Heterogeneous multi-material (HM) is composed of different materials with continuous or discontinuous structures distribution, and can obtain properties that cannot be achieved by a single material [43,44]. The CPF has a piezoresistivity and lower elastic modulus, while the elastomer has a high elastic modulus. Therefore, combining the CPF and the elastomer to design an HM is a potential solution to achieve high sensitivity and a wide measurement range in one sensor.

To solve the problem, a heterogeneous multi-material flexible piezoresistive sensor (HMFPS) with high sensitivity and a wide measurement range is proposed. The HMFPS consists of a graphene foam (GF) sensing layer, a polydimethylsiloxane (PDMS) supporting layer and an interdigital electrode. Among them, the GF is fixed at the center of the PDMS supporting layer, and the flexible interdigital electrode is placed under the surface of the GF, as shown in Figure 1a. We prepared the HMFPS with different PDMS supporting layer areas, namely HMFPS-3, HMFPS-5, HMFPS-10. The influence of the HM structure on the performance of the sensor and its working principle is investigated, proving that the HM structure is an effective way to improve the piezoresistivity of the FPS. Among them, HMFPS-10 has the best comprehensive performance, with a sensitivity of 0.695 KPa^{-1}, a measurement range of 0–14,122 KPa, fast response/recovery (83 ms and 166 ms) and excellent stability (2000 cycles). In addition, the flexible sensor can be applied to human movement monitoring.

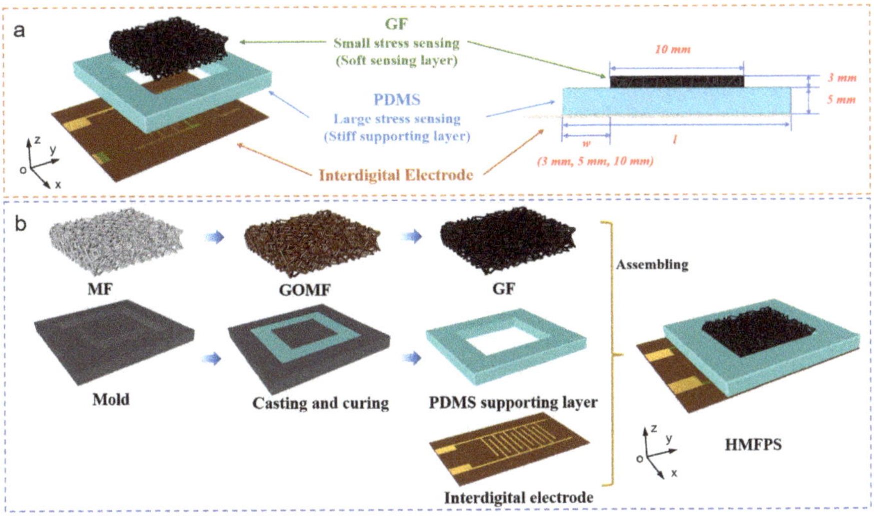

Figure 1. (**a**) Schematic of the HMFPS; (**b**) Schematic of the fabrication process of the HMFPS.

2. Experimental

2.1. Materials

PDMS (Sylgard 184) and curing agent were provided by Dow Corning Co., Ltd. (Midland, MI, USA). Graphene oxide (GO) powder was purchased from Nanjing XFNANO Materials Tech. Co., Ltd. (Nanjing, China). Melamine foams (MF) were obtained from Shenzhen Siyuan Rubber and Plastic Products Factory (Shenzhen, China). Vitamin C (VC) was purchased from Sinopharm Chemical Reagent Co., Ltd. (Shanghai, China). An interdigital electrode with polyimide (PI) substrate was purchased from Rigorous Technology Co., Ltd. (Shenzhen, China). All the materials and chemicals were used as received.

2.2. Preparation of HMFPS

The HMFPS consists of a GF sensing layer, a PDMS supporting layer and an interdigital electrode. Among them, the GF is fixed at the center of the PDMS supporting layer, and the flexible interdigital electrode is placed under the surface of the GF, as shown in Figure 1a. The preparation process of the HMFPS is shown in Figure 1b. First, the GFs were prepared by a dip-coating method. MFs (10 mm × 10 mm × 8 mm) were washed in deionized water and ethanol for 10 min then dip-coated in GO solution (2 mg mL^{-1}) under vacuum for 10 min. The resulting GOMFs were immersed in vitamin C (VC) solution (12 mg mL^{-1}) for 6 h at 90 °C to reduce the GO to rGO. The obtained GFs were washed in deionized water to remove the excess VC, and dried at 80 °C. We prepared five groups of GOMFs with different dip-coating times, respectively. The samples were chemically reduced to prepare the GF, expressed as GF_x (where x = 1, 2, 3, 4, and 5). Second, the PDMS supporting layers were prepared. PDMS and curing agent were mixed in a ratio of 10:1, then poured into a Teflon mold after removing bubbles. The mixture was cured at 80 °C for 4 h to obtain a PDMS supporting layer. We prepared three different sizes of PDMS supporting layers with widths of 3 mm, 5 mm, and 10 mm, respectively. Finally, the GF, PDMS supporting layer and interdigital electrode were assembled to obtain the HMFPS. The GF is fixed on the surface of the interdigital electrode using a conductive silver paste. Additionally, the copper wire is fixed to the interdigital electrode by soldering tin. The interdigital electrode enables the sensor to be compact and avoid exposed wire connections, increasing the stability of the sensor. The HMFPSs with three different supporting layer sizes are defined as HMFPS-3, HMFPS-5, and HMFPS-10, respectively.

2.3. Characterizations and Measurement

The structures of MF and GF were characterized by scanning electron microscopy (SEM) (Gemini SEM500, Zeiss, Oberkochen, Germany) at an operating voltage of 2 kV. Raman spectrums were recorded using a Raman spectrometer (Jobin Yvon LavRam HR800, Horiba, Japan) with laser excitation of 514 nm.

The mechanical and electrical property tests were conducted on HMFPS-3, HMFPS-5 and HMFPS-10 using a high-precision stress testing machine (TianYuan Test Instrument TY8000, Shanghai, China) and a precision source meter (2450, Keithley, Cleveland, OH, USA), respectively. The sensors were compressed to strain at a constant speed of 20 mm min^{-1}, and the loading rate remained at 60 mm min^{-1} for the endurance cycle tests. All the tests were conducted at room temperature.

3. Results and Discussion

3.1. Structure and Morphology

The HMFPS consists of three parts: the GF sensing layer, the PDMS supporting layer and an interdigital electrode. The GF is 3 mm higher than the PDMS supporting layer, which is fixed at the center of the PDMS supporting layer. The GF is connected to the interdigital electrode by silver paste. Three HMFPSs with different PDMS supporting layer areas (HMFPS-3, HMFPS-5 and HMFPS-10) are prepared. The photographs of the samples are shown in Figure 2a. We took the SEM images to observe the morphology of the GF sensing layer. Figure 2b–d and Figure 2e–g are the SEM images of MF and GF under different magnifications, respectively. The microscopic porous structures of the MF and GF are similar, indicating that the GF prepared by dip-coating method preserves the original pore structure of the MF. In addition, the surface of the MF is flat and smooth, while the surface of the GF presents an obvious fold structure, indicating that graphene sheets are successfully assembled on the MF skeleton.

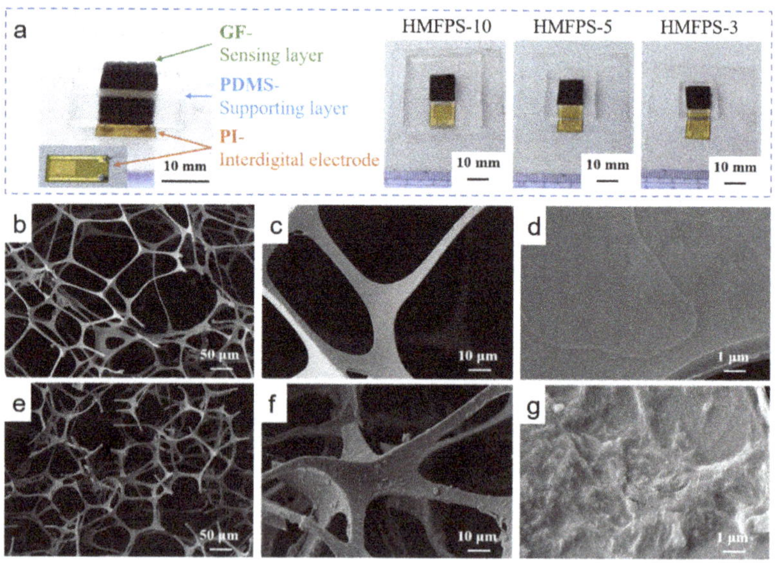

Figure 2. (**a**) Photographs of the HMFPS-3, HMFPS-5 and HMFPS-10; SEM images of (**b–d**) the MF and (**e–g**) the GF.

The porosity of the GF can be expressed as $Q = \frac{V_p}{V_0} \times 100\%$, where V_0 represents the volume of GF in its natural state ($V_0 = 0.8$ cm^{-3}), and V_p represents the porous volume of GF). We tested the porosity of the GF by drainage method. The quality of the GF is 6.75 mg.

The density of the GF can be calculated as 8.43 kg m^{-3}. Additionally, when the GF is full of water, its quality is 635.71 mg. Therefore, the quality of water absorbed by GF pores is 628.96 mg, which is equivalent to the volume of GF pores ($V_p = 0.629$ cm^{-3}). Hence, the porosity of the GF is 78.62%.

Figure S2 shows the Raman spectra of the GOMF and the GF contain D and G valence bands at 1343 cm^{-1} and 1580 cm^{-1}, respectively. The D band is attributed to sp3 defects in the graphene sheet, and the G band is attributed to the E2g phonon mode in the sp2 carbon atoms plane. The ratio of the intensity of the D valence band to the G valence band (I_D/I_G) of the GOMF was 0.82. The I_D/I_G of GF was 1.28 after chemical reduction by VC, indicating that the GO sheets attached to MF are efficiently reduced to G.

3.2. Sensing Mechanism and Piezoresistivity of the HMFPS

The sensing mechanism of the HMFPS is shown in Figure 3. Applying stress makes the GF skeleton of the HMFPS contact, which decreases the resistance by establishing new conductive paths. The sensitivity of the stress sensor (*S*) can be defined as $S = (\Delta I/I_0)/P$, where ΔI is the current variation, I_0 is the original current, and P is the applied stress. Due to the PDMS supporting layer and protruding GF, the piezoresistive effect of HMFPS presents two modes, as shown in Figure 3a. Without stress, the GF is higher than the PDMS supporting layer. With a small stress (<1.8 kPa), the GF above the PDMS supporting layer is compressed, while the PDMS supporting layer is not compressed. At this time, the overall equivalent elastic modulus of the sensor is equal to the GF, leading to a high *S*. However, with the stress increasing, the GF above the PDMS supporting layer is completely compressed. With more compressive strain, the GF and PDMS supporting layer are compressed simultaneously. In this case, the equivalent elastic modulus of the sensor is approximately equal to the PDMS supporting layer. Compared to only compressed GF, greater stress is required at the same strain, which greatly improves the measurement range of the sensor. In conclusion, the GF above the PDMS supporting layer allows the sensor to sense small stresses (<1.8 kPa). When the PDMS supporting layer is compressed, the sensor can sense large stress (14,122 kPa). Based on the HM design method, the low modulus of the GF can be combined with the high modulus of the PDMS, producing a sensor with high sensitivity and a large measurement range, simultaneously.

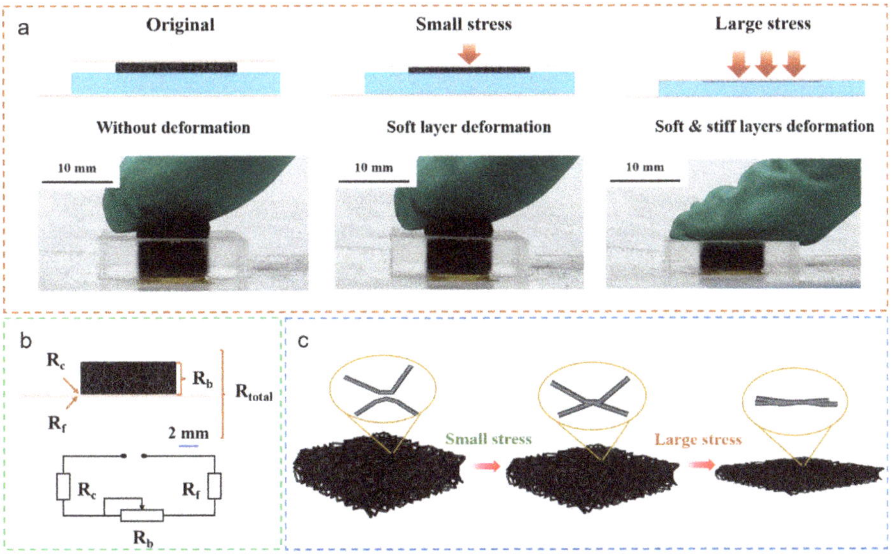

Figure 3. (**a**) Schematic and pictures of different amounts of compressive strain of HMFPS; (**b**) The equivalent circuit model of the HMFPS; (**c**) Schematics of the sensing mechanism of the HMFPS.

To further analyze the piezoresistivity of the sensor, we established the equivalent circuit model of the sensor, as shown in Figure 3b. The total resistance (R_{total}) of the HMFPS consists of the GF bulk resistance (R_b), the interdigital electrode resistance (R_f) and the contact resistance between the GF and the interdigital electrode (R_c). The total resistance can be calculated as follows: $R_{total} = R_b + R_f + R_c$. Generally, the R_f and R_c can be neglected, and the piezoresistivity of the sensor is mainly determined by R_b. In this study, five groups of GF samples were prepared by changing the dip-coating times. The conductivity of GF increases with the increase in dip-coating times, as shown in Figure S1. The conductivity of the GF increased rapidly for the first three dip-coating times, while it increased slowly for the fourth and fifth dip-coating times. According to the threshold effect, higher sensitivity can be obtained by keeping the conductivity of piezoresistive composites near the threshold value. Therefore, in order to obtain better pressure-sensitive characteristics, GF_3 foam samples with the largest change in conductivity should be selected. The conductivity of GF_3 is about 0.086 S m^{-1}. In the follow-up studies of this experiment, GF_3 samples were used. R_b depends on the number of conductive paths of the GF and can be equivalent to a sliding rheostat. Without stress, HMFPS remains in its original state, with a resistance of R_0. Applying stress compresses the GF, resulting in a deformation (Δh). The R_b of the GF decreases according to $R_b \propto 1/\Delta h$. The sensitivity and measurement range of the sensor are determined by the relationship between the Δh and the P, which can be expressed as $E = P/\Delta h$ (where E represents elastic modulus of the sensor). Within a certain stress range, the applied stress only works on the GF, and the E of the sensor is small. The smaller P can produce a larger Δh, indicating a large variation rate of the R_b, so that the sensor has high sensitivity. Beyond a certain range, the existence of the PDMS supporting layer increases the E of the sensor, which generates a certain amount of the Δh that requires a larger P. The variation rate of the R_b decreases, increasing the measurement range of the sensor.

The sensing mechanism of the HMFPS is shown in Figure 3c. The entire sensing range can be divided into two stages, which are the low stress range and the high stress range. In the original state, graphene is uniformly covered on the MF surface, forming a stable conductive network. Applying stress contacts the adjacent skeleton of the GF, establishing new conductive paths and decreasing the R_b. With the increase in the stress, the foam skeletons further contact each other to establish conductive paths, and the R_b reduced. The changes in the entire sensing process are attributed to the "contact effects".

3.3. Mechanical Properties of the HMFPS

The size of the PDMS supporting layer determines the equivalent elastic modulus of the sensor, thereby affecting the sensitivity and measurement range. Figure 4a–d are the compressive stress–strain curves of the GF, the HMFPS-3, the HMFPS-5, and the HMFPS-10 under 40%, 60%, and 80% compressive strain.

Figure 4e shows the compressive stress–strain curves of the GF, the HMFPS-3, the HMFPS-5, the HMFPS-10 and the PDMS. The stress required at the same strain for the HMFPS is sharply increased with the presence of the PDMS supporting layer. The stress of the HMFPS increases nonlinearly with the strain greater than 37.5%. When the strain is less than 37.5%, only the GF is compressed. Due to the lower elastic modulus of the GF, the sensor exhibits a lower elastic modulus (0.0633 kPa). When the strain is greater than 37.5%, the GF and the PDMS supporting layer need to be compressed simultaneously, which equivalently increases the stress required for the deformation of the GF. Figure 4f compares the E of the GF, the HMFPS-3, the HMFPS-5, the HMFPS-10 and the PDMS when the strain is less than 37.5% and greater than 37.5%. When the strain is less than 37.5%, the E of the sample decreases with the increase in the PDMS supporting layer area, which is because the compressed area of the sample (S) is equivalent to the size of the entire sensor during the experiment, that is, $S_{GF} < S_{\text{HMFPS}-3} < S_{\text{HMFPS}-5} < S_{\text{HMFPS}-10}$. The applied stress ($P$) can be defined as $P = F/S$, where F represents the applied stress, and S represents the area of the applied stress. The same F ($F_{GF} = F_{\text{HMFPS}-3} = F_{\text{HMFPS}-5} = F_{\text{HMFPS}-10}$) leads to the same Δh ($\Delta h_{GF} = \Delta h_{\text{HMFPS}-3} = \Delta h_{\text{HMFPS}-5} = \Delta h_{\text{HMFPS}-10}$), resulting in

$P_{GF} > P_{\text{HMFPS}-3} > P_{\text{HMFPS}-5} > P_{\text{HMFPS}-10}$. Additionally, as $E = P/\Delta h$, it can be seen that $E_{GF} > E_{\text{HMFPS}-3} > E_{\text{HMFPS}-5} > E_{\text{HMFPS}-10}$. During this process, the sensitivity of the sensor is equal to that of the GF, which enables the small stress (<1.8 kPa) sensation. When the strain is greater than 37.5%, the E of the sample increases with the increase in the PDMS supporting layer area, which increases the measurement range of the sensor. Therefore, the HMFPS has a larger measurement range compared with the GF, and has high sensitivity compared with the CC-based sensors.

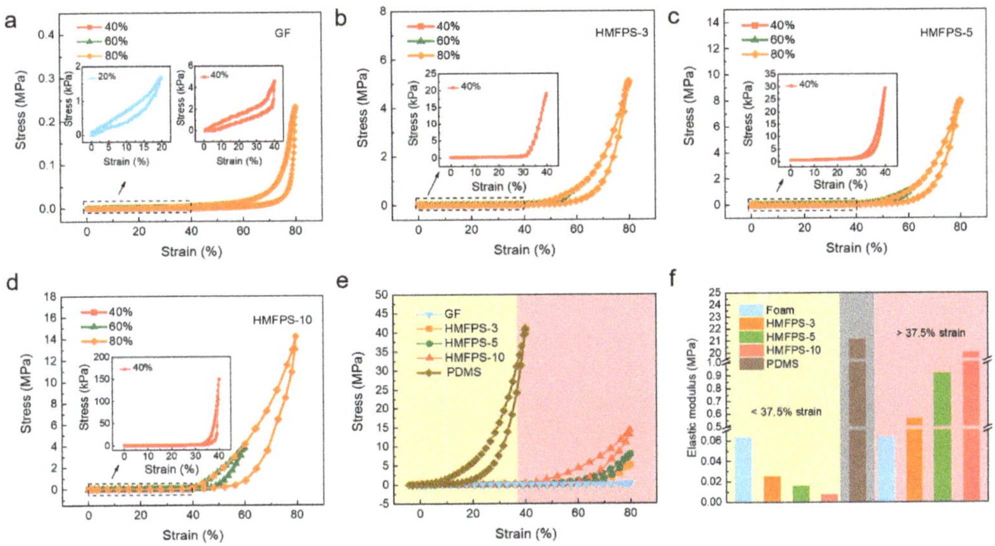

Figure 4. Compressive stress–strain curves of (**a**) GF, (**b**) HMFPS-3, (**c**) HMFPS-5, and (**d**) HMFPS-10 under 40%, 60%, and 80% strain; (**e**) Compressive stress–strain curves, and (**f**) elastic modulus of Gf, HMFPS-3, HMFPS-5, HMFPS-10, and PDMS bulk.

3.4. Piezoresistivity of the HMFPS

The current response curves of the GF, the HMFPS-3, the HMFPS-5 and the HMFPS-10 under different stresses are shown in Figure 5a,b. The sensitivity (S) of the HMFPS can be defined as $S = (\frac{\Delta I}{I_0})/P$, where ΔI is the current variation under a certain stress, I_0 is the original current of the GF, and P is the applied stress. Under small stress (<1.8 kPa), the sensor has ultra-high sensitivity of 0.695 kPa^{-1} due to the compression of only the GF sensing layer. With the increase in the stress, the GF sensing layer and the PDMS supporting layer are compressed simultaneously. The required stress increases with the current increases, which is affected by the PDMS supporting layer. The sensitivity of the HMFPS decreases, while the measurement range of the increases. This is consistent with the sensor mechanism of the sensor. Figure 5c shows the sensitivity and measurement range of the GF, the HMFPS-3, the HMFPS-5 and the HMFPS-10. In the small stress range (<1.8 kPa), the sensitivity of the HMFPS is 0.695 kPa^{-1}. HMFPS-10 has a wide stress measurement range of 0–14122 kPa.

The current response–strain curve of the HMFPS-10 is shown in Figure S3. The current variation rate of the HMFPS increases with the increase in the strain. The current variation rate of the HMFPS increases relatively slowly when the strain is less than 76%, while the current variation rate of the HMFPS increases rapidly when the strain is more than 76%. This is related to the porosity of the GF sensing layer. With the compressing deformation of the GF, the foam skeletons contact each other and establish new conductive paths, resulting in the current increasing. When the compressing deformation is smaller than the range of the GF porosity, the contact between the foam skeleton is limited, the new conductive path is less, and the current increases slowly. When the compressing deformation is close to or greater than the porosity of the GF, the deformation leads to direct contact of the foam skeleton, resulting in a large number of conductive paths and a sharp increase in the current.

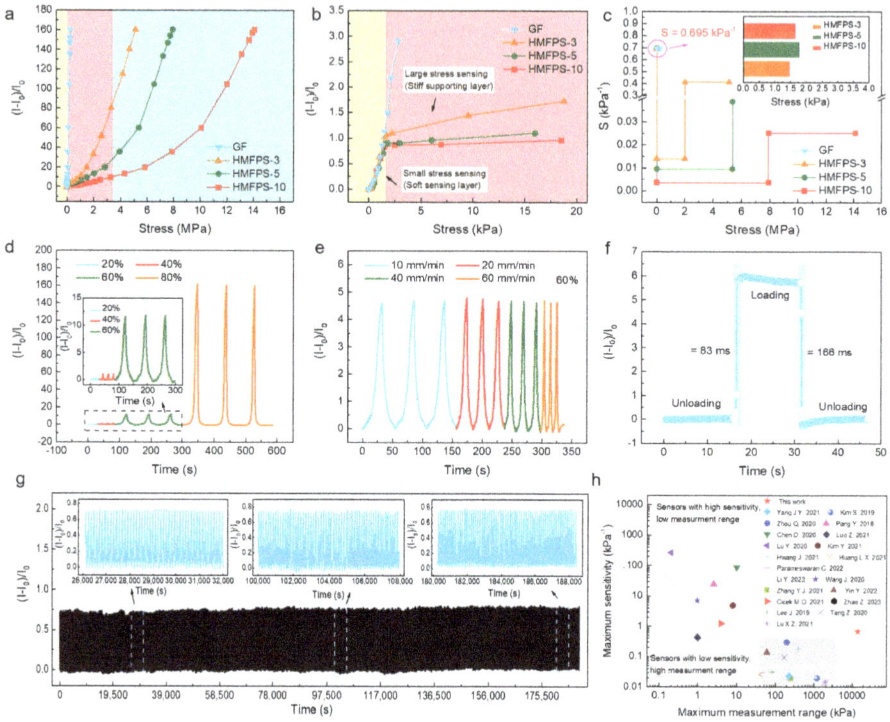

Figure 5. (**a**,**b**) Current responses of the Gf, HMFPS-3, HMFPS-5, and HMFPS-10; (**c**) Sensitivity and detection range of the Gf, HMFPS-3, HMFPS-5, and HMFPS-10; Current variation rates of the HMFPS-3 to cyclic compression (**d**) from 20% strain to 80% strain and (**e**) to different stress rates; (**f**) Response and recovery time of the HMFPS-3; (**g**) Durability test of the HMFPS over 2000 compression cycles; (**h**) Comparison of this sensor with state-of-the-art counterparts.

In order to study the dynamic performance of the sensor, we tested the dynamic response of the sensor under 20%, 40%, 60% and 80% compressive strain, as shown in Figure 5d. The response of the sensor can be clearly distinguished with increasing stress, and the three loading cycles under the same stress exhibit good repeatability. In addition, the dynamic responses of the sensor at four different loading rates of 10, 20, 40, and 60 mm min^{-1} under 60% strain are shown in Figure 5e. The results show that the response of the sensor is independent for different compression speeds. Figure 5f shows the response and recovery time of the HMFPS. The HMFPS has a fast response speed to the applied stress, and its response time and recovery time are 83 ms and 166 ms, respectively.

Repeatability and stability are important indicators in the practical application of stress sensors. In order to investigate the cyclic repeatability and stability of the HMFPS, 2000 cyclic loading/unloading tests were performed on the HMFPS, as shown in Figure 5g. The current variation rate of the sample remains stable during multiple cycles. The illustration in Figure 5g shows that the current variation is almost the same for each cycle. Experiments show that our proposed HMFPS has good repeatability and stability. Furthermore, we compared the performance of this work with existing literature on piezoresistive stress sensors, as shown in Figure 5h and Table S1 [17,18,37,45–62]. Benefiting from the HM design of the sensor, the prepared HMFPS has high sensitivity and wide measurement range.

3.5. Applications of the HMFPS

In order to verify the applicability of the HMFPS in human motion detection, the fabricated HMFPS was attached to different parts of the human body with breathable medical bandage. Additionally, the upper surface of the sensor is covered with the same size paper. Figure 6a is the current response curve of the finger pressing the HMFPS, which shows that the HMFPS can respond to the stress in real time and the current variation rate is proportional to the compressive stress. Figure 6b–f shows the response results of the sensors fixed on the finger, the neck, the elbow, the wrist, and the knee, respectively. The results show that the HMFPS can respond to the motion of the human body through the current variation rate, indicating that the proposed sensor can be used for human motion detection.

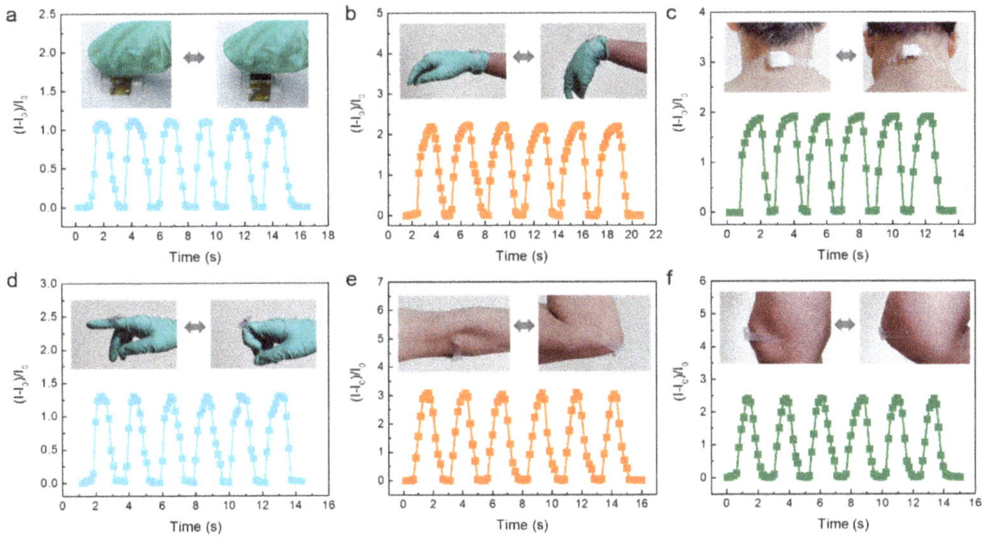

Figure 6. Monitoring human motion using HMFPS; (**a**) Finger pressing, (**b**) wrist bending, (**c**) nodding, (**d**) finger movement, (**e**) arm bending, and (**f**) knee bending.

4. Conclusions

In summary, we propose a HMFPS with high sensitivity and wide measurement range through HM design. The HMFPS consists of a GF sensing layer, a PDMS supporting layer and an interdigital electrode. We prepared three HMFPS with different area (HMFPS-3, HMFPS-5, HMFPS-10), and investigated the effect and mechanism of the PDMS supporting layer area on the piezoresistivity of the sensor. Moreover, we analyzed the sensing mechanism of the HMFPS, and proved that the HM design is an effective method to achieve high sensitivity and a wide measurement range. Among them, HMFPS-10 has the best overall performance, with a sensitivity of 0.695 kPa^{-1}, a measurement range of 0–14,122 kPa, fast response/recovery (83 ms and 166 ms) and excellent stability (2000 cycles). In addition, the application of the HMFPS in human motion monitoring was demonstrated.

Supplementary Materials: The following supporting information can be downloaded at: https://www.mdpi.com/article/10.3390/mi14040716/s1, Figure S1: Electrical conductivity of GFs with different dip-coating times, Figure S2: Raman spectrums of the GOMF and the GF, Figure S3: Current responses-strain curves of the HMFPS-10, Table S1: Comparison between the results of the proposed foam and its counterparts.

Author Contributions: Conceptualization, T.Y. and J.Y.; methodology, T.Y.; software, T.Y.; validation, T.Y. and J.Y.; formal analysis, T.Y.; investigation, T.Y.; resources, J.Y., D.Z. and G.G.; data curation, Y.W.; writing—original draft preparation, T.Y.; writing—review and editing, T.Y.; visualization, T.Y. and Y.T.; supervision, G.G.; project administration, J.Y. and D.Z.; funding acquisition, J.Y. All authors have read and agreed to the published version of the manuscript.

Funding: This research was funded by the National Natural Science Foundation of China, grant number 62101432, the China Postdoctoral Science Foundation, grant number 2022M722558, the Fundamental Research Program of Shanxi Province, grant number 20210302123182, the Research Project of Shanxi Scholarship Council of China, grant numbers 2020-036, 2022-069, the Scientific Research Fund of Zhejiang Provincial Education Department, grant number Y202146138 and the University Enterprise Cooperation Project, grant number FG2022314.

Data Availability Statement: Data available on request from the authors.

Conflicts of Interest: The authors declare no conflict of interest.

References

1. Zhu, M.L.; He, T.Y.Y.; Lee, C.K. Technologies toward next generation human machine interfaces: From machine learning enhanced tactile sensing to neuromorphic sensory systems. *Appl. Phys. Rev.* **2020**, *7*, 031305. [CrossRef]
2. Yang, J.C.; Mun, J.; Kwon, S.Y.; Park, S.; Bao, Z.; Park, S. Electronic Skin: Recent Progress and Future Prospects for Skin-Attachable Devices for Health Monitoring, Robotics, and Prosthetics. *Adv. Mater.* **2019**, *31*, e1904765. [CrossRef] [PubMed]
3. Ge, G.; Wang, Q.; Zhang, Y.Z.; Alshareef, H.N.; Dong, X. 3D Printing of Hydrogels for Stretchable Ionotronic Devices. *Adv. Funct. Mater.* **2021**, *31*, 2107437. [CrossRef]
4. Ge, G.; Mandal, K.; Haghniaz, R.; Li, M.; Xiao, X.; Carlson, L.; Jucaud, V.; Dokmeci, M.R.; Ho, G.W.; Khademhosseini, A. Deep Eutectic Solvents-Based Ionogels with Ultrafast Gelation and High Adhesion in Harsh Environments. *Adv. Funct. Mater.* **2023**, *33*, 2207388. [CrossRef]
5. Xu, J.; Wang, Z.M.; Wang, X.X.; Wu, Y.L.; Xing, R.Z.; Yu, T.T.; Li, Y.; Ao, J.P.; Tao, Y.B.; Bai, B.W.; et al. Breathable Encapsulated Liquid Metal Foam-Based Soft Stress Sensor. *Adv. Mater. Technol.* **2023**, 2201193. [CrossRef]
6. White, E.L.; Yuen, M.C.; Case, J.C.; Kramer, R.K. Low-Cost, Facile, and Scalable Manufacturing of Capacitive Sensors for Soft Systems. *Adv. Mater. Technol.* **2017**, *2*, 1700072. [CrossRef]
7. Shintake, J.; Piskarev, E.; Jeong, S.H.; Floreano, D. Ultrastretchable Strain Sensors Using Carbon Black-Filled Elastomer Composites and Comparison of Capacitive Versus Resistive Sensors. *Adv. Mater. Technol.* **2018**, *3*, 1700284. [CrossRef]
8. You, I.; Mackanic, D.G.; Matsuhisa, N.; Kang, J.; Kwon, J.; Beker, L.; Mun, J.; Suh, W.; Kim, T.Y.; Tok, J.B.; et al. Artificial multimodal receptors based on ion relaxation dynamics. *Science* **2020**, *370*, 961–965. [CrossRef]
9. Mannsfeld, S.C.; Tee, B.C.; Stoltenberg, R.M.; Chen, C.V.; Barman, S.; Muir, B.V.; Sokolov, A.N.; Reese, C.; Bao, Z. Highly sensitive flexible pressure sensors with microstructured rubber dielectric layers. *Nat. Mater.* **2010**, *9*, 859–864. [CrossRef]
10. Xue, X.T.; Zhang, D.G.; Wu, Y.L.; Xing, R.Z.; Li, H.; Yu, T.T.; Bai, B.W.; Tao, Y.B.; Dickey, M.D.; Yang, J.Y. Segregated and Non-Settling Liquid Metal Elastomer via Jamming of Elastomeric Particles. *Adv. Funct. Mater.* **2022**, *33*, 2210553. [CrossRef]
11. Yan, Y.; Hu, Z.; Yang, Z.; Yuan, W.; Song, C.; Pan, J.; Shen, Y. Soft magnetic skin for super-resolution tactile sensing with force self-decoupling. *Sci. Robot* **2021**, *6*, eabc8801. [CrossRef]
12. Ge, G.; Zhang, Y.Z.; Zhang, W.; Yuan, W.; El-Demellawi, J.K.; Zhang, P.; Di Fabrizio, E.; Dong, X.; Alshareef, H.N. $Ti_3C_2T_x$ MXene-Activated Fast Gelation of Stretchable and Self-Healing Hydrogels: A Molecular Approach. *ACS Nano* **2021**, *15*, 2698–2706. [CrossRef]
13. Jin, Y.; Lin, Y.; Kiani, A.; Joshipura, I.D.; Ge, M.; Dickey, M.D. Materials tactile logic via innervated soft thermochromic elastomers. *Nat. Commun.* **2019**, *10*, 4187. [CrossRef]
14. Ji, B.; Zhou, Q.; Hu, B.; Zhong, J.; Zhou, J.; Zhou, B. Bio-Inspired Hybrid Dielectric for Capacitive and Triboelectric Tactile Sensors with High Sensitivity and Ultrawide Linearity Range. *Adv. Mater.* **2021**, *33*, e2100859. [CrossRef]
15. Ge, G.; Lu, Y.; Qu, X.; Zhao, W.; Ren, Y.; Wang, W.; Wang, Q.; Huang, W.; Dong, X. Muscle-Inspired Self-Healing Hydrogels for Strain and Temperature Sensor. *ACS Nano* **2020**, *14*, 218–228. [CrossRef]
16. Yu, T.T.; Zhang, D.G.; Wu, Y.L.; Guo, S.Z.; Lei, F.; Li, Y.; Yang, J.Y. Graphene foam pressure sensor based on fractal electrode with high sensitivity and wide linear range. *Carbon* **2021**, *182*, 497–505. [CrossRef]
17. Zhang, D.G.; Wang, X.X.; Wu, Y.L.; Song, H.L.; Ma, Z.; Zhang, X.Y.; Yang, X.F.; Xing, R.Z.; Li, Y.; Yang, J.Y. Passive Particle Jamming Variable Stiffness Material-Based Flexible Capacitive Stress Sensor with High Sensitivity and Large Measurement Limit. *Adv. Mater. Technol.* **2021**, *6*, 2100106. [CrossRef]
18. Yang, J.Y.; Kwon, K.Y.; Kanetkar, S.; Xing, R.Z.; Nithyanandam, P.; Li, Y.; Jung, W.; Gong, W.; Tuman, M.; Shen, Q.C.; et al. Skin-Inspired Capacitive Stress Sensor with Large Dynamic Range via Bilayer Liquid Metal Elastomers. *Adv. Mater. Technol.* **2022**, *7*, 2101074. [CrossRef]
19. Lin, W.; Wang, B.; Peng, G.; Shan, Y.; Hu, H.; Yang, Z. Skin-Inspired Piezoelectric Tactile Sensor Array with Crosstalk-Free Row+Column Electrodes for Spatiotemporally Distinguishing Diverse Stimuli. *Adv. Sci.* **2021**, *8*, 2002817. [CrossRef]

20. Lai, Y.C.; Lu, H.W.; Wu, H.M.; Zhang, D.G.; Yang, J.Y.; Ma, J.; Shamsi, M.; Vallem, V.; Dickey, M.D. Elastic Multifunctional Liquid-Metal Fibers for Harvesting Mechanical and Electromagnetic Energy and as Self-Powered Sensors. *Adv. Energy Mater.* **2021**, *11*, 2100411. [CrossRef]
21. Yang, R.; Zhang, W.; Tiwari, N.; Yan, H.; Li, T.; Cheng, H. Multimodal Sensors with Decoupled Sensing Mechanisms. *Adv. Sci.* **2022**, *9*, e2202470. [CrossRef] [PubMed]
22. Trung, T.Q.; Lee, N.E. Flexible and Stretchable Physical Sensor Integrated Platforms for Wearable Human-Activity Monitoringand Personal Healthcare. *Adv. Mater.* **2016**, *28*, 4338–4372. [CrossRef] [PubMed]
23. Yang, J.Y.; Tang, D.; Ao, J.P.; Ghosh, T.; Neumann, T.V.; Zhang, D.G.; Piskarev, Y.; Yu, T.T.; Truong, V.K.; Xie, K.; et al. Ultrasoft Liquid Metal Elastomer Foams with Positive and Negative Piezopermittivity for Tactile Sensing. *Adv. Funct. Mater.* **2020**, *30*, 2002611. [CrossRef]
24. Zhang, D.G.; Zhong, Y.Q.; Wu, Y.L.; Zhang, X.F.; Dickey, M.D.; Yang, J.Y. Liquid metal elastomer with flytrap-inspired pillar structure for stress sensing. *Compos. Sci. Technol.* **2021**, *216*, 109066. [CrossRef]
25. Choi, S.; Han, S.I.; Kim, D.; Hyeon, T.; Kim, D.H. High-performance stretchable conductive nanocomposites: Materials, processes, and device applications. *Chem. Soc. Rev.* **2019**, *48*, 1566–1595. [CrossRef]
26. Wu, Q.; Qiao, Y.; Guo, R.; Naveed, S.; Hirtz, T.; Li, X.; Fu, Y.; Wei, Y.; Deng, G.; Yang, Y.; et al. Triode-Mimicking Graphene Pressure Sensor with Positive Resistance Variation for Physiology and Motion Monitoring. *ACS Nano* **2020**, *14*, 10104–10114. [CrossRef]
27. Zhang, Z.M.; Zhang, Y.X.; Jiang, X.; Bukhari, H.; Zhang, Z.X.; Han, W.H.; Xie, E.Q. Simple and efficient pressure sensor based on PDMS wrapped CNT arrays. *Carbon* **2019**, *155*, 71–76. [CrossRef]
28. Han, S.; Alvi, N.U.H.; Granlof, L.; Granberg, H.; Berggren, M.; Fabiano, S.; Crispin, X. A Multiparameter Pressure-Temperature-Humidity Sensor Based on Mixed Ionic-Electronic Cellulose Aerogels. *Adv. Sci.* **2019**, *6*, 1802128. [CrossRef]
29. Wang, Y.L.; Wang, Y.S.; Wan, B.L.; Han, B.G.; Cai, G.C.; Li, Z.Z. Properties and mechanisms of self-sensing carbon nanofibers/epoxy composites for structural health monitoring. *Compos. Struct.* **2018**, *200*, 669–678. [CrossRef]
30. Zhao, S.; Zhang, G.; Gao, Y.; Deng, L.; Li, J.; Sun, R.; Wong, C.P. Strain-driven and ultrasensitive resistive sensor/switch based on conductive alginate/nitrogen-doped carbon-nanotube-supported Ag hybrid aerogels with pyramid design. *ACS Appl. Mater. Interfaces* **2014**, *6*, 22823–22829. [CrossRef]
31. Zhang, S.; Liu, H.; Yang, S.; Shi, X.; Zhang, D.; Shan, C.; Mi, L.; Liu, C.; Shen, C.; Guo, Z. Ultrasensitive and Highly Compressible Piezoresistive Sensor Based on Polyurethane Sponge Coated with a Cracked Cellulose Nanofibril/Silver Nanowire Layer. *ACS Appl. Mater. Interfaces* **2019**, *11*, 10922–10932. [CrossRef]
32. Yang, J.Y.; Ye, Y.S.; Li, X.P.; Lu, X.Z.; Chen, R.J. Flexible, conductive, and highly pressure-sensitive graphene-polyimide foam for pressure sensor application. *Compos. Sci. Technol.* **2018**, *164*, 187–194. [CrossRef]
33. Ai, Y.F.; Hsu, T.H.; Wu, D.C.; Lee, L.; Chen, J.H.; Chen, Y.Z.; Wu, S.C.; Wu, C.; Wang, Z.M.M.; Chueh, Y.L. An ultrasensitive flexible pressure sensor for multimodal wearable electronic skins based on large-scale polystyrene ball@reduced graphene-oxide core-shell nanoparticles. *J. Mater. Chem. C* **2018**, *6*, 5514–5520. [CrossRef]
34. Wang, L.L.; Jackman, J.A.; Tan, E.L.; Park, J.H.; Potroz, M.G.; Hwang, E.T.; Cho, N.J. High-performance, flexible electronic skin sensor incorporating natural microcapsule actuators. *Nano Energy* **2017**, *36*, 38–45. [CrossRef]
35. Davoodi, E.; Montazerian, H.; Haghniaz, R.; Rashidi, A.; Ahadian, S.; Sheikhi, A.; Chen, J.; Khademhosseini, A.; Milani, A.S.; Hoorfar, M.; et al. 3D-Printed Ultra-Robust Surface-Doped Porous Silicone Sensors for Wearable Biomonitoring. *ACS Nano* **2020**, *14*, 1520–1532. [CrossRef]
36. Wang, L.; Wang, D.; Wu, Z.; Luo, J.; Huang, X.; Gao, Q.; Lai, X.; Tang, L.C.; Xue, H.; Gao, J. Self-Derived Superhydrophobic and Multifunctional Polymer Sponge Composite with Excellent Joule Heating and Photothermal Performance for Strain/Pressure Sensors. *ACS Appl. Mater. Interfaces* **2020**, *12*, 13316–13326. [CrossRef]
37. Kim, S.; Amjadi, M.; Lee, T.I.; Jeong, Y.; Kwon, D.; Kim, M.S.; Kim, K.; Kim, T.S.; Oh, Y.S.; Park, I. Wearable, Ultrawide-Range, and Bending-Insensitive Pressure Sensor Based on Carbon Nanotube Network-Coated Porous Elastomer Sponges for Human Interface and Healthcare Devices. *ACS Appl. Mater. Interfaces* **2019**, *11*, 23639–23648. [CrossRef]
38. Kweon, O.Y.; Lee, S.J.; Oh, J.H. Wearable high-performance pressure sensors based on three-dimensional electrospun conductive nanofibers. *NPG Asia Mater.* **2018**, *10*, 540–551. [CrossRef]
39. Guo, Y.; Guo, Z.; Zhong, M.; Wan, P.; Zhang, W.; Zhang, L. A Flexible Wearable Pressure Sensor with Bioinspired Microcrack and Interlocking for Full-Range Human-Machine Interfacing. *Small* **2018**, *14*, e1803018. [CrossRef]
40. Li, Y.; Luo, S.D.; Yang, M.C.; Liang, R.; Zeng, C.C. Poisson Ratio and Piezoresistive Sensing: A New Route to High-Performance 3D Flexible and Stretchable Sensors of Multimodal Sensing Capability. *Adv. Funct. Mater.* **2016**, *26*, 2900–2908. [CrossRef]
41. Huang, X.; Guo, W.; Liu, S.Y.; Li, Y.Y.; Qiu, Y.Q.; Fang, H.; Yang, G.G.; Zhu, K.H.; Yin, Z.P.; Li, Z.; et al. Flexible Mechanical Metamaterials Enabled Electronic Skin for Real-Time Detection of Unstable Grasping in Robotic Manipulation. *Adv. Funct. Mater.* **2022**, *32*, 2109109. [CrossRef]
42. Cao, K.L.; Wu, M.; Bai, J.B.; Wen, Z.; Zhang, J.W.; Wang, T.Y.; Peng, M.W.; Liu, T.; Jia, Z.; Liang, Z.Q.; et al. Beyond Skin Pressure Sensing: 3D Printed Laminated Graphene Pressure Sensing Material Combines Extremely Low Detection Limits with Wide Detection Range. *Adv. Funct. Mater.* **2022**, *32*, 2202360. [CrossRef]
43. Rafiee, M.; Farahani, R.D.; Therriault, D. Multi-Material 3D and 4D Printing: A Survey. *Adv. Sci.* **2020**, *7*, 1902307. [CrossRef] [PubMed]

44. Ravanbakhsh, H.; Karamzadeh, V.; Bao, G.Y.; Mongeau, L.; Juncker, D.; Zhang, Y.S. Emerging Technologies in Multi-Material Bioprinting. *Adv. Mater.* **2021**, *33*, 2104730. [CrossRef] [PubMed]
45. Zhou, Q.; Ji, B.; Hu, B.; Li, S.B.; Xu, Y.; Gao, Y.B.; Wen, W.J.; Zhou, J.; Zhou, B.P. Tilted magnetic micropillars enabled dual-mode sensor for tactile/touchless perceptions. *Nano Energy* **2020**, *78*, 105382. [CrossRef]
46. Pang, Y.; Zhang, K.; Yang, Z.; Jiang, S.; Ju, Z.; Li, Y.; Wang, X.; Wang, D.; Jian, M.; Zhang, Y.; et al. Epidermis Microstructure Inspired Graphene Pressure Sensor with Random Distributed Spinosum for High Sensitivity and Large Linearity. *ACS Nano* **2018**, *12*, 2346–2354. [CrossRef]
47. Chen, D.; Liu, Z.; Li, Y.; Sun, D.; Liu, X.; Pang, J.; Liu, H.; Zhou, W. Unsymmetrical Alveolate PMMA/MWCNT Film as a Piezoresistive E-Skin with Four-Dimensional Resolution and Application for Detecting Motion Direction and Airflow Rate. *ACS Appl. Mater. Interfaces* **2020**, *12*, 30896–30904. [CrossRef]
48. Luo, Z.; Chen, J.; Zhu, Z.; Li, L.; Su, Y.; Tang, W.; Omisore, O.M.; Wang, L.; Li, H. High-Resolution and High-Sensitivity Flexible Capacitive Pressure Sensors Enhanced by a Transferable Electrode Array and a Micropillar-PVDF Film. *ACS Appl. Mater. Interfaces* **2021**, *13*, 7635–7649. [CrossRef]
49. Lu, Y.; He, Y.; Qiao, J.; Niu, X.; Li, X.; Liu, H.; Liu, L. Highly Sensitive Interlocked Piezoresistive Sensors Based on Ultrathin Ordered Nanocone Array Films and Their Sensitivity Simulation. *ACS Appl. Mater. Interfaces* **2020**, *12*, 55169–55180. [CrossRef]
50. Kim, Y.; Yang, H.; Oh, J.H. Simple fabrication of highly sensitive capacitive pressure sensors using a porous dielectric layer with cone-shaped patterns. *Mater. Des.* **2021**, *197*, 109203. [CrossRef]
51. Hwang, J.; Kim, Y.; Yang, H.; Oh, J.H. Fabrication of hierarchically porous structured PDMS composites and their application as a flexible capacitive pressure sensor. *Compos. Part B-Eng.* **2021**, *211*, 108607. [CrossRef]
52. Huang, L.X.; Wang, H.; Zhan, D.H.; Fang, F.Y. Flexible Capacitive Pressure Sensor Based on Laser-Induced Graphene and Polydimethylsiloxane Foam. *IEEE Sens. J.* **2021**, *21*, 12048–12056. [CrossRef]
53. Parameswaran, C.; Chaudhary, R.P.; Prutvi, S.H.; Gupta, D. Rapid One Step Fabrication of Hydrophilic Hierarchical Porous PDMS with Negative Piezopermittivity for Sensing and Energy Storage Applications. *ACS Appl. Polym. Mater.* **2022**, *4*, 2047–2056. [CrossRef]
54. Li, Y.; Cui, Y.; Zhang, M.; Li, X.; Li, R.; Si, W.; Sun, Q.; Yu, L.; Huang, C. Ultrasensitive Pressure Sensor Sponge Using Liquid Metal Modulated Nitrogen-Doped Graphene Nanosheets. *Nano Lett.* **2022**, *22*, 2817–2825. [CrossRef]
55. Wang, J.; Lou, Y.; Wang, B.; Sun, Q.; Zhou, M.; Li, X. Highly Sensitive, Breathable, and Flexible Pressure Sensor Based on Electrospun Membrane with Assistance of AgNW/TPU as Composite Dielectric Layer. *Sensors* **2020**, *20*, 2459. [CrossRef]
56. Zhang, Y.J.; Zhao, Y.; Zhai, W.; Zheng, G.Q.; Ji, Y.X.; Dai, K.; Mi, L.W.; Zhang, D.B.; Liu, C.T.; Shen, C.Y. Multifunctional interlocked e-skin based on elastic micropattern array facilely prepared by hot-air-gun. *Chem. Eng. J.* **2021**, *407*, 127960. [CrossRef]
57. Yin, Y.; Wang, Y.; Li, H.; Xu, J.; Zhang, C.; Li, X.; Cao, J.; Feng, H.; Zhu, G. A flexible dual parameter sensor with hierarchical porous structure for fully decoupled pressure–temperature sensing. *Chem. Eng. J.* **2022**, *430*, 133158. [CrossRef]
58. Cicek, M.O.; Doganay, D.; Durukan, M.B.; Gorur, M.C.; Unalan, H.E. Seamless Monolithic Design for Foam Based, Flexible, Parallel Plate Capacitive Sensors. *Adv. Mater. Technol.* **2021**, *6*, 2001168. [CrossRef]
59. Zhao, Z.; Guo, Q.K.; Sun, Y.; An, N.L.; Hui, P.Z.; Yang, L.H.; Chen, X.F. Bioinspired Hierarchical Structure for an Ultrawide-Range Multifunctional Flexible Sensor Using Porous Expandable Polyethylene/Loofah-Like Polyurethane Sponge Material. *Adv. Intell. Syst.* **2023**, *5*, 2200295. [CrossRef]
60. Lee, J.; Kim, J.; Shin, Y.; Jung, I. Ultra-robust wide-range pressure sensor with fast response based on polyurethane foam doubly coated with conformal silicone rubber and CNT/TPU nanocomposites islands. *Compos. Part B-Eng.* **2019**, *177*, 107364. [CrossRef]
61. Tang, Z.; Jia, S.; Zhou, C.; Li, B. 3D Printing of Highly Sensitive and Large-Measurement-Range Flexible Pressure Sensors with a Positive Piezoresistive Effect. *ACS Appl. Mater. Interfaces* **2020**, *12*, 28669–28680. [CrossRef] [PubMed]
62. Lu, X.Z.; Yu, T.T.; Meng, F.C.; Bao, W.M. Wide-Range and High-Stability Flexible Conductive Graphene/Thermoplastic Polyurethane Foam for Piezoresistive Sensor Applications. *Adv. Mater. Technol.* **2021**, *6*, 2100248. [CrossRef]

micromachines

Article

A 7.6-nW 1-kS/s 10-Bit SAR ADC for Biomedical Applications

Yunfeng Hu [1,*], Bin Tang [1], Lexing Hu [1], Haibo Liang [1], Bin Li [2], Zhaohui Wu [2] and Xiaojia Liu [1]

[1] University of Electronic Science and Technology of China, Zhongshan Institute, Zhongshan 528402, China
[2] School of Microelectronics, South China University of Technology, Guangzhou 510640, China
* Correspondence: huyf@zsc.edu.cn

Abstract: This paper presents a 10-bit successive approximation register analog-to-digital converter with energy-efficient low-complexity switching scheme, automatic ON/OFF comparator and automatic ON/OFF SAR logic for biomedical applications. The energy-efficient switching scheme achieves an average digital-to-analog converter switching energy of 63.56 CV_{ref}^2, achieving a reduction of 95.34% compared with the conventional capacitor switching scheme for CDACs. With the switching scheme, the ADC can lower the dependency on the accuracy of V_{cm} and complexity of DAC control logic and DAC driver circuit. Moreover, dynamic circuits and automatic ON/OFF technology are used to reduce power consumption of comparator and SAR logic. The prototype is designed and fabricated in a 180 nm CMOS with a core size of 500 μm × 300 μm (0.15 mm^2). It consumes 7.6 nW at 1 kS/s sampling rate and 1.8-V supply with an achieved signal-to-noise-and distortion ratio of 45.90 dB and a resulting figure of merit of 51.7 fJ/conv.-step.

Keywords: analog-to-digital converter (ADC); energy-efficient; successive approximation register (SAR)

check for updates

Citation: Hu, Y.; Tang, B.; Hu, L.; Liang, H.; Li, B.; Wu, Z.; Liu, X. A 7.6-nW 1-kS/s 10-Bit SAR ADC for Biomedical Applications. *Micromachines* **2022**, *13*, 2110. https://doi.org/10.3390/mi13122110

Academic Editors: Xiao Xiao and Gang Ge

Received: 22 November 2022
Accepted: 28 November 2022
Published: 29 November 2022

1. Introduction

The development of low-power integrated circuits (ICs) will help bring portable and implantable biomedical devices and biosensors to the market. Analog front end (AFE) circuits in these products may consume most of the total power budget because they usually need to remain online to sense input signals continuously [1]. Various biomedical signals and their frequency ranges are shown in Table 1. Most biomedical signals have frequencies below 1 kHz. Figure 1 shows the basic processing units in a biomedical implantable device [2]; ADC is an intermediate unit that converts analog signals into digital signals. Successive-approximation register (SAR) analog-to-digital converter (ADC) has become an appropriate choice for low-power biomedical applications in recent years due to its low-power characteristics [2–5]. Figure 2 shows the basic components in an SAR ADC. Among the building blocks in an SAR ADC, a capacitive DAC always consumes a significant part of the total power consumption [6–8]. Recently, some energy-efficient switching schemes have been proposed to reduce the energy consumption of DAC capacitor arrays. [8–10]. Compared to conventional techniques [11], Charge-Recovery [8], Charge-Sharing [8], Capacitor-Splitting [8], Set-and-down [10], and V_{cm}-based [9] techniques reduce the switching energy by 12.52%, 24.99%, 37.48%, 81.26%, and 87.52%, respectively. However, these schemes have various drawbacks. Capacitor-Splitting [8] and V_{cm}-based [9] schemes have complex DAC drive circuits, the Set-and-down scheme [10] has large common-mode voltage shift, and the V_{cm}-based scheme [9] has a high dependence on the middle reference voltage (V_{cm}).

Table 1. Frequency ranges of various biomedical signals.

Biomedical Signals	Frequency Range
ECG [2]	0.05–100 Hz
ECoG [12]	70–110 Hz
EMG [13]	50–150 Hz
EEG [14]	0–100 Hz

Figure 1. Block diagram of biomedical implantable device.

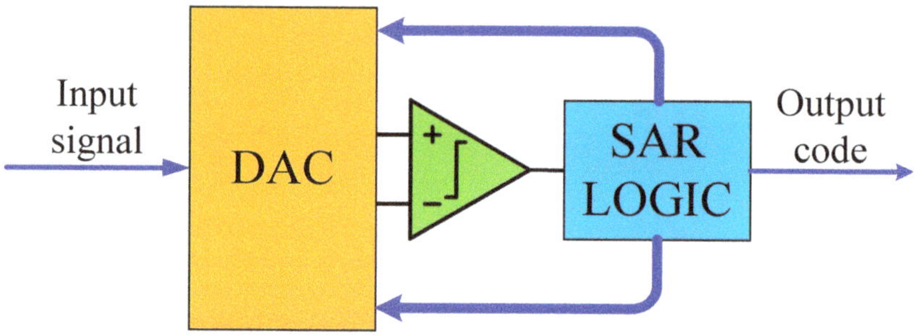

Figure 2. Building blocks of SAR ADC.

In this paper, the energy-efficient and low-complexity switching scheme [15] is used to realize successive approximation conversion. In the first comparison, no switching energy was consumed due to the use of top-plate sampling technology [10]. In the second comparison, no switching energy was consumed due to the closed-loop charge recycling method [16]. From the third comparison to the $(N-1)$th comparison, the reference voltage of the corresponding capacitor in the lower voltage capacitor array changes from gnd to V_{ref}. In the last comparison, the reference voltage of the last capacitor in the lower voltage capacitor array changes from gnd to V_{cm}. From the third comparison to the last comparison, since there is only one capacitor-switching reference voltage for each comparison, the power consumption is low. As a result, the energy-efficient and low-complexity switching scheme achieves an average switching energy of 63.56 CV_{ref}^2. Compared with the conventional switching scheme [11], this switching scheme reduces the switching energy by 95.34%. In addition, only the least significant bit (LSB) depends on the accuracy of V_{cm}, and each capacitor only uses two reference voltages, which reduces the dependence on the accuracy of V_{cm} and the complexity of DAC control logic and DAC driver circuit. An automatic ON/OFF comparator is used to achieve low power consumption. The comparator consists of three parts: the automatic ON/OFF clock circuit, the dynamic preamplifier stage, and the dynamic latch stage. The automatic ON/OFF clock circuit allows the comparator to work only during comparing. Automatic ON/OFF SAR Logic consists of three parts: automatic ON/OFF clock circuit, shift control, data latch. The automatic ON/OFF clock signal is generated by the comparator output signal, and the clock is output only when the comparator is active. In order to simplify the DAC control logic and DAC driver circuit, the latch of SAR logic uses a dynamic latch with differential output. When the proposed SAR ADC uses 180 nm CMOS process and operates at a sampling rate of 1 kS/s, the ADC achieves 45.90 dB SNDR and 58.79 dB SFDR and consumes only 7.6 nW [17]. The proposed SAR ADC is suitable for portable and implantable medical sensors.

This paper is organized as follows. Section 2 describes the ADC architecture and low-power circuits. Section 3 shows the measurement results and the comparison with other ADCs. Finally, Section 4 concludes.

2. Proposed ADC Architecture

As shown in Figure 3, the proposed SAR ADC consists of comparator, SAR logic, capacitor array DAC and DAC drive circuit. Because each capacitor of the capacitor array DAC has only two reference voltages, the DAC drive circuit and DAC control logic are simple. In addition, because the last capacitor uses V_{cm} as the reference voltage, the number of unit capacitors is reduced by half.

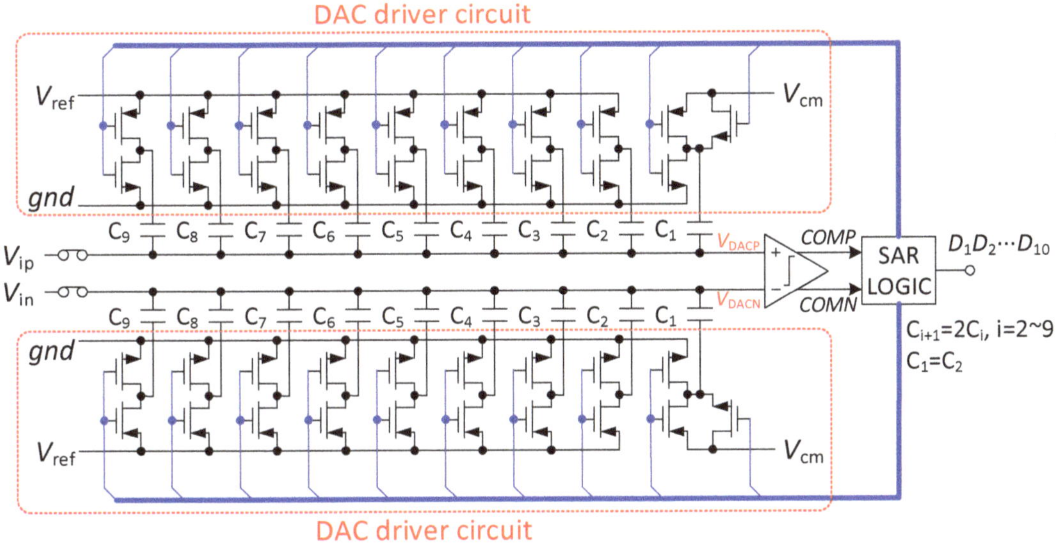

Figure 3. Proposed 10-bit SAR ADC architecture.

2.1. Switching Scheme

As shown in Figure 4, the operation of the switching scheme can be performed in five phases: sampling, the 1st comparison, the 2nd comparison, the 3rd to $(N-1)$th comparison, and the Nth comparison.

Sampling: In the sampling phase, the input signals are sampled on the top-plates of all capacitors via sampling switch, with the bottom-plates of the largest capacitors connecting to V_{ref} and other capacitors to *gnd*.

The 1st comparison: After sampling, the sampling switches are turned off. The output voltages of the DAC capacitor array are found to be

$$\begin{cases} V_{DACP}(1) = V_{ip} \\ V_{DACN}(1) = V_{in} \end{cases} \tag{1}$$

The comparator compares the sampling signals (V_{ip} and V_{in}) and gets D_1(MSB). No switching energy is consumed in the first comparison.

$$E_1 = 0 \tag{2}$$

The 2nd comparison (level-shift-*gnd*): If $D_1 = 1$, the reference voltage of the largest capacitor in the positive capacitor array changes from V_{ref} to *gnd*. If $D_1 = 0$, the reference voltage of the largest capacitor in the negative capacitor array becomes *gnd*. As a result, the voltage of the higher side is decreased by $V_{ref}/2$, and the output voltages are found to be

$$\begin{cases} V_{DACP}(2) = V_{ip} - D_1\frac{V_{ref}}{2} \\ V_{DACN}(2) = V_{in} - (1 - D_1)\frac{V_{ref}}{2} \end{cases} \tag{3}$$

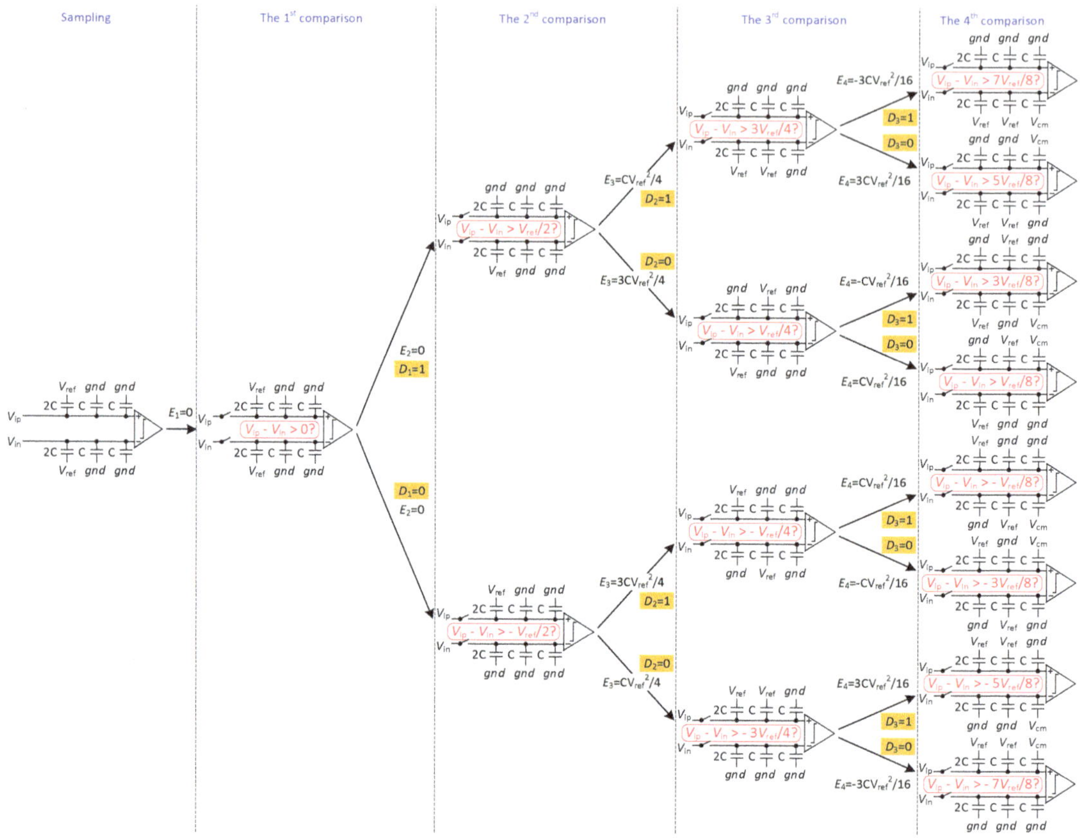

Figure 4. Switching procedure of 4-bit SAR DAC.

The comparator compares $V_{DACP}(2)$ with $V_{DACN}(2)$ and gets D_2. Due to the closed-loop charge recycling method [16], there is no switching energy consumption in the second comparison.

$$E_2 = 0 \tag{4}$$

The 3rd to $(N-1)$th comparison ("up" operation): According to the previous comparison results, the reference voltage of the corresponding capacitor in the lower voltage capacitor array is switched from *gnd* to V_{ref}, while the other one (in the higher voltage capacitor array) remains unchanged. For example, in the third comparison, if $D_2 = 1$, the reference voltage of the second largest capacitor in the negative capacitor array is switched from *gnd* to V_{ref}. If $D_2 = 0$, the reference voltage of the second largest capacitor in the positive capacitor array is switched from *gnd* to V_{ref}. The ADC repeats the procedure until the $(N-1)$th comparison is completed. The output voltages of each comparison are found to be

$$\begin{cases} V_{DACP}(i) = V_{ip} - D_1\frac{V_{ref}}{2} + \sum\limits_{j=2}^{i-1}(1-D_j)\frac{V_{ref}}{2^j} \\ V_{DACN}(i) = V_{in} - (1-D_1)\frac{V_{ref}}{2} + \sum\limits_{j=2}^{i-1}D_j\frac{V_{ref}}{2^j} \end{cases} \tag{5}$$

The comparator compares $V_{DACP}(i)$ with $V_{DACN}(i)$ and gets D_i. During the switching procedure, there is only one capacitor switch for each comparison, resulting in less switching activity and lower energy. Based on the switching energy calculation method in [6], the switching energy of each comparison is found to be

$$E_i = \left\{ \begin{array}{l} 2^{N-i-1} - 2^{N-2i} - D_{i-1} \sum\limits_{j=1}^{i-2} D[j] 2^{N-j-i-1} \\ -(1 - D_{i-1}) \sum\limits_{j=1}^{i-2} (1 - D[j]) 2^{N-j-i-1} \end{array} \right\} CV_{ref}^2 \quad (6)$$

*N*th comparison: In the *N*th comparison, the reference voltage of the last capacitor in the lower side is switched from *gnd* to V_{cm} while the other one (on the higher side) remains unchanged. The output voltages and switching energy are found to be

$$\left\{ \begin{array}{l} V_{DACP}(N) = V_{ip} - D_1 \frac{V_{ref}}{2} + \sum\limits_{j=2}^{N-1} (1 - D_j) \frac{V_{ref}}{2^j} \\ V_{DACN}(N) = V_{in} - (1 - D_1) \frac{V_{ref}}{2} + \sum\limits_{j=2}^{N-1} D_j \frac{V_{ref}}{2^j} \end{array} \right. \quad (7)$$

$$E_N = \left\{ \begin{array}{l} D_1(1 - D_{N-1}) \left[2^{-2} - 2^{-N} - \sum\limits_{j=1}^{N-2} (1 - D[j]) 2^{-j-1} \right] \\ +(1 - D_1) D_{N-1} \left[2^{-2} - 2^{-N} - \sum\limits_{j=1}^{N-2} D[j] 2^{-j-1} \right] \end{array} \right\} CV_{ref}^2 \quad (8)$$

The average switching energy of the switching scheme is derived as

$$E_{average} = \overline{\sum_{D_1 D_2 \cdots D_N = 00\cdots0}^{11\cdots1} \left(\sum_{i=1}^{N} E_i \right)} = \left(2^{N-4} - 2^{-1} + 2^{-4} \right) CV_{ref}^2 \quad (9)$$

Figure 5 shows switching energy at each output code for different switching schemes. The average switching energy of the switching scheme used for 10-bit SAR ADC is 63.56 CV_{ref}^2. Compared with the conventional switching scheme [11], the used switching scheme [15] and Capacitor-Splitting [8], Set-and-down [10], and V_{cm}-based [9] schemes reduce the switching energy by 95.34%, 37.48%, 81.26%, and 87.52%, respectively. Figure 6 presents the 500-run Monte Carlo simulation results of the proposed DAC switching scheme with unit capacitor mismatch of σ_u/C_u = 1%. The RMS DNL and the RMS INL of the proposed DAC switching scheme are 0.325 LSB and 0.326 LSB, respectively.

2.2. Automatic ON/OFF Comparator

A low-power two-stage full dynamic comparator is reported in [18]. In order to save more power consumption of the comparator, an automatic ON/OFF clock circuit is added to the comparator. As shown in Figure 7a, the comparator consists of automatic ON/OFF clock circuit, dynamic preamplifier stage, and dynamic latch stage. In the dynamic preamplifier stage, V_{DACP} and V_{DACN} are the output signals of the DAC capacitor array and are connected to the differential inputs of the comparator. *AP* and *AN* are differential outputs of the dynamic preamplifier stage. In the dynamic latch stage, *COMP* and *COMN* are the comparison results, which are obtained by *AP*, *AN*, and *CCLK* driving the latch. In the process of result latching, no power-to-ground current path is formed, so the comparator only has a dynamic power supply. The automatic ON/OFF clock circuit generates the clock for comparator operation. When the $\overline{RST}$ is high or P_{10} + N_{10} (P_{10} and N_{10} are the 10th comparison result that latched in SAR logic) is high, there is no clock output, the comparator is in the OFF state, and the comparator has no power consumption. Figure 7b shows the timing diagram of the comparator.

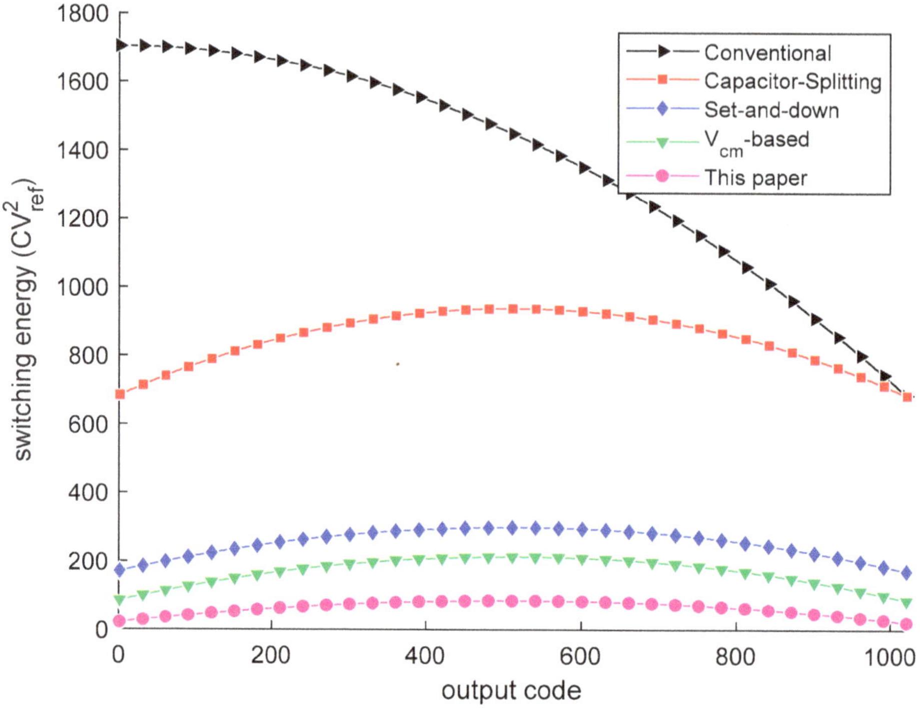

Figure 5. Switching energy against output codes. The black [11], red [8], blue [10], green [9], and magenta curves are switching energy.

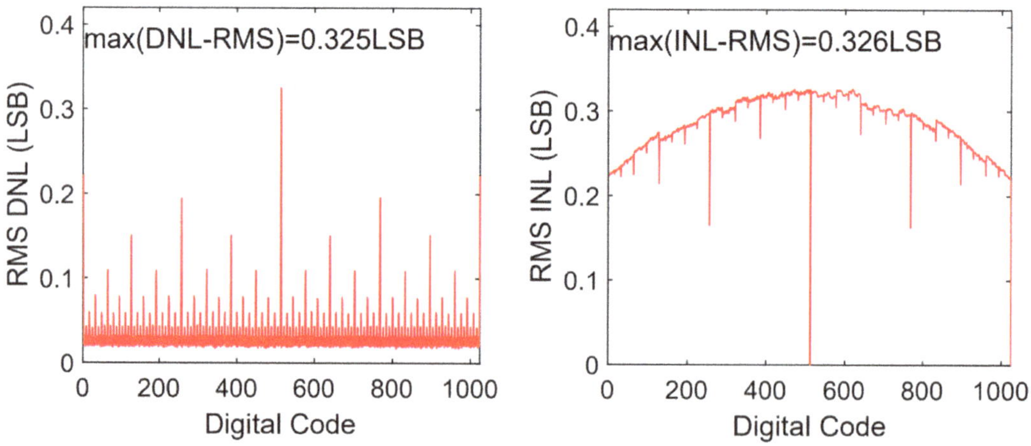

Figure 6. DNL and INL versus output code of the proposed switching scheme.

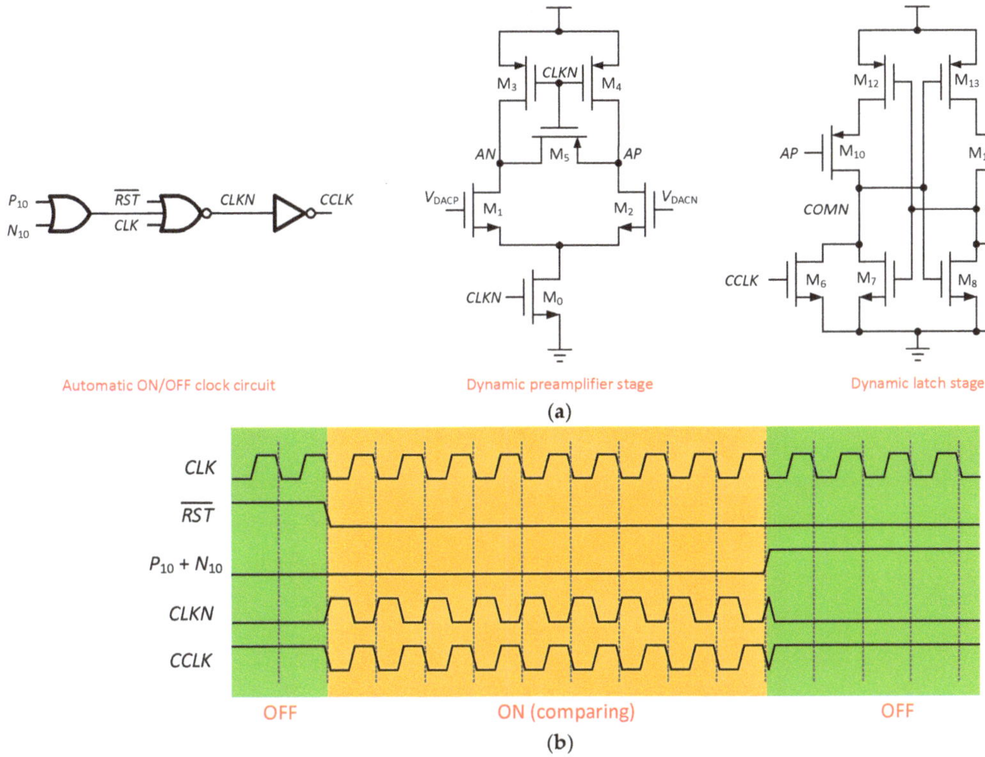

Figure 7. Automatic ON/OFF comparator. (**a**) Schematic diagram; (**b**) Timing diagram.

2.3. Automatic ON/OFF SAR Logic

As shown in Figure 8, the automatic ON/OFF SAR logic consists of an automatic ON/OFF clock circuit, a sequencer, and a data register. The sequencer is a shift register that shifts the set signal through a series of D flip-flops. The set signal is then used to activate the Latch in the data register. When the last D flip-flop in the sequencer is triggered, the sequencer will be reset and await the next conversion cycle. The data register is composed of dynamic latches, which can latch the differential outputs of the comparator and output differential data. Differential output makes DAC logic circuit simpler. The automatic ON/OFF clock circuit is used to provide the drive clock signal for the shift register. Drive clock is only ON while comparison results are being latched, thus reducing SAR logic power consumption. Figure 8b shows the timing diagram of the SAR logic.

2.4. DAC Driver Circuit

As shown in Figure 9, each capacitor requires two reference voltages. The reference voltages of C_2 to C_9 capacitors are V_{ref} and *gnd*, and the drive circuit can be realized by CMOS inverter. The reference voltage of C_1 capacitor is V_{cm} and *gnd*, and the driving circuit adopts a hybrid structure of CMOS transmission gate and CMOS inverter circuit.

2.5. Capacitor Array

Figure 10 illustrates the floorplan of the capacitor array DAC for a single side. Both sides have identical layout design. DAC capacitors and dummy capacitors are represented by squares (unit capacitors) in different colors. DAC capacitors are surrounded by dummy capacitors to minimize the proximity effects and second-order lithographic errors. Additionally, a common centroid layout is used to reduce parasitic effects.

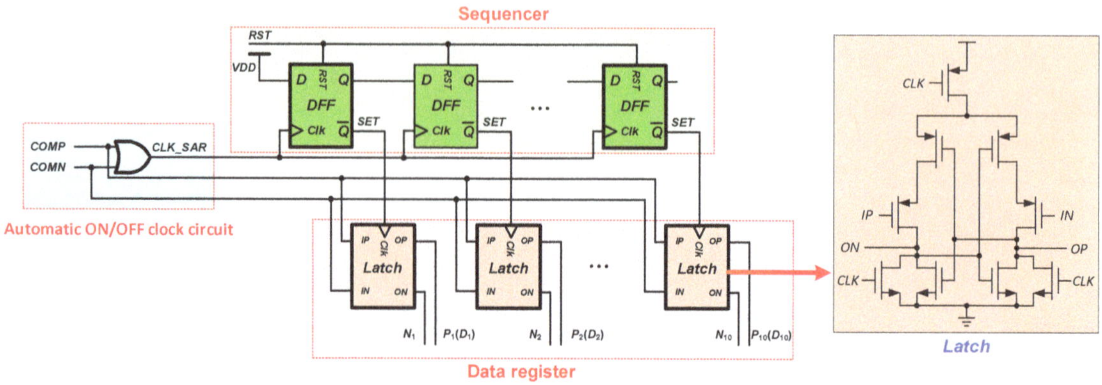

(a)

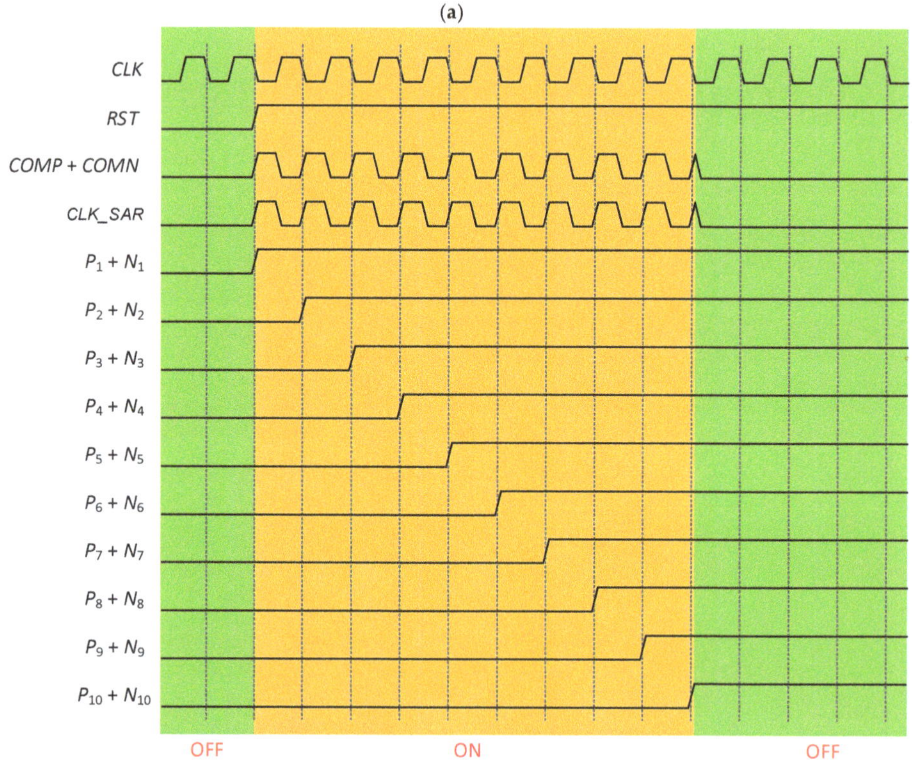

(b)

Figure 8. Automatic ON/OFF SAR logic. (**a**) Block diagram; (**b**) Timing diagram.

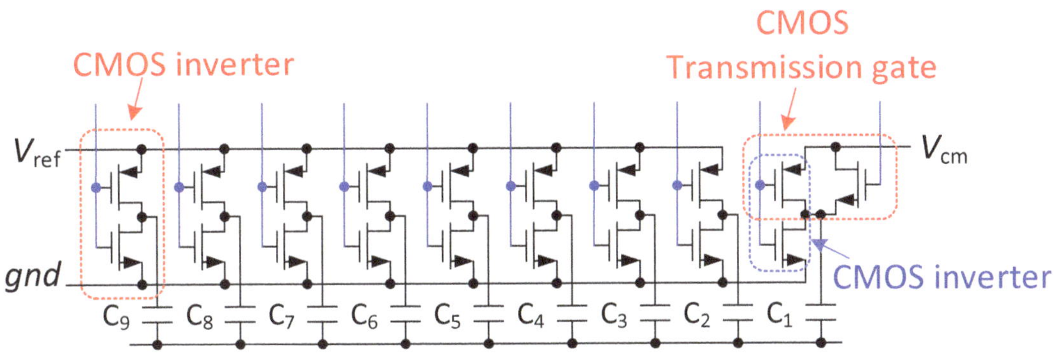

Figure 9. DAC driver circuit.

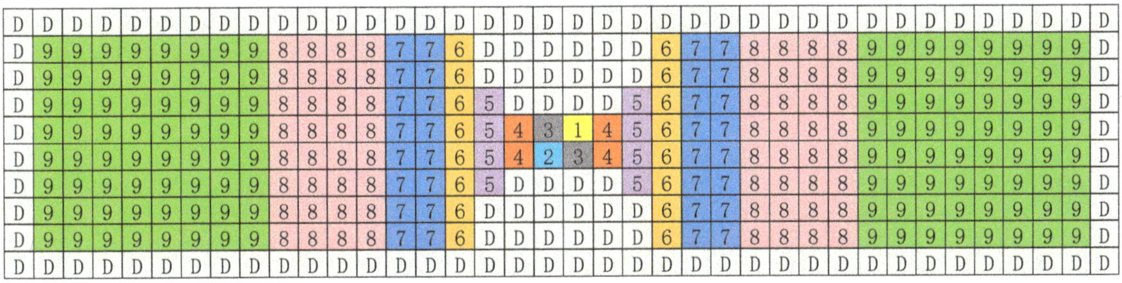

Figure 10. The layout floorplan of the DAC capacitor array.

3. Results

Figure 11 shows the chip micrograph of the ADC prototype fabricated in 180 nm CMOS process with a core area of 0.15 mm^2 (500 μm × 300 μm). Figure 12 shows the measurement environment. The power supply device provides power and reference voltage for the SAR ADC chip. The FPGA generates timing signals to control the work of SAR ADC chip. The signal generator generates a differential sinusoidal signal and connects to the analog signal input of the SAR ADC chip. The logic analyzer is connected to the digital signal output of the SAR ADC chip and collects the output data. The PC takes out the measured data from the logic analyzer and then analyzes various performance indicators of the chip. As shown in Figure 13, the signal-to-noise-and-distortion ratio (SNDR) is 45.15 dB, and the effective number of bits (ENOB) is 7.2 bit at 1.8 V supply and 1 kS/s sampling rate. The SAR ADC consumes 7.6 nW, and the calculated FoM is 51.7 fJ/conv.-step. The Figure-of-Merit (*FoM*) was calculated from the following equation:

$$FoM = \frac{Power}{2^{ENOB} \times f_{samping}} \tag{10}$$

Table 2 shows the performance comparison between the proposed ADC and other ADCs. Compared with another 180 nm ADC, the proposed ADC has a convincing performance. If the proposed SAR ADC uses the 65 nm process, it may have better performance.

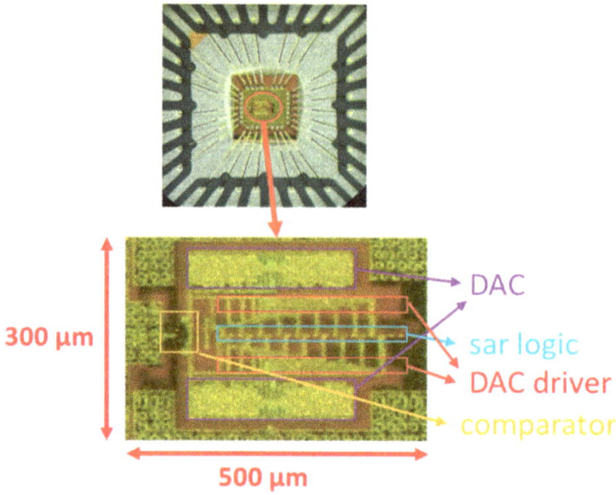

Figure 11. Chip micrograph.

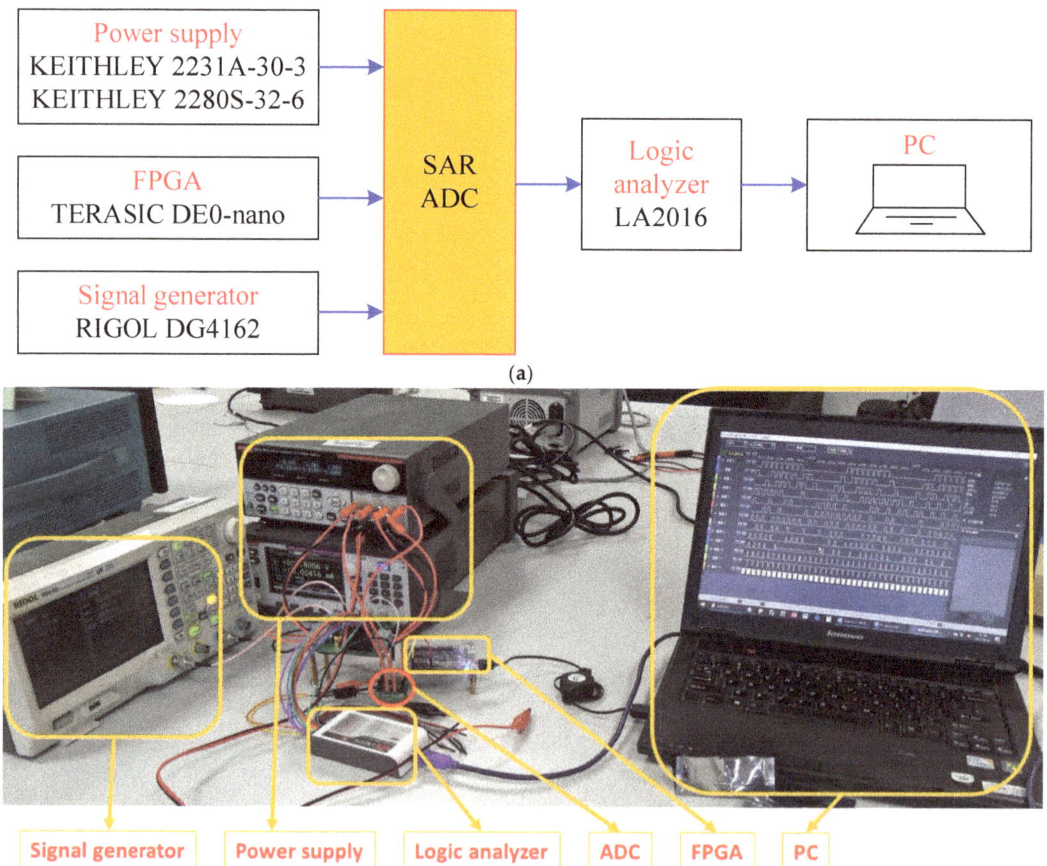

(a)

(b)

Figure 12. Measurement environment. (**a**) Block diagram; (**b**) Photo.

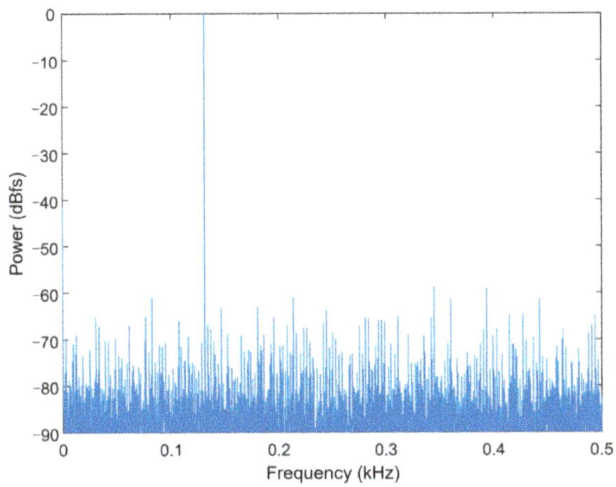

Figure 13. FFT spectrum.

Table 2. Performance comparison.

Specification	[19]	[20]	[21]	This Work
Technology (nm)	65	180	65	180
Resolution (bit)	14	10	8	10
Supply Voltage (V)	0.8	1	0.6	1.8
Sampling Rate (kS/s)	10	1	0.5	1
Power (nW)	1980	120	1.8	7.6
ENOB (bit)	12.5	9.76	7.14	7.2
FoM (fJ/conv.-step)	34.2	138.4	25.5	51.7

4. Conclusions

This paper has presented a low-power SAR ADC for biomedical applications. The ADC uses an energy-efficient, low-complexity switching scheme to reduce power consumption. Because of the top-plate sampling and level-shift-gnd operations, the switching scheme did not consume energy in the first and second comparisons. Thanks to the use of V_{cm} for the last capacitor, the total capacitance is reduced by half, so the energy consumption of DAC is also reduced. In addition, because of the automatic ON/OFF technology, the comparator and SAR logic only generate energy consumption during operation. The proposed SAR ADC achieves FoM of 51.7 fJ/conv.-step at 1.8 V supply and 1 KS/s sampling rate. If the ADC adopts a low-voltage design method, more energy consumption will be saved. The proposed low-power SAR ADC is suitable for biomedical applications.

Author Contributions: Conceptualization, Y.H.; methodology, Y.H.; software, Y.H. and B.T.; validation, Y.H., B.T. and L.H.; formal analysis, Y.H.; investigation, Y.H. and H.L.; resources, Y.H., B.L. and Z.W.; data curation, Y.H., B.T. and L.H.; writing—original draft preparation, Y.H.; writing—review and editing, L.H., B.L. and X.L.; visualization, Y.H., B.T. and L.H.; supervision, Y.H.; project administration, Y.H. and Z.W.; funding acquisition, B.L. and Y.H. All authors have read and agreed to the published version of the manuscript.

Funding: This research was funded by the National Natural Science Foundation of China (no. 60976026), the Key Field Project of Colleges and Universities in Guangdong Province (no. 2021ZDZX1081), the Key Project of Social Welfare and Basic Research Project in Zhongshan City (2021B2020), the Construction Project of Professional Quality Engineering in 2020 (no. YLZY202001), and the Construction Project of Professional Quality Engineering in 2021 (no. JD202101).

Data Availability Statement: Not applicable.

Conflicts of Interest: The authors declare no conflict of interest.

References

1. Weibo, H.; Yen-Ting, L.; Tam, N.; Lie, D.Y.C.; Ginsburg, B.P. An 8-Bit Single-Ended Ultra-Low-Power SAR ADC With a Novel DAC Switching Method and a Counter-Based Digital Control Circuitry. *IEEE Trans. Circuits Syst. I Regul. Pap.* **2013**, *60*, 1726–1739. [CrossRef]
2. Aneesh, K.; Manoj, G.; Sam, S.S. Design Approaches of Ultra-Low Power SAR ADC for Biomedical Systems—A Review. *J. Circuits Syst. Comput.* **2022**, *31*, 2230009. [CrossRef]
3. Mao, W.; Li, Y.; Heng, C.; Lian, Y. A Low Power 12-bit 1-kS/s SAR ADC for Biomedical Signal Processing. *IEEE Trans. Circuits Syst. I Regul. Pap.* **2019**, *66*, 477–488. [CrossRef]
4. Hou, Y.; Qu, J.; Tian, Z.; Atef, M.; Yousef, K.; Lian, Y.; Wang, G. A 61-nW Level-Crossing ADC with Adaptive Sampling for Biomedical Applications. *IEEE Trans. Circuits Syst. II Express Briefs* **2019**, *66*, 56–60. [CrossRef]
5. Kamata, T.; Ueda, M.; Hirai, Y.; Tani, S.; Asano, T.; Isami, S.; Kurata, T.; Tatsumi, K.; Matsuoka, T. An analog front-end employing 87 dB SNDR stochastic SAR-ADC for a biomedical sensor. In Proceedings of the 2017 15th IEEE International New Circuits and Systems Conference (NEWCAS), Strasbourg, France, 25–28 June 2017; pp. 301–304. [CrossRef]
6. Zhangming, Z.; Yuhua, L. A 0.6-V 38-nW 9.4-ENOB 20-kS/s SAR ADC in 0.18-um CMOS for Medical Implant Devices. *IEEE Trans. Circuits Syst. I Regul. Pap.* **2015**, *62*, 2167–2176. [CrossRef]
7. Byung-Geun, L. Power and Bandwidth Scalable 10-b 30-MS/s SAR ADC. *IEEE Trans. Very Large Scale Integr. VLSI Syst.* **2015**, *23*, 1103–1110. [CrossRef]
8. Ginsburg, B.P.; Chandrakasan, A.P. An energy-efficient charge recycling approach for a SAR converter with capacitive DAC. In Proceedings of the 2005 IEEE International Symposium on Circuits and Systems, Kobe, Japan, 23–26 May 2005; pp. 184–187. [CrossRef]
9. Yan, Z.; Chi-Hang, C.; Chio, U.F.; Sai-Weng, S.; Seng-Pan, U.; Martins, R.P.; Maloberti, F. A 10-bit 100-MS/s Reference-Free SAR ADC in 90 nm CMOS. *IEEE J. Solid-State Circuits* **2010**, *45*, 1111–1121. [CrossRef]
10. Liu, C.C.; Chang, S.J.; Huang, G.Y.; Lin, Y.Z. A 10-bit 50-MS/s SAR ADC With a Monotonic Capacitor Switching Procedure. *IEEE J. Solid-State Circuits* **2010**, *45*, 731–740. [CrossRef]
11. Johns, D.; Martin, K. *Analog Integrated Circuit Design*; John Wiley & Sons, Inc.: New York, NY, USA, 1997.
12. Hill, N.J.; Gupta, D.; Brunner, P.; Gunduz, A.; Adamo, M.A.; Ritaccio, A.; Schalk, G. Recording Human Electrocorticographic (ECoG) Signals for Neuroscientific Research and Real-time Functional Cortical Mapping. *J. Vis. Exp.* **2012**, e3993. [CrossRef] [PubMed]
13. Wang, J.; Tang, L.; Bronlund, J.E. Surface EMG Signal Amplification and Filtering. *Int. J. Comput. Appl.* **2013**, *82*, 15–22. [CrossRef]
14. Kumar, J.S.; Bhuvaneswari, P.T.V. Analysis of Electroencephalography (EEG) Signals and Its Categorization—A Study. *Procedia Eng.* **2012**, *38*, 2525–2536. [CrossRef]
15. Hu, Y.; Yi, Z.; He, Z.; Li, B. Energy-efficient, area-efficient, high-accuracy and low-complexity switching scheme for SAR ADC. *IEICE Electron. Expr.* **2017**, *14*, 20170428. [CrossRef]
16. Hu, Y.; Liu, A.; Li, B.; Wu, Z. Closed-loop charge recycling switching scheme for SAR ADC. *Electron. Lett.* **2017**, *53*, 66–68. [CrossRef]
17. Hu, Y.; Chen, L.; Chen, H.; Wen, Y.; Zhang, H.; Wu, Z.; Li, B. A 7.6-nW 1-kS/s 10-bit SAR ADC for Biomedical Applications. In Proceedings of the 2019 IEEE Asia Pacific Conference on Circuits and Systems (APCCAS), Bangkok, Thailand, 11–14 November 2019; pp. 129–132. [CrossRef]
18. Harpe, P.; Cantatore, E.; van Roermund, A. A 10b/12b 40 kS/s SAR ADC With Data-Driven Noise Reduction Achieving up to 10.1b ENOB at 2.2 fJ/Conversion-Step. *IEEE J. Solid-State Circuits* **2013**, *48*, 3011–3018. [CrossRef]
19. Zhang, D.; Alvandpour, A. A 12.5-ENOB 10-kS/s Redundant SAR ADC in 65-nm CMOS. *IEEE Trans. Circuits Syst. II Express Briefs* **2016**, *63*, 244–248. [CrossRef]
20. Wenliang, G.; Guoxing, W.; Kuan-Ting, L.; Kea-Tiong, T. A 10-bit 1kS/s-30kS/s successive approximation register analog-to-digital converter for biological signal acquisition. In Proceedings of the 2013 6th International Conference on Biomedical Engineering and Informatics, Hangzhou, China, 16–18 December 2013; pp. 403–407. [CrossRef]
21. Yen-Po, C.; Dongsuk, J.; Yoonmyung, L.; Yejoong, K.; Zhiyoong, F.; Inhee, L.; Langhals, N.B.; Kruger, G.; Oral, H.; Berenfeld, O.; et al. An Injectable 64 nW ECG Mixed-Signal SoC in 65 nm for Arrhythmia Monitoring. *IEEE J. Solid-State Circuits* **2015**, *50*, 375–390. [CrossRef]

micromachines

MDPI

Article

Simple Immunosensor Based on Carboxyl-Functionalized Multi-Walled Carbon Nanotubes @ Antimony-Doped Tin Oxide Composite Membrane for Aflatoxin B_1 Detection

Guanglei Chu [1,2,3,4], Zengning Liu [3], Yanyan Zhang [1,3,*], Yemin Guo [1,3], Xia Sun [1,3] and Ming Li [1,2,3,*]

1 School of Agriculture Engineering and Food Science, Shandong University of Technology, No. 12 Zhangzhou Road, Zibo 255049, China; chuguanglei1994@163.com (G.C.); gym@sdut.edu.cn (Y.G.); sunxia2151@sina.com (X.S.)
2 Hunan Agricultural Equipment Research Institute, Hunan Academy of Agricultural Sciences, No. 120 Donghu Road, Changsha 410125, China
3 Shandong Provincial Engineering Research Center of Vegetable Safety and Quality Traceability, No. 12 Zhangzhou Road, Zibo 255049, China; liuzengning@163.com
4 College of Design and Engineering, National University of Singapore, No. 21 Lower Kent Ridge Road, Singapore 119077, Singapore
* Correspondence: zyyan1104@163.com (Y.Z.); limingcs@hotmail.com (M.L.); Tel.: +86-533-2786558 (Y.Z. & M.L.)

Abstract: This paper presents a novel nano-material composite membrane for detecting aflatoxin B_1 (AFB_1). The membrane is based on carboxyl-functionalized multi-walled carbon nanotubes (MWCNTs-COOH) @ antimony-doped tin oxide (ATO)-chitosan (CS). To prepare the immunosensor, MWCNTs-COOH were dissolved in the CS solution, but some MWCNTs-COOH formed aggregates due to the intertwining of carbon nanotubes, blocking some pores. ATO was added to the solution containing MWCNTs-COOH, and the gaps were filled by adsorbing hydroxide radicals to form a more uniform film. This greatly increased the specific surface area of the formed film, resulting in a nano-composite film that was modified on screen-printed electrodes (SPCEs). The immunosensor was then constructed by immobilizing anti-AFB_1 antibodies (Ab) and bovine serum albumin (BSA) on an SPCE successively. The assembly process and effect of the immunosensor were characterized using scanning electron microscopy (SEM), differential pulse voltammetry (DPV), and cyclic voltammetry (CV). Under optimized conditions, the prepared immunosensor exhibited a low detection limit of 0.033 ng/mL with a linear range of 1×10^{-3}–1×10^{3} ng/mL. The immunosensor demonstrated good selectivity, reproducibility, and stability. In summary, the results suggest that the MWCNTs-COOH@ATO-CS composite membrane can be used as an effective immunosensor for detecting AFB_1.

Keywords: immunosensor; aflatoxin B_1; multi-walled carbon nanotubes; antimony-doped tin oxide; field real-time detection

check for updates

Citation: Chu, G.; Liu, Z.; Zhang, Y.; Guo, Y.; Sun, X.; Li, M. Simple Immunosensor Based on Carboxyl-Functionalized Multi-Walled Carbon Nanotubes @ Antimony-Doped Tin Oxide Composite Membrane for Aflatoxin B_1 Detection. *Micromachines* **2023**, *14*, 996. https://doi.org/10.3390/mi14050996

Academic Editors: Xiao Xiao and Gang Ge

Received: 20 March 2023
Revised: 6 April 2023
Accepted: 8 April 2023
Published: 3 May 2023

1. Introduction

Aflatoxin (AFT), a potent natural carcinogen, is a biologically active secondary metabolite mainly produced by Aspergillus flavus, Aspergillus parasiticus, and Aspergillus oryzae [1]. Among them, Aflatoxin B_1 (AFB_1) is the most toxic and widespread, posing a significant threat to human health [2]. It is almost impossible to avoid AFB1 in agricultural products, and its continuous consumption, even at low levels, can lead to a dramatic increase in the incidence of cancer and other diseases, particularly in developing countries, where AFB1 is associated with many types of cancers. Therefore, the maximum residue levels of AFB1 in foods must be strictly controlled. Traditional methods of AFB1 detection are time-consuming and not suitable for on-site monitoring of agricultural products. Hence, there is an urgent need to develop rapid and quantitative methods and instruments for AFB1 detection to achieve effective control of its levels in agricultural products.

To safeguard food safety and protect human health, a range of instrument detection techniques have been developed, such as thin-layer chromatography (TLC) [3], high-performance liquid chromatography (HPLC) [4], and HPLC-mass spectrometry (MS) [5]. Although these methods offer high detection accuracy and stability, their complex and time-consuming operations have hindered their advancement in the realm of rapid detection [6]. Hence, the need of the hour is to devise a simple, swift, and efficient means of detecting AFB1. In recent years, immunosensors have emerged as a promising detection method that leverages the specific binding of antigens and antibodies to detect small molecules [7]. Owing to their high detection accuracy, low cost, time efficiency, and ease of use, a range of immunosensors have been deployed to detect AFTs [8]. Among the electrochemical signal transduction techniques, electrochemiluminescence immunosensors [9] and electrochemical immunosensors [10] have exhibited high sensitivity and are widely applied in diverse immune reactions using voltammetry and amperometry, which are the most prevalent and widely adaptable methods [11]. Enhancing the current response of the immunosensor through the incorporation of nanomaterials represents an effective means of heightening its sensitivity. Nanomaterials such as gold nanoparticles [12], Fe_3O_4 [13], and graphene oxide [14] are typically employed to modify electrodes. This incorporation not only improves the transmission efficiency on the surface of the electrode but also amplifies the specific surface area.

Chitosan, a naturally-occurring alkaline polysaccharide compound, is also known as polyglucosamine (1-4)-2-amino-B-D glucose. It is obtained by the deacetylation of chitin, and is highly versatile due to its abundant amino groups. When chitosan is dissolved in acidic solution, these amino groups become free and protonated, giving it polycationic electrolyte properties. This remarkable feature allows chitosan to serve as a highly effective dispersant, capable of chelating various heavy metal ions. In addition, chitosan exhibits exceptional anion adsorption capabilities and is naturally biodegradable, making it a highly attractive and eco-friendly option [15].

Multi-walled carbon nanotubes (MWCNTs) possess remarkable electrical conductivity, thermal conductivity, and mechanical properties, rendering them ideal for developing high-performance polymer composites with multifunctional capabilities. The unique structure and high specific surface area of MWCNTs allow for precise control of electronic properties through molecular adsorption, doping, and charge transfer [16]. Furthermore, the deposition of MWCNTs onto electrode surfaces can create a continuous, ordered arrangement on the electrode, establish a favorable micro-environment for immobilizing enzymes, aptamers, or antibodies, and provide a pathway for electron transfer [17]. However, nano-films of MWCNTs have the tendency to non-specifically adsorb small molecules, which may lead to inaccuracies in small molecule detection.

Nanostructured doped oxides have become a focal point in the realm of research due to their remarkable electrochemical properties [18]. Among them, antimony-doped tin oxide (ATO) stands out as a promising alternative, as it enhances the conductivity of the doping agent without compromising on high optical transparency [19]. Additionally, ATO also exhibits exceptional electrocatalytic performance [20]. When added to a solution of carboxyl-functionalized multi-walled carbon nanotubes (MWCNTs-COOH), ATO engulfs the aggregates of MWCNTs-COOH to generate novel nanocomposites. The integration of ATO results in a denser and more uniform nanocomposite film, which in turn reduces the interaction between MWCNTs-COOH and the reaction substrate, thus lowering the rate of electron transfer.

In this work, taking advantage of the combination of MWCNTs-COOH and ATO nanomaterials, we developed nanocomposite membranes and used them for the construction of the AFB_1 immunosensor. The aim of this work is to provide a sensitive and stable membrane for quantitative determination of AFB_1 in food and agricultural products.

2. Experimental

2.1. Apparatus

The cyclic voltammetry (CV) and differential pulse voltammetry (DPV) measurements were conducted using a CHI660D electrochemical workstation purchased from Shanghai Chenhua Co., Shanghai, China. The screen-printed carbon electrodes (SPCEs, TE100, d = 3 mm) used in the experiment were purchased from Zensor R&D (Taiwan, China). The morphology of the modified electrodes was observed using a scanning electron microscope (SEM) from the Netherlands. Ultrasonication was performed using a SK3300H ultrasonic cleaner from Shanghai, China, and the solution was blended using a PTR-35 SPC vortex mixer from Britain. All experiments were performed at room temperature.

2.2. Reagents and Materials

The antibodies (Abs) were procured from the Oil Crops Research Institute of the Chinese Academy of Agricultural Sciences (Wuhan, China). The antimony-doped tin oxide (ATO) with an OD of 20 nm and purity of 99% and carboxyl-functionalized multi-walled carbon nanotubes (MWCNTs-COOH) with an OD range of 10–20 nm were purchased from Beijing Gold Deco Island Co., Ltd. (Beijing, China). Potassium ferrocyanide ($K_3[Fe(CN)_6]$) and potassium ferricyanide ($K_4[Fe(CN)_6]$) were purchased from Yongda Chemical Reagent Co., Ltd. (Tianjin, China). The 0.01 M pH 7.2–7.4 phosphate buffer solutions (PBS) were procured from Beijing Solarbio Science & Technology Co., Ltd., Beijing, China. All other chemicals used were of analytical reagent grade. Ultrapure water (18.2 MΩ·cm) purified with an LS MK_2 PALL-water purification system was used for the preparation of all solutions. Chitosan (CS) was obtained from Sangon Biotech Co., Ltd. (Shanghai, China).

2.3. Preparation of MWCNTs-COOH @ ATO-CS Composites

A total of 0.2 g of CS was dissolved in 100 mL of 1% acetic acid solution, and magnetic stirring was continued for more than 8 h to prepare a 0.1% (*w*/*v*) CS solution. Then, 2 mg of MWCNTs-COOH powder and 1.2 mg of ATO powder were accurately weighed and dissolved in 4 mL of CS solution using 2 h of ultrasonic treatment and 2 h of mixing with a mixer until the suspension became uniform and stable. The resulting highly dispersed and dark grey suspension was the MWCNTs-COOH @ ATO-CS solution, which was stored in a refrigerator (4 °C) for the experiment.

2.4. Preparation of Immunosensor Based on SPCEs

2.4.1. Preparation of SPCEs

To prepare the electrode for modification, it was thoroughly cleaned by ultrasonic cleaning in NaOH and HCl solutions for 5 min each. The electrode was then washed with ultrapure water and dried with nitrogen gas. To further ensure cleanliness, the electrode was washed with anhydrous ethanol and dried again with nitrogen gas. The treated electrodes were then immersed in pH 5.0 PBS solution for 300 s with a potential of 1.75 V and subsequently scanned from 0.3 V to 1.25 V and from 0.3 V to −1.3 V until a steady state CV curve was obtained [21].

2.4.2. Preparation of AFB_1/BSA/Ab/MWCNTs-COOH @ ATO-CS/SPCEs Immunosensor

A total of 6 μL MWCNTs-COOH @ ATO-CS solution was applied to the pre-treated SPCE surface and allowed to dry at room temperature, creating the MWCNTs-COOH @ ATO-CS/SPCE. Next, 6 μL of Ab was added to the modified electrode surface to form Ab/MWCNTs-COOH@ATO-CS/SPCE. To create the BSA/Ab/MWCNTs-COOH@ATO-CS/SPCE, 6 μL of BSA was added to the nanomaterial-modified electrode surface and dried at −4 °C. The electrode was then stored at 4 °C for AFB_1 detection. The preparation process of the immunosensor is illustrated in Figure 1.

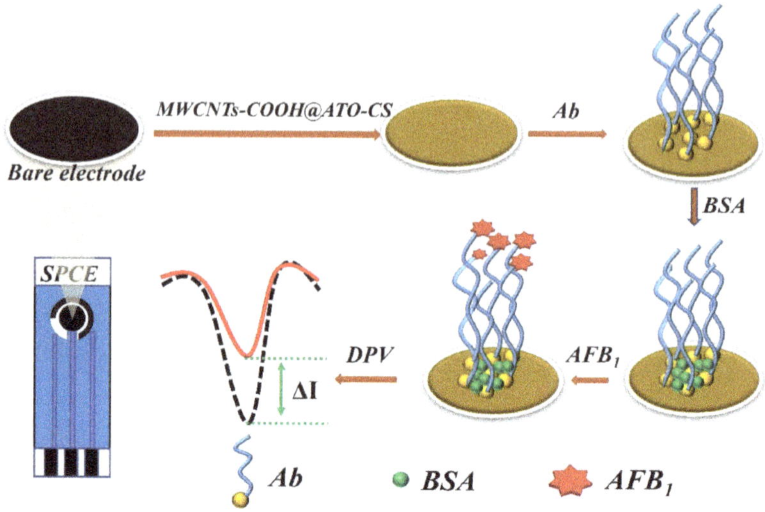

Figure 1. Schematic diagram of the preparation process of the immunosensor.

2.4.3. Electrochemical Measurements

The composites of MWCNTs-COOH @ ATO-CS were subjected to scanning electron microscopy (SEM) to analyze their properties. All electrochemical measurements were carried out in 15 mL of 0.1 M PBS (pH 7.0) that contained a 1:1 mixture of 5 mM $K_3[Fe(CN)_6]/K_4[Fe(CN)_6]$ as a redox probe and 0.1 M KCl at ambient temperature. CV was conducted within a potential range from −1.0 V to 1.0 V at a scan rate of 100 mV/s. Electrochemical differential pulse voltammetry (DPV) measurements were performed under the following conditions: voltage was scanned from −0.30 V to 0.60 V with a pulse height of 100 mV, while the step height and frequency were maintained at 4 mV and 15 Hz, respectively. To investigate the sensitivity and specificity of the proposed immunosensor, DPV was utilized. Additionally, parameters that affected the immunosensor response, such as the Ab concentration and response time, were optimized. Following optimization, the proposed immunosensor was used to detect AFB_1.

2.5. Immunosensor Specificity Analysis

In order to validate the specificity of the immunosensor, control experiments were conducted using five different fungal toxins, namely, AFM_1, α-zearalenone (α-ZEN), zearalenone (ZEN), ochratoxin A (OTA), and fumonisin B_1 (FB_1) at a concentration of 5 μg/mL. In addition, AFB_1 at a concentration of 100 ng/mL was also detected using the same sensor under unchanged experimental conditions. The detected currents from the control experiments were compared to observe any differences in response.

2.6. Peanut Oil Sample Pretreatment Method

Weigh precisely 5 g of peanut oil into a 100 mL triangular flask, and then add 25 mL of 10% methanol PBS solution. Different concentrations of AFB_1 should be added and the mixture should be shaken vigorously with an oscillator for 30 min to ensure thorough mixing. Next, the mixture should be centrifuged in a centrifuge for 10 min at 10,000 r/min. After centrifugation, 1 mL of the supernatant should be taken and 4 mL of methanol PBS solution should be added. Vigorous shaking for 5 s is necessary to ensure proper mixing, and then the sample should be stored in a refrigerator for further use.

3. Results and Discussion

3.1. Characterizations of Modified Electrodes

The modified electrodes were characterized using scanning electron microscopy (SEM) to investigate the morphology and structure of the prepared materials. The SEM images in Figure 2A revealed the presence of nano-scale cracks and holes on the electrode surface after the MWCNTs-COOH-CS composite film was modified onto the electrode. Figure 2B showed uniform density and size distribution of ATO-CS nanoparticles. The SEM image in Figure 2C depicted the morphology of MWCNTs-COOH @ ATO-CS. The MWCNTs-COOH formed clusters that were wrapped by a layer of ATO through electrostatic interaction, which increased the specific surface area of the nanocomposite material and made the material distribution more uniform. Compared to MWCNTs-COOH-CS or ATO-CS alone, MWCNTs-COOH @ ATO-CS was denser, which helped to increase the transfer rate of electrons. During the early stage of the formation of MWCNTs-COOH, they were prone to stacking together layer by layer, and many cracks or gaps appeared during the stacking process, resulting in the surface of MWCNTs-COOH being full of small holes. However, the addition of ATO compensated for these gaps and improved the uniformity of the nanocomposite, while also greatly enlarging its specific surface area. The EDS image in Figure 2D showed the analysis of the composition of C, O, Sn, Sb, and other elements in MWCNTs-COOH @ ATO-CS.

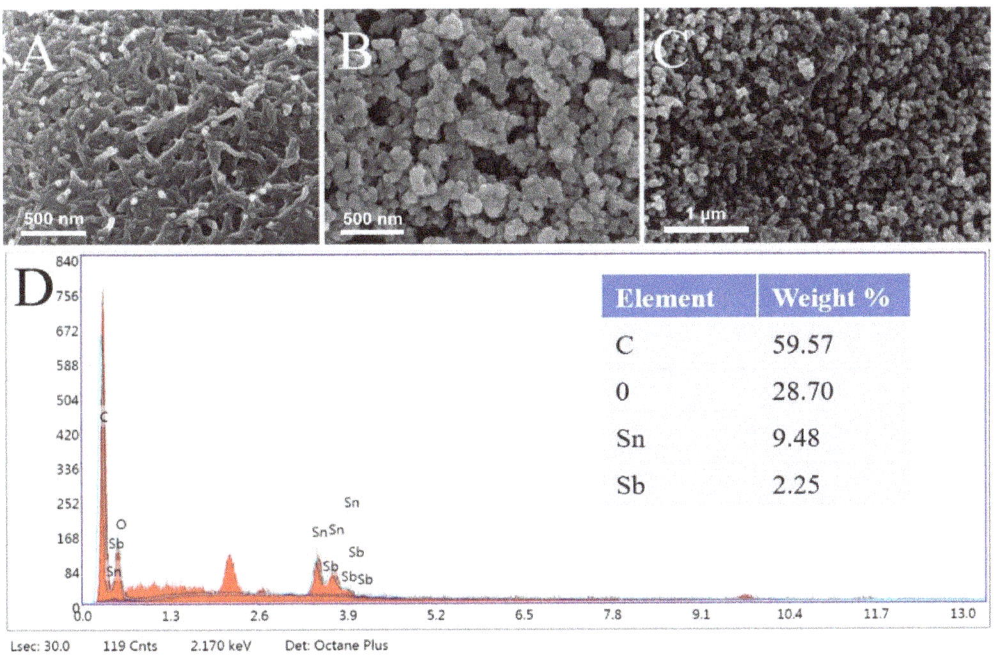

Figure 2. SEM images of (**A**) MWCNTs-COOH-CS, (**B**) ATO-CS, (**C**) MWCNTs-COOH @ ATO-CS, (**D**) EDS image of MWCNTs-COOH @ ATO-CS.

In addition, the stability of the MWCNTs-COOH @ ATO modified electrode was also evaluated. Six consecutive tests were conducted on the modified SPCE using the nanomaterials. As depicted in Figure 3, the six curves exhibited a high degree of consistency, suggesting that the nanocomposites demonstrated excellent stability.

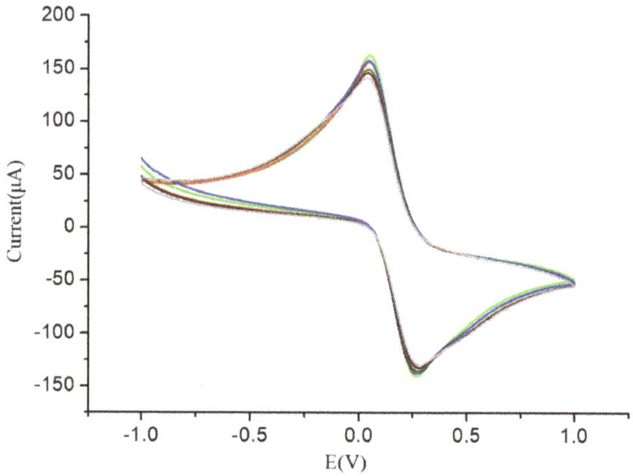

Figure 3. Stability of MWCNTs-COOH @ ATO.

3.2. *Electrochemical Behavior of the Modified Electrodes*

The assembly process of the immunosensor was investigated using CV and DPV techniques, where the bottom liquid contained 5 mM $[Fe(CN)_6]^{3-/4-}$ and 0.1 M KCl. As illustrated in Figure 4, the redox peaks of bare SPCEs were evident. Upon modification of the electrode with MWCNTs-COOH @ ATO, the peak current increased substantially to 160 μA (curve b), suggesting that the MWCNTs-COOH @ ATO possessed a large surface area and good electronic conductivity. Since antibodies lack conductivity as macromolecular proteins, their addition impeded electron transfer between $[Fe(CN)_6]^{3-/4-}$ and the electrode surface. BSA acted as a blocker, eliminating non-specific binding by closing specific sites on the surface of the electrode. Furthermore, since BSA had no conductivity, electron transfer was further obstructed, resulting in a rapid decrease in peak current. These results imply that the antibodies and BSA were successfully immobilized on the electrode surface.

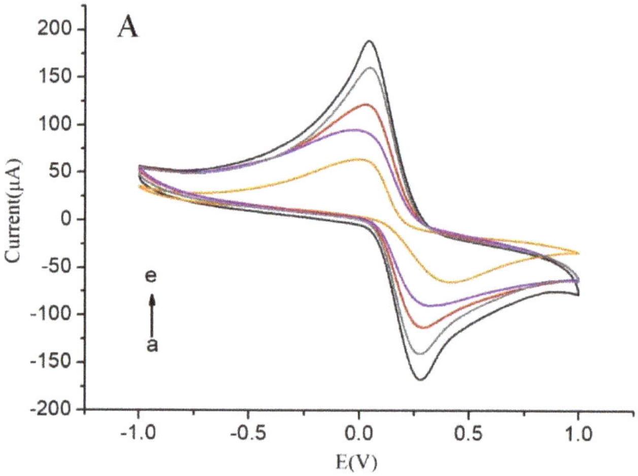

Figure 4. *Cont.*

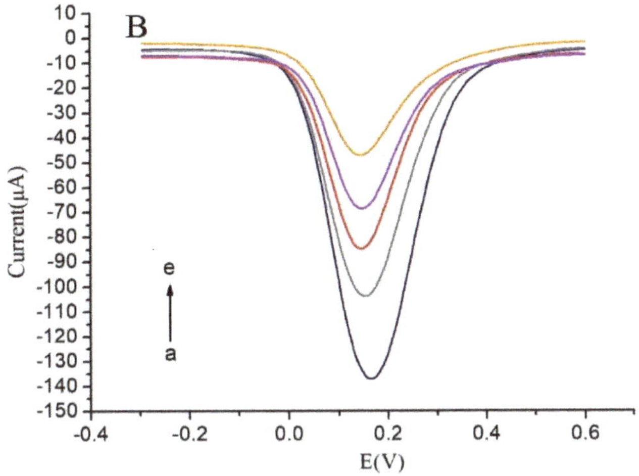

Figure 4. (**A**) CVs of modified screen-printed carbon electrodes; (**B**) DPVs of modified screen-printed carbon electrodes a: MWCNTs-COOH @ ATO-CS/SPCE, b: Ab/MWCNTs-COOH @ ATO-CS/SPCE, c: BSA/Ab/MWCNTs-COOH @ ATO-CS/SPCE, d: SPCE, e: AFB_1/BSA/Ab/MWCNTs-COOH @ ATO-CS/SPCE.

3.3. Optimization Parameters of the Immunosensor Performance

The impact of Ab concentration on the immunosensor's response was investigated and the results are presented in Figure 5. As shown in the figure, the peak current increased gradually with increasing Ab concentration until it reached a maximum value at 20 ng/mL. Subsequently, the response remained relatively stable as the Ab concentration continued to increase, indicating that the maximum number of Ab molecules had already been immobilized on the electrode surface. Thus, 20 ng/mL was identified as the optimal Ab concentration for fabrication of the immunosensor.

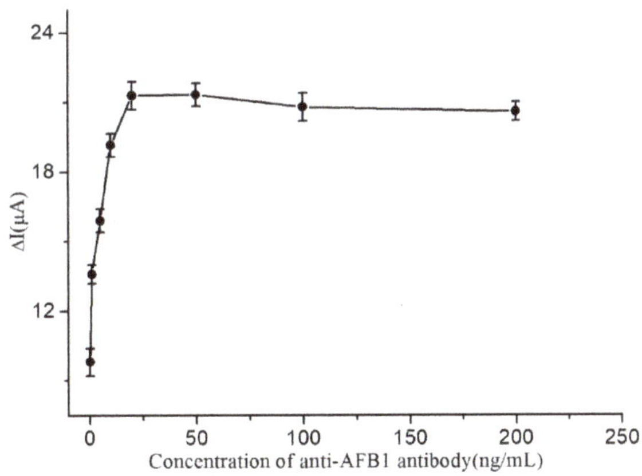

Figure 5. Influence of immunosensor concentration.

A study was conducted to investigate the effects of reaction time on the response of the immunosensor, and the findings are illustrated in Figure 6. It was observed that the peak current increased in a gradual manner with the increase in reaction time and reached its highest value at 40 min. Furthermore, the response became almost stable as the reaction time was further extended, indicating that the reaction time had reached a saturation point on the immunosensor. Based on these observations, 40 min was deemed the most suitable reaction time for the fabrication of the immunosensor.

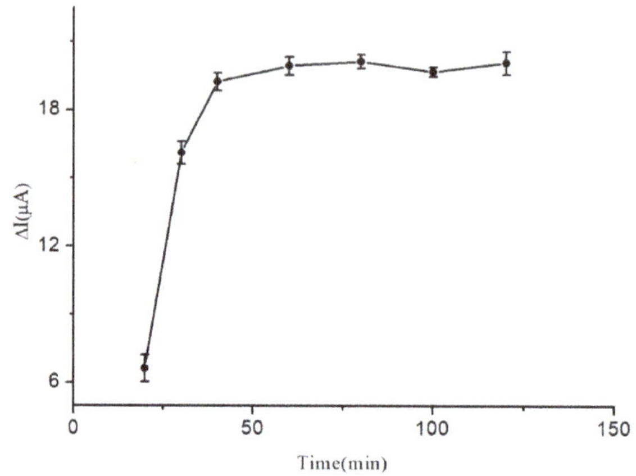

Figure 6. Influence of reaction time.

3.4. *Calibration Curve*

The unknown solution's AFB_1 concentration was deduced by analyzing the current response of the immunosensor at various AFB_1 concentrations immobilized on the electrode surface. A correlation between different AFB_1 concentrations and the immunosensor was established, revealing that as the AFB_1 concentration increased, the current response difference decreased.

Figure 7 demonstrates a remarkable linear relationship between the current variance and the corresponding AFB_1 concentration. With the ideal experimental conditions, the AFB_1 concentration could be accurately determined in the range of 1×10^{-3} to 1×10^{3} ng/mL using the linear equation y = 15.771 + 3.318x (R^2 = 0.995). The immunosensor also exhibited a remarkable detection limit of 0.03 ng/mL (S/N = 3).

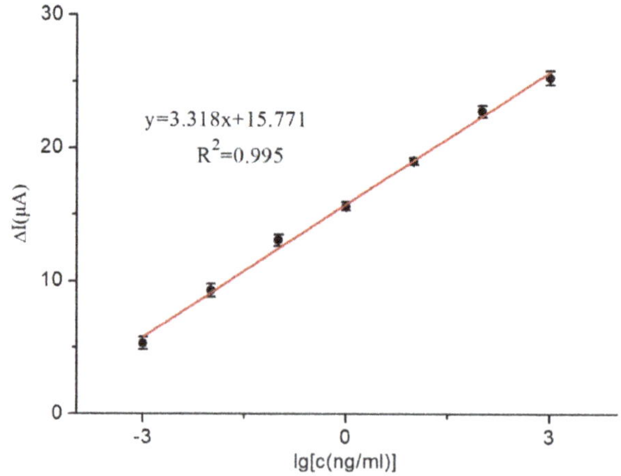

Figure 7. Relationship between current change and AFB_1 concentrations.

3.5. *Selectivity and Stability of Immunosensor*

The selectivity of the immunosensor was evaluated by comparing the sensing results of five different mycotoxins: α-zearalenone (α-ZEN), ochratoxin A (OTA), fumonisin B1 (FB_1), zearalenone (ZEN), and AFM1, all at concentrations of 5 μg/mL, while the concentration of

AFB_1 was maintained at 100 ng/mL. As illustrated in Figure 8, except for AFM_1, the current response of other toxins was lower than 5 μA. Although AFM_1 exhibited a current response of 7 μA, its concentration was 50 times higher than that of AFB_1. These findings indicate that the developed immunosensor possessed an excellent selectivity for practical applications.

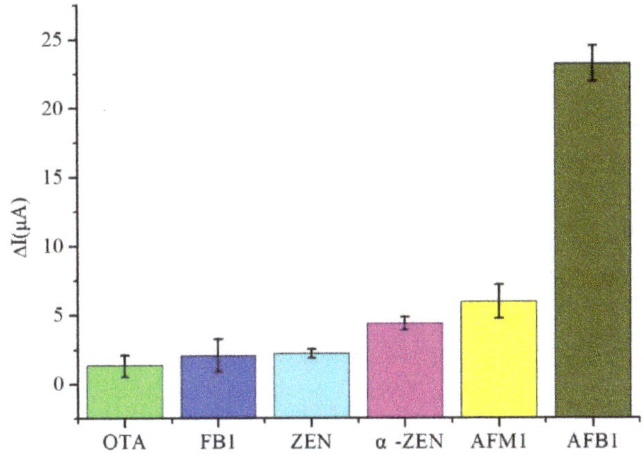

Figure 8. The selectivity for five kinds of mycotoxins (5 μg/mL): α-zearalenone (α-ZEN), ochratoxin A (OTA), fumonisin B_1 (FB_1), zearalenone (ZEN) and AFM_1.

To ensure the success of the immunosensor preparation, the stability of the prepared immunosensors was assessed as a crucial factor. Following optimized conditions, a total of 12 immunosensors were fabricated and stored at 4 °C for 15 days. Subsequently, these immunosensors were used to detect AFB_1 at a concentration of 100 ng/mL. The results revealed a slight decrease of approximately 9.2% in the measured current difference, indicating that the immunosensor exhibited good stability.

3.6. Determination of Spiked Recovery of AFB1 in Peanut Oil

To assess the accuracy of the immunosensor, spiked recoveries were measured in peanut oil samples after pretreatment. AFB_1 was added to the peanut oil samples at spiked concentrations of 10^{-1} ng/mL, 10 ng/mL, and 10^2 ng/mL. These experiments were conducted under optimal conditions, and the results are presented in Table 1. The results demonstrated spiked recoveries ranging from 95.15 to 111.60%, with relative standard deviations ranging from 2.3 to 5.3%. These findings suggest that the immunosensor exhibited good detection accuracy.

Table 1. Detection of peanut oil by immunosensor.

Sample	AFB_1 Addition(ng/mL)	Standard Current Difference (μA)	ΔI (μA)	RSD (%, *n* = 5)	Recovery Rate (%)
1	10^{-1}	5.82	5.6	5.2	95.15
2	10	9.14	10.2	3.2	111.60
3	10^2	12.45	12.08	2.3	96.10

3.7. Sensor Performance Comparison

Referring to Table 2, to conduct a comprehensive evaluation of the sensor developed in this study, we compared and assessed various performance indicators of immune sensors. Notably, our sensor boasts a wider detection range and a superior detection limit, which suggests its great potential for practical applications.

Table 2. Comparison of the performance of the sensor prepared in this study to other immunosensors.

Sensors	Detection Method	LOD	Linear Rang	Practical Samples	Ref.
Porous AuNPs/GCE	DPV	0.94 ng/mL	0.01–20 ng/mL	Glutinous rice/Corn/Rice	[22]
Au/Bi_2S_3/ERGO/CF	DPV	8 pg/mL	10 pg–20 ng/mL	Cornflour	[23]
MWCNTs/RTIL/Ab/AFB_1/GCE	EIS	0.03 ng/mL	0.1–10 ng/mL	Olive oils	[24]
BSA/Ab/MWCNTs-COOH @ ATO-CS/SPCEs	DPV	0.033 ng/mL	1×10^{-3}–1×10^{3} ng/mL	Peanut oil	This Work

GCE: glassy carbon electrode, ERGO: electrochemically reduced graphene oxide, CF: carbon fiber, RTIL: room temperature ionic liquid, EIS: electrochemical impedance spectroscopy.

4. Conclusions

This study aimed to develop an AFB_1 immunosensor using a nanocomposite material, MWCNTs @ ATO-CS, via an experimental approach. The nanocomposite film was created by incorporating multi-walled carbon nanotubes with nano-stannic oxide antimony and CS as a dispersant, resulting in a film with excellent adhesion and biocompatibility. SEM and CV were used to characterize the nanocomposite film, which exhibited a high level of uniformity, allowing for amplified current signal detection in the immunosensor and facilitating the immobilization of Ab. The immunosensor was constructed by immobilizing MWCNTs @ ATO-CS, Ab, and BSA on SPCE through the layer-wise self-assembly method. Each step of sensor fabrication was characterized using CV and DPV. Parameters, such as the loading of antibodies and the incubation time of the immune reaction, were optimized. The calibration curve was established under optimal conditions, showing a linear relationship for AFB_1 in the range of 10^{-3} to 10^3 ng/L with a linear equation of y = 3.318x + 15.771 (R^2 = 0.995). The immunosensor also demonstrated a limit of detection of 0.03 ng/L (S/N = 3). The specificity and stability of the immunosensor were verified, demonstrating excellent performance. Furthermore, the spiked recovery detection test in peanut oil samples illustrated the immunosensor's potential for practical sample detection.

Author Contributions: Conceptualization, Z.L.; methodology, Z.L.; software, G.C. and Z.L.; Data curation, G.C. and Z.L.; formal analysis, Y.Z. and Y.G.; investigation, X.S.; resources, X.S., Y.G. and M.L.; Visualization, G.C. and Z.L.; writing—original draft preparation, Z.L. and G.C.; writing—review and editing, Y.Z., Y.G., X.S. All authors have read and agreed to the published version of the manuscript.

Funding: This research received no external funding.

Institutional Review Board Statement: Not applicable.

Informed Consent Statement: Not applicable.

Data Availability Statement: Not applicable.

Conflicts of Interest: The authors declare no conflict of interest.

References

1. Sengul, U. Comparing determination methods of detection and quantification limits for aflatoxin analysis in hazelnut. *J. Food Drug Anal.* **2016**, *24*, 56–62. [CrossRef] [PubMed]
2. Liu, Y.; Liu, D.; Li, C.; Cui, S.; Yun, Z.; Zhang, J.; Wei, Y.; Sun, F. Chromatographic methods for rapid aflatoxin B_1 analysis in food: A review. *Crit. Rev. Food Sci. Nutr.* **2022**, 1–18. [CrossRef] [PubMed]
3. Li, Y.; Zhao, C.; Lu, C.; Zhou, S.; Tian, G.; He, L.; Bao, Y.; Fauconnier, M.L.; Xiao, H.; Zheng, J. Simultaneous determination of 14 bioactive citrus flavonoids using thin-layer chromatography combined with surface enhanced Raman spectroscopy. *Food Chem.* **2021**, *338*, 128115. [CrossRef]
4. Andrade, P.D.; Gomes da Silva, J.L.; Caldas, E.D. Simultaneous analysis of aflatoxins B_1, B_2, G_1, G_2, M_1 and ochratoxin A in breast milk by high-performance liquid chromatography/fluorescence after liquid-liquid extraction with low temperature purification (LLE-LTP). *J. Chromatogr. A* **2013**, *1304*, 61–68. [CrossRef]
5. Wang, L.; Wang, Z.; Gao, W.; Chen, J.; Yang, M.; Kuang, Y.; Huang, L.; Chen, S. Simultaneous determination of aflatoxin $B_{(1)}$ and ochratoxin A in licorice roots and fritillary bulbs by solid-phase extraction coupled with high-performance liquid chromatography-tandem mass spectrometry. *Food Chem.* **2013**, *138*, 1048–1054. [CrossRef]
6. Huang, Z.; Shu, Z.; Xiao, A.; Pi, F.; Li, Y.; Dai, H.; Wang, J. Determination of aflatoxin B_1 in rice flour based on an enzyme-catalyzed Prussian blue probe. *LWT* **2022**, *162*, 113500. [CrossRef]

7. Myndrul, V.; Viter, R.; Savchuk, M.; Koval, M.; Starodub, N.; Silamikelis, V.; Smyntyna, V.; Ramanavicius, A.; Iatsunskyi, I. Gold coated porous silicon nanocomposite as a substrate for photoluminescence-based immunosensor suitable for the determination of Aflatoxin B_1. *Talanta* **2017**, *175*, 297–304. [CrossRef]
8. Ma, H.; Sun, J.; Zhang, Y.; Xia, S. Disposable amperometric immunosensor for simple and sensitive determination of aflatoxin B 1 in wheat. *Biochem. Eng. J.* **2016**, *115*, 38–46. [CrossRef]
9. Yue, Q.; Li, X.; Fang, J.; Li, M.; Zhang, J.; Zhao, G.; Cao, W.; Wei, Q. Oxygen Free Radical Scavenger PtPd@PDA as a Dual-Mode Quencher of Electrochemiluminescence Immunosensor for the Detection of AFB1. *Anal. Chem.* **2022**, *94*, 11476–11482. [CrossRef]
10. Kunene, K.; Sayegh, S.; Weber, M.; Sabela, M.; Voiry, D.; Iatsunskyi, I.; Coy, E.; Kanchi, S.; Bisetty, K.; Bechelany, M. Smart electrochemical immunosensing of aflatoxin B_1 based on a palladium nanoparticle-boron nitride-coated carbon felt electrode for the wine industry. *Talanta* **2023**, *253*, 124000. [CrossRef]
11. Wen, W.; Yan, X.; Zhu, C.; Du, D.; Lin, Y. Recent Advances in Electrochemical Immunosensors. *Anal. Chem.* **2017**, *89*, 138–156. [CrossRef] [PubMed]
12. Abad-Gil, L.; Gismera, M.J.; Sevilla, M.T.; Procopio, J.R. Electrochemical sensing platform with gold nanoparticles capped by PDDA for benzyl alcohol determination. *Mikrochim. Acta* **2023**, *190*, 115. [CrossRef] [PubMed]
13. Taherimaslak, Z.; Amoli-Diva, M.; Allahyary, M.; Pourghazi, K. Magnetically assisted solid phase extraction using Fe_3O_4 nanoparticles combined with enhanced spectrofluorimetric detection for aflatoxin M1 determination in milk samples. *Anal. Chim. Acta* **2014**, *842*, 63–69. [CrossRef]
14. Sharma, A.; Kumar, A.; Khan, R. A highly sensitive amperometric immunosensor probe based on gold nanoparticle functionalized poly (3, 4-ethylenedioxythiophene) doped with graphene oxide for efficient detection of aflatoxin B_1. *Synth. Met.* **2018**, *235*, 136–144. [CrossRef]
15. Annu; Raja, A.-N. Recent development in chitosan-based electrochemical sensors and its sensing application. *Int. J. Biol. Macromol.* **2020**, *164*, 4231–4244. [CrossRef] [PubMed]
16. Gao, W.; Guo, J.; Xiong, J.; Smith, A.T.; Sun, L. Improving thermal, electrical and mechanical properties of fluoroelastomer/amino-functionalized multi-walled carbon nanotube composites by constructing dual crosslinking networks. *Compos. Sci. Technol.* **2018**, *162*, 49–57. [CrossRef]
17. Chen, D.; Liu, Z.; Fu, J.; Guo, Y.; Sun, X.; Yang, Q.; Wang, X. Electrochemical acetylcholinesterase biosensor based on multi-walled carbon nanotubes/dicyclohexyl phthalate modified screen-printed electrode for detection of chlorpyrifos. *J. Electroanal. Chem.* **2017**, *801*, 185–191. [CrossRef]
18. Rahman, M.M.; Ahmed, J.; Asiri, A.M. Development of Creatine sensor based on antimony-doped tin oxide (ATO) nanoparticles. *Sens. Actuators B Chem.* **2017**, *242*, 167–175. [CrossRef]
19. Tao, T.; Chen, C.; Qi, W.; Liang, B.; Yao, Y.; Lu, S.-G. Antimony doped tin oxide-coated $LiNi_{0.5}Co_{0.2}Mn_{0.3}O_2$ cathode materials with enhanced electrochemical performance for lithium-ion batteries. *J. Alloy. Compd.* **2018**, *765*, 601–607. [CrossRef]
20. Xu, Q.-C.; Zhang, Q.-Q.; Sun, X.; Guo, Y.-M.; Wang, X.-Y. Aptasensors modified by antimony tin oxide nanoparticle-chitosan based on interdigitated array microelectrodes for tetracycline detection. *RSC Adv.* **2016**, *6*, 17328–17335. [CrossRef]
21. Jiao, Y.; Jia, H.; Guo, Y.; Zhang, H.; Wang, Z.; Sun, X.; Zhao, J. An ultrasensitive aptasensor for chlorpyrifos based on ordered mesoporous carbon/ferrocene hybrid multiwalled carbon nanotubes. *RSC Adv.* **2016**, *6*, 58541–58548. [CrossRef]
22. Liu, B.; Peng, J.; Wu, Q.; Zhao, Y.; Shang, H.; Wang, S. A novel screening on the specific peptide by molecular simulation and development of the electrochemical immunosensor for aflatoxin B_1 in grains. *Food Chem.* **2022**, *372*, 131322. [CrossRef] [PubMed]
23. Ou, G.; Zhao, A.; Liao, H.; Zhang, Z.; Xiao, F. Au nanopartics decorated urchin-like Bi_2S_3 on graphene wrapped carbon fiber microelectrode: Towards electrochemical immunosensor for sensitive determination of aflatoxin B_1. *J. Electroanal. Chem.* **2023**, *909*, 117124. [CrossRef]
24. Yu, L.; Zhang, Y.; Hu, C.; Wu, H.; Yang, Y.; Huang, C.; Jia, N. Highly sensitive electrochemical impedance spectroscopy immunosensor for the detection of AFB1 in olive oil. *Food Chem.* **2015**, *176*, 22–26. [CrossRef] [PubMed]

micromachines

MDPI

Article

Cyclic Voltammetric-Paper-Based Genosensor for Detection of the Target DNA of Zika Virus

Anirudh Bishoyi [1], Md. Anish Alam [2], Mohd. Rahil Hasan [2], Manika Khanuja [3], Roberto Pilloton [4,*] and Jagriti Narang [2,*]

1 Department of Biotechnology & Microbiology, National College (Autonomous), Tiruchirapalli 620001, India
2 Department of Biotechnology, School of Chemical and Life Sciences, Jamia Hamdard, Hamdard Nagar, New Delhi 110062, India
3 Centre for Nanoscience and Nanotechnology, Jamia Millia Islamia, New Delhi 110025, India
4 Institute of Crystallography, National Research Council (IC-CNR), 00118 Rome, Italy
* Correspondence: roberto.pilloton@cnr.it (R.P.); jags_biotech@yahoo.co.in (J.N.)

Highlights:

- Recent report on paper-based genosensor for the diagnosis of ZIKV target DNA.
- Silver nanoparticles were characterized by XRD/UV-Vis/TEM.
- Cross-reactivity of genosensor was deduced.
- Electrochemical validation was performed through CV.

Abstract: Zika virus (ZIKV), a positive-sense single-stranded RNA virus, has been declared as the cause of a 'worldwide public health emergency' by the WHO since the year 2016. In cases of acute infections, it has been found to cause Guillain–Barre syndrome and microcephaly. Considering the tropical occurrence of the infections, and the absence of any proper treatments, accurate and timely diagnosis is the only way to control this infectious disease. Currently, there are many diagnostic methods under investigation by the scientific community, but they have some major limitations, such as high cost, low specificity, and poor sensitivity. To overcome these limitations, we have presented a low-cost, simple-to-operate, and portable diagnosis system for its detection by utilizing silver nanoparticles. silver nanoparticles were synthesized via chemical methods and characterization was confirmed by UV/TEM and XRD. The paper platform was synthesized using a graphene-based conductive ink, methylene blue as the redox indicator, and a portable potentiostat to perform the cyclic voltammetry to ensure true point-of-care availability for patients in remote areas.

Keywords: Zika virus; genosensor; paper electrodes; cross-reactivity; synthetic serum

Citation: Bishoyi, A.; Alam, M.A.; Hasan, M.R.; Khanuja, M.; Pilloton, R.; Narang, J. Cyclic Voltammetric-Paper-Based Genosensor for Detection of the Target DNA of Zika Virus. *Micromachines* **2022**, *13*, 2037. https://doi.org/10.3390/mi13122037

Academic Editors: Xiao Xiao, Gang Ge and Anas AlAzzam

Received: 22 August 2022
Accepted: 17 November 2022
Published: 22 November 2022

1. Introduction

A mosquito-borne flavivirus, ZIKV, has been the cause of multiple public health concerns and a long-running epidemic since 1947. Previously limited to occasional instances in Africa and Asia, the virus' outbreak in 2015 in Brazil suggested an immediate spread throughout the US. The WHO declared it a worldwide public health emergency in 2016 [1]. ZIKV has recently emerged as a major worldwide threat because it causes microcephaly in infants born to infected mothers. Epidemiological monitoring of infection has been hampered by the lack of reliable tests capable of distinguishing between Zika and other Flavivirus infections, especially the dengue virus. Various Zika-based biosensors have been developed, which offer advantageous features such as low detection limit, linear range, and stability, but have some limitations such as the use of the bulky three-electrode setup and high reagent/sample volume, which restrict their application as POC-based devices. ZIKV, like other Flaviviruses, is primarily an enveloped virus with a single-stranded RNA (ssRNA) genome of nearly 11Kb +ve polarity-based gene encoding a single polyprotein, bound by an icosahedral enclosure comprising 180 copies within every envelope (E) glycoprotein with about 500aa (amino acids), and approx. 75aa membrane protein and approx. 165aa precursor-membrane 'prM' protein [2,3] Symptoms include moderate influenza, neurological symptoms, and subclinical presentations in a child born to an infected mother, and Guillain–Barre syndrome (GBS) in adults [4]. There is currently no cure for ZIKV. The ZIKV may be detected quickly and easily, lowering the risk of infection and improving its management. Traditional ZIKV testing methods, such as virus isolation, identification of ZIKV-specific antibodies, ELISA, and PCR, have some drawbacks. There is a requirement for an immediate and uncomplicated technique for the diagnosis of ZIKV [5–10]. Scientists are very interested in developing an effortless and portable approach for diagnosing ZIKV. Screen-printed electrodes (SPEs) can be fabricated from a variety of materials, including plastic, paper, and tattoos, and they provide a lot of flexibility in terms of size, shape, dimensionality, and customizing options. SPEs could be especially useful in the development of new approaches and research. Screen printing is distinguished by its ease of production and application, as it does not necessitate the usage of cleanrooms and can be performed with low-cost components and even homemade setups. Conductive inks, an electrode-patterned mesh screen, a squeegee, and an oven are all that are needed for the screen-printing process [11].

Nanobioconjugates are nanomaterials that have been combined with biomolecules to form hybrid nano (bio) materials. Hundreds of functional biomolecules and tags immobilized on nanomaterials(nanobioconjugates) have been proven to boost signal enhancement in a variety of biosensors. Such hybrid systems outperform their individual components, with each component contributing a unique trait or function to the hybrid that the other does not [12]. The conjugation of AgNPs with single-stranded DNA (ssDNA) provides a simple way to create nanobioconjugates with exceptional capabilities for signal extension, improved sensitivity, and reduced LOD in bioassays [13]. Alves et al. [14] developed a genosensor based on the graphite carbon electrode. A DNA probe was immobilized covalently onto the sensing interface and hybridization was detected by the differential current response. The major advantage of the developed sensor was that it was applied in an experimental sample and interference with other flaviviruses was also studied. Another Zika-based genosensor was developed on SPEs by Cajigas et al. [12]. SPEs were modified with gold nano-architecture and Ru was used as a labeling agent. The sensor was also applied in serum samples, but the SPEs are very expensive and corrode easily, which limits their application in POC-based devices. Faria et al. [15] developed a disposable electrode, which offers many advantageous features such as being economical, having a low reagent volume, and no bulky electrode set up. The developed biosensor has some limitations, such as higher response time (1.5 h), and the sensor was not tested to an experimental sample. Paper-based carbon conductive SPEs modified with AgNPs were utilized as a platform for immobilization of the oligonucleotide probe of ZIKV. This substance is simple and economical, has a high reaction yield, an extensive surface area, and the ability to be functionalized, and is biocompatible. Thus, taking into account things such as enhanced performance, this effort represents a novel platform for detecting ZIKV, including the potential to fill the gap in the lack of approaches that sense the target at very low concentrations, while also delivering ZIKV analysis in the sample with stability, portability, and quick analysis. Therefore, PBG is a suitable substitute for easy, stable, low-cost, and quick identification. As a consequence of the high specificity of DNA, electrochemical DNA sensing approaches have received a lot of attention. The precision and sensitivity of DNA-based electrochemical biosensors has improved due to the use of nanomaterials for signal improvement. Specifically, the working electrode of PBG was modified with silver nanoparticles on account of its high electrical and thermal conductivity. Furthermore, the biosensor specific to the target TDNA of ZIKV was immobilized on the surface-modified working electrode of PBG and an electrochemical investigation was executed to verify the respective hybridization.

The current study aimed to develop an electrochemical paper-based device for the detection of ZIKA virus target DNA, which has many advantages such as low sample/reagent volume, ease of preparation, and more stable printing on paper versus plastic substrate, and the sensor was also tested in artificial serum samples to ensure sensor applicability in complex matrices. The use of a cutting-edge portable potentiostat limits the device's LOD, but this is mitigated by the fact that it can be replicated in any remote corner of the world, effectively making the device truly capable of providing POC diagnosis, which is critical taking the tropical presence of the disease under consideration.

The analytical performance of AgNP/PBG-based biosensor for the selective recognition of ZIKV TDNA was confirmed in this study, establishing the created sensor as a possible candidate for precise and sensitive Zika virus diagnosis.

2. Methods

2.1. Chemicals, Reagents, Apparatus

Methylene blue was purchased from Sigma Aldrich, India, and all the other chemicals were of AR grade: silver nitrate (Qualigens) and sodium borohydride (GLR Innovations). ZIKA target was dissolved in sterile water (100 μM). For the making of synthetic serum: NaCl, $CaCl_2$, KCl, $MgSO_4$, $NaHCO_3$, Na_2HPO_4, and NaH_2PO_4 were used. A 23-base oligonucleotide probe DNA(PDNA) was obtained from Merck (Rahway, NJ, USA) for capturing the 23-base target DNA (TDNA), which it was complimentary to. The used sequence of ZIKV target DNA and probe DNA was acquired from the work of Zhang et al. [16].

Probe Sequence: AGCCATGACCGACACCACACCGT

Target Sequence: TCGGTACTGGCTGTGGTGTGGCA

The yellow fever oligonucleotide, used for testing the specificity of the device, was also obtained from Merck (Rahway, NJ, USA).

Target Sequence: CGATTAACTCCACATAACCAGACG

For the preparation of silver NPs: chloroauric acid (John Baker Inc.), trisodium citrate (LOBA), NaBH4 (GRL Innovations), CTAB (HIMEDIA), and ascorbic acid (SRL) were used. All other chemicals are purchased from LOBA (Colaba, Mumbai, India).

The electrochemical measurements including cyclic voltammetry (CV) were performed on Metrohm Dropsens (Asturias, Spain) (μStat-I 400s). Photometric analysis of absorbance was performed using UV-Visdouble beam spectrophotometer (HALO DB-20R) (Dynamica Scientific, Livingston, United Kingdom). The microstructural characterizations were executed via using X-ray diffraction (XRD) (Smart Lab guidance, Rigaku, Tokyo, Japan) to check the phase and crystallinity. AgNP morphology was inspected via TEM on Talos L120C (Thermo Fisher Scientific, Waltham, MA, USA).

2.2. Synthesis of Silver NPs

Silver NPs were synthesized by the chemical reductant method. A total of 10 mL of freshly prepared 1mM $AgNO_3$ solution was added in a dropwise fashion (1 drop/second) to 30 mL of 2 mM ice-cold $NaBH_4$ solution, while continuously stirring the solution, to form a bright yellow solution when all the silver nitrate was added; this bright yellow solution contains the AgNPs.

$$AgNO_3 + NaBH_4 \rightarrow Ag + 1/2H_2 + 1/2B_2H_6 + NaNO_3$$

The bright yellow solution containing the Ag nanoparticles was analyzed by photometric analysis and showed peak absorbance at 387.5 nm using a HALO DB-20R UV-Vis beam spectrophotometer.

2.3. Preparation of Synthetic Serum

For the preparation of synthetic serum, 6.8 g of NaCl, 0.2 g $CaCl_2$, 0.4 g KCl, 0.1 g $MgSO_4$, 2.2 g $NaHCO_3$, 0.126 g Na_2HPO_4, and 0.026 g of NaH_2PO_4 were dissolved in distilled water, and the pH was maintained at 7.4.

2.4. Development of PBGs

A silk screen with a laser-cut patterned solid skin adhered to it, according to predetermined dimensions for a two-electrode system, was used for hand printing. Carbon conductive ink was pressed onto cellulose papers through the open regions of the overhead specified screen using a squeegee. The dimensions of the electrode were pre-fixed and framed on the silk screen, which was further used as a stencil for the preparation of the electrodes. The printed electrodes were a two-electrode system: a counter electrode (CE) and a working electrode (WE). This led to the construction of the PBGs.

2.5. Deposition of the Silver Nanoparticles and Immobilization on the PBG

The synthesized silver nanoparticles were drop-deposited (20 μL) on the circular working area of the paper-based biosensor. The paper-based biosensor was then dried on a hot plate at 60 °C. After modification with silver nanoparticles, the working area was immobilized with the consensus biosensor of the probe. For this, 20 μL of the probe DNA was dropped over the working area. This biosensor-modified electrode was further used for the recognition of the target DNA of Zika.

2.6. Stages for Electrochemical Detection

In order to make a functioning and selective biosensor for detecting Zika virus cDNA, the probe DNA needs to be deposited onto the working electrode with the help of the nanoparticles. For this, cyclic voltammetry (CV) values of bare electrodes with no deposition were analyzed. Next, the AgNPs were deposited onto the paper-based biosensor and dried over night after which each voltammetry was repeated as before; for the next step, the PDNA was deposited on a paper-based biosensor containing dried AgNPs, and CV values were recorded at each fabrication stage. For the last phase of the biosensor design, the analyte molecule, the TDNA, was deposited onto electrodes containing both AgNPs and PDNA; CV was performed and values were recorded.

2.7. Optimization of Physicochemical Parameters

The detection capacity of the genosensor was optimized by observing the alterations in the voltammograms caused by changes to the different detection parameters. This was achieved by manipulating the required parameters in the TDNA sample. Concentrations of 0.1, 1, 10, and100 μM were made, the TDNA was incubated at different temperatures (15, 25, 30, and 45 °C), and the time given for hybridization by the potentiostat (tcond) was altered.

2.8. Binding of the Analyte on AgNPs/Probe/Target/PBGs

The different concentrations of the biosensor (0.1, 1, 10, and100 μM) were detected by dropping a mixture of target DNA and MB. The CV measurements were performed so that the hybridization between the biosensor and target DNA could be confirmed. TDNA was drop-deposited over PDNA/AgNPs/PBGs with different concentrations of TDNA.

2.9. Procedure for Experimental Sample Analysis, Repeatability, and Stability Analysis

The capability of the genosensor to perform in experimental samples was checked by adding a known concentration of the target DNA in the synthetic serum. This solution along with the hybridization indicator was added over a probe biosensor-modified paper-based electrode system. Electrochemical assessments were performed to check the results.

2.10. Principle behind Sensing

The proposed PBG works on the principle that the hybridization of DNA into the double-helical structure causes a lowering of the peak current value of the CV curve, as demonstrated by Pan et al. [17]. This can be attributed to the molecular interaction of MB with ssDNA and dsDNA, while with ssDNA it can only have electrostatic interactions; however, when the DNA is present in its double-stranded helical form, it can also intercalate between the structure in addition to the electrostatic interactions. Therefore, in ssDNA, MB has a significantly lower binding constant, thus allowing for a better passage of current. This results in a greater negative formal potential with dsDNA than with ssDNA.

3. Results

The fabrication method of the electrode involves the deposition of AgNPs onto the working surface of the paper electrode. The AgNPs were drop-deposited onto the surface of the electrode. After the drop-deposition of the AgNPs, the DNA probe was also put onto the working surface of the paper electrodes. The AgNPs provided a biocompatible environment to the biological identification component and accelerated the electron transfer.

3.1. Morphological Characterization of Silver Nanoparticles

The XRD patterns of silver nanoparticles are shown in Figure 1a. The XRD pattern revealed that it has a face-centered cubic crystal structure. According to the obtained XRD result, the 111, 200, 220, and 311 crystallographic planes of silver nanoparticles were responsible for the 2thata values of 38.15°, 44.06°, 66.20°, and 76.61° peaks, respectively. The XRD pattern demonstrated that the Ag-NPs synthesized in this work were crystalline in nature. The formation of Ag-NPs was also confirmed by measuring the SPR over the wavelength range of 300–800 nm using UV-Vis spectroscopy. According to the reported data, the spherical Ag-NPs contribute to the absorption bands at approximately 400–450 nm in the UV-visible spectra. The UV-visible absorption spectrum revealed that the wide SPR band had a single peak at 445 nm, which was located in the broad SPR band. This peak confirmed the presence of a uniform distribution of Ag-NPs in the synthesized sample as shown in Figure 1b. TEM images (Figure 1c) depicted that the silver nanoparticles are spherical in shape and 30 nm in size. Figure 1d shows the average particle size of the nanoparticles to be around 25nm.

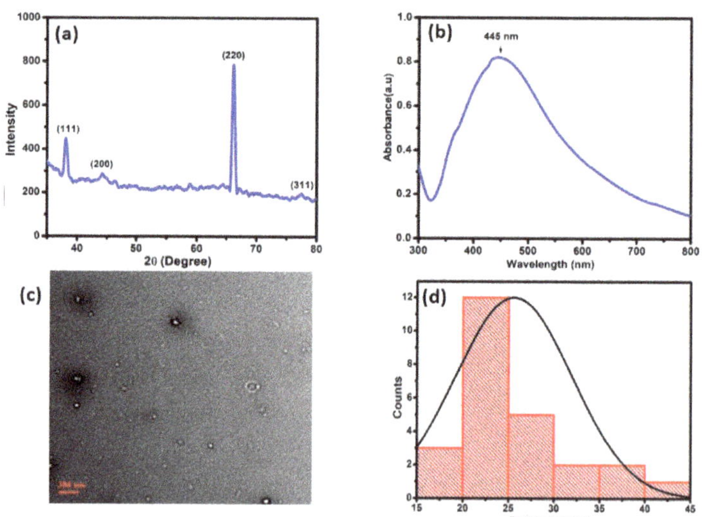

Figure 1. (**a**) XRD diffractogram and (**b**) UV-Vis spectroscopy; (**c**) TEM image of silver nanoparticles; (**d**) average particle size.

3.2. Electrochemical Properties of DNA Probe/AgNPs/PBGs

The electrochemical characterization of the DNA probe/AgNP/PBG-modified electrodes was performed by employing the electrochemical CV technique. Figure 2 shows the differential current response at different stages of the electrode. As the bare electrode showed a smaller peak current response (0.26 μA), which is attributed to the lower electron transfer kinetics. Upon deposition of the AgNPs onto the working surface, there was a significant two-fold increase in current response (1.59 μA), which was due to the fast electron transfer kinetics provided by silver nanoparticles. After immobilization of the biological recognition element (DNA probe) onto the working surface, the current (1.07 μA) was drastically decreased due to the nonconductive nature of DNA. After the introduction of target DNA, the current response (0.81 μA) was further decreased due to the well-known MB principle. MB was intercalated between the bases, which significantly decreased the current response.

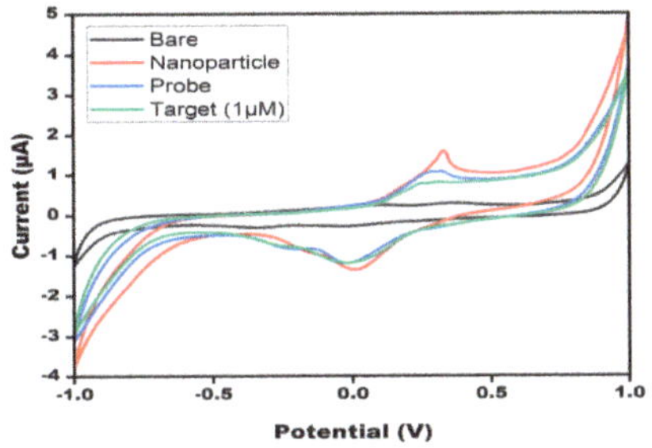

Figure 2. Cyclic voltammograms of 10 mM methylene blue in 0.1MKCl at bare PBGS, AgNPs/PBGS, DNA probe/AgNPs/PBGS and TDNA/PDNA/AgNPs/PBGS at 50 mV s^{-1} in the potential range from −1 V to +1 V.

3.3. Effect of Different Target DNA Concentrations on the DNA Probe/AgNPs/PBGs

Different concentrations of TDNA were analyzed to depict the quantitative performance of the developed sensor. Different concentrations varying from 0.1 to 100 μM were employed for the hybridization of the ssDNA probe. The results concluded that TDNA is showing hybridization with the ssDNA probe, and at different concentrations a varying current response was observed, which confirmed the quantitative performance of the developed sensor. The results obtained were in line with the earlier reported sensors. Upon increasing the concentrations of TDNA, there was a decreased current response as more insulating layers of biological recognition element retarded the electron transfer. A linear relationship was found between the log value of the TDNA concentration and the anodic current value, and a good r^2 value was found. The detection limit was found to be 0.1 μM (Figure 3).

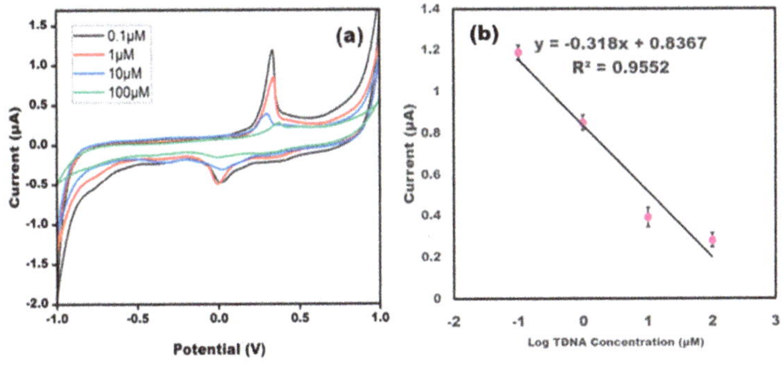

Figure 3. (**a**) CV of DNA probe/AgNP/PBG-modified electrode using different concentrations of ZIKV-TDNA (0.1 μM, 1 μM, 10 μM, and 100 μM) in 10 mM methylene blue in 0.1 M KCl at 50 mV s^{-1} in the potential range from −1 V to +1 V. (**b**) Linear curve of the current value and log of the target DNA concentration. The error bar represents the standard deviation of the sensor for each concentration repeated 5 times.

3.4. Optimization of DNA Probe/AgNP/PBG Platform in Terms of Temperature and Time

Optimization of the sensor is very important for the smooth functioning of the developed sensor. The performance of the sensor depends on pH, temperature, and time. Therefore, the sensor was optimized in terms of these experimental variables so that maximum response could be achieved (Figure 4). The performance of the developed sensor DNA probe/AgNPs/PBGswas extensively studied in varying different ranges of temperature and time. The cyclic voltammogram of the DNA probe/AgNPs/PBGs was observed at different temperatures ranging from 15 to 45 °C at a scan rate of 50 mV/s. The highest response was observed at 45 °C, with its peak almost approaching the peak of the CV without TDNA; this could be due to instability of the double helical structure at high temperatures, as the high-energy TDNA would not easily bind to the PDNA. Therefore, the sensor was optimized at 25 °C. The created sensor was tuned at various times to provide the best possible response at any given moment. After 20 s, there was no significant change in anodic current value as current values of 20 s and 30 s were overlapping with each other, implying that 20 s is the optimal reaction time. An ideal incubation time of 20 s was maintained prior to the anticipated sweep runs.

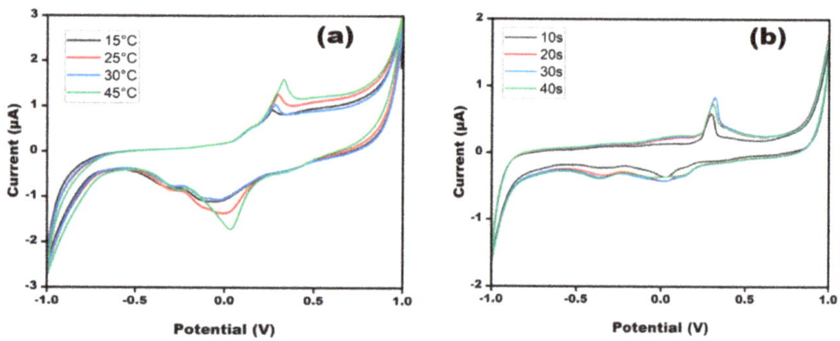

Figure 4. Cyclic voltammetry obtained at DNA probe/AgNPs/PBGs for different (**a**) temperature (15–45 °C) and (**b**) time (10–40 s) in 10 mM methylene blue in 0.1 M KCl at 50 mV s^{-1} in the potential range from −1 V to +1 V.

3.5. Evaluation Parameters

The developed sensor was evaluated in terms of its reproducibility, recovery, and stability (Figure 5 and Table 1). The recovery of the developed sensor was checked by adding 4 μM to the initial concentration 1 μM. The current response of the original concentration (5 μM) was similar to the current response obtained after addition. The developed sensor was able to show good recovery. CV was taken to check the reproducibility of the modified sensor. The developed sensor was checked at five replicates' determination. The target DNA was estimated on a single day five times and it was observed that the current remained unchanged at different determinations. The response of the sensor was also estimated after one week and the sensor response was found to be almost the same. The observations demonstrated that the sensor showed high reproducibility. The developed sensor was tested for its long-term stability. The paper electrodes were checked after 15 days. The sensor showed almost the same response as it did on day one.

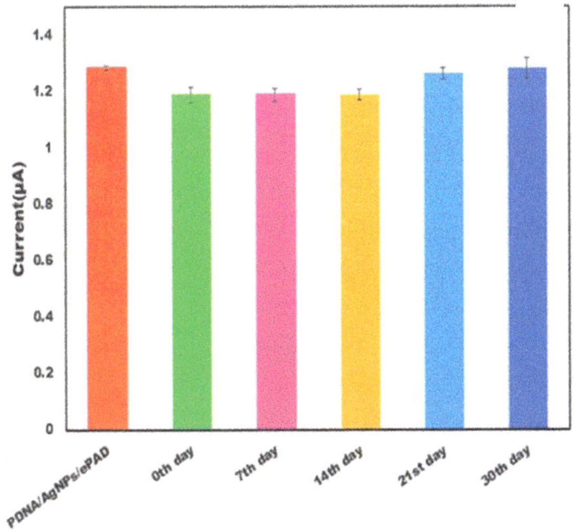

Figure 5. Bar graph depicting the stability of the modified electrode DNA probe/AgNPs/PBGS for TDNA on the 7th, 14th, 21st, and 30th day in 10 mM methylene blue in 0.1MKCl at 50 mV s^{-1} in the potential range from −1 V to +1 V.

Table 1. Recovery test of constructed biosensor of ZIKV.

Initial Concentration (μM)	Concentration Added (μM)	Concentration Found (μM)	Recovery (Percentage)
1	4	5.1	102

3.6. Specificity and Reliability

This PBG selectivity was estimated via analysis with the target yellow fever virus (YFV) DNA. Reactivity was observed only in the target DNA of ZIKV. Results show that there was no major change in the current peak of the DNA probe of ZIKV before or after the addition of the target DNA for YFV, respectively; the current peak remained unchanged (Figure 6). This study revealed that the target DNA YFV had no specific binding to the specific probe DNA of ZIKV. However, in presence of the target ZIKV DNA, binding occurred between the DNA probe and the target ZIKV TDNA. The bar graph in Figure 6 represents the confirmation of the above results.

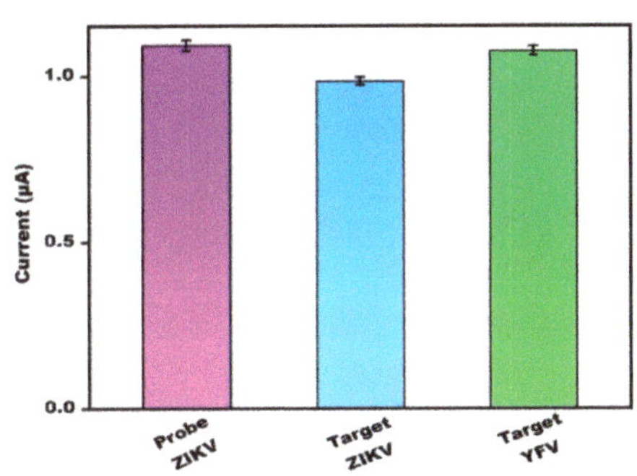

Figure 6. Bar graph depicting the CV confirmation with crossbar with the current value of probe ZIKV/target ZIKV and probe YFV.

3.7. Analysis of Experimental Sample

The CV response of the sensor when the TDNA ZIKV was spiked into a synthetic serum sample was found to be similar to the target DNA alone (Figure 7). The developed sensor was able to detect the target DNA in synthetic serum as the current response was found to be almost the same as the target DNA alone.

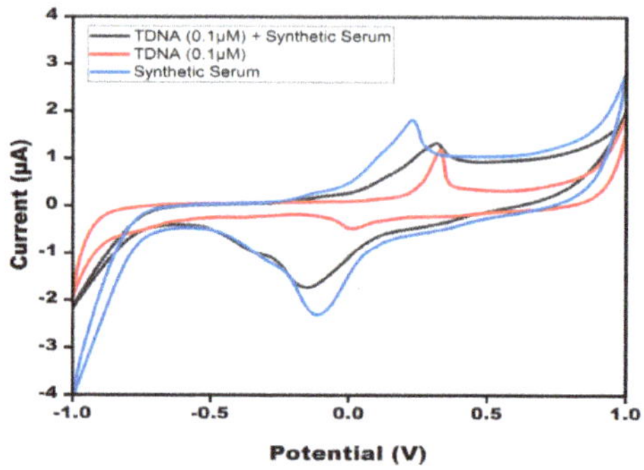

Figure 7. CV peak current study of TDNA of ZIKV in synthetic serum at modified DNA probe/AgNPs/PBGs in10 mM methylene blue in 0.1 M KCl at 50 mV s^{-1} in the potential range from -1 V to +1 V. CV peak current was compared with the synthetic serum, TDNA, and TDNA in synthetic serum.

4. Discussion

The present work aimed to develop an electrochemical paper-based device for the detection of ZIKA virus target DNA, which offered many advantages, such as low sample/reagent volume, ease of preparation, and more stable printing on paper, unlike plastic substrate, and the sensor was also applied in artificial serum samples so that sensor applicability could be checked in complex matrices. The use of a state-of-the-art portable potentiostat acts as a limiting factor for the device's LOD, as can be seen in Table 2, but this diminishes when the fact that this can be replicated in any remote corner of the world is taken into consideration, effectively making the device truly capable of providing POC diagnosis, which is crucial considering the tropical presence of the disease in question. Still, the developed genosensor has major disadvantages in that the sensor was not able to detect the direct antigen, which can be overcome by employing aptamers a biological recognition elements. The employment of aptamers as biological or structural recognition elements can be considered as a future perspective of the developed study. The advancement in technology and the development of more sensitive potentiostat devices also open the scope of lower LODs and better detection capabilities for such devices.

Table 2. Comparative study of various ZIKV-based genosensors.

Biosensors	Linear Range	LOD	References
Electrochemical genosensor for Zika virus based on a poly-(3-amino4-hydroxybenzoic acid)-modified pencil carbon graphite electrode	84.0 pM to 1.41 nM	25.4 pM	[14]
Gold nanoparticle/DNA-based nanobioconjugate for electrochemical detection of Zika virus	10 to 600 fM and from 500 fM to 10 pM of the target	0.2 and 33 fM at the SPAuE and SPCE/Au	[12]
Label-free electrochemical DNA biosensor for Zika virus identification	-	25.0 ± 1.7 nM	[15]
A sensitive label-free impedimetric DNA biosensor based on silsesquioxane-functionalized gold nanoparticles for Zika virus detection.	1.0×10^{-12}–1.0×10^{-6} M	0.82 pM	[18]
Electrochemical biosensor based on surface imprinting for Zika virus detection in serum	10 fM–1 μM	9.4 fM	[19]
Acrylic-based genosensor utilizing metal salphen labeling approach for reflectometric dengue virus detection	1×10^{-15} M to 1×10^{-3} M	1.21×10^{-16} M	[20]
Rapid, point-of-care, paper-based plasmonic biosensor for Zika virus diagnosis	10–105 nM	1 nM	[21]
Diagnosis of Zika infection using a ZnO nanostructure-based rapid electrochemical biosensor	0.1 nM to 100 nM	1.00 pM	[22]
Cyclic voltammetric PBG-based detection of the target DNA of Zika virus	0.1 to 100 μM	0.1 μM	This present work

5. Conclusions

The current research describes the synthesis of electrochemical PBGs for the identification of ZIKV TDNA. AgNPs deposited on paper-based electrodes were utilized for boosting the signal response. The sensor displayed an enhanced linear range from 0.1 to 100 μM and the lower limit of detection was 0.1 μM. The usage of paper-based electrodes also put on the benefit of being disposable and lower-cost, as well as having the capability of being fabricated in-house. All of these features together make the current effort ideal for developing point-of-care systems. With a few adjustments, this approach could develop into a technology that could be employed at the bedside of ZIKV patients.

Author Contributions: Conceptualization, J.N. and R.P.; methodology, A.B., M.A.A. and M.R.H.; software, A.B., M.A.A. and M.R.H.; validation, A.B., M.A.A. and M.R.H. formal analysis, J.N., M.K. and R.P.; investigation, J.N., M.K. and R.P.; resources, J.N. and M.K.; data curation, A.B., M.A.A. and M.R.H.; writing—original draft preparation, A.B. and M.R.H.; writing—review and editing, A.B., M.A.A. and M.R.H.; visualization, J.N., M.K. and R.P.; supervision, J.N., M.K. and R.P.; project administration, J.N., M.K. and R.P.; funding acquisition, J.N., M.K. and R.P. All authors have read and agreed to the published version of the manuscript.

Funding: This research received no external funding.

Data Availability Statement: Data will be made available on request.

Conflicts of Interest: The authors declare no conflict of interest.

References

1. Plourde, A.R.; Bloch, E.M. A literature review of Zika virus. *Emerg. Infect. Dis.* **2016**, *22*, 1185. [CrossRef] [PubMed]
2. Gorshkov, K.; Shiryaev, S.A.; Fertel, S.; Lin, Y.W.; Huang, C.T.; Pinto, A.; Farhy, C.; Strongin, A.Y.; Zheng, W.; Terskikh, A.V. Zika virus: Origins, pathological action, and treatment strategies. *Front. Microbiol.* **2019**, *7*, 9. [CrossRef] [PubMed]
3. Lei, J.; Hansen, G.; Nitsche, C.; Klein, C.D.; Zhang, L.; Hilgenfeld, R. Crystal structure of Zika virus NS2B-NS3 protease in complex with a boronate inhibitor. *Science* **2016**, *503*, 5. [CrossRef] [PubMed]
4. Abushouk, A.I.; Negida, A.; Ahmed, H. An updated review of Zika virus. *J. Clin. Virol.* **2016**, *53*, 8. [CrossRef] [PubMed]
5. Kaushik, A.; Yndart, A.; Kumar, S.; Jayant, R.D.; Vashist, A.; Brown, A.N.; Li, C.Z.; Nair, M. A sensitive electrochemical immunosensor for label-free detection of Zika-virus protein. *Sci. Rep.* **2018**, *1*, 5. [CrossRef] [PubMed]
6. Noorbakhsh, F.; Abdolmohammadi, K.; Fatahi, Y.; Dalili, H.; Rasoolinejad, M.; Rezaei, F.; Salehi-Vaziri, M.; Shafiei-Jandaghi, N.Z.; Gooshki, E.S.; Zaim, M.; et al. Zika virus infection, basic and clinical aspects: A review article. *Iran. J. Public Health* **2019**, *48*, 20. [CrossRef]
7. Priyamvada, L.; Quicke, K.M.; Hudson, W.H.; Onlamoon, N.; Sewatanon, J.; Edupuganti, S.; Pattanapanyasat, K.; Chokephaibulkit, K.; Mulligan, M.J.; Wilson, P.C.; et al. Human antibody responses after dengue virus infection are highly cross-reactive to Zika virus. *Proc. Natl. Acad. Sci. USA* **2016**, *7852*, 7. [CrossRef]
8. Wong, S.J.; Furuya, A.; Zou, J.; Xie, X.; Dupuis, A.P.; Kramer, L.D.; Shi, P.Y. A multiplex microsphere immunoassay for Zika virus diagnosis. *eBioMedicine* **2017**, *36*, 40. [CrossRef]
9. Bosch, I.; Puig, d.; Hiley, M.; Carré-Camps, M.; Perdomo-Celis, F.; Narváez, C.F.; Salgado, D.M.; Senthoor, D.; O'Grady, M.; Phillips, E.; et al. Rapid antigen tests for dengue virus serotypes and Zika virus in patient serum. *Sci. Transl. Med.* **2017**, *27*, 9. [CrossRef]
10. Shetti, N.P.; Bukkitgar, S.D.; Reddy, K.R.; Reddy, C.V.; Aminabhavi, T.M. ZnO-based nanostructured electrodes for electrochemical sensors and biosensors in biomedical applications. *Biosens. Bioelectron.* **2019**, *15*, 141. [CrossRef]
11. Sher, M.; Faheem, A.; Asghar, W.; Cinti, S. Nano-engineered screen-printed electrodes: A dynamic tool for detection of viruses. *TrAC Trends Anal. Chem.* **2021**, *1*, 143. [CrossRef] [PubMed]
12. Cajigas, S.; Alzate, D.; Orozco, J. Gold nanoparticle/DNA-based nanobioconjugate for electrochemical detection of Zika virus. *Microchim. Acta* **2020**, *187*, 11. [CrossRef]
13. Moço, A.C.; Guedes, P.H.; Flauzino, J.M.; da Silva, H.S.; Vieira, J.G.; Castro, A.C.; Gomes, É.V.; Tolentino, F.M.; Soares, M.M.; Madurro, J.M.; et al. Electrochemical detection of zika virus in biological samples: A step for diagnosis point-of-care. *Electroanalysis* **2019**, *580*, 7. [CrossRef]
14. da Fonseca Alves, R.; Franco, D.L.; Cordeiro, M.T.; de Oliveira Junior, E.M.; Dutra, R.A.; Sotomayor, M.D. Novel electrochemical genosensor for Zika virus based on a poly-(3-amino-4-hydroxybenzoic acid)-modified pencil carbon graphite electrode. *Sens. Actuators B Chem.* **2019**, *296*, 126681. [CrossRef]
15. Faria, H.A.; Zucolotto, V. Label-free electrochemical DNA biosensor for zika virus identification. *Biosens. Bioelectron.* **2019**, *131*, 149–155. [CrossRef]
16. Li, C.; Xu, D.; Ye, Q.; Hong, S.; Jiang, Y.; Liu, X.; Zhang, N.; Shi, L.; Qin, C.F.; Xu, Z. Zika Virus Disrupts Neural Progenitor Development and Leads to Microcephaly in Mice. *Cell Stem Cell* **2016**, *19*, 120–126. [CrossRef] [PubMed]
17. Pan, D.; Zuo, X.; Wan, Y.; Wang, L.; Zhang, J.; Song, S.; Fan, C. Electrochemical Interrogation of Interactions between Surface-Confined DNA and Methylene Blue. *Sensors* **2007**, *7*, 2671–2680. [CrossRef]
18. Steinmetz, M.; Lima, D.; Viana, A.G.; Fujiwara, S.T.; Pessôa, C.A.; Etto, R.M.; Wohnrath, K. A sensitive label-free impedimetric DNA biosensor based on silsesquioxane-functionalized gold nanoparticles for Zika Virus detection. *Biosens Bioelectron.* **2019**, *141*, 111351. [CrossRef]
19. Tancharoen, C.; Sukjee, W.; Thepparit, C.; Jaimipuk, T.; Auewarakul, P.; Thitithanyanont, A.; Sangma, C. Electrochemical Biosensor Based on Surface Imprinting for Zika Virus Detection in Serum. *ACS Sens.* **2019**, *4*, 69–75. [CrossRef] [PubMed]
20. Mazlan, N.F.; Tan, L.L.; Karim, N.H.A.; Heng, L.Y.; Jamaluddin, N.D.; Yusof, N.Y.M.; Quay, D.H.X.; Khalid, B. Acrylic-based genosensor utilizing metal salphen labeling approach for reflectometric dengue virus detection. *Talanta* **2019**, *198*, 358–370. [CrossRef] [PubMed]
21. Jiang, Q.; Chandar, Y.J.; Cao, S.; Kharasch, E.D.; Singamaneni, S.; Morrissey, J.J. Rapid, Point-of-Care, Paper-Based Plasmonic Biosensor for Zika Virus Diagnosis. *Adv. Biosyst.* **2017**, *1*, e1700096. [CrossRef]
22. Faria, A.M.; Mazon, T. Early diagnosis of Zika infection using a ZnO nanostructures-based rapid electrochemical biosensor. *Talanta* **2019**, *203*, 153–160. [CrossRef]

micromachines

Article

A Multi-Parameter Fusion Method for Cuffless Continuous Blood Pressure Estimation Based on Electrocardiogram and Photoplethysmogram

Gang Ma [1,2], Jie Zhang [2], Jing Liu [3], Lirong Wang [2,3,*] and Yong Yu [2,*]

1 School of Biomedical Engineering, Division of Life Sciences and Medicine, University of Science and Technology of China, Hefei 230026, China
2 Suzhou Institute of Biomedical Engineering and Technology, China Academy of Sciences, Suzhou 215163, China
3 School of Electronics and Information Technology, Soochow University, Suzhou 215031, China
* Correspondence: wanglirong@suda.edu.cn (L.W.); yuyong@sibet.ac.cn (Y.Y.)

Abstract: Blood pressure (BP) is an essential physiological indicator to identify and determine health status. Compared with the isolated BP measurement conducted by traditional cuff approaches, cuffless BP monitoring can reflect the dynamic changes in BP values and is more helpful to evaluate the effectiveness of BP control. In this paper, we designed a wearable device for continuous physiological signal acquisition. Based on the collected electrocardiogram (ECG) and photoplethysmogram (PPG), we proposed a multi-parameter fusion method for noninvasive BP estimation. An amount of 25 features were extracted from processed waveforms and Gaussian copula mutual information (MI) was introduced to reduce feature redundancy. After feature selection, random forest (RF) was trained to realize systolic BP (SBP) and diastolic BP (DBP) estimation. Moreover, we used the records in public MIMIC-III as the training set and private data as the testing set to avoid data leakage. The mean absolute error (MAE) and standard deviation (STD) for SBP and DBP were reduced from 9.12 ± 9.83 mmHg and 8.31 ± 9.23 mmHg to 7.93 ± 9.12 mmHg and 7.63 ± 8.61 mmHg by feature selection. After calibration, the MAE was further reduced to 5.21 mmHg and 4.15 mmHg. The result showed that MI has great potential in feature selection during BP prediction and the proposed multi-parameter fusion method can be used for long-term BP monitoring.

Keywords: blood pressure; wearable device; feature extraction; GCMI

Citation: Ma, G.; Zhang, J.; Liu, J.; Wang, L.; Yu, Y. A Multi-Parameter Fusion Method for Cuffless Continuous Blood Pressure Estimation Based on Electrocardiogram and Photoplethysmogram. *Micromachines* **2023**, *14*, 804. https://doi.org/10.3390/mi14040804

Academic Editor: Aiqun Liu

Received: 24 February 2023
Revised: 27 March 2023
Accepted: 29 March 2023
Published: 31 March 2023

1. Introduction

Blood pressure (BP) is one of the most important parameters for monitoring the status of the body and can be used in the diagnosis and treatment of many diseases. Unfortunately, in recent years, the number of hypertension (HPT) patients continued to increase and became an urgent global health problem [1]. HPT could increase the pressure of blood on the walls of the arteries, which gradually affects several organs, including the heart, brain, and kidneys. HPT even has the potential to cause the blockage or rupture of blood vessels that supply oxygen to the brain, which greatly increases the risk of accidents [2–4]. BP regulation is influenced by a variety of factors, such as cardiac ejection, peripheral resistance, and vessel wall elasticity, and also has a relation with mood and age. As HPT is a chronic disease that requires long-term monitoring, a single BP measurement is not able to provide an accurate view of the body's condition. Therefore, it is very important to propose a noninvasive continuous blood pressure estimation method, which can reflect the physical condition of the human body in real time and reduce the incidence of cardiovascular diseases [5].

The mainstream cuffless BP measurement methods analyze the features of physiological signals such as photoplethysmogram (PPG) [6], electrocardiogram (ECG) [7], ballistocardiogram (BCG) [8], etc., and predict BP by machine learning (ML). The key technologies are physiological signals acquisition [9] and feature extraction [10,11]. In the last decade, with the development of semiconductor technology and the popularity of wearable health monitoring devices, people were found to use such devices to achieve a variety of physiological signal acquisition without interfering with normal human activities. The quality of signals directly affects the accuracy of prediction. This poses a challenge to the design of acquisition hardware but also makes continuous BP monitoring possible. In [12], Rachim et al. designed a multimodal biosensor to measure PPG and impedance plethysmography (IPG) from the participant's wrist. Then, 14 PTT-features were calculated between the IPG peak-point and PPG. The comparative experiment showed that the extracted PTT-features had a certain correlation with BP, but its correlation coefficient for SBP was weaker than that of conventional PAT, which was located by ECG. In addition, compared with the fingertip, the wrist was more active and had a bad influence on the collected PPG signals. Therefore, the author only analyzed the time characteristics and did not use other amplitude characteristics. Bui et al. [13] chose the ear as the measuring location to minimize motion interference and ensure comfort. Benefiting from the good contact between the ear and the sensor, the author collected the PPG signal with high quality. However, because of the lack of ECG as a reference, the features provided by a single PPG are still limited. Multiple types of signals are conducive to obtaining richer features and improving the accuracy of results. Hence, researchers must consider comfort, signal diversity, and quality while using wearable devices to collect physiological signals.

For feature extraction, scholars try to synthesize multiple physiological signals and select information in multiple dimensions such as the time domain and frequency domain. Among them, pulse transit time (PTT) [14], pulse arrival time (PAT) [15], pulse wave velocity (PWV) [16], etc. were proven to be effective. Chen W et al. [17] adopted that PTT was highly correlated with the high-frequency components of BP. Through calibration, more accurate blood pressure estimation can be achieved. Meanwhile, age, gender, cardiovascular disease, and other factors determine arterial stiffness, which is significantly correlated with PTT [18]. Zhang et al. [19] considered how the autonomic nervous system affects the heart and blood vessels. He extracted nine features related to heart rate variability (HRV) and combined them with other PPG features. The result proved the effectiveness of HRV features. Geerthy et al. [20] extracted informative features such as SDI, Womersley, and QRS from PPG and ECG signals and used a genetic algorithm (GA) to select features. The best optimal feature set reduced the mean absolute error (MAE) from 13.20 mmHg to 9.54 mmHg for systolic BP (SBP) and 9.91 mmHg to 5.48 mmHg for diastolic BP (DBP), respectively. Wang et al. [21] collected 30 sets of PPG and ECG signals by Finometer. Then, 39 features were extracted and 10 features were finally retained by comparing their correlations with BP. In [22], 32 features from PPG were extracted to estimate BP. These features almost comprehensively contained multiple scale information such as amplitude, time, and frequency. Shuo et al. [23] introduced the mean impact value (MIV) to investigate the impact of each feature and the genetic algorithm (GA) to implement parameter optimization. After optimization, the MAE of SBP and DBP reached 3.27 mmHg and 1.16 mmHg. Through the joint analysis of the PPG signal and its derivatives, Chowdhury et al. [24] extracted up to 101 features, among which 75 features were from the time domain, 16 were from the frequency domain, and the remaining 20 were computed using statistical analysis. It was observed that the number of features extracted from the signal increased, which not only aggravated the workload of feature extraction but also made model training more difficult. Compared with the feature extraction itself, how to screen out the optimal solution from many features became another issue for scholars to research.

Different from traditional ML methods, deep learning (DL) methods avoid the drawbacks of manual feature extraction. It can automatically learn more abstract and high-dimensional features from the input signal, which enables it to have stronger adaptability for nonlinear system complexity [11]. Benefiting from its powerful data mining capabilities, machine learning can be combined with biological signal processing and achieved good performance [25]. Yu et al. [26] introduced an attention-based residual block to U-Net to predict BP. The result showed that the combination of PPG raw signal, first derivative, and second derivative as model input was helpful for the network to extract more information. Wang et al. [27] proposed an end-to-end model to measure BP. The model consisted of one-dimensional convolutional layers, depth-separable convolutional layers, and a gated recurrent unit (GRU). The average absolute error (MAE) was 3.95 mmHg for SBP and 2.14 mmHg for DBP, which met the international standard. Senturk et al. [28] analyzed the performance of dynamic learning methods, such as recurrent neural networks (RNN), nonlinear autoregressive networks with exogenous inputs neural networks (NARX-NN), and long short-term memory neural networks (LSTM-NN). The experiment showed that NARX had the most potential.

Despite the great performance in BP prediction, DL still has the following shortcomings. On the one hand, it relies on a large amount of data during the training process, and it is difficult to achieve good performance on small batches of data. On the other hand, neural network features are not easily understood at the semantic level, which makes it more challenging for scholars to interpret the experimental results.

Above all, as ECG and PPG signals have irreplaceable effects on BP assessment, we designed a wearable physiological signal acquisition device that could continuously collect and transmit two signals in real time. In order to make the acquisition process comfortable, we adopted a separate design for the host and the sensor. Meanwhile, combining individual features with waveform features, a multi-parameter fusion method for non-invasive continuous BP estimation was proposed and mutual information (MI) [29] was introduced for feature selection. The main contributions in this paper are as follows:

(1) We used a split design where the acquisition host and sensor were connected via a type-C interface. This not only ensured the quality of the collected signal but also facilitated the replacement of the sensor. The host can either transmit data wirelessly in real time or store data locally as a backup. The whole host size was only 42 mm $\times$ 29 mm $\times$ 13 mm and caused a little burden on the human body.

(2) We constructed a private database by self-collected data. Then, the model was trained on the MIMIC III dataset [30] (accessed on 10 January 2020) and tested on the private dataset, which avoided data leakage. After training, we calibrated the model with a quarter of the records in the testing set.

(3) Gaussian copula MI (GCMI) was used to rank the initial 25 features. Then, 11 and 15 features were retained for SBP and DBP prediction, respectively. The results showed that the optimal feature set improved performance.

2. Data Preparation

In this section, we will describe the wearable device structure and the data acquisition process for noninvasive continuous BP estimation.

2.1. Cuff-Less Continuous BP Measurement System

The hardware block diagram of the BP monitoring system designed in this paper is shown in Figure 1a. The host system consisted of five parts: the main control unit, the signal acquisition front-end (AFE) circuit, the storage circuit, the wireless transmission circuit, and the power supply management circuit. The STM32F205 chip was chosen as the main control, which had an Arm Cortex-M3 core with a maximum main frequency of 120 MHz. The acquisition circuit contained the synchronous record of PPG and ECG signals. For ECG signals, ADS1298 was selected as the AFE chip to collect the lead II signals. It could provide 24-bit analog-to-digital conversion accuracy and up to 12$\times$ signal amplification. At the same time, it had a 500 Mohm input impedance and $\pm$400 mV dynamic input range, which met the requirements of ECG acquisition. For PPG signals, AFE4490 was used as the analog AFE. It had two programmable 8-bit resolution current output functions. The maximum output current was 200 mA and could drive red light and infrared light. For the convenience of acquisition, we designed a peripheral sensor that integrated a finger-clip blood oxygen probe and ECG electrodes, which were connected to the main circuit through a type-C interface. The sampling rate was set to 125 Hz. We chose nRF52832 as the wireless transmission chip, which was equipped with Bluetooth Low-power Energy (BLE) wireless communication protocol and can communicate with other Bluetooth devices. Meanwhile, considering the instability of wireless transmission, we selected a 32 GB eMMC as the local storage, which was suitable for long-term collection.

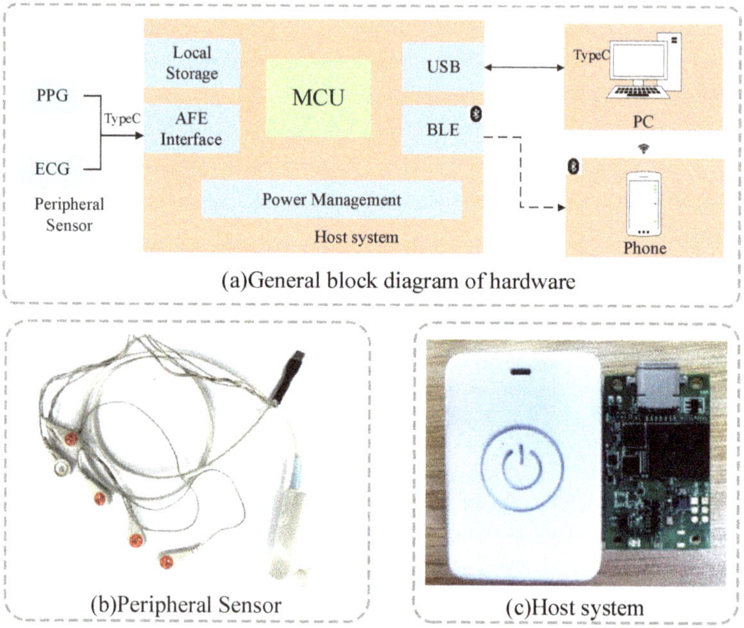

Figure 1. Hardware system. (**a**) General block diagram of hardware; (**b**) Peripheral sensor; (**c**) Host system.

The whole hardware system is shown above. Figure 1b exhibits the customized ECG electrodes and oximetry probe and Figure 1c exhibits the final host device and the internal PCB circuit.

2.2. Data Collection

A total of 15 volunteers participated in the data collection of this experiment, including 13 males and 2 females, aged from 20 to 30 years old. These experiments were conducted according to the Helsinki declaration. All subjects volunteered to participate in the study and signed a written consent form before participation. Participants did not take any specific medications. During the acquisition, RA and LL electrodes were connected to the subject to collect II-lead ECG. In order to reduce the noise effect, an additional RL electrode was added as the reference ground. The PPG signal was collected at the index finger by the finger clip-type oximetry probe. Figure 2a illustrates an example of wearing the device during acquisition.

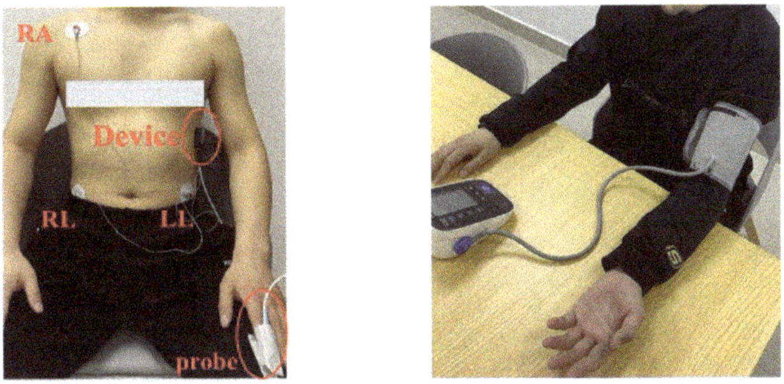

(a) (b)

Figure 2. Data acquisition process. (**a**) Device wearing display, (**b**) BP measured by OMROM HEM-7211.

After the connection, volunteers were asked to sit in the chair with a relaxed posture and PPG and ECG signals were simultaneously collected for 30 s. Next, SBP and DBP values were obtained by measuring the left arm of the volunteers using a traditional cuff BP monitor (OMROM HEM-7211) after two minutes of rest, as shown in Figure 2b. This process was repeated five times with a 2 min interval between each measurement. The lowest and highest values of the five measurements were removed, and the mean value of the remaining three measurements was referred to as ground truth. Finally, a total of 90 records were saved as a private dataset.

2.3. MIMIC-III Dataset

MIMIC-III is an extensive and freely available database comprising health-related data associated with over forty thousand patients who stayed in critical care units of the Beth Israel Deaconess Medical Center between 2001 and 2012 (Johnson et al., 2016) [30]. It consists of a clinical database and a waveform database. The clinical database is a relational database composed of 26 tables and the clinical records can be queried by SQL language. The waveform database contains the waveform records, and the sampling rate was 125 Hz. Although the original database contained a large number of data records, many patients had incomplete information, which did not meet the experimental requirement, and so, we conducted a preliminary screening of the data. Furthermore, since the data in the collected database were from young people, we only retained the data of patients aged between 20 and 40 years to avoid bias in model training. Finally, we kept 900 records as a training data set.

2.4. Preprocessing

The same preprocessing procedure was performed for both the data in MIMIC-III and the collected data. ECG signal acquisition process was mixed with various interferences, such as baseline drift, electrode motion, muscle artifact, and power line interference [31]. We used a bidirectional infinite impulse response filter to denoise the ECG signal. Firstly, the mean value of the ECG signal was subtracted to remove the direct current component. Then, the fourth-order bidirectional low-pass filter was used to remove the high-frequency noise with a cut-off frequency of 35 Hz [32]. Finally, the second-order bidirectional high pass filter was used to remove the baseline drift with a cut-off frequency of 0.9 Hz.

To prevent waveform distortion caused by denoising, we used a low-pass filter and cubic spline interpolation to denoise the PPG signal. Firstly, the mean value of the PPG signal was subtracted to remove the direct current component, and then, the high-frequency noise was removed by using a low-pass filter with a cut-off frequency of 10 Hz. Finally, we used cubic spline interpolation to remove the baseline drift.

MIMIC-III contains continuous ambulatory blood pressure waveforms, and so, we located the peaks and valleys of each segment as SBP and DBP labels, respectively. In order to facilitate subsequent feature extraction, we extracted corresponding feature points in ECG and PPG, respectively. Feature point detection mainly includes R-wave peak detection of the ECG signal, peak, valley, and maximum slope point detection of the PPG signal. In Figure 3, point A is the R-wave peak of the ECG signal. Points B, C, and D are the valley, maximum slope point, and the peak of the PPG signal, respectively. We used the Pan-Tompkin algorithm [33] to realize R-wave peak detection. Peak detection of the PPG signal was realized by the double-threshold detection algorithm [34]. Once the peak was determined, the search area of other feature points could be obtained. Starting from the peak, we searched forward the 0.4 times of the peak interval to find the zero-crossing point of the first derivative of the PPG signal from negative to positive, to realize the valley detection. The peak point of the first derivative was found in the area from the valley to the peak of the ascending branch of the PPG signal to detect the maximum slope point.

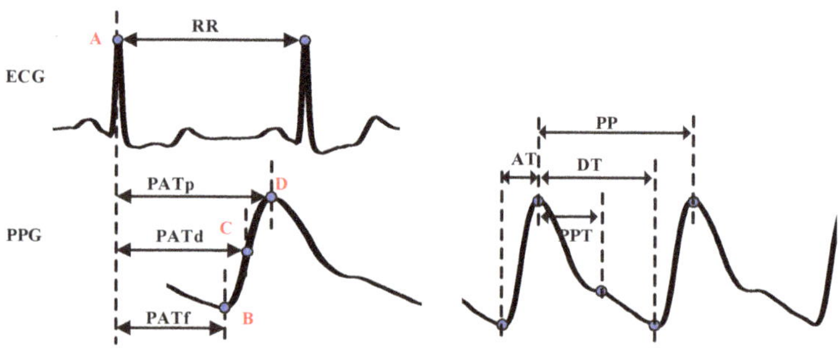

Figure 3. Schematic diagram of feature point detection.

3. Proposed Method

In this section, we will introduce the process of feature extraction, selection, and BP prediction. Figure 4 demonstrates the block diagram of the proposed method.

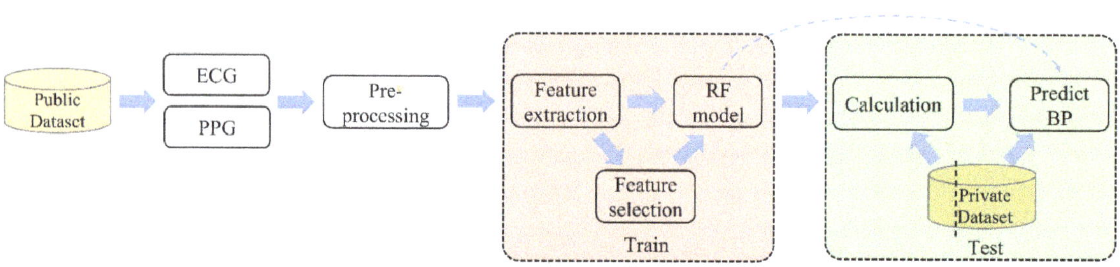

Figure 4. Block diagram of the proposed method.

3.1. Feature Extraction

We extracted 3 individual features and 22 waveform features. Table 1 lists the abbreviation and definition of features.

Table 1. Feature abbreviation and definition.

Index	Feature	Definition
1	Gender	Gender of subjects
2	Age	Age of subjects
3	Weight	Weight of subjects
4	PATp	Pulse transit time from the R peak of ECG to the peak point of PPG
5	PATf	Pulse transit time from the R peak of ECG to the valley point of PPG
6	PATd	Pulse transit time from the R peak of ECG to the maximum difference point of PPG
7	HR	Heart rate
8	PP	The peak-to-peak interval of PPG
9	PPT	The time interval from the first peak point to the second peak or the inflection point of PPG
10	PIRp	PPG intensity ratio of the peak point to valley point
11	PIRmd	PPG intensity ratio of the maximum difference point to the valley point
12	PIavg	The average amplitude of PPG
13	PIsd	The standard deviation of PPG
14	PImax	The maximum amplitude of PPG
15	PImin	The minimum amplitude of PPG
16	SA	The sum of ascending branch value of PPG
17	AA	The average of ascending branch value of PPG
18	SD	The sum of descending branch value of PPG
19	AD	The average descending branch value of PPG
20	SR	The sum ratio of ascending branch value to descending branch value of PPG
21	K	K value of PPG
22	AT	Ascending time of PPG
23	DT	Descending time of PPG
24	AS	The ascending slope of PPG
25	DS	The descending slope of PPG

1. Individual features

It was proved that age and gender were correlated with arterial stiffness and PWV [35,36], and so, we took individual information into account. During the collection process, we recorded the gender, age, and weight information of each volunteer. Meanwhile, by querying the MIMIC-III clinical database, this information can also be obtained.

2. PAT

PAT is defined as the time required for blood to flow from the beginning of the electrical activation of the heart to the distal point [15]. As it contains the PTT feature, PAT is often used as a critical parameter for BP estimation [14]. Generally, the R- peak of the ECG signal is taken as the starting point, and the feature points of the PPG signal are taken as the ending point to calculate PAT. We extracted PAT_p, PAT_f, and PAT_d, which means the distance from the R-peak of the ECG signal to the peak, the valley, and the maximum slope point of the PPG signal, which is shown in Figure 3.

3. Other time-related features

Heart rate (*HR*): *HR* reflects the heart cycle and can be calculated by the *RR* interval of the ECG signal. Cardiac output can be correlated with PTT through HR, and so, there is also a correlation between heart rate and blood pressure. HR can be calculated by the following formula:

$$HR = \frac{60 \times fs}{RR} \tag{1}$$

The peak-to-peak interval of the PPG signal (PP): PP is the time interval between two peaks. Studies found that participants with HPT or arteriosclerosis had "longer PP interval" than healthy participants [37].

The PPG waveform consists of ascending branches and descending branches. The ascending branch time (AT) is the time from the valley to the peak of the PPG signal, which was proven to be a useful feature for classifying the PWV. The descending branch time (DT) is the time from the peak to the next valley of the PPG signal.

Peak-to-peak time (PPT) is defined as the time between the first peak and the second peak or inflection point of the PPG signal. Its definition depends on the contour of the PPG waveform [38]. The second peak or inflection point is generated by reflection waves, which is related to the time required for the PPG signal to transit from the heart to the peripheral and return, so it can be used to evaluate the artery's stiffness and the PWV.

4. Intensity-related features

From the perspective of the formation of BP, it is mainly affected by five factors: cardiac output, peripheral resistance, arterial wall elasticity, circulating blood volume, and blood volume ratio. PPG intensity ratio (PIR) is related to changes in arterial diameter, which is the main cause of peripheral resistance and blood volume. PIR can be used to evaluate the smooth muscle tension that regulates arterial blood pressure in the low-frequency range and improve the accuracy of blood pressure estimation [39,40]. We selected the ratio of peak intensity to valley intensity (PIRp) and the ratio of maximum slope point intensity to valley intensity (PIRmd), which can be calculated by the Formula (2).

$$PIR_x = \frac{PI_x}{PI_v} \tag{2}$$

where PIR_x is the PPG intensity ratio, PI_x is the PPG intensity of the peak or the maximum slope point, and PI_v is the PPG intensity of the valley.

To reflect the change in the PPG intensity, we also added the statistical features of the PPG intensity, including the average value of the PPG intensity (PIavg), the standard deviation of the PPG intensity (PIsd), the maximum value of the PPG intensity (PImax), the minimum value of the PPG intensity (PImin).

5. *K* value

K value can reflect the physiological factors of the human cardiovascular system, such as vascular peripheral resistance, vascular wall elasticity, and blood viscosity. It is an important physiological index for the clinical examination of cardiovascular disease and has important clinical application value. It was calculated according to Formula (3):

$$K = \frac{P_m - PI_v}{PI_p - PI_v} \tag{3}$$

where P_m is the average intensity of the PPG signal in one cardiac cycle, which was calculated by Formula (4):

$$P_m = \frac{1}{T}\int_0^T PI(t)dt \tag{4}$$

$PI(t)$ is the intensity of the PPG signal at time t, and T is the time of one cycle of the PPG signal.

6. Other waveform features

The slope of ascending branch and descending branch can be calculated by PI_p, AT, and DT. The area of ascending branch and descending branch can be obtained by integrating the PPG intensity. We added seven features to the model, including the ascending slope of the PPG signal (AS), the descending slope of the PPG signal (DS), the sum of ascending branch value of the PPG signal (SA), the average of ascending branch value of the PPG signal (AA), the sum of descending branch value of the PPG signal (SD), the average of descending branch value of the PPG signal (AD), and the sum ratio of the ascending branch value to the descending branch value of the PPG signal (SR).

3.2. Regression Model

Random forest (RF) [41] uses decision trees as weak learners and combines multiple decision trees to make predictions through random sampling with replacement. It can be used in classification and regression tasks, in which the final result is obtained by voting in classification and the average value is taken as the final result in regression. Even if there is a non-linear relationship between the input and label, it can still maintain good performance. Each decision tree constructed is different to reduce the deviation and variance of the prediction results. Because of the multiple combinations of the prediction results, RF is not sensitive to outliers and has a better ability for anti-overfitting and stability.

Compared with random search, grid search is time-consuming and easy to result in dimension disaster, and so, we used the bootstrap method to train the random forest. The number of decision trees was set to 100, the maximum depth of trees was set to 50, the min_samples_split was set to 2, and the min_samples_leaf was set to 1.

3.3. GCMI

MI is a measure of the mutual dependence between two random variables [42]. It measures the degree of information about one variable that is learned through observing the other [43]. It is a non-negative value, with higher values indicating stronger dependence between the variables. When two random variables follow the Gaussian distribution and X is a multidimensional vector, the calculated MI is the GCMI [29]. It can be obtained by the following formula:

$$I(X,Y) = 0.5 \times \log_2[\frac{|\sum X||\sum Y|}{|\sum XY|}] \quad (5)$$

where $\sum X$ and $\sum Y$ are the covariance matrices of X and Y, respectively, and $\sum XY$ is the covariance matrix of the joint variables (X, Y). The detailed feature selection process is shown in Algorithm 1. First, each feature is calculated GCMI with SBP and DBP. If the value is zero, the corresponding feature will be removed. Due to the redundancy between features, the GCMI of the combined group may not necessarily be the highest. Therefore, we took the approach of deleting features and gradually eliminating the features that contributed the least to the group. Meanwhile, we set a threshold. When the minimum MI loss of the group exceeded the threshold after eliminating a feature, the feature selection process was stopped. The rest features were the filtered feature set.

Algorithm 1. GCMI-based feature selection method.

Input : Training set D : $F = \{f_1, f_2, \cdots, f_n\}$, the label C
Output: Selected features F'
Steps:
(1) For i = 1 to n, do
(2) If $I(f_i, C) = 0$, then $F1 = F - f_i$
(3) end
(4) Repeat
(5) Calculate $I_1 = I(F', C)$
(6) For i = 1 to n, do
(7) $F2 = F' - f_i$, Calculate $I_2 = I(F_2, C)$
(8) end
(9) Find min (I_2) with f_i, then
(10) $F' = F' - f_i \mid \min(I_2)$
(11) Until $I_1 - \min(I_2) > 0.002$
(12) Return F'

4. Results

In this paper, the mean absolute error (*MAE*) and standard deviation (*STD*) are used to evaluate the prediction results, given by

$$MAE = \frac{1}{N}\sum_{i}^{N}\left|\widetilde{y}_{BP}^{i} - y_{BP}^{i}\right| \quad (6)$$

$$STD = \sqrt{\frac{1}{N}\sum_{i=1}^{N}\left(\widetilde{y}_{BP}^{i} - \overline{y}_{BP}\right)^{2}} \quad (7)$$

where N is the number of samples, i is one of the samples, $\widetilde{y}_{BP}^{i}$, y_{BP}^{i}, and $\overline{y}_{BP}$ are the reference *BP*, estimated *BP*, and average *BP*, respectively.

4.1. Waveform Display

Figure 5 shows the raw and filtered waveforms recorded within 30 s. It was observed that the raw ECG contained significant baseline drift due to respiration and body movement. After processing, the noise was removed, and the characteristic waveforms of the ECG, such as the P wave, QRS complex, and T wave, were well preserved. For PPG, with the use of the finger-clip blood oxygen probe, the shape of the original PPG signal was already very good, and the peak points, valley points, and dicrotic notch points were all very clear. High-quality signals provided a guarantee for the accurate extraction of more feature information. Moreover, comparing the ECG and PPG signals, they contained the same number of peaks during the same period, indicating that the recording of the signals was synchronous. It should be noted that due to PAT, there was a slight delay in their peak value of them. The waveform demonstrated that the entire hardware acquisition system was working well.

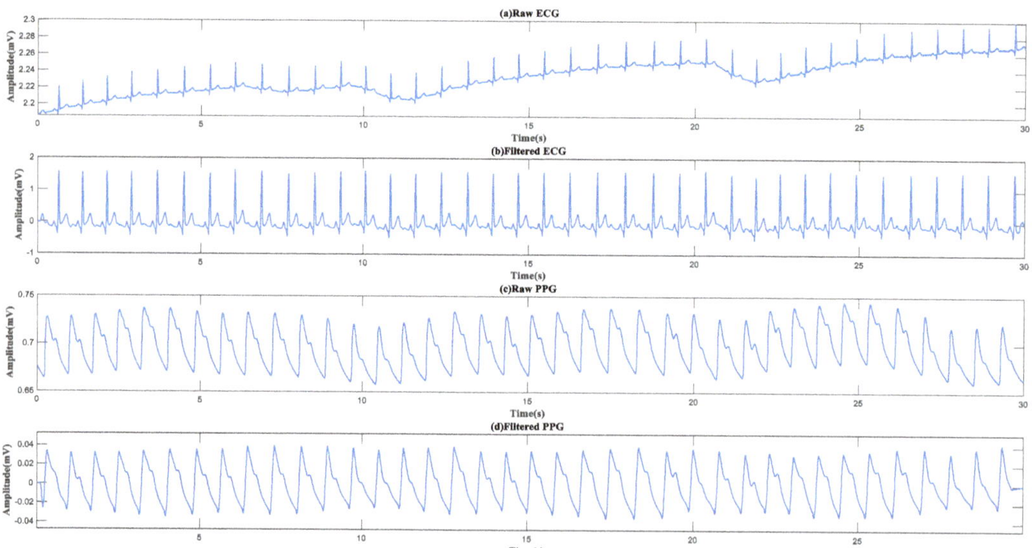

Figure 5. Raw and filtered PPG and ECG signals acquired from eMMC within 30 s.

4.2. Feature Selection

Figure 6 shows the GCMI between BP values and different feature sets. It can be seen that there was almost no change in GCMI when the features were initially deleted. This indicates that these features were redundant and the information they contained could be replaced by other features. As the number of features decreased, the GCMI started to decrease gradually. When the number of features of SBP and DBP was less than 11 or 15, respectively, the value of GCCI decreased significantly. Therefore, we chose these two values as thresholds for the feature subset. Unfortunately, if the number of features was less than 10, the GCMI values decreased rapidly for both SBP and DBP, suggesting that the feature set was no longer favorable for BP prediction at this time, because most information was lost.

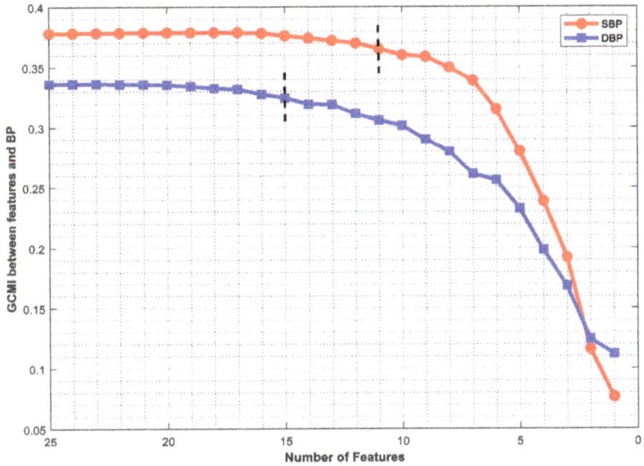

Figure 6. GCMI between different numbers of features and BP (the black dotted line represents the cut-off point for feature selection).

4.3. *Model Performance*

According to the threshold, 11 and 15 optimal feature sets were retained for SBP and DBP, respectively, which are described in Table 2. It was found that almost all individual information was retained, which reflects that individual information plays an essential role in BP values. Meanwhile, from the definition, we can conclude that there was a certain similarity between the feature "PP" and "HR". Only one of them was retained in the filtered feature set after screening.

Table 2. Optimal feature subset for SBP and DBP.

	SBP			DBP		
Feature	Age	AA	DS	Gender	HR	AD
	Weight	SD		Age	K	SR
	PATf	AD		Weight	PIsd	AT
	HR	SR		PATf	AA	AS
	PIsd	AT		PATd	SD	DS

The predicted results before and after feature selection are compared in Table 3. After feature selection, the MAE and STD were 7.93 ± 9.12 mmHg and 7.63 ± 8.61 mmHg, which was smaller than before. This indicates that there were abnormal features in the initial feature set, which negatively affected the robustness of the system. The selected feature set not only accelerated the training speed of the model but also improved its performance This also demonstrated the effectiveness of feature screening.

Table 3. Results before and after feature selection.

	Feature Selection	MAE ± STD (mmHg)
SBP	No	9.12 ± 9.83
	Yes	7.93 ± 9.12
DBP	No	8.31 ± 9.23
	Yes	7.63 ± 8.61

Since the records in the training and testing sets were from different individuals, calibration was necessary for BP prediction. Based on the model trained on the training sets, we further selected a quarter of the testing sets to fine-tune the model, and used the calibrated model to predict the remaining data [44]. This operation is also known as personal calibration and is beneficial to reduce error. The result after calibration is shown in Table 4 and Figure 7. It can be seen that after calibration, the MAE and STD for SBP and DBP were significantly reduced, and both reached grade "A" on the criteria of British Hypertension Society Standard (BHS). This proves that after calibration, the optimal feature set achieved great performance.

Table 4. The MAE and STD for SBP and DBP after calibration.

	MAE ± STD (mmHg)	The Proportion of MAE ≤5 mmHg	≤10 mmHg	≤15 mmHg	Grade
SBP	5.21 ± 5.98	65%	86%	97%	A
DBP	4.15 ± 5.66	72%	89%	98%	A
BHS	-	60%	85%	95%	A
	-	50%	75%	90%	B
	-	40%	65%	85%	C

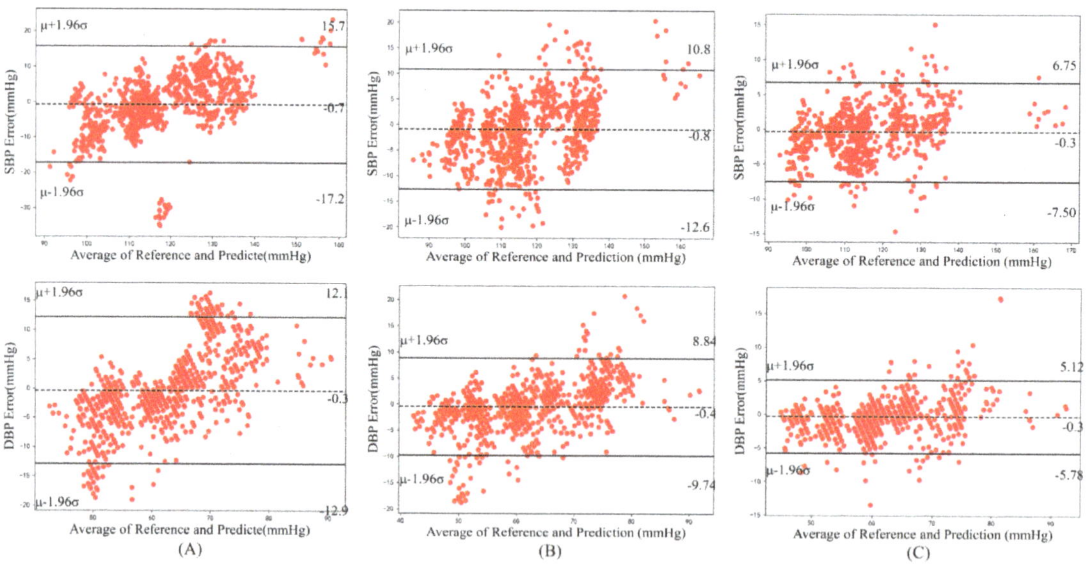

Figure 7. Bland–Altman plots showing the mean difference and 95% limits of agreement for SBP and DBP. The dotted line represents the mean value and the solid line represents the mean value ± 1.96 standard deviation. ((**A**) all features; (**B**) after feature selection; (**C**) after calculation).

To further analyze the results of this experiment, we displayed the Bland–Altman plots of the predicted SBP and DBP under different conditions. As shown in Figure 6, the feature selection proposed in this paper was effective by comparing A and B. The low and upper limits of agreement reduced from [−17.2, 15.7] mmHg to [−12.6, 10.8] mmHg for SBP and [−12.9, 12.1] mmHg to [−9.74, 8.84] mmHg for DBP. The calibration process helped it down further to [−7.5, 6.75] mmHg for SBP and [−5.78, 5.12] mmHg for DBP, respectively.

4.4. Comparison with Related Works

Many scholars conducted research on BP prediction based on feature extraction and selection from PPG and ECG signals. Table 5 provides details and the performance of these methods. Before calculation, although the prediction for DBP was slightly worse, the error for SBP in this paper was smaller than that of other methods. Furthermore, the calibration operation made up for this deficiency, and the results were further improved. After feature screening, the number of features retained were the least. However, in [45], the MAE for SBP and DBP was much smaller. On the one hand, the author trained and tested the model on a single dataset and the distribution difference of the data was small. On the other hand, the performance of Bi-GRU and attention mechanism was superior to traditional machine learning methods but also increased the computational complexity.

Table 5. Performance comparison with related works.

Calculation	Works	Feature	Methods	SBP (mmHg)		DBP (mmHg)	
		Num.*		MAE	STD	MAE	STD
No	Ours	11/15 *	RF	7.93	9.12	7.63	8.61
	Kachuee [15]	15/15	Adaboost	11.17	10.09	5.35	6.14
	Zhang [19]	20/20	Adaboost	10.03	14.43	5.35	8.33
Yes	Ours	11/15	RF	5.21	5.98	4.15	5.66
	Zhang [19]	20/20	Adaboost	7.73	7.96	4.30	4.50
	Geerthy [20]	23/17	RF	9.00	-	5.48	-
	Kachuee [15]	15/15	Adaboost	8.21	5.45	4.31	3.52
	El-Hajj [45]	22/22	Bi-GRU	2.58	3.35	1.26	1.63

*: The number before and after "/" is the feature number for SBP and DBP respectively.

5. Discussion

Recently, wearable devices made great achievements in human multi-physiological signal detection [46,47]. The research of noninvasive continuous BP estimation based on the ECG signal and PPG signal attracted people's attention, which makes the BP measurement equipment more portable and convenient. BP changes with time and is affected by multiple factors, such as exercise, age, emotion, diet, and temperature. It is easy to manifest the "white coat hypertension effect" when measuring BP in the hospital, which makes the accurate value more difficult to obtain [48]. Most databases such as UCI [15] contain multiple physiological signals, but information about age, gender, and weight is not available. Previous research had found that BP is highly related to these factors, which brings disadvantages to the research.

The designed device had the necessary functions for physiological signals acquisition, and the overall size was very small and convenient for daily signal acquisition. This provides a technical guarantee for our subsequent research without having to be limited by the public database. Meanwhile, as the number of extracted features grows, it is critical to obtain the most optimal set of features. Since the MI is nonlinear and simple to calculate [49], it can effectively represent the dependence between features and is suitable for BP prediction.

6. Conclusions

In this paper, we designed a wearable device to collect ECG and PPG signals. Then, we proposed a multi-parameter fusion method for noninvasive continuous BP estimation. GCMI method was introduced to reduce features from 25 to 11 and 15 for SBP and DBP, respectively. This procedure effectively avoided the influence of redundant features on the model. We trained the model on a public dataset and tested it on a private dataset. The results showed that after calibration, the proposed method achieved good results.

In the future, we intend collect more signals from the hospital or community, especially from the elderly over 60 years old with hypertension. The amount of current private datasets is still small and the source of data is concentrated in younger people. The old are at high risk for HPT and have a more complex physical situation. It is meaningful to improve the accuracy of continuous BP monitoring for the old. Furthermore, we attempted to combine the traditional features with the CNN and LSTM networks. The 25 features extracted in this paper were still limited. For information mining of big data, these networks have more advantages than manual extraction.

Author Contributions: Conceptualization, G.M. and Y.Y.; methodology, G.M. and J.Z.; software, J.L.; validation and formal analysis, G.M., J.Z. and J.L.; investigation and resources, G.M., J.L. and L.W.; data curation, Y.Y. writing—original draft preparation, G.M., J.Z. and J.L.; writing—review and editing, G.M., L.W and Y.Y.; visualization and supervision, G.M. and L.W. All authors have read and agreed to the published version of the manuscript.

Funding: This work was supported by the National Key Research and Development Program of China (2022YFC3601003), the National Natural Science Foundation of China (62001470) and the Natural Science Foundation of Shandong Province, China (ZR2020QF021).

Data Availability Statement: The data presented in this study are available from the corresponding author upon reasonable request.

Conflicts of Interest: The authors declare no conflict of interest.

References

1. Kannel, W.B. Blood Pressure as a Cardiovascular Risk Factor. *JAMA* **1996**, *275*, 1571. [CrossRef] [PubMed]
2. Macmahon, S.; Peto, R.; Collins, R.; Godwin, J.; Cutler, J.; Sorlie, P.; Abbott, R.; Neaton, J.; Dyer, A.; Stamler, J. Blood pressure, stroke, and coronary heart disease: Part 1, prolonged differences in blood pressure: Prospective observational studies corrected for the regression dilution bias. *Lancet* **1990**, *335*, 765–774. [CrossRef] [PubMed]
3. Group, S.R. A randomized trial of intensive versus standard blood-pressure control. *N. Engl. J. Med.* **2015**, *373*, 2103–2116. [CrossRef]
4. Thomopoulos, C.; Parati, G.; Zanchetti, A. Effects of blood pressure lowering on outcome incidence in hypertension: Effects of more vs. less intensive blood pressure lowering and different achieved blood pressure levels–updated overview and meta-analyses of randomized trials. *J. Hypertens.* **2016**, *34*, 613–622. [CrossRef]
5. Olesen, T.B.; Pareek, M.; Stidsen, J.V.; Blicher, M.K.; Rasmussen, S.; Vishram-Nielsen, J.; Maagaard, L.; Olsen, M.H. Association between antecedent blood pressure, hypertension-mediated organ damage and cardiovascular outcome. *Blood Press.* **2020**, *29*, 232–240. [CrossRef]
6. Natarajan, K.; Block, R.C.; Yavarimanesh, M.; Chandrasekhar, A.; Mestha, L.K.; Inan, O.T.; Hahn, J.-O.; Mukkamala, R. Photoplethysmography Fast Upstroke Time Intervals Can Be Useful Features for Cuff-Less Measurement of Blood Pressure Changes in Humans. *IEEE Trans. Biomed. Eng.* **2022**, *69*, 53–62. [CrossRef]
7. Sharifi, I.; Goudarzi, S.; Khodabakhshi, M.B. A novel dynamical approach in continuous cuffless blood pressure estimation based on ECG and PPG signals. *Artif. Intell. Med.* **2019**, *97*, 143–151. [CrossRef] [PubMed]
8. Gupta, K.; Bajaj, V.; Ansari, I.A. A support system for automatic classification of hypertension using BCG signals. *Expert Syst. Appl.* **2023**, *214*, 119058. [CrossRef]
9. Agham, N.D.; Chaskar, U.M. Learning and non-learning algorithms for cuffless blood pressure measurement: A review. *Med. Biol. Eng. Comput.* **2021**, *59*, 1201–1222. [CrossRef]
10. Martinez-Ríos, E.; Montesinos, L.; Alfaro-Ponce, M.; Pecchia, L. A review of machine learning in hypertension detection and blood pressure estimation based on clinical and physiological data. *Biomed. Signal Process. Control* **2021**, *68*, 102813. [CrossRef]
11. Maqsood, S.; Xu, S.; Tran, S.; Garg, S.; Springer, M.; Karunanithi, M.; Mohawesh, R. A survey: From shallow to deep machine learning approaches for blood pressure estimation using biosensors. *Expert. Syst. Appl.* **2022**, *197*, 116788. [CrossRef]
12. Rachim, V.P.; Chung, W.Y. Multimodal Wrist Biosensor for Wearable Cuff-less Blood Pressure Monitoring System. *Sci. Rep.* **2019**, *9*, 7947. [CrossRef] [PubMed]

13. Bui, N.; Pham, N.; Barnitz, J.J.; Zou, Z.; Nguyen, P.; Truong, H.; Kim, T.; Farrow, N.; Nguyen, A.; Xiao, J.; et al. eBP: A Wearable System For Frequent and Comfortable Blood Pressure Monitoring From User's Ear. In Proceedings of the 25th Annual International Conference on Mobile Computing and Networking, Association for Computing Machinery, Los Cabos, Mexico, 21–25 October 2019.
14. Mukkamala, R.; Hahn, J.-O.; Inan, O.T.; Mestha, L.K.; Kim, C.-S.; Toreyin, H.; Kyal, S. Toward Ubiquitous Blood Pressure Monitoring via Pulse Transit Time: Theory and Practice. *IEEE Trans. Biomed. Eng.* **2015**, *62*, 1879–1901. [CrossRef] [PubMed]
15. Kachuee, M.; Kiani, M.M.; Mohammadzade, H.; Shabany, M. Cuffless Blood Pressure Estimation Algorithms for Continuous Health-Care Monitoring. *IEEE Trans. Biomed. Eng.* **2017**, *64*, 859–869. [CrossRef]
16. Nabeel, P.M.; Kiran, V.R.; Joseph, J.; Abhidev, V.V.; Sivaprakasam, M. Local Pulse Wave Velocity: Theory, Methods, Advancements, and Clinical Applications. *IEEE Rev. Biomed. Eng.* **2020**, *13*, 74–112. [CrossRef]
17. Chen, W.; Kobayashi, T.; Ichikawa, S.; Takeuchi, Y.; Togawa, T. Continuous estimation of systolic blood pressure using the pulse arrival time and intermittent calibration. *Med. Biol. Eng. Comput.* **2000**, *38*, 569–574. [CrossRef]
18. Jia, L. Biomechanics and Blood Pressure with Modeling of Pulse Wave Velocity Based on Multiple Linear Regression. In *Advanced Materials Research*; Trans Tech Publications: Zurich, Switzerland, 2014; pp. 261–264.
19. Zhang, Y.; Zhou, C.; Huang, Z.; Ye, X. Study of cuffless blood pressure estimation method based on multiple physiological parameters. *Physiol. Meas.* **2021**, *42*, 055004. [CrossRef]
20. Thambiraj, G.; Gandhi, U.; Mangalanathan, U.; Jose, V.J.M.; Anand, M. Investigation on the effect of Womersley number, ECG and PPG features for cuff less blood pressure estimation using machine learning. *Biomed. Signal Process. Control* **2020**, *60*, 101942. [CrossRef]
21. Wang, B.; Huang, Z.; Wu, J.; Liu, Z.; Liu, Y.; Zhang, P. *Continuous Blood Pressure Estimation Using PPG and ECG Signal*; Springer International Publishing: Berlin, Germany, 2019; pp. 61–74.
22. Hsu, Y.C.; Li, Y.H.; Chang, C.C.; Harfiya, L.N. Generalized Deep Neural Network Model for Cuffless Blood Pressure Estimation with Photoplethysmogram Signal Only. *Sensors* **2020**, *20*, 5668. [CrossRef]
23. Chen, S.; Ji, Z.; Wu, H.; Xu, Y. A Non-Invasive Continuous Blood Pressure Estimation Approach Based on Machine Learning. *Sensors* **2019**, *19*, 2585. [CrossRef]
24. Chowdhury, M.H.; Shuzan, N.I.; Chowdhury, M.E.; Mahbub, Z.B.; Uddin, M.M.; Khandakar, A.; Reaz, M.B.I. Estimating Blood Pressure from the Photoplethysmogram Signal and Demographic Features Using Machine Learning Techniques. *Sensors* **2020**, *20*, 3127. [CrossRef] [PubMed]
25. Tobore, I.; Liew, M.S.; Yuhang, L.; Al-Handarish, Y.; Kandwal, A.; Nie, Z.; Wang, B. Deep Learning Intervention for Health Care Challenges: Some Biomedical Domain Considerations. *JMIR Mhealth Uhealth* **2019**, *7*, e11966. [CrossRef] [PubMed]
26. Yu, M.; Huang, Z.; Zhu, Y.; Zhou, P.; Zhu, J. Attention-based residual improved U-Net model for continuous blood pressure monitoring by using photoplethysmography signal. *Biomed. Signal Process. Control* **2022**, *75*, 103581. [CrossRef]
27. Wang, C.; Yang, F.; Yuan, X.; Zhang, Y.; Chang, K.; Li, Z. *An End-to-End Neural Network Model for Blood Pressure Estimation Using PPG Signal*; Springer: Singapore, 2020; pp. 262–272.
28. Senturk, U.; Polat, K.; Yucedag, I. A non-invasive continuous cuffless blood pressure estimation using dynamic Recurrent Neural Networks. *Appl. Acoust.* **2020**, *170*, 107534. [CrossRef]
29. Ince, R.A.; Giordano, B.L.; Kayser, C.; Rousselet, G.A.; Gross, J.; Schyns, P.G. A statistical framework for neuroimaging data analysis based on mutual information estimated via a gaussian copula. *Hum. Brain Mapp.* **2017**, *38*, 1541–1573. [CrossRef]
30. Johnson, A.E.W.; Pollard, T.J.; Shen, L.; Lehman, L.-W.H.; Feng, M.; Ghassemi, M.; Moody, B.; Szolovits, P.; Celi, L.A.; Mark, R.G. MIMIC-III, a freely accessible critical care database. *Sci. Data* **2016**, *3*, 160035. [CrossRef]
31. Ziarani, A.K.; Konrad, A. A nonlinear adaptive method of elimination of power line interference in ECG signals. *IEEE Trans. Biomed. Eng.* **2002**, *49*, 540–547. [CrossRef]
32. Yin, S.; Li, G.; Luo, Y.; Lin, L. Cuff-less continuous blood pressure measurement based on multiple types of information fusion. *Biomed. Signal Process. Control* **2021**, *68*, 102549. [CrossRef]
33. Member, S.; IEEE; Pan, J.; Tompkins, W.J. A Real-Time QRS Detection Algorithm. *IEEE Trans. Biomed. Eng.* **2007**, *32*, 230–236.
34. Elgendi, M.; Norton, I.; Brearley, M.; Abbott, D.; Schuurmans, D. Systolic peak detection in acceleration photoplethysmograms measured from emergency responders in tropical conditions. *PLoS ONE* **2013**, *8*, e76585. [CrossRef]
35. Mitchell, G.F.; Parise, H.; Benjamin, E.J.; Larson, M.G.; Keyes, M.J.; Vita, J.A.; Vasan, R.S.; Levy, D. Changes in arterial stiffness and wave reflection with advancing age in healthy men and women: The Framingham Heart Study. *Hypertens. Off. J. Am. Heart Assoc.* **2004**, *43*, 1239–1245. [CrossRef] [PubMed]
36. Schiffrin, E.L. Vascular stiffening and arterial compliance: Implications for systolic blood pressure. *Am. J. Hypertens.* **2004**, *17*, 39S–48S. [CrossRef] [PubMed]
37. Dillon, J.B.; Hertzman, A.B. The form of the volume pulse in the finger pad in health, arteriosclerosis, and hypertension. *Am. Heart J.* **1941**, *21*, 172–190. [CrossRef]
38. Alty, S.R.; Angarita-Jaimes, N.; Millasseau, S.C.; Chowienczyk, P.J. Predicting arterial stiffness from the digital volume pulse waveform. *IEEE Trans. Biomed. Eng.* **2007**, *54*, 2268–2275. [CrossRef]
39. Ding, X.-R.; Zhang, Y.-T. Photoplethysmogram intensity ratio: A potential indicator for improving the accuracy of PTT-based cuffless blood pressure estimation. In Proceedings of the 2015 37th Annual International Conference of the IEEE Engineering in Medicine and Biology Society (EMBC), Milan, Italy, 25–29 August 2015; pp. 398–401.

40. Ding, X.-R.; Zhang, Y.-T.; Liu, J.; Dai, W.-X.; Tsang, H.K. Continuous Cuffless Blood Pressure Estimation Using Pulse Transit Time and Photoplethysmogram Intensity Ratio. *IEEE Trans. Biomed. Eng.* **2016**, *63*, 964–972. [CrossRef]
41. Breiman, L. Random forests. *Mach. Learn.* **2001**, *45*, 5–32. [CrossRef]
42. Kraskov, A.; Stogbauer, H.; Grassberger, P. Estimating mutual information. *Phys. Rev. E Stat. Nonlin Soft Matter Phys.* **2004**, *69*, 066138. [CrossRef]
43. Liu, H.; Sun, J.; Liu, L.; Zhang, H. Feature selection with dynamic mutual information. *Pattern Recognit.* **2009**, *42*, 1330–1339. [CrossRef]
44. Pandit, J.A.; Lores, E.; Batlle, D. Cuffless Blood Pressure Monitoring: Promises and Challenges. *Clin. J. Am. Soc. Nephrol.* **2020**, *15*, 1531–1538. [CrossRef]
45. El-Hajj, C.; Kyriacou, P.A. Deep learning models for cuffless blood pressure monitoring from PPG signals using attention mechanism. *Biomed. Signal Process. Control* **2021**, *65*, 102301. [CrossRef]
46. Moon, J.H.; Kang, M.-K.; Choi, C.-E.; Min, J.; Lee, H.-Y.; Lim, S. Validation of a wearable cuff-less wristwatch-type blood pressure monitoring device. *Sci. Rep.* **2020**, *10*, 19015. [CrossRef] [PubMed]
47. Guo, Y.; Liu, X.; Peng, S.; Jiang, X.; Xu, K.; Chen, C.; Wang, Z.; Dai, C.; Chen, W. A review of wearable and unobtrusive sensing technologies for chronic disease management. *Comput. Biol. Med.* **2021**, *129*, 104163. [CrossRef] [PubMed]
48. Mancia, G.; De Backer, G.; Dominiczak, A.; Cifkova, R.; Fagard, R.; Germano, G.; Grassi, G.; Heagerty, A.M.; Kjeldsen, S.E.; Laurent, S.; et al. 2007 ESH-ESC practice guidelines for the management of arterial hypertension: ESH-ESC task force on the management of arterial hypertension. *J. Hypertens.* **2007**, *25*, 1751–1762. [CrossRef]
49. Hoque, N.; Bhattacharyya, D.K.; Kalita, J.K. MIFS-ND: A mutual information-based feature selection method. *Expert Syst. Appl.* **2014**, *41*, 6371–6385. [CrossRef]

micromachines

MDPI

Article

Evaluating and Visualizing the Contribution of ECG Characteristic Waveforms for PPG-Based Blood Pressure Estimation

Gang Ma [1,2], Yuhang Chen [1,2], Wenliang Zhu [1,2], Lesong Zheng [1], Hui Tang [3], Yong Yu [2] and Lirong Wang [2,3,*]

[1] School of Biomedical Engineering (Suzhou), Division of Life Sciences and Medicine, University of Science and Technology of China, Hefei 230026, China
[2] Suzhou Institute of Biomedical Engineering and Technology, China Academy of Sciences, Suzhou 215163, China
[3] School of Electronics and Information Technology, Soochow University, Suzhou 215006, China
* Correspondence: wanglirong@suda.edu.cn

Abstract: Non-invasive continuous blood pressure monitoring is of great significance for the preventing, diagnosing, and treating of cardiovascular diseases (CVDs). Studies have demonstrated that photoplethysmogram (PPG) and electrocardiogram (ECG) signals can effectively and continuously predict blood pressure (BP). However, most of the BP estimation models focus on the waveform features of the PPG signal, while the peak value of R-wave in ECG is only used as a time reference, and few references investigated the ECG waveforms. This paper aims to evaluate the influence of three characteristic waveforms in ECG on the improvement of BP estimation. PPG is the primary signal, and five input combinations are formed by adding ECG, P wave, QRS complex, T wave, and none. We employ five common convolutional neural networks (CNN) to validate the consistency of the contribution. Meanwhile, with the visualization of Gradient-weighted class activation mapping (Grad-CAM), we generate the heat maps and further visualize the distribution of CNN's attention to each waveform of PPG and ECG. The heat maps show that networks pay more attention to the QRS complex and T wave. In the comparison results, the QRS complex and T wave have more contribution to minimizing errors than P wave. By separately adding P wave, QRS complex, and T wave, the average MAE of these networks reaches 7.87 mmHg, 6.57 mmHg, and 6.21 mmHg for systolic blood pressure (SBP), and 4.27 mmHg, 3.65 mmHg, and 3.73 mmHg, respectively, for diastolic blood pressure (DBP). The results of the experiment show that QRS complex and T wave deserves more attention and feature extraction like PPG waveform features in the continuous BP estimation.

Keywords: ECG characteristic waveforms; Grad-CAM; PPG; CNN; blood pressure

check for updates

Citation: Ma, G.; Chen, Y.; Zhu, W.; Zheng, L.; Tang, H.; Yu, Y.; Wang, L. Evaluating and Visualizing the Contribution of ECG Characteristic Waveforms for PPG-Based Blood Pressure Estimation. *Micromachines* **2022**, *13*, 1438. https://doi.org/10.3390/mi13091438

Academic Editor: Xiao Xiao

Received: 14 August 2022
Accepted: 27 August 2022
Published: 31 August 2022

1. Introduction

Hypertension is one of the most serious but potential public health problems in the world, which will threaten the patients' life and can even result in many diseases such as heart failure, stroke, coronary heart disease, etc. [1]. Modern medical evidence shows that early detection and strict control of hypertension could slow its further development and reduce the risk of disease [2]. Nevertheless, as it is often referred to as a "silent killer", most people with pre-hypertension have no significant symptoms and always ignore it. *The Lancet* reports that the number of people with hypertension between the ages of 30 and 79 doubled from 1990 to 2019, and nearly 580 million people are unaware of it [3]. Therefore, continuous BP monitoring is an effective and necessary solution for the early detection and control of these diseases [4].

In general, there are two basic kinds of BP measurement methods: direct and indirect. Direct measurement, as the gold standard, is carried out by the insertion of a catheter, but is often restricted in critical situations such as intensive care units and clinical surgeries; its result is the most accurate and continuous in nature [5]. As this method is invasive and carries multiple risks, clinically people tend to use indirect methods. Traditionally they use an inflatable cuff to compress the blood vessel and observe changes in pulse wave signal during this process. However, the measurement mechanism of inflation and deflation is cumbersome and not suitable for long-term observation [6]. What is more, due to the pressure on blood vessels, the time interval between two monitoring readings needs to be greater than 2 min. In the past decade, with the emergence of wearable devices that are convenient for physiological signal collection [7], continuous blood pressure monitoring has gradually shown a lot of promise. Based on PPG and ECG, the researchers summarized three basic physiological indicators highly related to BP: pulse arrival time (PAT), pulse transit time (PTT) [8,9], and pulse wave velocity (PWV) [10]. Many studies have proved the effectiveness of these three indicators [11]. Once the PTT, PAT, or PWV are determined, BP can be calculated from a mathematical model.

The above three indicators are essentially based on the blood flow velocity in the blood vessel, and another research trend is the shape of the waveform. The propagation of pulse waves along the artery is affected by heart, vascular consistency, vascular stiffness, vascular resistance [12], etc. Changes in these factors will also be reflected in the waveform shape. Therefore, researchers focus on morphology and are keen to extract characteristic features from these signals [13,14]. Kruylyak [15] extracted 21 physiological indices from PPG features and used an artificial neural network (ANN) model to verify the improvement in accuracy. Liu et al. [16] extracted 14 feature parameters from the second derivative of the PPG(SDPPG) and combined 21 regular time-domain PPG features to establish a Support Vector Regression (SVM). In addition to using PPG and ECG signal features, Yin [17] also added personal information, such as age, weight, and gender from 186 volunteers. From PPG and its derivatives, Lin et al. [18] proposed 19 new physiological parameters and the predicted BP value achieved a decrease when tested on 22 subjects. In the literature [19], the author combined up to 107 features from each PPG signal, including 75 time-domain, 16 frequency-domain and 10 statistical features, as well as 6 demographic data. Then, the feature selection technique was used to reduce the amount of computation, and the Gaussian Process Regression (GPR) method is used. The estimation performance of the ReliefF feature selection algorithm outperformed other algorithms in estimating. The authors above have made full use of PPG waveform characteristics, while only a few articles discussed ECG features. To the best of our knowledge, Jamal [20] firstly explored the importance of the ECG waveform in a BP model. They applied the CNNs to extract 56 features from PPG and ECG segments. By adding and removing ECG features, they proved that ECG waveforms contained important information which is helpful to improve accuracy. Then Geerthy [21] compared the model trained with PPG features fused with 18 ECG features and PPG features alone. They investigated the influence of these features and used random forest-based genetic algorithm (GA) and confirmed that significant features like QRS complex, QT interval, SDI, and heart rate could improve the results.

Compared with classical machine learning models, neural network models have gradually been applied to the field of continuous BP monitoring based on physiological signals and have become a new trend. This method eliminates the step of manual feature extraction and takes advantage of its powerful data mining and feature extraction capabilities [22]. In [23], a four-layer long short-term memory (LSTM) network was employed to estimate systolic blood pressure (SBP) and diastolic blood pressure (DBP). The model contains a bidirectional structure for accessing larger-scale context information of input sequence and residual connections to allow efficient gradient in the LSTM network to propagate. Tanveer et al. [24] combined the preprocessed PPG and ECG waveforms into the hierarchical ANN-LSTM to obtain the value of SBP and DBP. Yu et al. [25] improved U-Net with an attention-based residual and added derivatives of the PPG signal as additional inputs, aiming to improve the effectiveness of information mining. Rong [26] proposed a multi-type features fusion (MTFF) neural network model. This model contains two basic blocks: one is CNN block, which was used to learn the morphological and frequency spectrum features, and the other is Bi-directional LSTM (BLSTM) block, which was used to focus on temporal features.

To sum up, it can be found that BP prediction models based on physiological signals are the mainstream trend of continuous BP monitoring. On the one hand, researchers focus on the extraction of new effective artificial features. However, most of the literature pays too much attention to the characteristics of PPG signal, with the R-peak of the ECG signal used as the time reference point, while few articles use the characteristics of ECG waveform. On the other hand, the CNN models have been continuously improved to extract features, and ECG and PPG signals are used as the original signal input. However, due to the uninterpretable nature of neural networks, it is difficult for people to know which feature of the signal the networks have learned.

In order to better understand and explain CNNs, a visualization technique known as Grad-CAM [27] was developed for visual interpretation. It utilizes gradient information flowing into the last convolutional layer of the CNNs to assign importance values to each neuron for a specific attention decision. Grad-CAM is an improvement of CAM [28] that overcomes the problem inherent in CAM. The obvious shortcoming of CAM is that the network structure needs to be modified, which means that the data need to be retrained. Grad-CAM solved this problem and is suitable for any CNN-based network without modifying the structure. Li et al. [29] applied Grad-CAM to the channel selection of Electroencephalography (EEG) signal and achieved great accuracy. Kin et al. [30] used Grad-CAM for the interpretation of the basis of the judgment of the ECG classification model. They proposed a visual-DenseNet for ECG model using Grad-CAM. These experiments show that Grad-CAM is also suitable for 1-dimensional signals.

Finally, the main contributions of this study are listed as follows:

(1) This paper specifically compares the contribution of the three characteristic waveforms of ECG to the BP prediction model. Each ECG signal is masked as separate P, QRS, and T waves. Then these signals are put into different networks combined with PPG signal for training, rather than directly adding or deleting whole ECG signals [20].
(2) This paper introduces Grad-CAM into a BP regression model so that we can more clearly understand which features of physiological signals will receive more attention or be ignored in the model, so as to provide a reference for artificial feature extraction or modification of the network model. Meanwhile, the visualization results further corroborate with (1), which strongly illustrate the different characteristics that ECG signals and PPG signals provide in the model.

2. Materials and Methods

2.1. Dataset

In this study, original PPG, II-lead ECG, and continuous arterial blood pressure (ABP) signals are from the MIMIC-III database [31]. MIMIC-III is an extensive, freely available database comprising health-related data from more than 40,000 patients who stayed in critical care units of the Beth Israel Deaconess Medical Center between 2001 and 2012 [32]. It contains thousands of distinct physiological signal records and vital sign time series collected from bedside patient monitors in the adult and neonatal intensive care unit, and the sampling rate is 125 Hz. Due to human operation factors, these records can last from a few seconds (usually abnormal) to several hours. For details of the MIMIC-III database, see https://mimic.physionet.org/ accessed on 13 January 2021. PPG and ECG signals are inputs, and the ABP signals are used for calculating the SBP and DBP values as the reference values in this experiment.

2.2. Preprocessing

2.2.1. Signal Selection and Filtering

Since the whole database is very large and the signal quality is even, we conduct a detailed cleaning procedure by the following criteria: First, we used raw PPG and ECG waveforms as network input, and ABP was used to make labels. Some records may lack one of these signals. Second, each waveform needs to be long enough to contain some changes in SBP and DBP. The minimal required record length was set to 1 min. Third, due to occasional human activity that led to sensor separation, a flat line appeared for a period of time in some records, which caused the loss of valid data [33]. Fourth, some records may contain abnormal ABP values (e.g., SBP $\geq$ 180 or DBP $\leq$ 60) [34]. Once a record has any of the above conditions, the corresponding record will be removed. The final database retained 938 subject records.

In order to remove the impacts of noise and artifacts during the collection process, some filters were further implemented to denoise the signals. PPG is susceptible to baseline drift, motion artifacts, and high-frequency noise, so a 4th order Butterworth bandpass filter with cutoff frequencies of 0.05 and 10 Hz was used to remove interfering signals [35]. For ECG, we used a 3rd order Butterworth bandpass filter with cutoff frequencies of 0.05 and 35 Hz [21]. Then each filtered signal was cut into 8 s fragments. In order to make blood pressure labels, we conducted peak detection on ABP signals, taking the mean peak value of each fragment as SBP and the valley value as DBP. The final dataset consists of 9280 fragments and is split into the training set, validating set, and testing set according to the ratio of 6:2:2.

2.2.2. ECG Detection

An ECG normal cardiac cycle contains three essential components: P wave, QRS complex, and T wave, representing different stages of cardiac activity. P wave is the first upward wave, which reflects the process of atrial depolarization and is an important detail in the diagnosis of arrhythmia. QRS complex is the most significant part, which is formed by ventricular contraction. T wave is the slow waveform that appears later, with a longer duration and gentle fluctuations representing the recovery process of ventricular excitation. Guillermo [36] applied the U-net architecture [37], which has been successfully applied in the field of biomedical image segmentation, for ECG delineation and achieved excellent results compared with traditional DSP methods. We adopted this architecture to extract three waveforms. Because the database in paper [36] is different from that of this paper, the accuracy of detection results cannot be guaranteed. However, the delineation results have a direct impact on the subsequent experiments, so we adopt a machine-based and manual-assisted strategy to manually verify the results of network segmentation. Because in an ECG cycle, the three characteristic waveforms have standard intervals, the automatic segmentation and manual inspection can reduce the workload and ensure the accuracy of the results. The extracted results of each ECG signal fragment are shown in Figure 1.

2.3. *Multiple Regression Models*

PPG has been widely used in BP prediction models. In order to verify the improvement of ECG waveform characteristics, we used five combinations as the input of the network: (1) PPG; (2) PPG + P wave; (3) PPG + QRS wave; (4) PPG + T wave; and (5) PPG + ECG. Meanwhile, to avoid accidental errors caused by a single model, we explored several well-known CNN architectures in our experiments, detailed as follows:

(1) AlexNet [38]: was one of the classic CNN networks in the past decade, which has laid an important foundation for the network and proved the effectiveness of CNN in complex models.
(2) GoogLeNet [39]: multiple convolutions or pooling operations were assembled into an inception module, and the network structure was constructed with this module. Then the network was concentrated in feature dimension, which reduced the number of parameters and further improved the network performance.
(3) ResNet18 [40]: proposed in 2015. It established the "shortcut connection" between the front and the back layer, which is useful for the back-propagation of gradients in the training process and deepens the number of layers.
(4) DenseNet121 [41]: established a dense connection between all previous layers and the back layer and realizes feature reuse by connecting features on the channel.

(5) DPN68 [42]: combined the fundamental ideas of ResNet and DenseNet. ResNet was the main framework to ensure low redundancy of features and a DenseNet branch was added to generate new features.

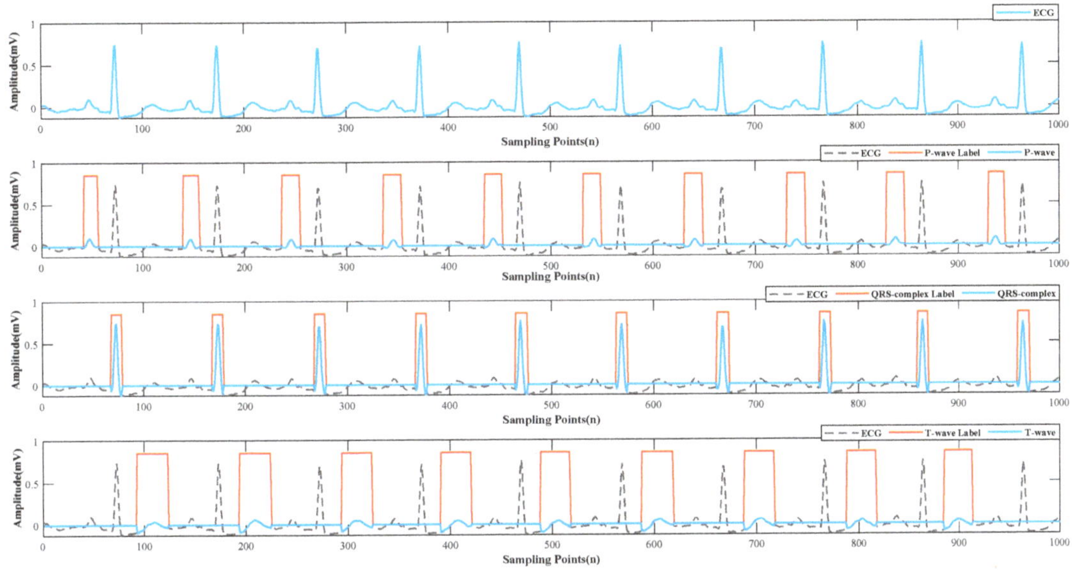

Figure 1. Raw ECG and P wave, QRS complex, T wave.

These original networks all take 2-dimensional images as input, while our experimental data are 1-dimensional signals. Therefore, we modified the first parameter of the convolution kernel to 1, so that the network structure could adapt to the experimental data. Other network parameters are not modified. The experiments were performed on a computer with 1 CPU at 2.6 GHz, 1 NVIDIA GeForce RTX2060 GPU and 64-Gb memory. All the models are run over highly efficient GPU using the PyTorch deep learning framework

2.4. Visualization with Grad-CAM

As a classification localization technology, Grad-CAM [27] can generate visual interpretation from any CNN-based network without an attention mechanism. Therefore, we can visualize the location of feature maps for these five models. For Grad-CAM, it uses the gradient to calculate the fusion weights of target feature images. The activation function Relu is then added to eliminate negative values and retain only those that have a positive effect on the results. As expressed by (1) and (2), A represents a feature map of size i, j, and dimension k. Then the importance weights are achieved by a global average pooling on i and j. Similar to CAM, it is possible to get $L^c_{(Grad-CAM)}$ by weighted sum.

$$\alpha_k^c = \frac{1}{Z}\sum_i \sum_j \frac{\partial y^c}{\partial A_{ij}^k} \tag{1}$$

$$L^c_{Grad-CAM} = \mathrm{ReLU}(\sum_k \alpha_k^c A^k) \tag{2}$$

For the selected five CNN models, we use the last convolution output before the fully connected layer as the input of Grad-CAM, and the output of Grad-CAM is a heat map of hot regions of a specific class that has the same size to the input signal [29]. The overall experimental process is shown in Figure 2.

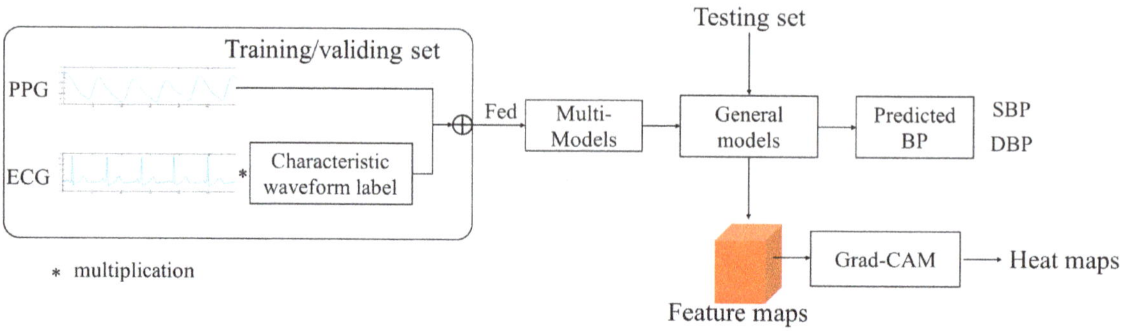

Figure 2. The overall framework of the experimental method.

3. Results

3.1. Metrics

In this study, the quantitative indicators for the performance evaluation of the blood pressure estimation model include the following [35,43]: mean error (*ME*), mean absolute error (*MAE*), standard deviation (*SD*), and root mean square error (*RMSE*). *ME* can reflect the relationship between the predicted value and the actual value. *SD* can reflect the dispersion degree of the prediction results. *MAE* and *RMSE* are also two important scales for evaluating models in machine learning. Compared with *MAE*, *RMSE* is more sensitive to outliers. These four indices are all expected to be as small as possible. Among them, N is the number of test segments, $\hat{y}^i$, y^i, and $\overline{y}^i$ represent the predicted BP, reference BP, and average BP, respectively.

$$ME = \frac{1}{N}\sum_{i}^{N}(\hat{y}^i - y^i) \quad (3)$$

$$MAE = \frac{1}{N}\sum_{i}^{N}\left|\hat{y}^i - y^i\right| \quad (4)$$

$$SD = \sqrt{\frac{1}{N}\sum_{i}^{N}(\hat{y}^i - \overline{y})^2} \quad (5)$$

$$RMSE = \sqrt{\frac{1}{N}\sum_{i}^{N}(\hat{y}^i - y^i)^2} \quad (6)$$

3.2. Analysis of Error Distribution

Table 1 presents the numerical comparison results of the mentioned five models with different signal combinations. Figure 3 compares the MAE estimated by different models for SBP and DBP under the five signal combinations. It can be found that the performance of these five CNN models was basically consistent, and the combination of PPG and ECG achieved the best results. This result is consistent with [20] that ECG contains important information related to BP, because the formation of BP is related to cardiac contractility, blood volume, and peripheral blood resistance. As an external manifestation of electrical signals of cardiac activity, ECG can be used to assess cardiac contractility, thus significantly improving the accuracy of blood pressure models.

Table 1. MAE and RMSE of five models with different input signal combinations.

Models	AlexNet				GoogLeNet				ResNet18			
	SBP/mmHg		DBP/mmHg		SBP/mmHg		DBP/mmHg		SBP/mmHg		DBP/mmHg	
	MAE	RMSE	MAE	RMSE	MAE	RMSE	MAE	RMSE	MAE	RMSE	MAE	RMSE
PPG	7.74	10.53	4.02	5.19	8.35	11.15	5.25	6.71	13.46	17.53	6.81	8.47
PPG + P wave	5.83	7.98	3.34	4.50	8.04	10.57	5.41	7.06	11.08	14.35	5.54	7.16
PPG + QRS wave	6.42	8.80	2.95	3.95	7.70	10.33	4.75	6.34	8.21	10.74	4.64	5.91
PPG + T wave	5.25	6.96	2.67	3.47	6.57	8.68	4.17	5.38	7.80	10.39	4.81	5.87
PPG + ECG	**4.93**	**6.2**	**2.44**	**4.19**	**4.75**	**6.44**	**2.97**	**3.97**	**7.63**	**10.26**	**4.58**	**6.21**

Models	DenseNet121				DPN68			
	SBP/mmHg		DBP/mmHg		SBP/mmHg		DBP/mmHg	
	MAE	RMSE	MAE	RMSE	MAE	RMSE	MAE	RMSE
PPG	8.75	11.81	4.52	5.91	8.48	11.41	4.46	5.98
PPG + P wave	7.25	9.80	3.54	4.71	7.16	9.64	3.69	4.95
PPG + QRS wave	7.70	10.33	4.75	6.34	6.22	8.50	3.40	4.55
PPG + T wave	4.80	6.87	3.37	4.35	6.62	9.14	3.62	4.90
PPG + ECG	**3.69**	**5.18**	**1.95**	**2.71**	**4.92**	**6.86**	**2.86**	**3.97**

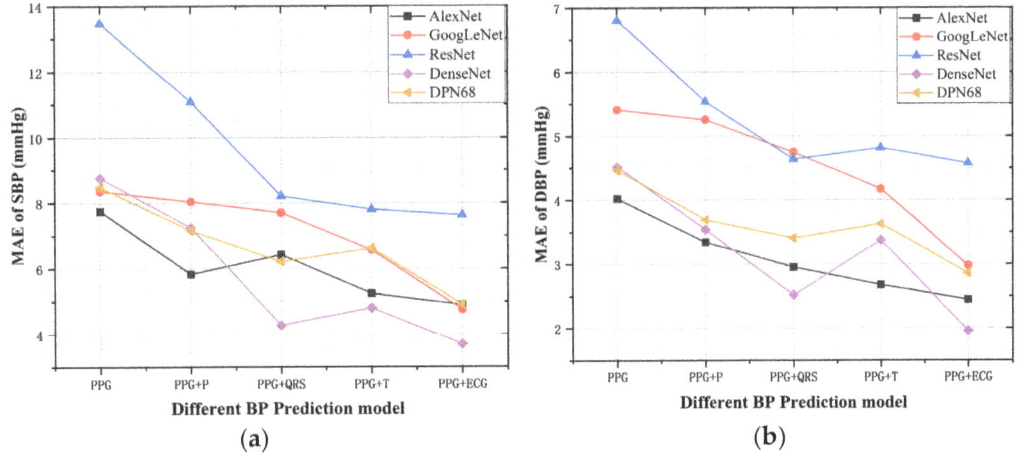

Figure 3. Estimation errors of five models with different input signal combinations. (**a**) MAE of SBP (**b**) MAE of DBP.

For the specific three ECG characteristic waveforms, no matter which one was added, it would have a positive effect on BP prediction, which is more obvious in Figure 3. Generally, the P wave contributes the least to the accuracy of the model, and the QRS complex and T wave have relatively high contributions. In the QRS complex, the R-wave morphology can reflect the intensity of ventricular contraction and then affect cardiac ejection. The contribution of the T wave to the results was unexpected; it followed the QRS complex with a small amplitude but the longest duration of the three characteristic waveforms. This enlightens us that we can pay more attention to these two characteristic waveforms when performing feature extraction.

Figure 4 is a Bland–Altman plot established between the estimated and reference values in the testing group. The mean and limits of agreement (mean ± 1.96SD) are illustrated with the solid dots and short solid lines. The error distribution range with PPG and ECG as inputs is more concentrated. By adding ECG characteristic waveforms, the error distribution range becomes obviously more concentrated and reaches an optimum when PPG and ECG are used as inputs. This result is consistent in all five models.

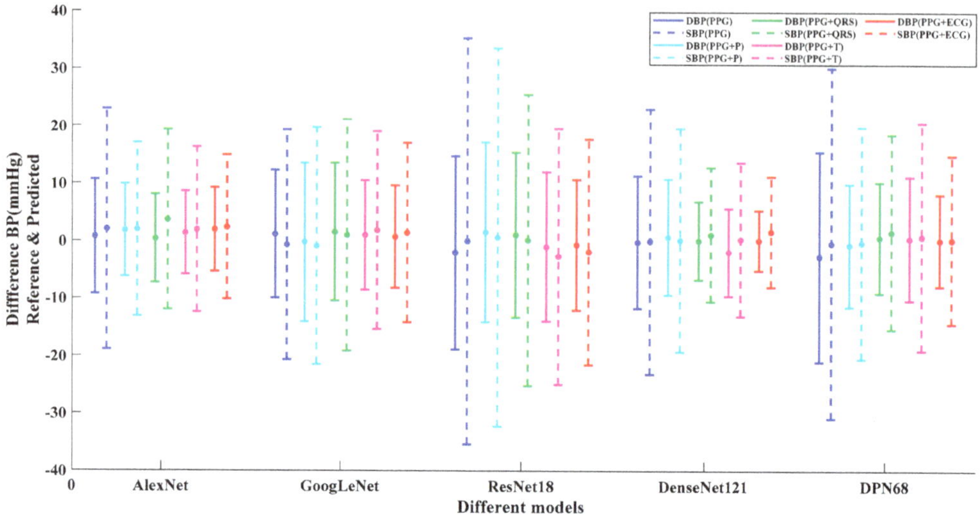

Figure 4. Mean and 95% CI plot for Different Models (solid dots represents mean value and short solid lines represent mean ± 1.96SD).

3.3. *Performance of Visualization*

Figure 5 shows the heat map of the features that each model learns when PPG and ECG are used as the signal input, in which the warm-colored parts indicate the more attention, and the cool-colored parts represent less attention on signal segments. For the ECG signal, it can be found that the networks show different degrees of highlight for each wave group. In Figure 5a,b,d,e, QRS and T waves receive more attention than P waves. Especially in Figure 5c, there is a significant periodic highlight in the QRS wave. This can be consistent with the results in Figure 3. Adding QRS and T waves greatly improves the accuracy of the prediction results because the network has learned more features in these two parts.

For PPG signal, we find that CNN networks are more likely to focus on the locations of peaks, valleys, and dicrotic notch, which are all the extreme points in a cycle of PPG. In traditional manual extraction of features [15–18], these points represent much time information and amplitude features, and scholars have deeply excavated them in both the time and frequency domains. This indicates that the features learned by the network inherit the advantages of artificial features.

It can be seen from Figure 3 that the prediction error of model Resnet18 is much larger than that of the other models when PPG and ECG are used as input signals. By comparing Figure 5c and others, we find that the focus of model Resnet18 is limited to the QRS complex, ignoring the rest waveform features; the characteristics learned in PPG are not significant, with only a narrow highlight at the peaks.

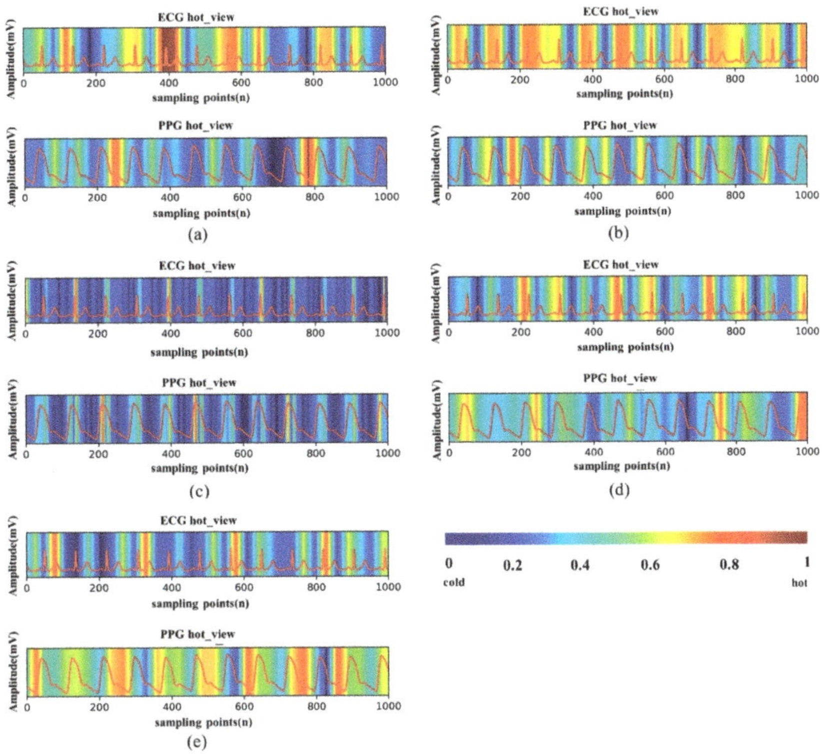

Figure 5. Heat maps generated by different models. (**a**) AlexNet. (**b**) GoogLeNet. (**c**) ResNet18. (**d**) DenseNet121. (**e**) DPN68.

4. Discussion

Literature [44] has proved that ECG signals can be used to evaluate the contractility of the heart, which has a direct impact on the formation of blood pressure. The external manifestation of cardiac activity is the ECG signal, so the study of ECG waveform has an important role in predicting blood pressure. In the past, the blood pressure prediction models were mainly based on PPG, supplemented by ECG, or the whole ECG waveform as input, without specific analysis of the influence of the three characteristic waveforms of ECG on blood pressure prediction. In this paper, two experimental methods are used to get a consistent conclusion: QRS complex and T wave have a better effect on BP than P wave. In the first method, different ECG characteristic waveforms are combined to evaluate the results directly. In the second method, we try to explain the black box of CNN network with the use of Grad-CAM.

The experimental results may be due to the following two reasons. As can be seen from Figure 1, the QRS complex is the most prominent characteristic waveform during the ECG cycle [36,45]. Its amplitude fluctuates the most, and the amount of information it can provide is higher than that of the P and T waves with smaller amplitude changes. The addition of T wave also significantly improved the results, because the T wave lasts the longest and accounts for a larger proportion in a cycle. This experimental result shows that we can focus on these two waveforms if we need the most effective feature extraction for ECG in the blood pressure regression task. These results suggest that we should focus on these two waveforms when we need to extract features most efficiently or add attention mechanisms to the ECG in blood pressure regression tasks.

5. Conclusions

In this study, to explore the importance of the ECG characteristic waveforms in the BP estimation model, we used a common ECG detection procedure supplemented by manual annotation and divided the ECG signal into three parts: P wave, QRS complex, and T wave. Then we adopted five CNN models with multi-combination of signals as input. The results showed that compared with the single PPG, adding ECG waveforms can effectively improve the accuracy and the contribution of QRS and T wave is better than that of P wave. Meanwhile, we introduced the Grad-CAM to generate heat maps. Comparing the heat maps of different models, we found that the networks have a response to almost all segments of the ECG but the QRS and T waves received more highlights. This means that the network has learned more valuable features at these two waveforms. At the same time, if the network focuses on too one-sided parts, it is not conducive to the final result.

Extracting multiple features from an entire ECG may be a bit blind, but this article points out important directions: QRS and T waves. In future work, based on the information contained in the ECG signal, we will try to extract several effective artificial features; QRS and T wave will be the focus of research. Then they will be combined with PPG features to build a BP regression model. Attention mechanisms or other modules can also be added to the structure to take advantage of the CNN networks feature mining.

Author Contributions: Conceptualization, G.M. and Y.C.; methodology, G.M. and W.Z.; software, L.Z.; validation and formal analysis, Y.C., L.Z. and H.T.; investigation and resources, G.M., Y.C. and W.Z.; data curation, H.T. writing—original draft preparation, G.M., Y.C. and W.Z.; writing—review and editing, L.W. and Y.Y.; visualization and supervision, G.M. and L.W. All authors have read and agreed to the published version of the manuscript.

Funding: This research received no external funding.

Data Availability Statement: The data presented in this study are available from the corresponding author upon reasonable request.

Conflicts of Interest: The authors declare no conflict of interest.

References

1. Kannel, W.B. Blood Pressure as a Cardiovascular Risk Factor. *JAMA* **1996**, *275*, 1571–1576. [CrossRef] [PubMed]
2. Sharifi, I.; Goudarzi, S.; Khodabakhshi, M.B. A novel dynamical approach in continuous cuffless blood pressure estimation based on ECG and PPG signals. *Artif. Intell. Med.* **2019**, *97*, 143–151. [CrossRef] [PubMed]
3. Zhou, B.; Carrillo-Larco, R.M.; Danaei, G.; Riley, L.M.; Paciorek, C.J.; Stevens, G.A.; Gregg, E.W.; Bennett, J.E.; Solomon, B.; Singleton, R.K.; et al. Worldwide trends in hypertension prevalence and progress in treatment and control from 1990 to 2019: A pooled analysis of 1201 population-representative studies with 104 million participants. *Lancet* **2021**, *398*, 957–980. [CrossRef]
4. Fischer, C.; Penzel, T. Continuous non-invasive determination of nocturnal blood pressure variation using photoplethysmographic pulse wave signals: Comparison of pulse propagation time, pulse transit time and RR-interval. *Physiol. Meas.* **2019**, *40*, 014001. [CrossRef] [PubMed]
5. Bugarini, A.; Young, A.J.; Griessenauer, C.J.; Dalal, S.S.; Schirmer, C.M.; Vaughan, S.; Goren, O. Perioperative Continuous Noninvasive Arterial Pressure Monitoring for Neuroendovascular Interventions: Prospective Study for Evaluation of the Vascular Unloading Technique. *World Neurosurg.* **2021**, *153*, e195–e203. [CrossRef] [PubMed]
6. Heydari, F.; Ebrahim, M.P.; Redoute, J.-M.; Joe, K.; Walker, K.; Yuce, M.R. A chest-based continuous cuffless blood pressure method: Estimation and evaluation using multiple body sensors. *Inf. Fusion* **2020**, *54*, 119–127. [CrossRef]
7. Guo, Y.; Liu, X.; Peng, S.; Jiang, X.; Xu, K.; Chen, C.; Wang, Z.; Dai, C.; Chen, W. A review of wearable and unobtrusive sensing technologies for chronic disease management. *Comput. Biol. Med.* **2021**, *129*, 104163. [CrossRef]
8. Geddes, L.A.; Voelz, M.H.; Babbs, C.F.; Bourland, J.D.; Tacker, W.A. Pulse Transit Time as an Indicator of Arterial Blood Pressure. *Psychophysiology* **1981**, *18*, 71–74. [CrossRef]
9. Park, Y.-S.; Kim, S.-H.; Lee, Y.S.; Choi, S.-H.; Ku, S.-W.; Hwang, G.-S. Real-Time Monitoring of Blood Pressure Using Digitalized Pulse Arrival Time Calculation Technology for Prompt Detection of Sudden Hypertensive Episodes During Laryngeal Microsurgery: Retrospective Observational Study. *J. Med. Internet Res.* **2020**, *22*, e13156. [CrossRef]
10. Najjar, S.S.; Scuteri, A.; Shetty, V.; Wright, J.G.; Muller, D.C.; Fleg, J.L.; Spurgeon, H.P.; Ferrucci, L.; Lakatta, E.G. Pulse wave velocity is an independent predictor of the longitudinal increase in systolic blood pressure and of incident hypertension in the Baltimore Longitudinal Study of Aging. *J. Am. Coll. Cardiol.* **2008**, *51*, 1377–1383. [CrossRef]

11. Sharma, M.; Barbosa, K.; Ho, V.; Griggs, D.; Ghirmai, T.; Krishnan, S.K.; Hsiai, T.K.; Chiao, J.-C.; Cao, H. Cuff-Less and Continuous Blood Pressure Monitoring: A Methodological Review. *Technologies* **2017**, *5*, 21. [CrossRef]
12. Awad, A.A.; Haddadin, A.S.; Tantawy, H.; Badr, T.M.; Stout, R.G.; Silverman, D.G.; Shelley, K.H. The relationship between the photoplethysmographic waveform and systemic vascular resistance. *J. Clin. Monit. Comput.* **2007**, *21*, 365–372. [CrossRef]
13. Ding, X.-R.; Zhang, Y.-T.; Liu, J.; Dai, W.-X.; Tsang, H.K. Continuous Cuffless Blood Pressure Estimation Using Pulse Transit Time and Photoplethysmogram Intensity Ratio. *IEEE Trans. Biomed. Eng.* **2016**, *63*, 964–972. [CrossRef]
14. Tan, X.; Ji, Z.; Zhang, Y. Non-invasive continuous blood pressure measurement based on mean impact value method, BP neural network, and genetic algorithm. *Technol. Health Care* **2018**, *26*, 87–101. [CrossRef]
15. Kurylyak, Y.; Lamonaca, F.; Grimaldi, D. A Neural Network-based method for continuous blood pressure estimation from a PPG signal. In Proceedings of the 2013 IEEE International Instrumentation and Measurement Technology Conference (I2MTC), Minneapolis, MN, USA, 6–9 May 2013; pp. 280–283.
16. Liu, M.; Po, L.-M.; Fu, H. Cuffless Blood Pressure Estimation Based on Photoplethysmography Signal and Its Second Derivative. *Int. J. Comput. Theory Eng.* **2017**, *9*, 202–206. [CrossRef]
17. Yin, S.; Li, G.; Luo, Y.; Lin, L. Cuff-less continuous blood pressure measurement based on multiple types of information fusion. *Biomed. Signal Process. Control* **2021**, *68*, 102549. [CrossRef]
18. Lin, W.-H.; Wang, H.; Samuel, O.W.; Liu, G.; Huang, Z.; Li, G. New photoplethysmogram indicators for improving cuffless and continuous blood pressure estimation accuracy. *Physiol. Meas.* **2018**, *39*, 025005. [CrossRef]
19. Chowdhury, M.H.; Shuzan, N.I.; Chowdhury, M.E.; Mahbub, Z.B.; Uddin, M.M.; Khandakar, A.; Reaz, M.B.I. Estimating Blood Pressure from the Photoplethysmogram Signal and Demographic Features Using Machine Learning Techniques. *Sensors* **2020**, *20*, 3127. [CrossRef]
20. Esmaelpoor, J.; Moradi, M.H.; Kadkhodamohammadi, A. Cuffless blood pressure estimation methods: Physiological model parameters versus machine-learned features. *Physiol. Meas.* **2021**, *42*, 035006. [CrossRef]
21. Thambiraj, G.; Gandhi, U.; Mangalanathan, U.; Jose, V.J.M.; Anand, M. Investigation on the effect of Womersley number, ECG and PPG features for cuff less blood pressure estimation using machine learning. *Biomed. Signal Process. Control* **2020**, *60*, 101942. [CrossRef]
22. Panwar, M.; Gautam, A.; Biswas, D.; Acharyya, A. PP-Net: A Deep Learning Framework for PPG-Based Blood Pressure and Heart Rate Estimation. *IEEE Sens. J.* **2020**, *20*, 10000–10011. [CrossRef]
23. Su, P.; Ding, X.R.; Zhang, Y.T.; Liu, J.; Miao, F.; Zhao, N. Long-term blood pressure prediction with deep recurrent neural networks. In Proceedings of the 2018 IEEE EMBS International Conference on Biomedical & Health Informatics (BHI), Las Vegas, NV, USA, 4–7 March 2018; pp. 323–328.
24. Tanveer, M.S.; Hasan, M.K. Cuffless blood pressure estimation from electrocardiogram and photoplethysmogram using waveform based ANN-LSTM network. *Biomed. Signal Process. Control* **2019**, *51*, 382–392. [CrossRef]
25. Yu, M.; Huang, Z.; Zhu, Y.; Zhou, P.; Zhu, J. Attention-based residual improved U-Net model for continuous blood pressure monitoring by using photoplethysmography signal. *Biomed. Signal Process. Control* **2022**, *75*, 103581. [CrossRef]
26. Rong, M.; Li, K. A multi-type features fusion neural network for blood pressure prediction based on photoplethysmography. *Biomed. Signal Process. Control* **2021**, *68*, 102772. [CrossRef]
27. Selvaraju, R.R.; Cogswell, M.; Das, A.; Vedantam, R.; Parikh, D.; Batra, D. Grad-CAM: Visual Explanations from Deep Networks via Gradient-Based Localization. In Proceedings of the 2017 IEEE International Conference on Computer Vision (ICCV), Venice, Italy, 22–29 October 2017.
28. Zhou, B.; Khosla, A.; Lapedriza, A.; Oliva, A.; Torralba, A. Learning deep features for discriminative localization. In Proceedings of the 2016 IEEE Conference on Computer Vision and Pattern Recognition (CVPR), Las Vegas, NV, USA, 27–30 June 2016; pp. 2921–2929.
29. Li, Y.; Yang, H.; Li, J.; Chen, D.; Du, M. EEG-based intention recognition with deep recurrent-convolution neural network: Performance and channel selection by Grad-CAM. *Neurocomputing* **2020**, *415*, 225–233. [CrossRef]
30. Kim, J.-K.; Jung, S.; Park, J.; Han, S.W. Arrhythmia detection model using modified DenseNet for comprehensible Grad-CAM visualization. *Biomed. Signal Process. Control* **2022**, *73*, 103408. [CrossRef]
31. Johnson, A.; Pollard, T.; Mark, R. MIMIC-III Clinical Database (version 1.4). *PhysioNet* **2016**. [CrossRef]
32. Johnson, A.E.W.; Pollard, T.J.; Shen, L.; Lehman, L.-W.H.; Feng, M.; Ghassemi, M.; Moody, B.; Szolovits, P.; Celi, L.A.; Mark, R.G. MIMIC-III, a freely accessible critical care database. *Sci. Data* **2016**, *3*, 160035. [CrossRef]
33. Slapničar, G.; Mlakar, N.; Luštrek, M. Blood Pressure Estimation from Photoplethysmogram Using a Spectro-Temporal Deep Neural Network. *Sensors* **2019**, *19*, 3420. [CrossRef]
34. Harfiya, L.; Chang, C.-C.; Li, Y.-H. Continuous Blood Pressure Estimation Using Exclusively Photopletysmography by LSTM-Based Signal-to-Signal Translation. *Sensors* **2021**, *21*, 2952. [CrossRef]
35. Lin, W.-H.; Chen, F.; Geng, Y.; Ji, N.; Fang, P.; Li, G. Towards accurate estimation of cuffless and continuous blood pressure using multi-order derivative and multivariate photoplethysmogram features. *Biomed. Signal Process. Control* **2021**, *63*, 102198. [CrossRef]
36. Jimenez-Perez, G.; Alcaine, A.; Camara, O. Delineation of the electrocardiogram with a mixed-quality-annotations dataset using convolutional neural networks. *Sci. Rep.* **2021**, *11*, 863. [CrossRef]

37. Ronneberger, O.; Fischer, P.; Brox, T. U-Net: Convolutional Networks for Biomedical Image Segmentation. In *Medical Image Computing and Computer-Assisted Intervention 2015*; Navab, N., Hornegger, J., Wells, W.M., Frangi, A.F., Eds.; Springer International Publishing: Cham, Switzerland, 2015; pp. 234–241.
38. Krizhevsky, A.; Sutskever, I.; Hinton, G.E. Imagenet classification with deep convolutional neural networks. *NIPS* **2012**, *25*, 84–90. [CrossRef]
39. Szegedy, C.; Liu, W.; Jia, Y.; Sermanet, P.; Reed, S.; Anguelov, D.; Erhan, D.; Vanhoucke, V.; Rabinovich, A. Going deeper with convolutions. In Proceedings of the 2015 IEEE Conference on Computer Vision and Pattern Recognition (CVPR), Boston, MA, USA, 7–12 June 2015; pp. 1–9.
40. He, K.; Zhang, X.; Ren, S.; Sun, J. Deep residual learning for image recognition. In Proceedings of the 2016 IEEE Conference on Computer Vision and Pattern Recognition, Seattle, WA, USA, 27–30 June 2016; pp. 770–778.
41. Huang, G.; Liu, Z.; Van Der Maaten, L.; Weinberger, K.Q. Densely connected convolutional networks. In Proceedings of the 2017 IEEE Conference on Computer Vision and Pattern Recognition, Honolulu, HI, USA, 21–26 July 2017; pp. 4700–4708.
42. Chen, Y.; Li, J.; Xiao, H.; Jin, X.; Yan, S.; Feng, J. Dual path networks. *Adv. Neural Inf. Process. Syst.* **2017**, *30*, 9.
43. Qin, K.; Huang, W.; Zhang, T. Deep generative model with domain adversarial training for predicting arterial blood pressure waveform from photoplethysmogram signal. *Biomed. Signal Process. Control* **2021**, *70*, 102972. [CrossRef]
44. Attia, Z.I.; Kapa, S.; Lopez-Jimenez, F.; McKie, P.M.; Ladewig, D.J.; Satam, G.; Pellikka, P.A.; Enriquez-Sarano, M.; Noseworthy, P.A.; Munger, T.M.; et al. Screening for cardiac contractile dysfunction using an artificial intelligence-enabled electrocardiogram. *Nat. Med.* **2019**, *25*, 70–74. [CrossRef]
45. Cheffer, A.; Savi, M.A.; Pereira, T.L.; de Paula, A.S. Heart rhythm analysis using a nonlinear dynamics perspective. *Appl. Math. Model.* **2021**, *96*, 152–176. [CrossRef]

micromachines

MDPI

Article

Performance Analysis of Electromyogram Signal Compression Sampling in a Wireless Body Area Network

Liangyu Zhang [1], Junxin Chen [1], Chenfei Ma [2], Xiufang Liu [3] and Lisheng Xu [1,*]

[1] College of Medicine and Biological Information Engineering, Northeastern University, 195 Innovation Road, Shenyang 110169, China
[2] Edinburgh Neuroprosthetics Laboratory, School of Informatics, The University of Edinburgh, 10 Crichton Street, Edinburgh EH8 9AB, UK
[3] Paul C. Lauterbur Research Center for Biomedical Imaging, Shenzhen Institutes of Advanced Technology, Chinese Academy of Sciences, Shenzhen 518055, China
* Correspondence: xuls@bmie.neu.edu.cn; Tel.: +86-024-83683200

Abstract: The rapid growth in demand for portable and intelligent hardware has caused tremendous pressure on signal sampling, transfer, and storage resources. As an emerging signal acquisition technology, compressed sensing (CS) has promising application prospects in low-cost wireless sensor networks. To achieve reduced energy consumption and maintain a longer acquisition duration for high sample rate electromyogram (EMG) signals, this paper comprehensively analyzes the compressed sensing method using EMG. A fair comparison is carried out on the performances of 52 ordinary wavelet sparse bases and five widely applied reconstruction algorithms at different compression levels. The experimental results show that the db2 wavelet basis can sparse EMG signals so that the compressed EMG signals are reconstructed properly, thanks to its low percentage root mean square distortion (PRD) values at most compression ratios. In addition, the basis pursuit (BP) reconstruction algorithm can provide a more efficient reconstruction process and better reconstruction performance by comparison. The experiment records and comparative analysis screen out the suitable sparse bases and reconstruction algorithms for EMG signals, acting as prior experiments for further practical applications and also a benchmark for future academic research.

Keywords: compressed sensing; electromyogram; reconstruction algorithm; wavelet basis

Citation: Zhang, L.; Chen, J.; Ma, C.; Liu, X.; Xu, L. Performance Analysis of Electromyogram Signal Compression Sampling in a Wireless Body Area Network. *Micromachines* **2022**, *13*, 1748. https://doi.org/10.3390/mi13101748

Academic Editors: Xiao Xiao and Gang Ge

Received: 2 September 2022
Accepted: 13 October 2022
Published: 15 October 2022

1. Introduction

Electromyography (EMG) is a technique for evaluating and recording the electrical activity produced by skeletal muscles [1]. Electromyography is performed using an instrument called an "electromyograph" to generate a record of electromyography. When myocytes are activated by electricity or nerves, electromyography can detect the potential that is generated. The signals can be analyzed to detect medical abnormalities, activation levels or recruitment sequences, or to analyze the biomechanics of human or animal movements. In computer science, EMG is also used as a middleware for gesture recognition to allow physical actions to be input into a computer as a form of human computer interaction [2]. The efficient acquisition, storage and transmission of electromyography (EMG) data is important for emerging applications such as telemedicine. However, the transmission and storage of EMG data is challenging due to limitations relating to Internet speed and hardware resources. These problems have been effectively solved with the emergence of wireless body sensor networks (WBSNs) [3–5].

Compressed sensing is widely implemented in the WBSN to reduce the power consumption of the sensors and to increase the security of the data. Xu et al. presented a compressed sensing-based approach to co-recognize human activities and sensor locations in a single framework [6]. Kimia et al. proposed a new method for EMG data compression using deep convolutional autoencoders (CAE) [7]. Shoaib et al. proposed a method to capture and detect EEG signals based on compressed perception, which reduced the energy consumption during communication and computation [8]. Zhang et al. came up with a method called block sparse Bayesian learning (BSBL) to solve the problem of compressed sensing and applied it for long-distance EEG monitoring [9]. Imtiaz et al. proposed a low-power MSP430 compressive sensing implementation, focusing mainly on the impact of the sensor node architecture on the compression performance [10]. Liu et al. solved the problem whereby EEG signals cannot be represented sparsely and shortened the computing time of single-channel EEG signals in compressed sampling. An optimization model with the l_0 norm and Schatten-0 norm was proposed to enforce sparsity and low-rank structures in the reconstructed multi-channel EEG signals [11]. Selin Aviyente applied a compression sensing framework for EEG signal compression [12], and the Gauss random matrix was utilized to compress the input signal in the measurement. In the reconstruction process, a discrete cosine matrix was selected as a sparse basis and the EEG signal recovery was completed by combining it with an orthogonal matching pursuit (OMP) algorithm [13]. Amir et al. [14] proposed a reconstruction scheme analysis framework, which employed a Gaussian random matrix to compress the original signal. Then, different sparse bases and several main reconstruction algorithms were combined separately. After this, they exploited the compressive sensing theory to analyze the refactored EEG signals and analyzed the performances of various combinations in detail. Zhao et al. [15] studied and improved the disadvantages of the high power consumption and large area of random dense binary measurement matrices. They proposed the QCAC matrix and random sparse binary matrix, and achieved a positive signal reconstruction effect while saving hardware computing resources. Moreover, they implemented the two measurement matrices in hardware based on a 65 nm CMOS processor. Emil Jovanov and Aleksandar Milenkovic designed the infrastructure for a wireless body area network based on an intelligent motion sensor for computer-aided physical rehabilitation [5]. In 2010, Hyunwoong Park put forward a space–time block coding (STBC) scheme based on a double transmitter and single receiver. They studied the influence of three communication channels on the body, through the body, diffraction around the body, and reflection of the body, and composed a propagation channel model of the body's surface [16].

Mamaghanian et al. applied the hardware platform of a shimmer wireless body area network to compare the compression performance for ECG based on a wavelet transform and compressed sensing. They indicated that the compressed sensing of ECG signals consumed less power and was suitable for the body area network with high real-time requirements [17]. Zhang et al. exploited the intra-block correlation of ECG signals under the framework of the compressed sensing theory and came up with a reconstruction algorithm based on block sparse Bayesian learning. Compared with other reconstruction algorithms, this algorithm has better ECG reconstruction performance and was beneficial in the data analysis and diagnosis of body area network telemonitoring centers [18]. Dixon et al. presented a 1-bit Bernoulli compressed sensing observation matrix for the dynamic threshold method of ECG signals, which reduced the transmission power consumption of body area network sensor nodes by increasing the compression ratio [19].

In this paper, we applied EMG compressed sensing on wireless body area networks. A fixed measurement matrix and reconstruction algorithm were selected and the most suitable wavelet base for the EMG signal was determined according to the reconstruction quality of the EMG signal. Our research was based on the existing research results, adding different sparse bases and reconstruction algorithms and evaluating the compression sensing performance of the EMG signals. The contributions can be summarized as follows: (1) The fixed measurement matrix (Bernoulli matrix) and the fixed BP reconstruction algorithm were employed [20]. Fifty-two kinds of sparse wavelet bases were selected, and the compression ratio range was set from 10% to 90%. Finally, after sparse decomposition and reconstruction, the average PRD value for 500 segments of EMG signals for each wavelet basis was calculated. From this, the most suitable wavelet basis for the EMG signals was evaluated. (2) A fixed measurement matrix (Bernoulli matrix) and wavelet basis were applied. Five reconstruction algorithms were evaluated via PRD, along with the time consumed. (3) The future development prospects were discussed and trends were identified.

The remainder of this paper is structured as follows. The basic concepts of CS and its application in WBSN are elucidated in the Methods section. In the results section, the compressed sampling results based on the EMG signals are shown and compared in detail. Finally, in the discussion section, the experimental results are summarized, the deficiencies are put forward, and the future implementation plan is preliminarily formulated.

2. Materials and Methods

2.1. WBSN

As a communication network for the long-distance, real-time continuous monitoring of a patient's physical conditions, important links applied via the Internet of Things (IoT) [21,22]. A WBSN is a wireless network for wearable devices [23–26]. It transmits physiological signal data between patients and hospitals far away via wireless networks based on wearable devices [27]. As a portable device, it has intimate connections with the surface of the human skin [28]. Although the acquisition device is small, the application management system in the data center and the user interface require a larger intelligent device and greater support [29]. WBSNs became popular at the end of the 20th century. The original idea was to connect individuals and surrounding devices based on a wireless personal area network (WPAN). To cover a more expansive space, WPAN technology is utilized as the gateway of the WBSN system so that the device placed in the human body can be connected to the Internet. Benefiting from this, medical staff can read the physiological data of remote patients in real time through the Internet.

As shown in Figure 1, there are five stages for the real-time monitoring and remote transmission of human physiological signals.

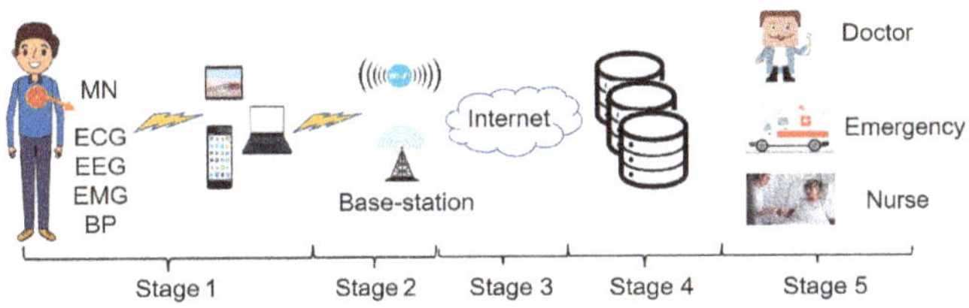

Figure 1. Real-time monitoring and remote transmission of human physiological signals.

Stage1: Physiological signal acquisition stage. Sensors for measuring ECG, EEG, EMG, and other physiological signals are placed on the body's surface or implanted under the skin. The data collected by the sensors are sent to the master node (MN) following IEEE standard 802.15.6. Then, the MN transmits the data to the local mobile devices (mobile phones, tablets, laptops) through wireless media and the local processing unit transmits the data to the next level.

Stage2: Wireless communication transmission stage. The further transmission of physiological signal data is achieved through the access point, WIFI, or cellular base station.

Stage3: Internet transmission stage. In this stage, long-distance data transmission is achieved through optical fiber technology.

Stage4: Data storage and analysis stage. The physiological signal data transmitted by long-distance are stored and analyzed in this stage. After the analysis, the data are classified and stored.

Stage5: Medical diagnosis and treatment stage. In this stage, the stored data are transmitted to the medical and health institutions. The doctor diagnoses the patient's condition according to the remote output data and gives the treatment plan [30].

Patient medical problems have caused various countries to bear many economic expenditures in recent years. WBSNs can be employed as a feasible solution to reduce medical expenses.

2.2. Compressed Sensing

Figure 2 shows the CS sampling process, in which the signal x to be collected is only non-zero at k times (k is the sparse degree). To manage the information in x, it is projected on a given set of sensing waveforms ϕ (that is to say, x is scented with a given set of waveforms). The most common measurement matrices are the Gaussian random matrix and the Bernoulli matrix, which do not correlate with the sparsest matrices. In addition, the measurement matrix can also use deterministic matrices such as a polynomial matrix, chaos matrix, and structured random matrix (e.g., the Toeplitz matrix).

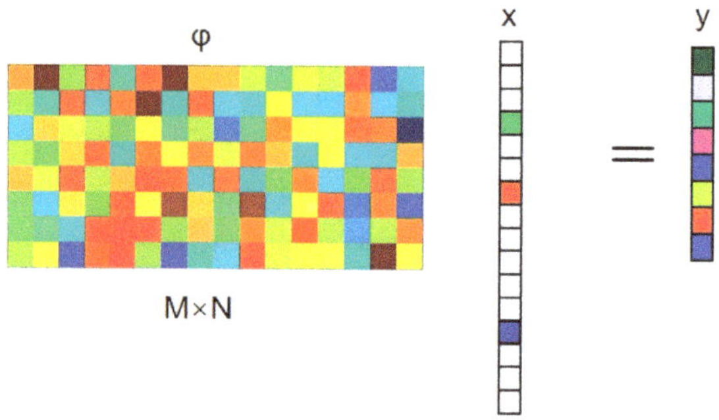

Figure 2. The CS sampling process.

The Bernoulli matrix is selected as a fixed measurement matrix in this paper. It constructs a matrix ϕ of size $M \times N$ in order that every element in ϕ independently obeys the Bernoulli distribution:

$$\Phi_{i,j} = \begin{cases} +\frac{1}{\sqrt{M}} & \mathrm{P} = \frac{1}{2} \\ -\frac{1}{\sqrt{M}} & \mathrm{P} = \frac{1}{2} \end{cases} = \frac{1}{\sqrt{M}} \begin{cases} +1 & \mathrm{P} = \frac{1}{2} \\ -1 & \mathrm{P} = \frac{1}{2} \end{cases} \tag{1}$$

and a set of measures y, which is far less than the original length of the signal that is obtained:

$$y = \phi x \tag{2}$$

When compressible, the method can also estimate the items with a larger amplitude. In practical applications, the signal x to be collected is usually not sparse, but in the transformation coefficient on an individual basis it is sparse or compressible.

The EMG is not sparse in the time domain, so it is necessary to design a suitable sparse transformation matrix to obtain the EMG signal projected onto the sparse matrix representation. Considering the embedded hardware system used in the wearable EMG monitoring device and the excellent performance of wavelets in signal compression [31], the sparse representation of EMG signals under a discrete wavelet transform (DWT) will be discussed. The Mallat decomposition of discrete sequences is performed for the myoelectric signal X. We construct a discrete wavelet transform matrix, i.e., the sparse matrix used in CS, as shown in Equations (3) and (4):

$$G_{j+1}(n) = G(n)X_j(n) = \sum_{k=-\infty}^{\infty} G(k)X_j(n-k) \tag{3}$$

$$H_{j+1}(n) = H(n)X_j(n) = \sum_{k=-\infty}^{\infty} H(k)X_j(n-k) \tag{4}$$

$H(n)$ and $G(n)$ are the high-pass and low-pass filter coefficient sequences corresponding to the selected wavelet function for both $G(n)$ and $H(n) \in R^{l\times 1}$. When the decomposition level $j = 0$, $X_0(n)$ is the EMG signal X to be compressed. After convolution, $G_1(n) \in R^{(N+L-1)\times N}$. We remove the front $L/2 = -1$ line and the rear $L/2 - 1$ line and select even rows from the remaining $N+1$ rows, giving a total of $N/2$ rows. Similarly, we select $N/2$ rows from $H_1(n)$ and superimpose them to form a decomposed wavelet transform matrix $\Psi \in R_{N\times N}$. The $N/2$ rows selected from $H_1(n)$ form the low-frequency part $X_1(n)$. The multi-layer decomposed wavelet transform matrix can be gained by continuing the iteration process according to Formulas (3) and (4).

A sparse basis needs to be exploited for sparse decomposition in compression measurements. To acquire a sparse basis suitable for EMG signals, a total of 52 types of wavelet bases in 6 wavelet functions are tested in this paper. They are haar, dbn, symn, coifn, bior, and rbio:

$$x = \psi\theta \tag{5}$$

The transformation coefficient θ is sparse, and a few coefficients contain almost all of the energy. Combining Equations (2) and (5), the relationship between the perceptual data and transform coefficients can be expressed as follows:

$$y = \phi x = \phi\psi\theta \tag{6}$$

If A (measurement matrix) = $\phi\psi$, then:

$$y = A\theta \tag{7}$$

Since Equations (2) and (7) have the same form and assumption, the transformation coefficient can also be estimated according to the optimization method. Then, the signal x to be collected can be estimated through transformation. The sparsity of x itself can be regarded as the sparsity of x on the unit matrix, and the measurement matrix is the perception matrix.

When matrix A satisfies RIP, sparse signals can be recovered by minimizing the l1 norm. The more uncorrelated the perception matrix (ϕ) and the transformation matrix (ψ), the less sparse the signal that can be recovered [32]. Moreover, the random matrix's probability that it is not related to any fixed transformation matrix is exceptionally high; that is to say, it is optimal to collect the unknown signal by collecting the random projection coefficient of the signal. In addition, the sparser the signal, the less perceptual data are needed to rebuild the original signal.

The compressed sensing method aims to restore all contents of the signal x through the measure y, which is far less than the amount of collected signal data.

There are two conditions under which recovery is possible [33]. Firstly, the signal is required to be sparse in a specific domain. The second one is incoherence, which is applied through the isometric property, which is sufficient for sparse signals [34].

Solving x from Formula (8) is an underdetermined problem, but on the other hand, the signal has only K unknown variables at unknown positions; that is, the signal has only K + 1 degrees of freedom. Therefore, under certain conditions, when the number of measurements exceeds the signal degrees of freedom, it can be recovered using some non-linear methods.

Obviously, when the K + 1 column of the perception matrix ϕ is selected arbitrarily and the linearity is independent, the signal with the sparsest characteristic found in all cases satisfying $y = \phi x$ is required, solving the following optimization problem:

$$\begin{gathered} min\|\mathrm{x}\|_0 \\ \text{subject to } y = \phi x \end{gathered} \tag{8}$$

where $\|\mathrm{x}\|_0$ represents the zero norm of x; that is, the number of non-zero elements.

For the CS reconstruction, some greedy algorithms are proposed to reduce the amount of computation. The greedy algorithms include the matching pursuit [35], compressive sampling matching pursuit (CoSaMP) [36], iteratively reweighted least squares (Irls) [37], subspace pursuit (SP) [38], and stage-wise orthogonal matching pursuit [39] algorithms.

The OMP algorithm is a representative greedy algorithm. As shown in Algorithm 1, its principle is to add the best column of each iteration to the estimation. Then, to achieve orthogonality between the estimated value and the residual, the optimization operation is implemented and the least square method is used in the subspace of the selected best fitting columns [40].

Algorithm 1 Orthogonal matching pursuit (OMP)

Input: matrix ϕ, measurements y, sparsity K
Output: sparse reconstruction x^K
1: $r^0 = y$ and $\Gamma^0 = \varnothing$
2: **for** $\mathrm{i} = 1 \ldots, K$ **do**
3: $\lambda^{\mathrm{i}} \Leftarrow argmax_j \left| \left\langle r^{i-1}, \phi_j \right\rangle \right|$ ⊳ Find best fitting column
4: $\Gamma^0 \Leftarrow \Gamma^{i-1} \cup \lambda^{\mathrm{i}}$
5: $x^i \Leftarrow argmin_x \left\| r^{i-1} - \phi_{\Gamma^i} x \right\|_2^2$ ⊳ LS optimization
6: $r^i \Leftarrow r^{i-1} - \phi_{\Gamma^i} x^i$ ⊳ Residual update
7: **end for**

In the fifth step of Algorithm 1, the least squares (LS) optimization problem will be generated. QR decomposition (QRD) is applied to solve this problem. This process will be decomposed into two matrices, the unitary matrix and upper triangular matrix, respectively represented by Q and R. In the last step of the OMP iteration, a new iterative decomposition is reused via QRD to calculate Q and R.

However, l_0 is an NP hard problem [41], which can replace non-convex problems via the solving convex. As a classical convex optimization method, the BP algorithm is often employed for CS reconstruction. The L1 minimum norm is equivalent to the L0 minimum norm under certain conditions, and the same solution can be obtained. Then, the above Equation (8) can be transformed into the optimization problem under the minimum norm of L1, as shown in Equation (9):

$$\begin{gathered} min\|\mathrm{x}\|_1 \\ \text{subject to } y = \phi x \end{gathered} \tag{9}$$

Under the condition of a restricted isometry property (RIP) [42], the minimum value of l_0 is equal to l_1, and l_1 can be found in polynomial time.

The theory of compressed sensing requires that the sensing matrix satisfy the RIP. For any k-sparse vector θ, if $\delta_k \in (0,1)$ is satisfied then:

$$(1-\delta_k)\|\theta\|_2 \le \|A\theta\|_2 \le (1+\delta_k)\|\theta\|_2 \tag{10}$$

Then, matrix A satisfies the RIP.

However, it is arduous to verify the RIP conditions. If confirmed, cross-correlation is demonstrated as the equivalent condition of the RIP conditions. The measurement matrix ϕ is uncorrelated with the sparse representation matrix ψ; that is, any one of ϕ's rows cannot be represented linearly by ψ's columns, nor can any column of ψ be represented linearly by ϕ's rows. Specifically, the correlation between the measurement matrix ϕ and sparse representation matrix ψ is defined as:

$$\mu(\phi,\psi) = \max\left|\left\langle \phi_i, \psi_j \right\rangle\right| = \max|\langle A(i,j)\rangle| = \max\frac{\left|\left\langle \mathrm{a}_i, \mathrm{a}_j \right\rangle\right|}{\|\mathrm{a}_i\|_2 \|\mathrm{a}_j\|_2} \quad 1 \le \mathrm{i} \le j \le N \tag{11}$$

In Formula (11), ϕ represents the measurement matrix and ψ is the sparse base. The correlation of ϕ and ψ is expressed by $\mu(\phi,\psi)$. Here, ϕ_i and ψ_j are row and column vectors of ϕ and ψ, respectively. In addition, a_i and a_j represents the ith row and jth column of matrix A, respectively. The more uncorrelated the matrix ϕ and sparse matrix ψ, the smaller the value of μ (ϕ, ψ), indicating the smaller the observation value that is required.

This paper is based on the existing compressed sensing theory for research. Although our theoretical innovations are limited, this ingenious sampling method is used to discuss the sparse basis and reconstruction algorithm of EMG signals in compressed sampling. Through the use of different compression ratios, a fixed measurement matrix, and a large number of calculations, the compressed sensing sparse basis and reconstruction algorithm most suitable for EMG signal are obtained.

2.3. Description of the EMG Datasets

A variety of physiological signals of the human body are stored in the MIT-BIH Polysomnographic Database [43], from which the EMG signals are extracted to analyze the compressed sensing accurately. We select the data for 5 channels here (slp32, slp37, slp41, slp45, slp48) and extract 100 fragments from the data for each channel, with 1024 points in each fragment. The sampling frequency is 250 Hz.

2.4. Performance Indicator

Several performance indicators are applied in this paper: the compression ratio (CR), percentage root mean square difference, and the time consumption.

2.5. Compression Ratio

Here, CR represents the measurement of the number of actual acquisition points required to collect the original signal x, which stands for the ratio of the number of actually collected signals to the number of original signals, along with the number of measurement M needed to obtain the accurately reconstructed signal. If M represents the dimensions of the measurement matrix and N represents the real length of the original signal x, CR can be expressed as:

$$CR = 1 - \frac{M}{N} \tag{12}$$

Since M is less than N, the value of CR is less than 1. Therefore, in actual measurements, only a small amount of data is needed to obtain a large amount of plain text information. In this paper, a total of 9 CR values are selected from 0.1 to 0.9 at equal intervals.

2.6. Percentage Root Mean Square Distortion

The percentage root mean square distortion (PRD) is an indicator of the difference between the reconstructed signal $\hat{X}$ and the original signal X, defined as:

$$PRD = \frac{||\hat{X} - X||_2}{||X||_2} \cdot 100 \tag{13}$$

Lower PRD numbers represent better reconstruction performance. Another indicator is the execution time, reflecting the reconstruction algorithm's efficiency.

2.7. Sparse Basis

Since EMG signals are not sparse in the time domain, a sparse basis needs to be utilized for sparse decomposition in the compression measurement. To acquire the sparse basis suitable for EMG signals, a total of 52 types of wavelet bases in 6 wavelet functions were tested in this paper. They were haar, dbn, symn, coifn, bior, and rbio, respectively.

2.8. Reconstruction Algorithms

The quality and efficiency of five kinds of commonly utilized reconstruction algorithms were tested to recover the signal. The five reconstruction algorithms were OMP, BP, CoSaMP, Irls, and SP.

2.9. Hardware Platform

The experiment was run on a laptop with a 1.9 GHz Intel Core i7-8650u CPU, 16 GB of memory, and a 512 GB hard disk. The program was run on MATLAB 2018b.

3. Results

This section may be divided by subheadings. It should provide a concise and precise description of the experimental results, their interpretation, as well as the experimental conclusions that can be drawn.

3.1. Compressed Sensing with Different Kinds of Wavelet Bases

Fifty-two different kinds of wavelet bases were applied for the sparse decomposition of EMG signals. The Bernoulli matrix and BP algorithm belonging to the convex optimization were exploited as fixed measurement matrices and reconstruction algorithms, respectively. Table 1 shows that the test effect, CR, and PRD values are inversely proportional. This means that the reconstruction quality is positively related to the number of measurements.

Table 1. PRD of reconstructed signals based on different wavelet bases. Values are means ± SD.

Wavelet	CR								
	10%	20%	30%	40%	50%	60%	70%	80%	90%
haar	30.67 ± 9.64	44.71 ± 15.35	56.62 ± 19.77	67.72 ± 21.97	78.64 ± 21.76	89.84 ± 21.33	100.41 ± 20.97	111.34 ± 20.98	120.4 ± 21.71
db2	23.74 ± 7.82	35.66 ± 10.63	45.99 ± 12.71	55.71 ± 13.83	65.23 ± 13.93	74.68 ± 13.09	84.34 ± 11.21	94.2 ± 7.93	103.48 ± 3.77
db3	34 ± 13.6	45.17 ± 15.65	55.86 ± 18.73	66.26 ± 20.84	77.4 ± 20.99	88.59 ± 19.82	99.2 ± 18.65	109.18 ± 17.97	115.23 ± 14.19
db4	24.15 ± 7.61	36.33 ± 10.45	46.9 ± 12.67	56.74 ± 13.94	66.32 ± 13.87	75.88 ± 12.83	85.49 ± 10.6	95.25 ± 7.36	103.65 ± 3.57
db5	34.51 ± 13.95	44.91 ± 14.47	54.38 ± 16.07	63.69 ± 16.75	73.42 ± 16.39	83.56 ± 14.94	93.56 ± 13.43	104.01 ± 11.12	112.27 ± 9.95
db6	26.25 ± 5.08	39.05 ± 6.1	50.01 ± 6.96	60.13 ± 7.2	69.63 ± 7.34	78.71 ± 6.83	87.81 ± 5.66	96.87 ± 3.88	104.78 ± 1.8
db7	26.4 ± 4.98	39.31 ± 5.96	50.19 ± 6.92	60.12 ± 7.26	69.65 ± 7	78.77 ± 6.6	87.91 ± 5.4	96.76 ± 3.74	104.87 ± 1.59
db8	26.5 ± 5.15	39.23 ± 6.17	50.09 ± 6.9	60.22 ± 7.02	69.8 ± 6.9	78.99 ± 6.15	88.17 ± 5.45	96.97 ± 3.83	105.06 ± 1.38
db9	26.55 ± 4.99	39.41 ± 6	50.43 ± 6.77	60.41 ± 7.09	69.92 ± 6.84	79.2 ± 6.34	88.16 ± 5.32	96.98 ± 3.71	104.93 ± 1.49
db10	26.65 ± 5.08	39.52 ± 6.05	50.42 ± 6.92	60.55 ± 6.97	69.82 ± 6.88	79.18 ± 6.27	88.12 ± 5.57	97.13 ± 3.66	104.99 ± 1.51
sym2	25.7 ± 5.1	38.51 ± 6.02	49.29 ± 7.08	59.37 ± 7.65	68.93 ± 7.56	78.25 ± 7.06	87.46 ± 6.07	96.44 ± 4.25	104.82 ± 1.61
sym3	25.81 ± 5.31	38.49 ± 6.26	49.4 ± 7.19	59.53 ± 7.45	68.94 ± 7.54	78.17 ± 6.89	87.44 ± 6.05	96.35 ± 4.19	104.64 ± 1.82
sym4	25.87 ± 5.2	38.58 ± 6.31	49.38 ± 7.16	59.44 ± 7.48	68.92 ± 7.71	78.27 ± 6.9	87.46 ± 6.16	96.39 ± 4.49	104.54 ± 1.81
sym5	25.93 ± 5.16	38.61 ± 6.2	49.49 ± 7.1	59.41 ± 7.46	69.03 ± 7.55	78.2 ± 6.99	87.41 ± 5.92	96.44 ± 4.17	104.63 ± 1.86
sym6	25.93 ± 5.16	38.67 ± 6.21	49.56 ± 7.17	59.54 ± 7.44	69.04 ± 7.74	78.33 ± 6.83	87.57 ± 6.03	96.36 ± 4.53	104.61 ± 1.8
sym7	26.03 ± 5.32	38.79 ± 6.21	49.55 ± 7.03	59.68 ± 7.36	69 ± 7.58	78.42 ± 6.84	87.42 ± 6.05	96.53 ± 4.24	104.68 ± 1.71
sym8	25.99 ± 5.13	38.73 ± 6.14	49.64 ± 7.15	59.62 ± 7.38	69.07 ± 7.7	78.35 ± 6.79	87.63 ± 5.98	96.4 ± 4.47	104.67 ± 1.7
coif1	25.65 ± 5.12	38.39 ± 6.07	49.19 ± 7.01	59.22 ± 7.68	68.78 ± 7.61	78.06 ± 7.03	87.33 ± 5.92	96.49 ± 4.24	104.5 ± 1.77
coif2	25.83 ± 5.21	38.51 ± 6.23	49.25 ± 7.13	59.41 ± 7.49	68.9 ± 7.62	78.06 ± 7	87.35 ± 6.08	96.22 ± 4.35	104.54 ± 1.8
coif3	25.95 ± 5.14	38.82 ± 6.17	49.55 ± 7.07	59.63 ± 7.39	69.15 ± 7.3	78.34 ± 6.98	87.54 ± 5.61	96.71 ± 4.09	104.47 ± 1.94
coif4	26.01 ± 5.12	38.77 ± 6.18	49.54 ± 7.12	59.64 ± 7.42	69.16 ± 7.47	78.29 ± 6.77	87.53 ± 5.86	96.39 ± 4.32	104.69 ± 1.57
coif5	26.05 ± 5.1	38.89 ± 6.12	49.73 ± 6.89	59.84 ± 7.32	69.25 ± 7.17	78.59 ± 6.57	87.69 ± 5.56	96.57 ± 4.06	104.7 ± 1.64
bior1.1	25.38 ± 5.22	38.16 ± 6.3	48.93 ± 7.15	59.22 ± 7.62	68.59 ± 7.79	77.89 ± 7.23	87.16 ± 6.47	96.3 ± 4.47	104.3 ± 2.35
bior1.3	26.13 ± 5.24	39.08 ± 6.3	50.02 ± 7.15	60.3 ± 7.49	69.84 ± 7.63	79.06 ± 7.09	88.03 ± 6.27	97.12 ± 4.09	104.76 ± 2.07
bior1.5	26.74 ± 5.32	39.89 ± 6.35	50.99 ± 7.15	61.28 ± 7.4	70.8 ± 7.56	79.95 ± 6.96	88.81 ± 6.19	97.68 ± 3.98	105.06 ± 1.99
bior2.2	27.19 ± 5.32	40.77 ± 6.25	52.45 ± 7.26	63.25 ± 7.7	73.29 ± 7.6	83.16 ± 6.78	92.33 ± 5.58	101.39 ± 3.46	108.44 ± 1.31
bior2.4	27.01 ± 5.31	40.46 ± 6.25	51.98 ± 7.13	62.56 ± 7.57	72.54 ± 7.48	82.12 ± 6.77	91.22 ± 5.52	100.33 ± 3.73	107.45 ± 1.33
bior2.6	27.12 ± 5.33	40.6 ± 6.26	52.09 ± 7.05	62.61 ± 7.49	72.56 ± 7.4	82.17 ± 6.75	91.22 ± 5.54	100.28 ± 3.79	107.41 ± 1.44
bior2.8	27.28 ± 5.33	40.83 ± 6.25	52.32 ± 7.02	62.83 ± 7.45	72.76 ± 7.32	82.4 ± 6.71	91.43 ± 5.51	100.39 ± 3.81	107.48 ± 1.45
bior3.1	31.7 ± 5.32	47.76 ± 6.35	61.99 ± 7.2	75.45 ± 7.31	88.62 ± 6.63	101.05 ± 5.78	114.82 ± 5.26	128.26 ± 3.62	135.14 ± 1.27
bior3.3	30.81 ± 5.07	45.97 ± 6.38	59.09 ± 7.18	71.06 ± 7.51	82.48 ± 6.86	93.13 ± 6.18	103.41 ± 4.95	113.08 ± 2.79	118.76 ± 0.9
bior3.5	30.59 ± 5.12	45.37 ± 6.41	58.1 ± 7.21	69.95 ± 7.44	80.92 ± 6.9	91.26 ± 6.02	100.76 ± 4.87	110.2 ± 2.87	115.69 ± 0.84
bior3.7	30.52 ± 5.2	45.12 ± 6.41	57.72 ± 7.17	69.49 ± 7.41	80.36 ± 6.94	90.51 ± 6.07	99.96 ± 4.93	109.31 ± 2.89	115.01 ± 0.85
bior3.9	28.26 ± 5.08	41.86 ± 6.01	53.02 ± 7.04	63.32 ± 7.44	72.8 ± 7.2	81.82 ± 6.59	90.73 ± 5.4	99.52 ± 3.82	106.96 ± 2.08
bior4.4	27.66 ± 5.05	40.99 ± 5.95	52.06 ± 6.98	62.27 ± 7.29	71.89 ± 7.03	81.06 ± 6.43	90.17 ± 5.35	99.19 ± 3.77	106.92 ± 1.97
bior5.5	27.71 ± 5.07	41.03 ± 5.97	52.08 ± 6.95	62.32 ± 7.19	71.96 ± 6.89	81.11 ± 6.29	90.26 ± 5.25	99.28 ± 3.64	107.03 ± 1.92
bior6.8	27.83 ± 5.08	41.19 ± 6.01	52.25 ± 6.94	62.53 ± 7.13	72.17 ± 6.8	81.29 ± 6.2	90.45 ± 5.16	99.43 ± 3.53	107.15 ± 1.87
rbio1.1	51.95 ± 6.79	67.99 ± 5.93	78.33 ± 5.85	85.43 ± 5.45	91.67 ± 4.53	97 ± 4.04	102.15 ± 3.49	107.48 ± 3.11	110.34 ± 2.02
rbio1.3	35.01 ± 5.42	52.02 ± 5.99	64.7 ± 6.58	74.85 ± 6.3	83.53 ± 5.58	91.01 ± 4.74	98.25 ± 3.97	105.47 ± 3.28	110.4 ± 2.1
rbio1.5	32.77 ± 5.26	48.84 ± 6	61.35 ± 6.63	71.9 ± 6.64	81.07 ± 5.98	89.16 ± 4.99	96.97 ± 4.23	104.85 ± 3.29	110.48 ± 2.09
rbio2.2	32.44 ± 5.2	48.1 ± 6.06	60.4 ± 6.69	70.97 ± 6.71	80.14 ± 6.1	88.41 ± 5.18	96.51 ± 4.25	104.61 ± 3.34	110.57 ± 2.08
rbio2.4	32.4 ± 5.19	47.84 ± 6.13	60.06 ± 6.72	70.61 ± 6.73	79.78 ± 6.13	88.18 ± 5.24	96.38 ± 4.21	104.55 ± 3.35	110.63 ± 2.11
rbio2.6	26.28 ± 5.16	39.22 ± 6.13	50.06 ± 7.12	60.13 ± 7.62	69.82 ± 7.49	79.08 ± 6.87	88.16 ± 5.78	97.04 ± 4.05	104.8 ± 1.8
rbio2.8	26.57 ± 5.3	39.77 ± 6.33	50.96 ± 7.18	61.36 ± 7.69	71.29 ± 7.64	80.76 ± 7.05	89.85 ± 5.76	98.73 ± 4	106.28 ± 1.68
rbio3.1	26.19 ± 5.13	39.06 ± 6.13	49.83 ± 6.98	59.83 ± 7.46	69.44 ± 7.28	78.71 ± 6.7	87.85 ± 5.63	96.82 ± 3.94	104.92 ± 1.8
rbio3.3	28.26 ± 5.08	41.86 ± 6.01	53.02 ± 7.04	63.32 ± 7.44	72.8 ± 7.2	81.82 ± 6.59	90.73 ± 5.4	99.52 ± 3.82	106.96 ± 2.08
rbio3.5	27.66 ± 5.05	40.99 ± 5.95	52.06 ± 6.98	62.27 ± 7.29	71.89 ± 7.03	81.06 ± 6.43	90.17 ± 5.35	99.19 ± 3.77	106.92 ± 1.97
rbio3.7	27.71 ± 5.07	41.03 ± 5.97	52.08 ± 6.95	62.32 ± 7.19	71.96 ± 6.89	81.11 ± 6.29	90.26 ± 5.25	99.28 ± 3.64	107.03 ± 1.92
rbio3.9	27.83 ± 5.08	41.19 ± 6.01	52.25 ± 6.94	62.53 ± 7.13	72.17 ± 6.8	81.29 ± 6.2	90.45 ± 5.16	99.43 ± 3.53	107.15 ± 1.87
rbio4.4	51.95 ± 6.79	67.99 ± 5.93	78.33 ± 5.85	85.43 ± 5.45	91.67 ± 4.53	97 ± 4.04	102.15 ± 3.49	107.48 ± 3.11	110.34 ± 2.02
rbio5.5	35.01 ± 5.42	52.02 ± 5.99	64.7 ± 6.58	74.85 ± 6.3	83.53 ± 5.58	91.01 ± 4.74	98.25 ± 3.97	105.47 ± 3.28	110.4 ± 2.1
rbio6.8	32.77 ± 5.26	48.84 ± 6	61.35 ± 6.63	71.9 ± 6.64	81.07 ± 5.98	89.16 ± 4.99	96.97 ± 4.23	104.85 ± 3.29	110.48 ± 2.09

3.2. Compressed Sensing with Different Kinds of Construction Algorithms

In this paper, we applied the PRD and execution time as performance indicators to evaluate five kinds of reconstruction algorithms. In this test, the Bernoulli matrix and coif5 wavelet basis were employed as the fixed measurement matrix and sparse basis, respectively.

From the perspective of the deviation between the original and reconstructed signals shown in Table 2 and Figure 3, the BP and Irls reconstruction algorithms were better than the other three algorithms. To be more precise, the BP reconstruction algorithm has a better recovery effect under all compression ratios.

Table 2. PRD values of the reconstruction signals for the different compression ratios (CRs) based on the same wavelet basis. Values are means ± SD.

Method	CR								
	0.1	0.2	0.3	0.4	0.5	0.6	0.7	0.8	0.9
OMP	48.24 ± 16.85	65.04 ± 21.35	79.3 ± 24.66	92.13 ± 25.64	104.56 ± 23.64	117.41 ± 20.89	128.26 ± 16.89	139.69 ± 11.11	146.37 ± 4.82
BP	23.87 ± 7.98	35.71 ± 10.75	46.1 ± 12.81	55.87 ± 13.86	65.31 ± 13.97	74.71 ± 13.07	84.48 ± 11.31	94.23 ± 7.95	103.51 ± 3.86
CoSaMP	55.61 ± 18.29	65.58 ± 21.42	77.07 ± 24.22	88.58 ± 25.93	101.44 ± 24.29	113.75 ± 20.46	123.59 ± 15.49	132.66 ± 10.19	132.89 ± 4.04
Irls	24.99 ± 8.44	37.53 ± 11.39	48.37 ± 13.65	58.42 ± 14.87	68.03 ± 14.85	77.49 ± 13.62	87 ± 11.47	96.1 ± 7.79	103.63 ± 3.59
SP	55.51 ± 17.55	62.37 ± 19.22	70.17 ± 20.76	78.89 ± 21.43	88.76 ± 20.44	98.88 ± 18.09	108.94 ± 14.9	119.03 ± 9.39	125.35 ± 4.43

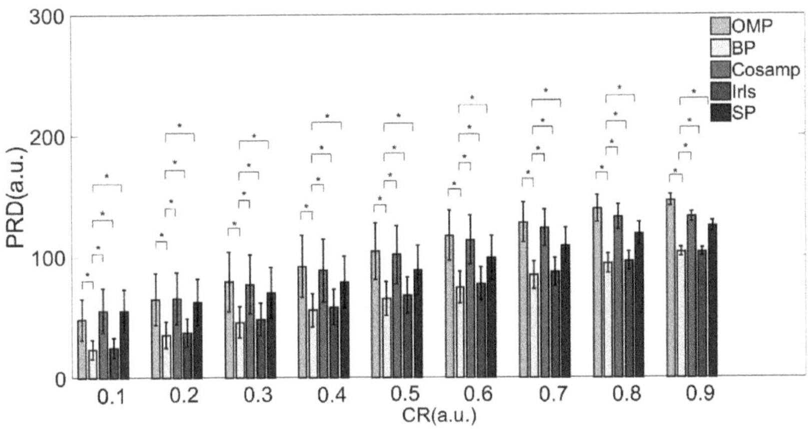

Figure 3. PRD values of various reconstruction algorithms. Note: * Indicates statistical significance via two-sample *t*-tests.

The execution times of the five reconstruction algorithms are shown in Table 3 and Figure 4. There were significant differences among each algorithm.

The IRLS algorithm consumed the most time under all compression ratios as compared with the other algorithms. When the CR increases from 0.1 to 0.9, the time consumed increases by more than 100 times. The reconstruction speed of the BP algorithm was the fastest under almost all compression ratios. When CR > 80%, the CoSaMP and SP algorithms took about the same time as the BP algorithm.

Combined with the results in Tables 2 and 3 and Figures 3 and 4, providing a comprehensive evaluation of the reconstruction quality and execution time, the BP reconstruction algorithm had superior reconstruction quality and high efficiency at all compression ratios. Although the Irls algorithm also showed high reconstruction quality, the execution efficiency was not satisfied. Conversely, the although CoSaMP and SP algorithms showed excellent efficiency when the CR was over 0.8, the reconstruction quality was poor. Therefore, considering these two factors, the BP algorithm was the most suitable for EMG signals in practical applications.

Table 3. Execution times (in seconds) of various reconstruction algorithms. Values are mean ± SD.

Method	CR								
	0.1	0.2	0.3	0.4	0.5	0.6	0.7	0.8	0.9
OMP	2236.66 ± 574.21	1368.17 ± 300.14	849.29 ± 169.05	496.98 ± 96.43	275.77 ± 57.91	141.54 ± 41.09	65.38 ± 32.34	23.88 ± 17.25	21.01 ± 0.82
BP	24.35 ± 3.71	20.49 ± 3.18	19.55 ± 6.75	13.43 ± 1.62	10.68 ± 1.75	9 ± 2.73	6.8 ± 2	5.43 ± 1.51	3.89 ± 0.82
CoSaMP	722.35 ± 173.09	519.86 ± 130.66	401.89 ± 152.8	221.14 ± 64.39	134.06 ± 54.4	63.89 ± 29.86	26.77 ± 15.09	9.69 ± 5.96	1.9 ± 1.4
Irls	3207.79 ± 817.23	2405.05 ± 578.46	1548.56 ± 346.9	1086.58 ± 230.09	705.22 ± 120.33	423.7 ± 83.59	206.53 ± 34.87	94.26 ± 20.81	26.17 ± 9.02
SP	431.65 ± 112.13	397.19 ± 221.28	194.22 ± 61.99	128.29 ± 55.02	68.97 ± 35.96	36.75 ± 20.47	18.9 ± 12.11	4.48 ± 1.7	1.33 ± 0.87

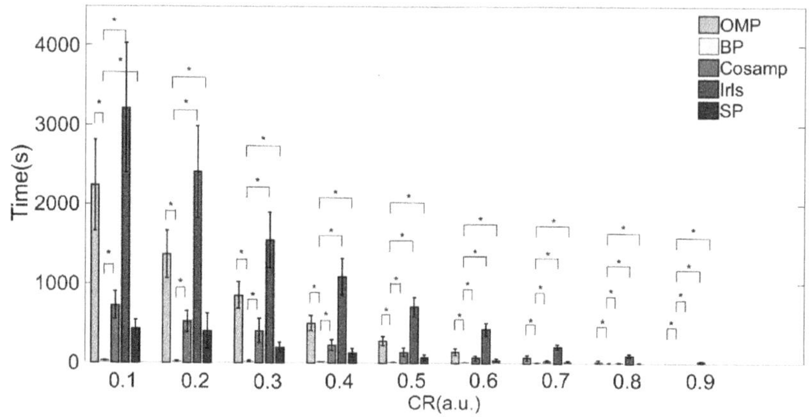

Figure 4. Execution times of various reconstruction algorithms. Note: * Indicates statistical significance via two-sample *t*-tests.

4. Discussion

The continuous and real-time monitoring of physiological signals play a significant role in disease diagnosis. However, in the long-term signal monitoring and wireless transmission of physiological signals, a large number of node resources will be required. Although the storage requirements can be satisfied due to Moore's law, the physiological signal collection systems are also supposed to ensure the energy consumption efficiency. As an effective post-acquisition processing method, CS can reduce the power consumption and improve the physiological signal sampling efficiency significantly.

The experimental results indicated that via CS one can transmit and compress EMG signals efficiently. The db2 wavelet base was the most suitable sparse basis for EMG signals under various compression ratios and can be utilized as a sparse transform basis for EMG. In addition, under most compression ratios, the BP algorithm provided better reconstruction quality and higher reconstruction efficiency.

However, although the most suitable wavelet basis and reconstruction algorithm was acquired for EMG compressed sensing, there was still a gap between the rebuilt signal and the original one according to the PRD performance parameter. This implied that the existing convex optimization and greedy reconstruction algorithms cannot reorganize the information from the EMG signals well. A similar conclusion was also presented in [44]. Casson et al. used the BP reconstruction algorithm and cubic B-spline dictionary sparse basis to conduct a compression sensing sampling test on EMG signals, which also showed that the EMG signals could not be restored satisfactorily (the PRD values were above 50% in all cases). In our study, although the results were not very satisfactory, compared with the conclusions of Casson et al., when a suitable compression ratio is selected, the PRD value will be reduced to less than 50%. In practical applications, the db2 and BP algorithms can be used as the sparse basis and reconstruction algorithm, respectively, for the compression sensing sampling of EMG signals, and EMG signals with minor errors can be obtained by reducing the compression ratio.

In other words, the CS approach is not fully functional at present in terms of the post-acquisition processing. Nevertheless, the wavelet basis most suitable for compressed

EMG sensing was found through a calculation and analysis process. Our future work will focus on EMG rebuilding process and deep learning applications in reconstruction.

5. Conclusions

This research aimed to evaluate the wavelet bases and reconstruction algorithms best suited for the compressive sampling of EMG signals. From the analysis based on the sampling quality and speed, it was concluded that the db2 wavelet basis provides a more balanced performance, while the BP reconstruction algorithm can achieve high accuracy in a short time.

Author Contributions: L.Z., J.C. and L.X. designed the project. L.Z. finished the manuscript. L.Z. and J.C. performed the experiments. L.Z. analyzed the results. J.C., X.L., C.M. and L.X. revised and spell-checked the manuscript. All authors have read and agree to the published version of the manuscript.

Funding: This research was supported by the National Natural Science Foundation of China (No. 62273082, and No. 61773110), the Natural Science Foundation of Liaoning Province (No. 2021-YGJC-14), the Basic Scientific Research Project (Key Project) of Liaoning Provincial Department of Education (LJKZ00042021), the Fundamental Research Funds for the Central Universities (No. N2119008), the Shenyang Science and Technology Plan Fund (No. 21-104-1-24, No. 20-201-4-10, and No. 201375), and the Member Program of Neusoft Research of Intelligent Healthcare Technology, Co., Ltd. (No. MCMP062002).

Data Availability Statement: Not applicable.

Conflicts of Interest: The authors declare no conflict of interest.

References

1. Paoletti, M.; Belli, A.; Palma, L.; Vallasciani, M.; Pierleoni, P.J.E. A Wireless Body Sensor Network for Clinical Assessment of the Flexion-Relaxation Phenomenon. *Electronics* **2020**, *9*, 1044. [CrossRef]
2. Kobylarz, J.; Bird, J.J.; Faria, D.R.; Ribeiro, E.P.; Intelligence, A.E.J.J.o.A.; Computing, H. Thumbs up, thumbs down: Non-verbal human-robot interaction through real-time EMG classification via inductive and supervised transductive transfer learning. *J. Ambient. Intell. Humaniz. Comput.* **2020**, *11*, 6021–6031. [CrossRef]
3. Akyildiz, I.F.; Su, W.; Sankarasubramaniam, Y.; Cayirci, E. Wireless sensor networks: A survey. *Comput. Netw.* **2002**, *38*, 393–422. [CrossRef]
4. Akyildiz, I.F.; Vuran, M.C.; Akan, O.B.; Su, W. Wireless sensor networks: A survey revisited. *Comput. Netw. J. Elsevier Sci.* **2004**, *45*, 1–39.
5. Jovanov, E.; Milenkovic, A.; Otto, C.; de Groen, P.C. A wireless body area network of intelligent motion sensors for computer assisted physical rehabilitation. *J. Neuroeng. Rehabil.* **2005**, *2*, 6. [CrossRef] [PubMed]
6. Xu, W.; Zhang, M.; Sawchuk, A.A.; Sarrafzadeh, M. Robust human activity and sensor location corecognition via sparse signal representation. *IEEE Trans. Biomed. Eng.* **2012**, *59*, 3169–3176. [PubMed]
7. Yildirim, O.; San Tan, R.; Acharya, U.R.J.C.S.R. An Efficient Compression of ECG Signals Using Deep Convolutional Autoencoders. *Cogn. Syst. Res.* **2022**, *52*, 198–211. [CrossRef]
8. Shoaib, M.; Jha, N.K.; Verma, N. A Compressed-Domain Processor for Seizure Detection to Simultaneously Reduce Computation and Communication Energy. In Proceedings of the IEEE 2012 Custom Integrated Circuits Conference, San Jose, CA, USA, 9–12 September 2012; pp. 1–4.
9. Zhang, Z.; Jung, T.-P.; Makeig, S.; Rao, B.D. Compressed sensing of EEG for wireless telemonitoring with low energy consumption and inexpensive hardware. *IEEE Trans. Biomed. Eng.* **2012**, *60*, 221–224. [CrossRef] [PubMed]
10. Imtiaz, S.A.; Casson, A.J.; Rodriguez-Villegas, E. Compression in wearable sensor nodes: Impacts of node topology. *IEEE Trans. Biomed. Eng.* **2013**, *61*, 1080–1090. [CrossRef]
11. Liu, Y.; de Vos, M.; van Huffel, S. Compressed sensing of multichannel EEG signals: The simultaneous cosparsity and low-rank optimization. *IEEE Trans. Biomed. Eng.* **2015**, *62*, 2055–2061. [CrossRef] [PubMed]
12. Aviyente, S. Compressed Sensing Framework for EEG Compression. In Proceedings of the IEEE 14th Workshop on Statistical Signal Processing, Madison, WI, USA, 26–29 August 2007; pp. 181–184.
13. Tropp, J.A.; Gilbert, A.C. Signal recovery from random measurements via orthogonal matching pursuit. *IEEE Trans. Inf. Theory* **2007**, *53*, 4655–4666. [CrossRef]
14. Abdulghani, A.M.; Casson, A.J.; Rodriguez-Villegas, E. Compressive sensing scalp EEG signals: Implementations and practical performance. *Med. Biol. Eng. Comput.* **2012**, *50*, 1137–1145. [CrossRef] [PubMed]

15. Zhao, W.; Sun, B.; Tong, W.; Zhi, Y. On-Chip Neural Data Compression based on Compressed Sensing with Sparse Sensing Matrices. *IEEE Trans. Biomed. Circuits Syst.* **2018**, *12*, 242–254. [CrossRef] [PubMed]
16. Quwaider, M.; Rao, J.; Biswas, S.K. Body-posture-based dynamic link power control in wearable sensor networks. *Commun. Mag. IEEE* **2010**, *48*, 134–142. [CrossRef]
17. Mamaghanian, H.; Khaled, N.; Atienza, D.; Vandergheynst, P. Compressed Sensing for Real-Time Energy-Efficient ECG Compression on Wireless Body Sensor Nodes. *IEEE Trans. Biomed. Eng.* **2011**, *58*, 2456–2466. [CrossRef] [PubMed]
18. Zhang, Z.; Jung, T.P.; Makeig, S.; Rao, B.D. Compressed sensing for energy-efficient wireless telemonitoring of noninvasive fetal ECG via block sparse Bayesian learning. *IEEE Trans. Biomed. Eng.* **2012**, *60*, 300–309. [CrossRef] [PubMed]
19. Dixon, A.M.; Allstot, E.G.; Gangopadhyay, D.; Allstot, D.J. Compressed sensing system considerations for ECG and EMG wireless biosensors. *IEEE Trans. Biomed. Circuits Syst.* **2012**, *6*, 156–166. [CrossRef]
20. Chen, S.S.; Donoho, D.L.; Saunders, M.A. Atomic decomposition by basis pursuit. *SIAM Rev.* **2001**, *43*, 129–159. [CrossRef]
21. Ashton, K. That 'internet of things' thing. *RFID J.* **2009**, *22*, 97–114.
22. Da Xu, L.; He, W.; Li, S. Internet of things in industries: A survey. *IEEE Trans. Ind. Inform.* **2014**, *10*, 2233–2243.
23. Hur, K.; Sohn, W.-S.; Kim, J.-K.; Lee, Y. A Real-Time Localization Platform Design in WUSB over WBAN Protocol for Wearable Computer Systems. In Proceedings of the International Conference on Future Generation Information Technology, Jeju Island, Korea, 8–10 December 2011; pp. 173–180.
24. Azhari, M.E.; Toumanari, A.; Latif, R. Performance analysis of IEEE 802.15.6 and IEEE 802.15.4 for wireless body sensor networks. In Proceedings of the International Conference on Multimedia Computing & Systems, Marrakech, Morocco, 14–16 April 2014.
25. Huque, M.; Munasinghe, K.S.; Jamalipour, A. A Probabilistic Energy-Aware Routing Protocol for Wireless Body Area Networks. In Proceedings of the 2014 IEEE 80th Vehicular Technology Conference (VTC2014-Fall), Vancouver, BC, Canada, 14–17 September 2014.
26. Chen, G.; Naganawa, J.-I.; Takada, J.-I.; Kim, M. Development of a tri-polarized dynamic channel sounder for wireless body area network. In Proceedings of the 2015 9th International Symposium on Medical Information and Communication Technology (ISMICT), Kamakura, Japan, 14 May 2015; pp. 30–34.
27. Jain, P.C. Wireless body area network for medical healthcare. *IETE Tech. Rev.* **2011**, *28*, 362–371. [CrossRef]
28. Gelogo, Y.E.; Hwang, H.J.; Kim, H.-K. Internet of things (IoT) framework for u-healthcare system. *Int. J. Smart Home* **2015**, *9*, 323–330. [CrossRef]
29. Gelogo, Y.E.; Kim, H.-K. Unified ubiquitous healthcare system architecture with collaborative model. *Int. J. Multimed. Ubiquitous Eng.* **2013**, *8*, 239–244.
30. Alkhayyat, A.; Thabit, A.A.; Al-Mayali, F.A.; Abbasi, Q.H. WBSN in IoT health-based application: Toward delay and energy consumption minimization. *J. Sens.* **2019**, *2019*, 2508452. [CrossRef]
31. Kumari, R.S.S.; Rajalakshmi, E. Very large-scale integration architecture for wavelet-based ECG signal adaptive coder. *IET Signal Process.* **2019**, *13*, 56–64. [CrossRef]
32. Candes, E.; Romberg, J. Sparsity and incoherence in compressive sampling. *Inverse Probl.* **2007**, *23*, 969. [CrossRef]
33. Erlich, Y.; Chang, K.; Gordon, A.; Ronen, R.; Navon, O.; Rooks, M.; Hannon, G.J. DNA Sudoku—Harnessing high-throughput sequencing for multiplexed specimen analysis. *Genome Res.* **2009**, *19*, 1243–1253. [CrossRef] [PubMed]
34. Donoho, D.L. For most large underdetermined systems of linear equations the minimal 1-norm solution is also the sparsest solution. *Commun. Pure Appl. Math.* **2006**, *59*, 797–829. [CrossRef]
35. Mallat, S.G.; Zhang, Z. Matching pursuits with time-frequency dictionaries. *IEEE Trans. Signal Process.* **1993**, *41*, 3397–3415. [CrossRef]
36. Gui, G.; Wan, Q.; Peng, W.; Adachi, F. Sparse multipath channel estimation using compressive sampling matching pursuit algorithm. *arXiv* **2010**, arXiv:1005.2270.
37. Ramani, S.; Fessler, J.A. An Accelerated Iterative Reweighted least squares algorithm for compressed sensing MRI. In Proceedings of the 2010 IEEE International Symposium on Biomedical Imaging: From Nano to Macro, Rotterdam, The Netherlands, 14–17 April 2010; pp. 257–260.
38. Liu, J.; Huang, K.; Yao, X. Common-innovation subspace pursuit for distributed compressed sensing in wireless sensor networks. *IEEE Sens. J.* **2018**, *19*, 1091–1103. [CrossRef]
39. Donoho, D.L.; Tsaig, Y.; Drori, I.; Starck, J.-L. Sparse solution of underdetermined systems of linear equations by stagewise orthogonal matching pursuit. *IEEE Trans. Inf. Theory* **2012**, *58*, 1094–1121. [CrossRef]
40. Bai, L.; Maechler, P.; Muehlberghuber, M.; Kaeslin, H. High-speed compressed sensing reconstruction on FPGA using OMP and AMP. In Proceedings of the 2012 19th IEEE International Conference on Electronics, Circuits, and Systems (ICECS 2012), Seville, Spain, 9–12 December 2012; pp. 53–56.
41. Nguyen, R.; Brown, M.S. Fast and Effective L0 Gradient Minimization by Region Fusion. In Proceedings of the IEEE International Conference on Computer Vision, Santiago, Chile, 13–16 December 2015; pp. 208–216.
42. Bandeira, A.S.; Fickus, M.; Mixon, D.G.; Wong, P. The Road to Deterministic Matrices with the Restricted Isometry Property. *J. Fourier Anal. Appl.* **2012**, *19*, 1123–1149. [CrossRef]

43. Goldberger, A.; Amaral, L.; Glass, L.; Hausdorff, J.; Ivanov, P.C.; Mark, R.; Mietus, J.; Moody, G.; Peng, C.; Stanley, H. Components of a new research resource for complex physiologic signals. *PhysioBank PhysioToolkit Physionet.* **2000**, *101*, 215–220.
44. Casson, A.J.; Rodriguez-Villegas, E. Signal agnostic compressive sensing for body area networks: Comparison of signal reconstructions. In Proceedings of the 2012 Annual International Conference of the IEEE Engineering in Medicine and Biology Society, San Diego, CA, USA, 28 August–1 September 2012; pp. 4497–4500.

micromachines

Article

An Arrhythmia Classification Model Based on Vision Transformer with Deformable Attention

Yanfang Dong [1,2], Miao Zhang [2], Lishen Qiu [1], Lirong Wang [2,3,*] and Yong Yu [2,*]

1 School of Biomedical Engineering, Division of Life Sciences and Medicine, University of Science and Technology of China, Hefei 230026, China
2 Suzhou Institute of Biomedical Engineering and Technology, China Academy of Sciences, Suzhou 215163, China
3 School of Electronics and Information Technology, Soochow University, Suzhou 215031, China
* Correspondence: wanglirong@suda.edu.cn (L.W.); yuyong@sibet.ac.cn (Y.Y.)

Abstract: The electrocardiogram (ECG) is a highly effective non-invasive tool for monitoring heart activity and diagnosing cardiovascular diseases (CVDs). Automatic detection of arrhythmia based on ECG plays a critical role in the early prevention and diagnosis of CVDs. In recent years, numerous studies have focused on using deep learning methods to address arrhythmia classification problems. However, the transformer-based neural network in current research still has a limited performance in detecting arrhythmias for the multi-lead ECG. In this study, we propose an end-to-end multi-label arrhythmia classification model for the 12-lead ECG with varied-length recordings. Our model, called CNN-DVIT, is based on a combination of convolutional neural networks (CNNs) with depthwise separable convolution, and a vision transformer structure with deformable attention. Specifically, we introduce the spatial pyramid pooling layer to accept varied-length ECG signals. Experimental results show that our model achieved an F1 score of 82.9% in CPSC-2018. Notably, our CNN-DVIT outperforms the latest transformer-based ECG classification algorithms. Furthermore, ablation experiments reveal that the deformable multi-head attention and depthwise separable convolution are both efficient in extracting features from multi-lead ECG signals for diagnosis. The CNN-DVIT achieved good performance for the automatic arrhythmia detection of ECG signals. This indicates that our research can assist doctors in clinical ECG analysis, providing important support for the diagnosis of arrhythmia and contributing to the development of computer-aided diagnosis technology.

Keywords: arrhythmia; deep learning; ECG signal; deformable attention transformer; depthwise separable convolution

check for updates

Citation: Dong, Y.; Zhang, M.; Qiu, L.; Wang, L.; Yu, Y. An Arrhythmia Classification Model Based on Vision Transformer with Deformable Attention. *Micromachines* **2023**, *14*, 1155. https://doi.org/10.3390/mi14061155

Academic Editor: Angeliki Tserepi

Received: 28 April 2023
Revised: 28 May 2023
Accepted: 29 May 2023
Published: 30 May 2023

1. Introduction

Cardiovascular disease (CVD) is the leading cause of death, accounting for over 32% of all deaths worldwide [1]. Cardiac arrhythmia (CA) serves as a warning signal for cardiovascular disease and enables clinicians to provide timely interventions through early diagnosis. Electrocardiogram (ECG) is an effective non-invasive tool for monitoring heart activity and diagnosing CA [2]. Precisely detecting arrhythmia has become a significant focus for biomedical researchers. However, accurately recognizing these complex CA-associated ECG rhythms requires considerable clinical experience and expertise. Manual detection of arrhythmia consumes considerable time for expert clinicians and cardiologists and it can be a task prone to errors even for these human experts. In fact, research shows that cardiologists or diagnosing doctors sometimes misjudge the type of arrhythmia [3,4]. The introduction of computer-aided diagnosis has aimed to enhance accurate ECG interpretation and reduce costs. Consequently, it has become increasingly essential to automatically detect arrhythmia using ECG signals, as this can assist clinical diagnosis during ECG analysis [5].

As digital ECG data become more widely available and deep learning algorithms continue to advance [6], an increasing number of researchers are turning to deep learning approaches for automated arrhythmia detection. These studies have demonstrated that automated feature extraction, as opposed to manual ECG morphological feature extraction, leads to more accurate prediction results when using deep-learning-based methods [7–9]. In the automatic ECG analysis algorithms based on deep learning, convolutional neural networks (CNNs), which can achieve data feature extraction through local receptive fields, weight sharing, downsampling, and other methods [10], are usually used as the backbone to extract features automatically. For instance, Kiranyaz et al. proposed an adaptive 1D CNN model that integrates feature extraction and classification into a single learning body for ECG classification. The model was trained using relatively small sets of common and patient-specific training data and achieved remarkable accuracy [11]. Rajpurkar et al. proposed a 34-layer CNN for classifying 14 types of cardiac arrhythmias. The model was trained end-to-end on a single-lead ECG signal and outperformed the average cardiologist in terms of both recall (sensitivity) and precision (positive predictive value) [12]. Acharya et al. utilized an 11-layer CNN algorithm for automated detection of normal and myocardial infarction ECG beats. Their model was able to accurately detect unknown ECG signals even in the presence of noise [13]. He et al. developed a 2D CNN for detecting atrial fibrillation (AF) episodes, achieving sensitivity, specificity, positive predictive value, and overall accuracy rates of 99.41%, 98.91%, 99.39%, and 99.23%, respectively [14]. Jun et al. proposed an effective ECG arrhythmia classification model that utilizes two-dimensional convolutional neural networks with ECG images as an input [15]. Then, in the detection of life-threatening cardiac arrhythmias, Elola et al. proposed two deep neural network (DNN) architectures to classify the rhythm into pulseless electrical activity (PEA) or pulse-generating rhythm (PR) using short ECG segments, and both architectures achieved excellent performance [16]. Dubatovka et al. explored deep neural networks (DNNs) for learning cardiac cycles and reliably detecting AF from single-lead electrocardiogram (ECG) signals with a superior performance [17]. Krasteva et al. reported the optimal hyperparameters of an end-to-end fully convolutional DNN architecture for detecting shockable and nonshockable rhythm using single-lead raw ECG signals with life-threatening arrhythmias [18]. Jekova et al. optimized the architecture of a computationally efficient end-to-end CNN models for ECG rhythm analysis during cardiopulmonary resuscitation [19].

Although the aforementioned methods have achieved great success in the detection of ECG arrhythmia, it is widely acknowledged that temporal information plays an important role in tackling even more complex arrhythmia detection problems [20]. As a time series of data, ECG signals inherently contain temporal dependencies within their waveform. Recurrent neural networks (RNNs) can capture temporal dependencies in sequential data more efficiently compared to CNNs [8]. For example, Wang et al. proposed a global and updatable classification scheme called the global RNN (GRNN) in which RNN was used for automatic feature learning and classification based on the morphological and temporal information of ECG [21]. Recognizing the unique characteristics of ECG signals, Chen et al. achieved excellent performance in arrhythmia classification by fusing the CNN and RNN models [22].

In recent years, the transformer has become a popular deep learning model alongside CNNs and RNNs. It utilizes an attention mechanism to capture temporal features and context vectors and was originally developed for natural language processing (NLP) tasks [23–26]. However, it has also demonstrated superior performance on various vision tasks [27,28], including image classification. One representative work is the vision transformer (ViT) [29], which processes images as sequences of patches using a standard transformer encoder used in NLP. Compared to CNNs, transformer-based models have larger receptive fields and excel at modeling long-range dependencies, resulting in better performance on many image classification data sets [30,31]. As ECG signals exhibit temporal dependencies in their waveforms, the ViT's mechanism can be applied to arrhythmia classification tasks. Yan et al. proposed a heartbeat classification model based on a transformer that utilized the encoder part to process segmented single-lead ECG signals [32]. Natarajan et al. developed a wide and deep transformer neural network that combined handcrafted ECG features determined by a random forest model with discriminative feature representations automatically learned from a transformer neural network to classify each 12-lead ECG sequence into 27 cardiac abnormality classes [33]. Che et al. embedded a transformer network in a CNN to capture the temporal information of ECG signals for arrhythmia classification while introducing a new link constraint to the loss function to enhance the classification ability of the embedding vector [34]. These studies demonstrate the effectiveness of utilizing the transformer network structure for solving arrhythmia classification problems. However, due to the interference on the ECG waveform morphology of different diseases and the complex relationship between them, existing transformer approaches still have several limitations.

In summary, the existing automatic ECG analysis algorithms that solely used CNN cannot fully exploit the temporal features of ECG signals. Compared to CNN, RNNs can more efficiently capture temporal dependencies in sequential data, and the models fusing CNN and RNN have demonstrated excellent performance in arrhythmia classification. However, these models typically use isolated heartbeat signals as input, which results in a failure to leverage inter-heartbeat dependencies and requires the explicit segmentation of heartbeats. The segmentation of heartbeats usually necessitates the use of QRS detection algorithms such as Wavedet [35] and Pan–Tompkins [36]. For real-world scenarios, this means that an additional preprocessing step is required. As mentioned in the previous paragraph, the transformer architecture can capture temporal features through an attention mechanism and research has demonstrated the effectiveness in solving arrhythmia classification problems. However, the attention mechanism in the current approaches based on the transformer has a wide receptive field that can include irrelevant information outside the region of interest, affecting the amplitude and local statistical information from ECG signals. Moreover, the majority of existing methods take equal-length ECG signal segments as input. To address these issues, we propose the CNN-DVIT model, an end-to-end multi-label classification model that combines CNN with depthwise separable convolution and a vision transformer structure with deformable attention.

The major contributions of our model are as follows: Firstly, our approach involves replacing the multi-head self-attention mechanism in the original vision transformer model with a more effective deformable self-attention module, which enables the self-attention module to focus on relevant regions and capture more informative features. Secondly, we introduce the spatial pyramid pooling layer to accept variable-length 12-lead ECG signals as input, which eliminates the explicit segmentation of heartbeats beforehand and can make use of inter-heartbeat dependencies to improve the classification performance. Thirdly, to further mine the information of every lead for multi-lead ECG signals, the depthwise separable convolutions replace the conventional convolution in the CNN backbone.

2. Methods

In this section, we will initially provide an overview of the model's overall structure, and then proceed to introduce the specific structure of each individual component.

2.1. Model Architecture

Our proposed model, illustrated in Figure 1, is able to take continuous 12-lead ECG signals as input and output the arrhythmia diagnosis result in an end-to-end manner. Concretely, the model is composed of three main components: (1) a CNN-based backbone for feature extraction from each lead; (2) a deformable attention transformer encoder module to combine the CNN-extracted features and the positional encoding; and (3) the classification layer to obtain the probability that each patient may have for each type of heart disease. The first part, the CNN-based backbone, is based on the original Inception module with residual connections, in which the depthwise separable convolutions replace the conventional convolutions. We attempt to extract details of waveform variation from every lead of ECG by using depthwise separable convolutions that operate independently on each lead. Following the CNN-based backbone, the extracted features combined with positional encoding pass through the deformable attention transformer encoder module. The output from the deformable transformer encoder is then fed into the classification layer, which generates the predicted probability distribution over the nine classes.

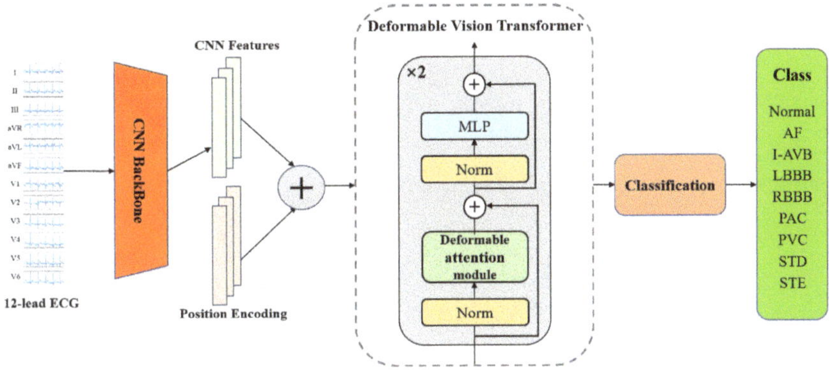

Figure 1. Overview of the proposed model.

2.2. CNN-Based Backbone

CNNs can extract data features using local receptive fields, weight sharing, downsampling, and other methods [10], which are commonly used as the backbone for feature extraction due to their ability to capture local features and translation invariance [37]. The depthwise separable convolutions were utilized in the Xception architecture developed by Google and demonstrated enhanced performance in image classification [38]. A depthwise separable convolution consists of a depthwise convolution and a pointwise convolution. The depthwise convolution is a spatial convolution performed independently over each channel of an input; and then the pointwise convolution is a 1×1 convolution projecting the channel output by the depthwise convolution onto a new channel space. For multi-lead ECG signals, the spatial convolutions are applied to each lead, and then the feature map of every channel is projected onto a new space, enabling ECG details to be extracted from every lead. Actually, our CNN-based backbone is built on the original Inception module with residual connections, where conventional convolutions are replaced by depthwise separable convolutions. Specific configuration details are shown in Figure 2. In the early stage of this part, we use a large convolution kernel with a size of 15, increasing the receptive field of the convolution network and facilitating the detection of large-scale waveforms. In the middle, we utilize the residual network structure and Inception structure. To learn information from different scales, we employ three different scales of convolution kernels. At the end of the CNN-based backbone, we introduce the spatial pyramid pooling layer to convert the different dimensions of the final output feature map into a fixed-dimensional CNN feature without considering the length of input signals [39]. As depicted in Figure 2, we divide the output feature graph from each lead in the first part into 36 blocks, 9 blocks, 4 blocks, and 1 block, and then compute the maximum pooling for each block individually. Therefore, the model can accommodate arbitrary-length ECG signals.

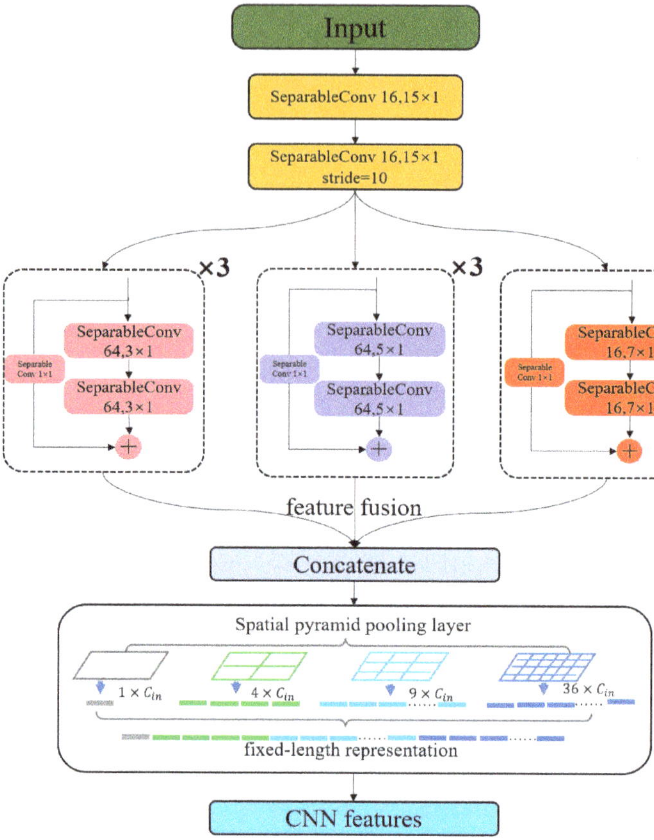

Figure 2. This is the structure of CNN Backbone.

2.3. *Deformable Vision Transformer*

The deformable vision transformer [40] was developed based on a deformable self-attention module, where the positions of key and value pairs in self-attention are selected in a data-dependent way. In the ECG signal classification problem, this flexible scheme enables the self-attention module to focus on relevant regions and capture more informative features to accurately identify the distinct characteristic wave types and recognize different arrhythmia categories. The structure is shown in the deformable vision transformer module in Figure 1. The deformable vision transformer network in our model contains two identical layer stacks and each layer has two sub-layers. The first sub-layer is the deformable attention module; the second is the multi-layer perceptron (MLP) block which adopts two linear transformations and a GELU activation. In addition, LayerNorm (LN) is applied before every block, and residual connections after every block [41,42].

Deformable Attention Module

The ability to flexibly model relevant features is a crucial aspect of our proposed model. This crucial characteristic is implemented by the deformable attention module described in the Ref. [40], and the structure is shown as Figure 3. Specifically, the deformable attention can be viewed as a spatial adaptive mechanism to effectively model the relationship between tokens under the guidance of important regions in the feature maps. These regions are computed from sets of deformed points which are learned from the queries by an offset network. After obtaining the regions, a bilinear interpolation method is applied to sample features from the feature maps, and then the sampled features are input into the key and value projections to obtain the deformed keys and values. Finally, standard multi-head attention is applied to attend queries to the sampled keys and aggregate features from the deformed values. This can be better understood mathematically using a set of theoretical notations, as follows.

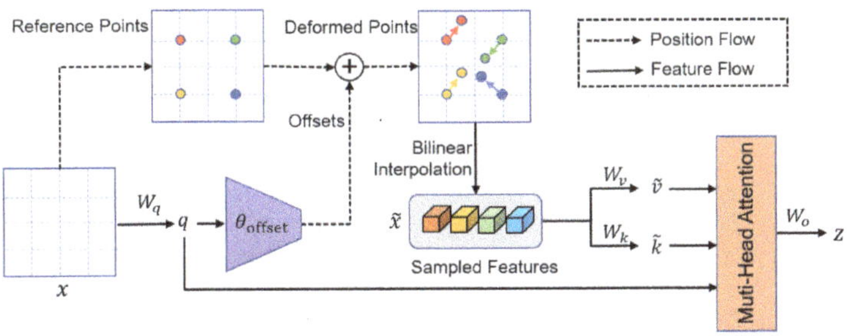

Figure 3. This is the structure of deformable attention module.

We take a feature map $x \in \mathbb{R}^{H\times W\times C}$ as the input, and then generate a uniform grid of reference points $p \in \mathbb{R}^{H/r\times W/r\times 2}$, where r is the factor by which the grid size is downsampled from the input feature map size. To obtain the offset for each reference point, the feature maps are projected linearly to obtain the query tokens q, and then fed into a lightweight sub-network offset network to generate the offsets Δp of the reference points p. Specifically, the sub-network contains two convolution modules with a nonlinear activation. First, the input features pass through a depthwise convolution to capture local features. Then, GELU activation and a 1×1 convolution are adopted to calculate the offsets. After generating the offsets for the reference points, the features at the locations of deformed points are sampled by the bilinear interpolation method and then projected as keys $\tilde{k}$ and values $\tilde{v}$. Next, we perform standard multi-head attention on q, $\tilde{k}$, and $\tilde{v}$, where a multi-head attention block with M heads is formulated as:

$$q = xW_q,\ \tilde{k} = \tilde{x}W_k,\ \tilde{v} = \tilde{x}W_v \tag{1}$$

$$z^{(m)} = \sigma\left(q^{(m)}\tilde{k}^{(m)\top}/\sqrt{d}\right)\tilde{v}^{(m)},\ m = 1,\ldots,M \tag{2}$$

$$z = \text{Concat}\left(z^{(1)},\ldots,z^{(M)}\right)W_o \tag{3}$$

where $W_q,\ W_k,\ W_v,\ W_o \in \mathbb{R}^{C\times C}$ are the projection matrices, $\tilde{x} = \phi(x; p + \Delta p)$, $\Delta p = \theta_{offset}(q)$, and the sampling function $\phi(\cdot;\cdot)$ is a bilinear interpolation as Equation (4). In Equation (2), $\sigma(\cdot)$ denotes the softmax function, and $d = C/M$ is the dimension of each head. $z^{(m)}$ denotes the output of the m-th attention head, and $q^{(m)}, \tilde{k}^{(m)}, \tilde{v}^{(m)}$ represent query, the deformed key, and value embeddings, respectively.

$$\phi(z; (p_x,p_y)) = \sum_{(r_x,r_y)} g(p_x,r_x)g(p_y,r_y)z[r_y,r_x,:] \tag{4}$$

where the $g(a,b) = \max(0, 1 - |a - b|)$ and (r_x, r_y) indexes all the locations on $z \in \mathbb{R}^{H\times W\times C}$.Finally, the features of each head are concatenated together and projected through W_o to obtain the final output z as Equation (3). To build up a deformable vision transformer, the normalization layer and an MLP block with two linear transformations and a GELU activation are adopted after the deformable attention module [43].

2.4. Classification Layer

The deformable vision transformer network is connected to the classification layer for multi-classification. The classification layer consists of a full connection layer and a softmax layer, which outputs the predicted probability indicating the likelihood that each patient has each type of cardiac arrhythmia.

3. Experiments and Results

In this section, we will delineate the data sources, expound on the experimental design, and elaborate on the analysis of experimental results.

3.1. Data Description and Experiment Setup

Our paper utilized data from the 1st China Physiological Signal Challenge (CPSC-2018), consisting of 6877 12-lead ECG recordings collected from 11 hospitals [44]. The signals contained normal heart rhythms and eight types of cardiac arrhythmia, namely, Normal (N), Atrial fibrillation (AF), First-degree atrioventricular block (I-AVB), Left bundle branch block (LBBB), Right bundle branch block (RBBB), Premature atrial contraction (PAC), Premature ventricular contraction (PVC), ST-segment depression (STD), and ST-segment elevated (STE). The recordings had one, two, or three labels, with some designated as First, Second, and Third. Thus, the ECG arrhythmia detection task in our paper became a multi-label classification problem.

The model being suggested is evaluated on every data category present in the CPSC-2018 data set using a train/validation/independent-test strategy. In the ten-fold cross-validation experiment, the training data set was divided into ten parts, with nine parts used as training data and one part as validation data. After iterations of training and validation, the model with the best performance on the validation set was evaluated on unseen test data for final performance evaluation. The Adam optimizer with default parameters and a learning rate of 0.0001 was used to train the model, and the Focal Loss function was adopted as the objective loss function [45].

3.2. Classification Performance

3.2.1. Evaluation Metrics

To measure the classification performance of our method from multiple perspectives, we introduce three performance indicators: the average *precision*, *recall* rate, and *F*1 score. The details are as follows:

$$Precision = \frac{TP}{TP + FP} \tag{5}$$

$$Recall = \frac{TP}{TP + FN} \tag{6}$$

$$F1 = \frac{2 \times Precision \times Recall}{Precision + Recall} \tag{7}$$

In our multi-label classification task, for a certain class, *TP* indicates the number of correctly classified samples in this class. *FN* indicates the number of samples belonging to this class that are misclassified as in other classes. *FP* indicates the number of samples misclassified as in this class when they belong to other classes. The averages of all nine categories of the *F*1 score were used to evaluate the final performance of the model.

3.2.2. Comparison with Existing Methods

To verify the performance of our proposed model, we compared our CNN-DVIT with recently proposed ECG classification methods as well as basic neural networks. Table 1 presents the results for the class-level F1 score and average F1 score of our model and six reference models. As shown in Table 1, our proposed CNN-DVIT has an average F1 score of 0.829, which is comparatively better than the other methods. Our model achieved the highest F1 score in six out of the nine ECG categories. Specifically, our approach outperformed the conventional Resnet [30] and LSTM [8] methods in all nine categories, with an average F1 score that was 0.092 and 0.165 higher than them, respectively. Moreover, the F1 scores of all categories performed better than the other methods except for PVC, where VGG-16 [10] obtained a slightly higher score. Notably, compared to the latest competitor [34], which also introduced the transformer, our CNN-DVIT outperformed in each category with an average F1 score improvement of 0.112.

Table 1. Classification performance on CPSC-2018 data set.

Type	F1 Score						
	ResNet [30]	LSTM [8]	VGG-16 [10]	Mostayed et al., 2018 [46]	Chen et al., 2020 [22]	Che et al., 2021 [34]	CNN-DVIT
N	0.730	0.730	0.750	0.702	0.795	0.817	0.831
AF	0.882	0.792	0.861	0.815	0.897	0.858	0.924
I-AVB	0.877	0.763	0.874	0.767	0.865	0.878	0.887
LBBB	0.786	0.848	0.857	0.847	0.821	0.800	0.905
RBBB	0.905	0.909	0.918	0.898	0.911	0.872	0.935
PAC	0.487	0.268	0.333	0.397	0.734	0.618	0.704
PVC	0.733	0.763	0.859	0.807	0.852	0.830	0.842
STD	0.784	0.800	0.814	0.768	0.788	0.711	0.823
STE	0.444	0.105	0.462	0.286	0.509	0.686	0.610
Average F1	0.737	0.664	0.748	0.699	0.797	0.786	0.829

Additionally, we present the confusion matrix, receiver operating characteristic (ROC) curve, and area under the curve (AUC) for CNN-DVIT in Figure 4a,b, respectively. The confusion matrix demonstrates our model's strong classification ability for almost all types of ECGs. However, in Figure 4a, we can observe that the PAC, STD, and STE categories exhibited relatively poor performance compared to other categories in the data set. This is attributed to a limited distribution of data within these categories. Meanwhile, the normal and STE classes demonstrated high similarity, particularly with regard to the morphology of their T waveforms. As a consequence, there was confusion and misclassification between these two classes. The ROC curve illustrates the network output at different classification criteria with FP and TP as the axis co-ordinates, effectively reflecting the classification performance of the network structure. As shown in Figure 4b, the AUC values of our CNN-DVIT model are greater than 95% for most classes, except for ST-segment elevated (STE). Therefore, it can be concluded that our model has excellent performance in classifying different cardiac diseases. These results align well with our F1 score assessment.

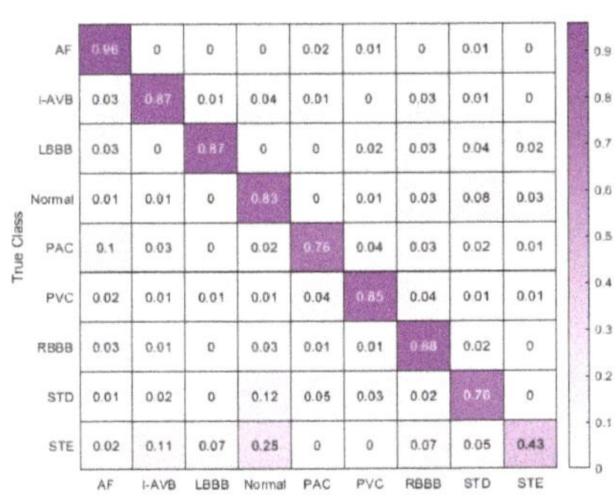

(a)

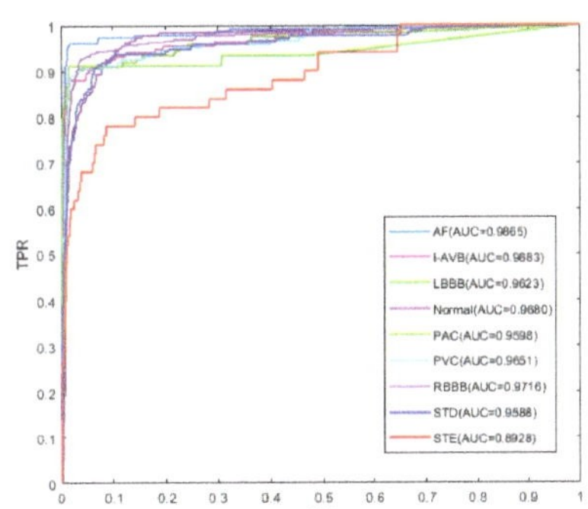

(b)

Figure 4. (**a**) Normalized confusion matrix of CNN-DVIT; and (**b**) the ROC curve and AUC of CNN-DVIT.

4. Discussion

In this section, we analyze the effectiveness of key components in our CNN-DVIT model through ablation experiments. We present the results for the ECG arrhythmia detection task in the CPSC-2018 data set.

First, we assess the effectiveness of the deformable attention module. In our CNN-DVIT model, we apply the deformable multi-head attention (DMHA) module to replace the multi-head self-attention (MHSA) block of the vision transformer. We compare the classification performance of both methods, and the results are recorded in Table 2. As shown in Table 2, the average F1 score of the models that use the DMHA module is higher than those using the multi-head self-attention (MHSA) block. These experimental results indicate that the deformable multi-head attention can more effectively extract information from ECG signals and exhibit outstanding performance in ECG classification on CPSC-2018 data sets.

Table 2. Results of ablation experiments.

DMHA	MHSA	DWS-CNN	CNN	Average		
				F1	Precision	Recall
	√	√		0.797	0.751	0.860
	√		√	0.789	0.775	0.825
√			√	0.819	0.814	0.830
√		√		0.829	0.819	0.849

Next, we investigate the impact of depthwise separable convolution (DWS-CNN) in our model. We conduct an ablation experiment by comparing the DWS-CNN with conventional convolution (CNN). As shown in Table 2, models that apply the DWS-CNN have a clear advantage over those using conventional convolution. These ablation experiments demonstrate that the depthwise separable convolution (DWS-CNN), which is applied to each lead and projects the feature map of every channel to a new space, can effectively extract features of heart disease from multi-lead ECG signals for diagnosis.

5. Conclusions

This paper presents an end-to-end model named CNN-DVIT for arrhythmia classification of multi-lead ECG signals. The model combines a CNN backbone and a transformer block to extract information from ECG signals through two steps: the first step is learning the details of waveform variation from every lead of ECG by the CNN backbone with depthwise separable convolutions; and the second step involves combining the features extracted by CNN with positional encoding using the deformable attention transformer encoder module. Particularly, in the first step, we employ a spatial pyramid pooling layer to convert the variable dimensions of the final output feature graph into fixed-dimensional CNN features. This enables our model to accept ECG signals of varying lengths. Our CNN-DVIT network architecture exhibits exceptional performance in ECG classification on CPSC-2018 data sets, achieving an average F1 score of 82.9% across eight types of arrhythmias and sinus rhythms. These results demonstrate that deformable attention is well-suited to the unique characteristics of ECG signals and can effectively perform ECG classification tasks.

In the task of arrhythmia classification, CNN-DVIT synthesizes feature data more effectively than using any single method alone, such as LSTM, ResNet, or the transformer with the multi-head self-attention. In recent years, advancements in hardware technology, information transmission, and computing capabilities have contributed to the increasing significance of wearable ECG devices as a diagnostic modality [47]. However, our model in this study is a multi-label arrhythmia classification model for the 12-lead ECG. As such, it may not be suitable for dynamic ECG data from wearable ECG devices, which are subject to greater interference. In a clinical setting, the timeliness of the auxiliary diagnosis system is generally required. Therefore, we plan to shift our focus towards developing lightweight models with fewer parameters that are better suited for analyzing dynamic ECG data from wearables in future work.

Author Contributions: Conceptualization, Y.D. and Y.Y.; methodology, Y.D. and L.W.; software, Y.D., and L.Q.; validation, Y.D., M.Z. and L.W.; formal analysis, Y.D. and Y.Y.; investigation and resources, Y.D. and L.W.; data curation, Y.Y.; writing—original draft preparation, Y.D., M.Z. and L.W.; writing—review and editing, Y.D., Y.Y. and L.W.; visualization and supervision, Y.D. and L.W. All authors have read and agreed to the published version of the manuscript.

Funding: This research was supported by the National Key Research and Development Program of China (2021YFC2501500).

Data Availability Statement: The data presented in this study are available from the corresponding author upon reasonable request.

Conflicts of Interest: The authors declare no conflict of interest.

References

1. Hu, R.; Chen, J.; Zhou, L. A transformer-based deep neural network for arrhythmia detection using continuous ECG signals. *Comput. Biol. Med.* **2022**, *144*, 105325. [CrossRef]
2. Kibos, A.S.; Knight, B.P.; Essebag, V.; Fishberger, S.B.; Tintoiu, I.C. *Cardiac Arrhythmias: From Basic Mechanism to State-of-the-Art Management*; Springer: London, UK, 2014; pp. 43–60.
3. Hanna, E.B.; Glancy, D.L. ST-segment depression and T-wave inversion: Classification, differential diagnosis, and caveats. *Clevel. Clin. J. Med.* **2011**, *78*, 404. [CrossRef] [PubMed]
4. Shiyovich, A.; Wolak, A.; Yacobovich, L.; Grosbard, A.; Katz, A. Accuracy of diagnosing atrial flutter and atrial fibrillation from a surface electrocardiogram by hospital physicians: Analysis of data from internal medicine departments. *Am. J. Med. Sci.* **2010**, *340*, 271–275. [CrossRef] [PubMed]
5. Schläpfer, J.; Wellens, H.J. Computer-Interpreted Electrocardiograms: Benefits and Limitations. *J. Am. Coll. Cardiol.* **2017**, *70*, 1183–1192. [CrossRef]
6. LeCun, Y.; Bengio, Y.; Hinton, G. Deep learning. *Nature* **2015**, *521*, 436–444. [CrossRef] [PubMed]
7. Baloglu, U.B.; Talo, M.; Yildirim, O.; Tan, R.S.; Acharya, U.R. Classification of Myocardial Infarction with Multi-Lead ECG Signals and Deep CNN. *Pattern Recognit. Lett.* **2019**, *122*, 23–30. [CrossRef]
8. Saadatnejad, S.; Oveisi, M.; Hashemi, M. LSTM-based ECG classification for continuous monitoring on personal wearable devices. *IEEE J. Biomed. Health Inform.* **2019**, *24*, 515–523. [CrossRef]
9. Qin, L.; Wang, X.; Li, S.; Wen, J.; Gao, X. An End-to-End 12-Leading Electrocardiogram Diagnosis System Based on Deformable Convolutional Neural Network with Good Antinoise Ability. *IEEE Trans. Instrum. Meas.* **2021**, *70*, 1–13. [CrossRef]
10. Simonyan, K.; Zisserman, A. Very deep convolutional networks for large-scale image recognition. *arXiv* **2014**, arXiv:1409.1556.
11. Kiranyaz, S.; Ince, T.; Gabbouj, M. Real-time patient-specific ecg classification by 1-D convolutional neural networks. *IEEE Trans. Biomed. Eng.* **2016**, *63*, 664–675. [CrossRef]
12. Rajpurkar, P.; Hannun, A.Y.; Haghpanahi, M.; Bourn, C.; Ng, A.Y. Cardiologist-level arrhythmia detection with convolutional neural networks. *arXiv* **2017**, arXiv:1707.01836.
13. Acharya, U.R.; Fujita, H.; Oh, S.L.; Hagiwara, Y.; Tan, J.H.; Adam, M. Application of deep convolutional neural network for automated detection of myocardial infarction using ECG signals. *Inf. Sci.* **2017**, *415*, 190–198. [CrossRef]
14. He, R.; Wang, K.; Zhao, N.; Liu, Y.; Yuan, Y.; Li, Q.; Zhang, H. Automatic detection of atrial fibrillation based on continuous wavelet transform and 2D convolutional neural networks. *Front. Physiol.* **2018**, *9*, 1206. [CrossRef]
15. Jun, T.J.; Nguyen, H.M.; Kang, D.; Kim, D.; Kim, D.; Kim, Y.-H. ECG arrhythmia classification using a 2-D convolutional neural network. *arXiv* **2018**, arXiv:1804.06812.
16. Elola, A.; Aramendi, E.; Irusta, U.; Picón, A.; Owens, P.; Idris, A. Deep neural networks for ECG-based pulse detection during out-of-hospital cardiac arrest. *Entropy* **2019**, *21*, 305. [CrossRef]
17. Dubatovka, A.; Buhmann, J.M. Automatic Detection of Atrial Fibrillation from Single-Lead ECG Using Deep Learning of the Cardiac Cycle. *BME Front.* **2022**, *2022*, 2765–8031. [CrossRef]
18. Krasteva, V.; Ménétré, S.; Didon, J.-P.; Jekova, I. Fully convolutional deep neural networks with optimized hyperparameters for detection of shockable and non-shockable rhythms. *Sensors* **2020**, *20*, 2875. [CrossRef]
19. Jekova, I.; Krasteva, V. Optimization of End-to-End Convolutional Neural Networks for Analysis of Out-of-Hospital Cardiac Arrest Rhythms during Cardiopulmonary Resuscitation. *Sensors* **2021**, *21*, 4105. [CrossRef]

20. Qiu, X.; Liang, S.; Meng, L.; Zhang, Y.; Liu, F. Exploiting Feature Fusion and Long-term Context Dependencies for Simultaneous ECG Heartbeat Segmentation and Classification. *Int. J. Data Sci. Anal.* **2021**, *11*, 181–193. [CrossRef]
21. Wang, G.; Zhang, C.; Liu, Y.; Yang, H.; Fu, D.; Wang, H.; Zhang, P. A Global and Updatable ECG Beat Classification System Based on Recurrent Neural Networks and Active Learning. *Inf. Sci.* **2019**, *501*, 523–542. [CrossRef]
22. Chen, T.-M.; Huang, C.-H.; Shih, E.S.; Hu, Y.-F.; Hwang, M.-J. Detection and Classification of Cardiac Arrhythmias by a Challenge-Best Deep Learning Neural Network Model. *iScience* **2020**, *23*, 100886. [CrossRef] [PubMed]
23. Vaswani, A.; Shazeer, N.; Parmar, N.; Uszkoreit, J.; Jones, L.; Gomez, A.N.; Kaiser, Ł.; Polosukhin, I. Attention Is All You Need. *NeurIPS* **2017**, *30*, 5998–6008.
24. Onan, A.; Korukolu, S. Exploring Performance of Instance Selection Methods in Text Sentiment Classification. In *Artificial Intelligence Perspectives in Intelligent Systems*; Springer: Berlin/Heidelberg, Germany, 2016; pp. 167–179.
25. Onan, A.; Korukolu, S. A Feature Selection Model Based on Genetic Rank Aggregation for Text Sentiment Classification. *J. Inf. Sci.* **2017**, *43*, 25–38. [CrossRef]
26. Onan, A. Sentiment Analysis on Product Reviews Based on Weighted Word Embeddings and Deep Neural Networks. *Concurr. Comput. Pract. Exp.* **2020**, *33*, e5909. [CrossRef]
27. Liu, Z.; Lin, Y.; Cao, Y.; Hu, H.; Wei, Y.; Zhang, Z.; Lin, S.; Guo, B. Swin Transformer: Hierarchical Vision Transformer using Shifted Windows. In Proceedings of the IEEE/CVF International Conference on Computer Vision, Montreal, QC, Canada, 10–17 October 2021; pp. 10012–10022. Available online: https://openaccess.thecvf.com/content/ICCV2021/html/Liu_Swin_Transformer_Hierarchical_Vision_Transformer_Using_Shifted_Windows_ICCV_2021_paper.html (accessed on 27 April 2023).
28. Wang, W.; Xie, E.; Li, X.; Fan, D.-P.; Song, K.; Liang, D.; Lu, T.; Luo, P.; Shao, L. Pyramid Vision Transformer: A Versatile Backbone for Dense Prediction Without Convolutions. In Proceedings of the IEEE/CVF International Conference on Computer Vision, Montreal, QC, Canada, 10–17 October 2021.
29. Dosovitskiy, A.; Beyer, L.; Kolesnikov, A.; Weissenborn, D.; Zhai, X.; Unterthiner, T.; Dehghani, M.; Minderer, M.; Heigold, G.; Gelly, S.; et al. An Image Is Worth 16 × 16 Words: Transformers for Image Recognition at Scale. *arXiv* **2020**, arXiv:2010.11929.
30. He, K.; Zhang, X.; Ren, S.; Sun, J. Deep Residual Learning for Image Recognition. In Proceedings of the IEEE Computer Society Conference on Computer Vision and Pattern Recognition (CVPR), Las Vegas, NV, USA, 27–30 June 2016.
31. Huang, G.; Liu, Z.; Pleiss, G.; Van Der Maaten, L.; Weinberger, K. Convolutional Networks with Dense Connectivity. *IEEE Trans. Pattern Anal. Mach. Intell.* **2019**, *44*, 8704–8716. [CrossRef]
32. Yan, G.; Liang, S.; Zhang, Y.; Liu, F. Fusing Transformer Model with Temporal Features for ECG Heartbeat Classification. In Proceedings of the 2019 IEEE International Conference on Bioinformatics and Biomedicine (BIBM), San Diego, CA, USA, 18–21 November 2019; IEEE: Piscataway, NJ, USA, 2019; pp. 898–905.
33. Natarajan, A.; Chang, Y.; Mariani, S.; Rahman, A.; Boverman, G.; Vij, S.; Rubin, J. A Wide and Deep Transformer Neural Network for 12-Lead ECG Classification. In Proceedings of the 2020 Computing in Cardiology, Rimini, Italy, 13–16 September 2020; IEEE: Piscataway, NJ, USA, 2020; pp. 1–4.
34. Che, C.; Zhang, P.; Zhu, M.; Qu, Y.; Jin, B. Constrained Transformer Network for ECG Signal Processing and Arrhythmia Classification. *BMC Med. Inform. Decis. Mak.* **2021**, *21*, 184. [CrossRef]
35. Martínez, J.P.; Almeida, R.; Olmos, S.; Rocha, A.P.; Laguna, P. A Wavelet-Based ECG Delineator: Evaluation on Standard Databases. *IEEE Trans. Biomed. Eng.* **2004**, *51*, 570–581. [CrossRef]
36. Pan, J.; Tompkins, W.J. A Real-Time QRS Detection Algorithm. *IEEE Trans. Biomed. Eng.* **1985**, *32*, 230–236. [CrossRef]
37. Shang, L.; Yang, Q.; Wang, J.; Li, S.; Lei, W. Detection of Rail Surface Defects Based on CNN Image Recognition and Classification. In Proceedings of the 20th International Conference on Advanced Communication Technology (ICACT), Chuncheon-si, Republic of Korea, 11–14 February 2018; IEEE: Piscataway, NJ, USA, 2018; pp. 45–51.
38. Chollet, F. Xception: Deep Learning with Depthwise Separable Convolutions. In Proceedings of the 2017 IEEE Conference on Computer Vision and Pattern Recognition (CVPR), Honolulu, HI, USA, 21–26 July 2017; IEEE: Piscataway, NJ, USA, 2017.
39. He, K.; Zhang, X.; Ren, S.; Sun, J. Spatial Pyramid Pooling in Deep Convolutional Networks for Visual Recognition. *IEEE Trans. Pattern Anal. Mach. Intell.* **2015**, *37*, 1904–1916. [CrossRef]
40. Xia, Z.; Pan, X.; Song, S.; Li, L.E.; Huang, G. Vision Transformer with Deformable Attention. In Proceedings of the IEEE/CVF Conference on Computer Vision and Pattern Recognition, New Orleans, LA, USA, 18–24 June 2022; pp. 4794–4803.
41. Wang, Q.; Li, B.; Xiao, T.; Zhu, J.; Li, C.; Wong, D.F.; Chao, L.S. Learning Deep Transformer Models for Machine Translation. *arXiv* **2019**, arXiv:1906.01787.
42. Baevski, A.; Auli, M. Adaptive Input Representations for Neural Language Modeling. *arXiv* **2018**. [CrossRef]
43. Ba, J.L.; Kiros, J.R.; Hinton, G.E. Layer Normalization. *arXiv* **2016**, arXiv:1607.06450.
44. Liu, F.; Liu, C.; Zhao, L.; Zhang, X.; Kwee, E.N.Y. An Open Access Database for Evaluating the Algorithms of Electrocardiogram Rhythm and Morphology Abnormality Detection. *J. Med. Imaging Health Inform.* **2018**, *8*, 1368–1373. [CrossRef]
45. Lin, T.Y.; Goyal, P.; Girshick, R.; He, K.; Dollar, P. Focal Loss for Dense Object Detection. *IEEE Trans. Pattern Anal. Mach. Intell.* **2019**, *99*, 2999–3007.

46. Mostayed, A.; Luo, J.; Shu, X.; Wee, W. Classification of 12-Lead ECG Signals with Bi-directional LSTM Network. *arXiv* **2018**, arXiv:1811.02090.
47. Tomasic, I.; Petrovic, N.; Linden, M.; Rashkovska, A. Comparison of Publicly Available Beat Detection Algorithms Performances on the ECGs Obtained by a Patch ECG Device. In Proceedings of the 2019 42nd International Convention on Information and Communication Technology, Electronics and Microelectronics (MIPRO), Opatija, Croatia, 20–24 May 2019; IEEE: Piscataway, NJ, USA, 2019; pp. 275–278.

micromachines

MDPI

Article

Dynamic Image Difficulty-Aware DNN Pruning

Vasileios Pentsos †, Ourania Spantidi † and Iraklis Anagnostopoulos *

School of Electrical, Computer and Biomedical Engineering, Southern Illinois University, Carbondale, IL 62901, USA
* Correspondence: iraklis.anagno@siu.edu
† These authors contributed equally to this work.

Abstract: Deep Neural Networks (DNNs) have achieved impressive performance in various image recognition tasks, but their large model sizes make them challenging to deploy on resource-constrained devices. In this paper, we propose a dynamic DNN pruning approach that takes into account the difficulty of the incoming images during inference. To evaluate the effectiveness of our method, we conducted experiments on the ImageNet dataset on several state-of-art DNNs. Our results show that the proposed approach reduces the model size and amount of DNN operations without the need to retrain or fine-tune the pruned model. Overall, our method provides a promising direction for designing efficient frameworks for lightweight DNN models that can adapt to the varying complexity of input images.

Keywords: Deep Neural Networks; pruning; embedded systems; image difficulty

1. Introduction

Deep Neural Networks (DNNs) have emerged as a popular choice for solving various challenging machine learning tasks, especially in the field of computer vision. Inside a DNN, matrix multiplication is a fundamental operation that forms the backbone of many common network layers, such as fully connected and convolutional layers. These operations are typically implemented using multiply accumulate (MAC) units, which perform both multiplication and addition in a single operation [1]. While MAC operations are essential for DNNs to function accurately, they also contribute significantly to the elevated energy consumption of these networks [2,3].

Therefore, even though DNN models have demonstrated remarkable accuracy on different image recognition benchmarks, their high computational and memory requirements limit their deployment on resource-constrained devices such as embedded systems. A commonly used technique is pruning, where the weight tensors of a DNN are modified to be sparse [4]. This approach involves removing a subset of network parameters, such as weights or neurons, that are deemed unnecessary or redundant. By reducing the number of parameters, DNN pruning reduces the number of MAC operations required for inference, thereby decreasing the energy consumption of the network. However, this removal of parameters from the original DNN model can lead to drops in the final accuracy, which is usually recovered through additional retraining or fine-tuning on the pruned model [5].

Previous research on DNN pruning has proposed various techniques for lightweight models. An approach is to zero weights whose magnitude exceeds a specified threshold [6]. A more aggressive approach is the removal of entire filters and neurons [7]. Instead of pruning individual weights, the work in [8] prunes groups of weights through the introduction of depth-wise and shape-wise sparsity instead. Through reinforcement learning, the work in [9] prunes convolution channels. To recover from the pruning-induced accuracy loss, these works fine-tune and retrain the compressed generated models. However, retraining is not always a feasible task. The required dataset might not even be available, or there

Citation: Pentsos, V.; Spantidi, O.; Anagnostopoulos, I. Dynamic Image Difficulty-Aware DNN Pruning. *Micromachines* **2023**, *14*, 908. https://doi.org/10.3390/mi14050908

Academic Editors: Xiao Xiao and Gang Ge

Received: 3 March 2023
Revised: 14 April 2023
Accepted: 21 April 2023
Published: 23 April 2023

might be privacy issues [10]. At the same time, retraining on embedded devices is a very time-consuming and computationally-intensive procedure [11].

In this work, we present a pruning framework that takes into account the difficulty of the input image during inference . In contrast to related works, we do not employ extensive offline explorations or retraining, and at the same time, we allow for a more data-driven and dynamic selection of pruning combinations. Image difficulty refers to the level of challenge posed by an input image for a DNN. Different images may require different levels of network complexity to achieve high accuracy. Some images may be relatively easy to classify, while others may be more challenging due to factors such as occlusion, lighting, or complexity. However, the term image difficulty can be subjective and arbitrary, and previous works have relied on human feedback alone to categorize an image as difficult for classification [12]. However, human perception may not always be an accurate measure of image difficulty, as DNNs extract features that are not visible to the human eye and process image information in a different way. Therefore, it is necessary to develop objective metrics that can quantify image difficulty and accurately assess the network's performance on different input images.

Overall, we propose a framework that:

1. Utilizes difficulty metrics that incorporate human observations as well as image quality scores to build a prediction model that can predict the difficulty of an image at run-time;
2. Introduces a lightweight exploration loop for pruning combinations; and
3. Adaptively prunes DNNs during inference time based on the predicted image difficulty scores.

2. Methodology

This section will present the proposed framework and its different components. First, we will present our employed metric for image difficulty and its utilization in a Support Vector Machine (SVM) classifier model. Afterwards, we will show an exploration loop that generates a pool of candidate pruned DNN models, and finally, we will present how to combine the SVM classifier and the pruned DNN models at run-time to achieve lightweight inference.

2.1. Quantifying and Classifying Image Difficulty

For this part of the methodology, we employed the minimum required viewing times dataset [13]. This dataset assesses the time required by human subjects to recognize objects from the ImageNet and ObjectNet datasets. The dataset provides valuable insights into how long it takes humans to classify images. We propose that this human-based information can be used toward building an effective prediction model. By analyzing the dataset in [13], we can identify patterns and characteristics of images that are easy or difficult for humans to classify. Intuitively, an image that requires a longer viewing time to classify is likely to have more complex features, occlusions, or ambiguities that make it challenging for humans to recognize. On the other hand, an image that can be classified quickly likely has clearer and more distinctive features that make it easier for humans to identify. Alongside the viewing times dataset, the work in [13] also proposed two difficulty metrics. The first metric is a difficulty score, which measures the proportion of humans who were able to correctly classify a given image. The second metric is the minimum amount of time required to reliably classify an image. This metric takes into account the time taken by the participants to accurately recognize the image, with longer times indicating greater complexity and difficulty.

To that end, we propose an image difficulty metric that combines both of these aspects into a single value. Therefore, for an image i, we define the difficulty as follows:

$$Difficulty_i = \frac{Correct\_Classifications}{Total\_Participants} \times (1 - \frac{Avg\_Time_i - min(Time_{total})}{max(Time_{total}) - min(Time_{total})}) \quad (1)$$

In Equation 1, the fraction $\frac{Correct\_Classifications}{Total\_Participants}$ corresponds to the number of correct classifications observed out of all of the participants. $Time_i$ is the average response time spent by all human subjects on the classification of image i, and $Time_{total}$ is a list that contains all $Time_i$ values for each image i recorded in the viewing times dataset [13]. With this metric, a value near 0 would mean that it is highly likely the specific image lies in the most difficult group of images in the dataset, while a value closer to 1 would indicate that the image was rarely misclassified and could be deemed an easier one. At this point, we want to mention that we do not conduct any further experiments on human subjects but solely rely on the dataset found in [13]. We use the recorded values from the participants as our only indicator of human-measured image difficulty.

We use the image difficulty metric defined in Equation (1) to build an SVM model that classifies any input image into one of the following three classes: Easy, Medium, and Hard. An overview of the SVM training process is shown in Figure 1. The input image is analyzed for its image quality (BRISQUE) [14], spatial information (SI) [15] and Gradient Sparsity Index (GSI) [16] metrics, which are used as the input to an SVM model with a polynomial kernel. As the output of the SVM, we combine two different pieces of information to acquire the final class: the difficulty metric defined in Equation (1) and the information of whether the DNN classifies correctly or not the image which can be obtained by running a single inference on the input image. Note that this procedure is being performed only for the training of the SVM predictor. Therefore, the final metric we use as the utilized difficulty is defined by:

$$Difficulty_i = 0.8 * \frac{Correct\_Classifications}{Total\_Participants} \times (1 - \frac{Avg\_Time_i - min(Time_{total})}{max(Time_{total}) - min(Time_{total})}) + 0.2 * DNN\_Classification \quad (2)$$

The $DNN\_Classification$ value in Equation (2) will either be 1 or 0 depending on whether the DNN can correctly classify the input image or not. We include the DNN's ability to classify the image to have a well-rounded difficulty metric that includes not only human perception abilities but also takes into consideration how difficult the image is from the DNN's perspective. However, we still consider the human-measured metric to be the most significant in terms of importance.

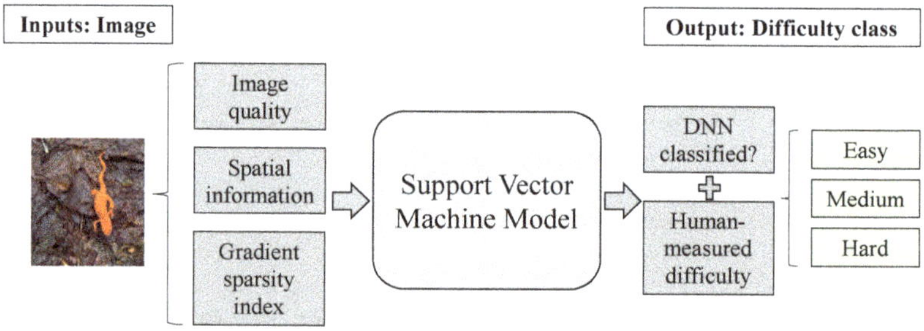

Figure 1. the SVM model to predict image difficulty of an input image for a target DNN.

Finally, instead of employing the SVM for a regression task where it would try to map an image's metrics to a single difficulty value as defined in Equation (2), we simply divide all the difficulty values we observe for the dataset in [13] to three equally sized categories (Easy, Medium, and Hard), and we use the SVM to classify the input image to one of these three classes. Therefore, at run-time, where we do not have the knowledge of the DNN's classification score for a previously unseen image, the SVM will simply return a classification score which will be either Easy, Medium, or Hard.

2.2. DNN Pruning Tolerance

As mentioned in Section 1, related works that utilize pruning toward more efficient DNN inference usually employ extensive explorations that also require retraining. In this framework, however, one of our main goals is to prune DNNs while completely avoiding retraining and further fine-tuning.

An overview of this step is shown in Figure 2. During the pruning stage, we avoid fine-grain time-consuming explorations and initially employ layer-wise pruning, i.e., we apply pruning, and specifically channel pruning [17] at a 10% consistent sparsity to one layer at a time. The proposed strategy involves adding pruning to each layer of the DNN individually, starting with the first layer and moving toward the last (Figure 2 ①). After pruning each layer, the accuracy of the resulting network is evaluated on the validation set of the target dataset, and the respective accuracy is monitored. This process allows us to evaluate the impact of pruning each layer of the DNN in isolation. This helps us to identify the relative importance of each layer and its contribution to the overall accuracy of the DNN.

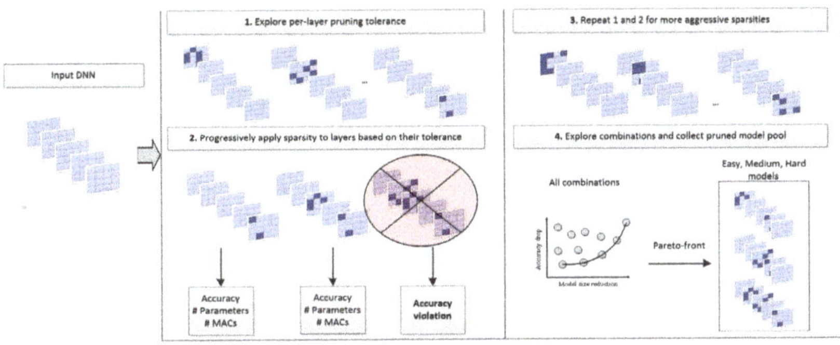

Figure 2. Overview of the pruning combination explorations on a given pretrained DNN model. No retraining is involved in this procedure, and the final outcome is three different combinations that will be invoked as the Easy, Medium, and Hard models.

Once each layer has been pruned individually, we can determine the resilience of each layer to pruning based on its accuracy drop. The layer with the least accuracy drop is considered the most resilient, while the layer with the highest accuracy drop is considered the least resilient. We then add pruning back to the network one layer at a time, starting with the most resilient layer and moving toward the least resilient (Figure 2, ②). Every combination is being evaluated on the validation set of the target dataset, and we monitor not only the accuracy drop but the reduction in parameters and MAC operations this time as well.

So far, steps ① and ② as described above and shown in Figure 2 correspond to a 10% sparsity ratio. For step ③, we repeat steps ① and ② for 20% sparsity. Note that we have already tested the layers for resilience in step ①, so their resilience is going to be similar if not identical for more aggressive sparsity ratios, too. Then, starting from the most resilient layer to the least resilient one, we swap the initially set 10% pruning ratio to 20% until an accuracy threshold is violated. We then take the Pareto front of all generated combinations and keep only three based on the reduction in the DNN's model size. All final three selected combinations should lie on the Pareto front and have increasing accuracy and model size reduction and will be used as the Easy, Medium, and Hard models during run-time.

2.3. Bringing It All Together: Run-Time Pruning Based on Image Difficulty

We combine the approaches presented in Sections 2.1 and 2.2 to achieve lightweight DNN inference without compromises in terms of accuracy at run-time. We consider run-time scenarios where real-time performance is prioritized [18] and therefore target DNN inference where each image is processed independently rather than in big batches.

Figure 3 shows how we combine the SVM from Section 2.1 with the pruning pool resulting from the exploration described in Section 2.2. For each incoming image in the input stream, we calculate the metrics that are needed by the SVM, which then classifies it as Easy, Medium, or Hard. From the pruned models pool that we have generated in Section 2.2, we select the respective model for inference based on the verdict of the SVM, and finally, we evaluate it on a DNN accelerator. We keep the resulting accuracy, model parameters, and total MAC operations to evaluate our model in Section 3.

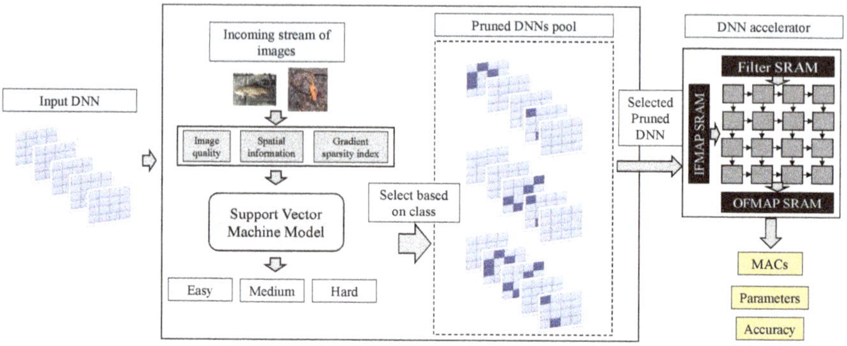

Figure 3. The execution flow of the proposed framework at run-time. For a target DNN model, each image during inference is being analyzed for its metrics and classified through the SVM. Then, the respective DNN model is selected for inference on a DNN accelerator.

3. Evaluation

The human-based dataset presented in [13] targets the ImageNet dataset which we also employ in this work [19]. The ImageNet dataset is a widely recognized benchmark in the field of computer vision and deep learning. It consists of more than a million images distributed over 1000 categories, making it a comprehensive and diverse dataset for evaluating the performance of DNNs. The accuracy achieved on this dataset by DNNs has become a standard metric of their evaluation. We evaluate the proposed framework on the entire validation dataset of ImageNet which is 50,000 images in total. We used pretrained models on the ImageNet from Pytorch [20], which is a popular open-source machine learning library heavily used in the development of DNNs. We evaluate a variety of DNNs with diverse structures which are the following: ResNet18 and ResNet50 [21], VGG11 and VGG16 [22], AlexNet [23] and GoogLeNet [24]. We use channel pruning [17] as the employed pruning technique and evaluate DNN inference on the systolic accelerator simulator in [25]. We simulate a systolic array following the Eyeriss [26] architecture paradigm, where the input feature map (IFMAP), filter, and output feature map (OFMAP) SRAM sizes are 108 kB, and the array has a height of 12 and a width of 14.

It is important to note that the aim of this paper is not to propose a new pruning method but utilize an already proposed approach such as channel pruning. In the case of channel pruning, there are various methods for retraining the pruned model to recover the lost accuracy, such as fine-tuning the model on the pruned architecture or using knowledge distillation to transfer the knowledge from the original model to the pruned model [27]. However, these methods can be computationally expensive and time-consuming, and they may not be feasible or practical in all scenarios [18]. On the contrary, the proposed method does not involve retraining or fine-tuning; therefore, to keep evaluations fair, we avoid directly comparing our results with other pruning methods that do involve such techniques. Instead, we provide comparisons in terms of the efficiency and effectiveness of the proposed methodology based on the compression achieved alongside the drop in final inference accuracy.

Our proposed approach is intended for lightweight Convolutional Neural Networks (CNNs) utilized in image classification tasks. While we evaluate our approach on several state-of-the-art CNNs on the ImageNet dataset, it can be applied to any CNN architecture. It is important to note that our evaluation is limited to the image classification task and convolutional neural networks. However, the same principles of layer sensitivity to pruning ratios can be applied to other neural network architectures. The ImageNet dataset is widely used in the computer vision community due to its large size and complexity, and evaluating on this dataset provides valuable insights into the effectiveness of our approach.

Table 1 contains all the results of the conducted evaluation of the proposed framework. We can observe that there are three different behavior patterns based on the type of DNN deployed with our framework.

1. The ResNet models: The ResNet models are overall more tolerant to pruning, allowing for significant speedups and a reduction in MAC operations and model size. Both of the examined ResNets comprise multiple convolutional layers, allowing for more pruning combinations to be explored. An illustrative example on the ResNet-18 specifically is shown in Figure 4. For all the combinations, we only consider solutions that lie on the Pareto front, and we select three different accuracy drop thresholds of increasing aggressiveness to select the Hard, Medium, and Easy DNNs: 1%, 4% and 8%, respectively. This way, we aim to have an overall average accuracy that does not fall more than 4% below the baseline accuracy on average. As shown in Table 1, we manage to stay below a 2% drop in accuracy for all DNNs considered in this work due to the adaptive nature of the framework at run-time.
2. AlexNet and GoogLeNet: Both of these models have very different structures. AlexNet has a small amount of convolution layers (just 8 in total), but its model requires tens of millions of parameters (61.1M) as opposed to the GoogLeNet model that has 22 convolution layers but requires just 6.62M parameters. The parameter amount for each DNN was calculated through [17]. Even though different in structure, convolution layers and parameters, they exhibit the same behavior, which is fairly conservative: they do not tolerate pruning as well, and the acceptable solutions found by the proposed framework led to lower reductions in MAC operations and parameters, which however came with a lower drop in the average accuracy.

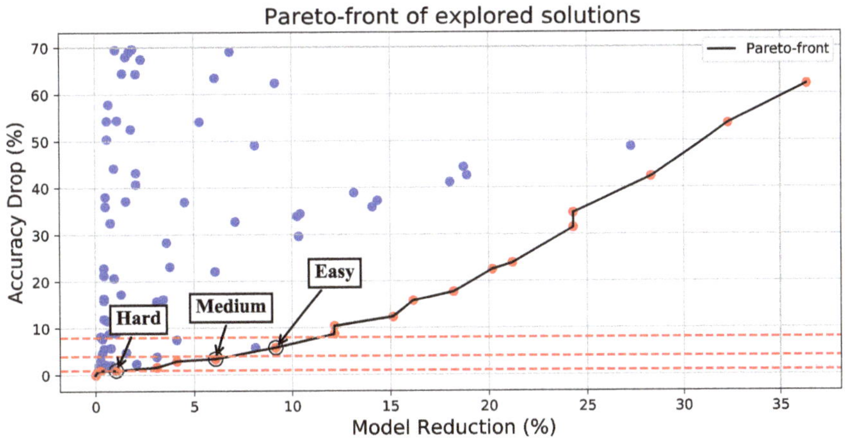

Figure 4. A visualization of all the explored pruning combinations of ResNet-18 alongside the underlying Pareto frontier. We select the best combination (maximum number of parameters reduced) for accuracy drops less than 1%, 4% and 8% for the Hard, Medium, and Easy DNNs, respectively.

We additionally evaluated two VGG models, as shown in Table 2. Even though the observed accuracy drops for the VGG-11 and VGG-16 models are similar to those of the ResNets, the reduction in parameter size and MAC operations is not as significant as the ResNet models. This is attributed to the fact that the VGG models comprise a smaller number of convolutional layers, which are however of higher depth. Therefore, applying a 10% sparsity to the multiple convolution layers can significantly lower the model size, which can lead to drastic drops in accuracy, as shown in the first two lines of

Table 2. Therefore, for these two cases only, instead of following the procedure described in Section 2.2 with an initial sparsity of 10% followed by the more aggressive 20%, we started the procedure with a 5% sparsity ratio followed by 10%.

Table 1. Evaluation results.

	Accuracy Drop (%)	MAC Reduction (%)	Parameter Reduction (%)	Speedup
ResNet-18	1.84	12.11	9.65	×1.28
ResNet-34	1.75	11.53	9.41	×1.36
ResNet-50	1.79	11	9.32	×1.35
AlexNet	0.58	5.61	4.23	×1.1
GoogLeNet	0.78	7.88	5.82	×1.15

Table 2. Evaluation results for VGG networks: At first, with no alterations in the initial pruning ratio (PR) (10%) and then with the adjusted initial PR (5%).

	Accuracy Drop (%)	MAC Reduction (%)	Parameter Reduction (%)	Speedup
VGG11 (10% PR)	3.25	8.4	6.6	×1.22
VGG16 (10% PR)	4.13	9.1	8.91	×1.26
VGG11 (5% PR)	1.61	6.67	3.5	×1.08
VGG16 (5% PR)	1.42	7.32	6.68	×1.14

Overall, the proposed framework always produced final solutions that achieved speedups compared to the baseline execution due to the reduction of MAC operations and parameters without surpassing 1.85% accuracy drop from the baseline. Therefore, this evaluation indicates that incorporating image difficulty and incoming data attributes during pruning can provide several benefits. Specifically, by pruning more aggressively on easy images, we can save more computational resources without sacrificing accuracy on the most challenging samples.

Cost Efficiency Analysis

The utilized SVM comprises a third-degree polynomial kernel and is only trained once. The main reason why the SVM is efficient for run-time deployment is that it can quickly classify new data points based on the decision boundary learned during training. To classify a new image at run-time, the SVM computes the dot product between the image's features and the weights of the support vectors. The resulting values are then passed through a decision function that maps them to the predicted class label. This process is relatively fast since it involves only a few computations and does not require iterative training steps.

The utilized SVM is only used to classify images into three difficulty levels (Easy, Medium, and Hard) at run-time, the classification time is expected to be very fast, and it should not be a significant bottleneck in the overall inference process. Additionally, computing the necessary inputs for the SVM (image quality, spatial information, and gradient sparsity index) is in the range of microseconds.

In our evaluation, we found that the additional time required to perform SVM inference and image property calculations was negligible compared to the overall time required for DNN inference. We observed that even with the added SVM inference step, the pruned DNNs still resulted in significantly faster inference times compared to the baseline unpruned DNNs. None of the evaluated DNNs showed a final inference time similar to the baseline, indicating that the pruning of the DNN more than compensated for the minimal overhead of SVM inference.

4. Conclusions

Deep Neural Networks (DNNs) are becoming increasingly complex, making them difficult to deploy on resource-constrained devices. While pruning mechanisms can result in lightweight DNNs, they typically require additional retraining to recover from accuracy losses induced by pruning. This paper proposes a dynamic DNN pruning approach that takes into account the difficulty of incoming images during inference on pretrained DNNs. Experimental results on the ImageNet dataset demonstrate that the proposed approach effectively reduces model size and DNN operations without the need for retraining or fine-tuning on the pruned model, resulting in speedups during inference. By considering the attributes of the input data, the proposed approach presents a promising direction for improving the deployment of DNNs on resource-constrained devices.

Author Contributions: Conceptualization, I.A.; Methodology, V.P. and O.S.; Software, O.S.; Data curation, V.P.; Writing—original draft, O.S.; Writing—review and editing, V.P. and I.A. All authors have read and agreed to the published version of the manuscript.

Funding: This research has been funded in part by the Consortium for Embedded Systems at SIUC.

Conflicts of Interest: The authors declare no conflict of interest.

References

1. Spantidi, O.; Zervakis, G. ; Anagnostopoulos, I; Amrouch, H.; Henkel. J. Positive/negative approximate multipliers for DNN accelerators. In Proceedings of the IEEE/ACM International Conference On Computer Aided Design (ICCAD), Munich, Germany, 1–4 November 2021; pp. 1–9.
2. Spantidi, O.; Anagnostopoulos, I. How much is too much error? Analyzing the impact of approximate multipliers on DNNs. In Proceedings of the 23rd International Symposium on Quality Electronic Design (ISQED), Virtual Event, 6–7 April 2022; pp. 1–6.
3. Amrouch, H.; Zervakis, G.; Salamin, S.; Kattan, H.; Anagnostopoulos, I.; Henkel, J. NPU Thermal Management. *IEEE Trans. Comput. Aided Des. Integr. Circuits Syst.* **2020**, *39*, 3842–3855. [CrossRef]
4. Yang, H.; Gui, S.; Zhu, Y.; Liu, J. Automatic neural network compression by sparsity-quantization joint learning: A constrained optimization-based approach. In Proceedings of the IEEE/CVF Conference on Computer Vision and Pattern Recognition, Virtual Event, 14–19 June 2020; pp. 2178–2188.
5. Han, S.; Han, S.; Mao, H.; Dally, W.J. Deep compression: Compressing deep neural networks with pruning, trained quantization and huffman coding. *arXiv* **2015**, arXiv:1510.00149 .
6. Han, S.; Pool, J.; Tran, J.; Dally, W. Learning both weights and connections for efficient neural network. In Proceedings of the 29th Annual Conference on Neural Information Processing Systems 2015, Montreal, QC, Canada, 7–12 December 2015.
7. Li, H.; Kadav, A.; Durdanovic, I.; Samet, H.; Graf, H.P. Pruning filters for efficient convnets. *arXiv* **2016**, arXiv:1608.08710.
8. Wen, W.; Wu, C.; Wang, Y.; Chen, Y.; Li, H. Learning structured sparsity in deep neural networks. In Proceedings of the 30th Annual Conference on Neural Information Processing Systems 2016, Barcelona, Spain, 5–10 December 2016.
9. He, Y.; Lin, J.; Liu, Z.; Wang, H.; Li, L.J.; Han, S. Amc: Automl for model compression and acceleration on mobile devices. In Proceedings of the European Conference on Computer Vision (ECCV), Munich, Germany, 8–14 September 2018; pp. 784–800.
10. Banner, R.; Nahshan, Y.; Soudry, D. Post training 4-bit quantization of convolutional networks for rapid-deployment. In Proceedings of the 33rd Conference on Neural Information Processing Systems (NeurIPS 2019), Vancouver, BC, Canada, 8–14 December 2019.
11. Spantidi, O.; Zervakis, G.; Alsalamin, S.; Roman-Ballesteros, I.; Henkel, J.; Amrouch, H.; Anagnostopoulos, I. Targeting DNN Inference via Efficient Utilization of Heterogeneous Precision DNN Accelerators. *IEEE Trans. Emerg. Top. Comput.* **2022**, *11*, 112–125. [CrossRef]
12. Tudor Ionescu, R.; Alexe, B.; Leordeanu, M.; Popescu, M.; Papadopoulos, D.P.; Ferrari, V. How hard can it be? Estimating the difficulty of visual search in an image. In Proceedings of the Conference on Computer Vision and Pattern Recognition, Las Vegas, NV, USA, 27–30 June 2016; pp. 2157–2166.
13. Mayo, D.; Cummings, J.; Lin, X.; Gutfreund, D.; Katz, B.; Barbu, A. How hard are computer vision datasets? Calibrating dataset difficulty to viewing time. In Proceedings of the 36th Conference on Neural Information Processing Systems, New Orleans, LA, USA, 28 November–9 December 2022.
14. Mittal, A.; Moorthy, A.K.; Bovik, A.C. No-reference image quality assessment in the spatial domain. *IEEE Trans. Image Process.* **2012**, *21*, 4695–4708. [CrossRef] [PubMed]
15. Yu, H.; Winkler, S. Image complexity and spatial information. In Proceedings of the 5th International Workshop on Quality of Multimedia Experience (QoMEX), Klagenfurt am Worthersee, Austria, 3–5 July 2013; pp. 12–17.
16. Li, L.; Cai, H.; Zhang, Y.; Lin, W.; Kot, A.C.; Sun, X. Sparse representation-based image quality index with adaptive sub-dictionaries. *IEEE Trans. Image Process.* **2016**, *25*, 3775–3786. [CrossRef] [PubMed]
17. Fang, G.; Ma, X.; Song, M.; Mi, M.B.; Wang, X. DepGraph: Towards Any Structural Pruning. *arXiv* **2023**, arXiv:2301.12900.

18. Spantidi, O.; Zervakis, G.; Anagnostopoulos, I.; Henkel, J. Energy-Efficient DNN Inference on Approximate Accelerators through Formal Property Exploration. *IEEE Trans. Comput. Aided Des. Integr. Circuits Syst.* **2022**, *41*, 3838–3849. [CrossRef]
19. Deng, J.; Dong, W.; Socher, R.; Li, L.J.; Li, K.; Fei-Fei, L. Imagenet: A large-scale hierarchical image database. In Proceedings of the Conference on Computer Vision and Pattern Recognition, Miami, FL, USA, 20–25 June 2009; pp. 248–255.
20. Paszke, A.; Gross, S.; Massa, F.; Lerer, A.; Bradbury, J.; Chanan, G.; Killeen, T.; Lin, Z.; Gimelshein, N.; Antiga, L.; et al. Pytorch: An imperative style, high-performance deep learning library. In Proceedings of the 33rd Conference on Neural Information Processing Systems (NeurIPS 2019), Vancouver, BC, Canada, 8–14 December 2019.
21. He, K.; Zhang, X.; Ren, S.; Sun, J. Deep residual learning for image recognition. In Proceedings of the Conference on Computer Vision and Pattern Recognition, Las Vegas, NV, USA, 27–30 June 2016; pp. 770–778.
22. Simonyan, K.; Zisserman, A. Very deep convolutional networks for large-scale image recognition. *arXiv* **2014**, arXiv:1409.1556.
23. Krizhevsky, A.; Sutskever, I.; Hinton, G.E. Imagenet classification with deep convolutional neural networks. *Commun. Acm* **2017**, *60*, 84–90. [CrossRef]
24. Szegedy, C.; Liu, W.; Jia, Y.; Sermanet, P.; Reed, S.; Anguelov, D.; Erhan, D.; Vanhoucke, V.; Rabinovich, A. Going deeper with convolutions. In Proceedings of the Conference on Computer Vision and Pattern Recognition, Boston, MA, USA, 7–12 June 2015; pp. 1–9.
25. Samajdar, A.; Zhu, Y.; Whatmough, P.; Mattina, M.; Krishna, T. Scale-sim: Systolic cnn accelerator simulator. *arXiv* **2018**, arXiv:1811.02883.
26. Chen, Y.H.; Krishna, T.; Emer, J.S.; Sze, V. Eyeriss: An energy-efficient reconfigurable accelerator for deep convolutional neural networks. *IEEE J. Solid-State Circuits* **2016**, *52*, 127–138. [CrossRef]
27. Zmora, N.; Jacob, G.; Zlotnik, L.; Elharar, B.; Novik, G. Neural Network Distiller: A Python Package For DNN Compression Research. *arXiv* **2019**, arXiv:1910.12232.

micromachines

MDPI

Article

Architecture Optimization of a Non-Linear Autoregressive Neural Networks for Mackey-Glass Time Series Prediction Using Discrete Mycorrhiza Optimization Algorithm

Hector Carreon-Ortiz, Fevrier Valdez, Patricia Melin and Oscar Castillo *

Tijuana Institute of Technology, TecNM, Tijuana 22379, Mexico
* Correspondence: ocastillo@tectijuana.mx

Abstract: Recurrent Neural Networks (RNN) are basically used for applications with time series and sequential data and are currently being used in embedded devices. However, one of their drawbacks is that RNNs have a high computational cost and require the use of a significant amount of memory space. Therefore, computer equipment with a large processing capacity and memory is required. In this article, we experiment with Nonlinear Autoregressive Neural Networks (NARNN), which are a type of RNN, and we use the Discrete Mycorrhizal Optimization Algorithm (DMOA) in the optimization of the NARNN architecture. We used the Mackey-Glass chaotic time series (MG) to test the proposed approach, and very good results were obtained. In addition, some comparisons were made with other methods that used the MG and other types of Neural Networks such as Backpropagation and ANFIS, also obtaining good results. The proposed algorithm can be applied to robots, microsystems, sensors, devices, MEMS, microfluidics, piezoelectricity, motors, biosensors, 3D printing, etc.

Keywords: optimization; nonlinear autoregressive neural networks; Mackey–Glass

Citation: Carreon-Ortiz, H.; Valdez, F.; Melin, P.; Castillo, O. Architecture Optimization of a Non-Linear Autoregressive Neural Networks for Mackey-Glass Time Series Prediction Using Discrete Mycorrhiza Optimization Algorithm. *Micromachines* **2023**, *14*, 149. https://doi.org/10.3390/mi14010149

Academic Editors: Xiao Xiao and Gang Ge

Received: 1 December 2022
Revised: 30 December 2022
Accepted: 30 December 2022
Published: 6 January 2023

1. Introduction

Optimization is not limited to applied mathematics, engineering, medicine, economics, computer science, operations research or any other science, but has become a fundamental tool in all fields, where constantly developing new algorithms and theoretical methods have allowed it to evolve in all directions, with a particular focus on artificial intelligence, such as deep learning, machine learning, computer vision, fuzzy logic systems, and quantum computing [1,2].

Optimization has grown steadily over the past 50 years. Modern society not only lives in a highly competitive environment, but is also forced to plan for growth in a sustainable manner and be concerned about resource conservation. Therefore, it is essential to optimally plan, design, operate and manage resources and assets. The first approach is to optimize each operation separately. However, the current trend is toward an integrated approach: synthesis and design, design and control, production planning, scheduling and control [3].

Theoretically, optimization has evolved to provide general solutions to linear, non-linear, unbounded and constrained network optimization problems. These optimization problems are called mathematical programming problems and are divided into two different categories: linear and nonlinear programming problems. Biologically derived genetic algorithms and simulated annealing are two equally powerful methods that have emerged in recent years. The development of computer technology has provided users with a variety of optimization codes with varying degrees of rigor and complexity. It is also possible to extend the capabilities of an existing method by integrating the features of two or more optimization methods to achieve more efficient optimization methodologies [4]; current optimization methods that can solve specific problems are still being developed, as we do not yet have a method that can solve them all, such as explained by the No Free Lunch (NFL) Algorithm [5], although the research trend is moving in that direction.

RNNs are a special class of neural network characterized by their inherent self-connectivity [6], and their variants are used in many contexts where temporal dependence of data is an important latent feature in model design [7]. The most important applications of RNNs include time series prediction [8], sequence transduction [9], language modeling [10–13], speech recognition [14], word embedding learning [15], sound modeling [16], handwriting recognition [17,18], and image generation [19]. A common variant of RNN called long short-term memory [20] is used in many of these studies.

One of the main advantages of this method with respect to others is that in general the NARNN-DMOA method is much easier to implement with better results, and with lower computation costs. Other methods use very robust Ensemble Neural Network architectures of several layers and of more than 2000 neurons and different architectures of Interval Type-2 Fuzzy Logic Systems (IT2FLSs), in addition to optimization algorithms such as PSO Genetic Algorithms [21–23], which implies a high computational cost.

The algorithm can be applied, as we previously mentioned, to robots, microsystems, sensors, devices, etc., in the optimization of the parameters of their models that are being experimented upon. The proposed algorithm can be used in the optimization of the architecture of a neural network or in the parameters of the membership functions of a fuzzy logic system; as we have seen in other articles [24–27], this type of experimentation with the DMOA is the subject of a future work that we plan to undertake in due course

The main contribution of this research is to use the DMOA algorithm to optimize the architecture of the NARNN neural network using the MG chaotic time series, which has not previously been done in the current literature.

The structure of this paper is as follows: (1) brief introduction of Optimization and Recurrent Neural Networks (RNNs), (2) we include a brief description of Nonlinear Autoregressive Neural Networks (NARNNN), (3) presentation of the Discrete Mycorrhiza Optimization Algorithm (DMOA) inspired by the symbiosis of plant roots and MN, (4) proposed method using the NARNN, the new DMOA algorithm and Mackey Glass chaotic time series, (5) results obtained from this research, such as statistical data, hypothesis testing and comparison of the DMOA-NARNN method with other methods, (6) in-depth discussion of the results and comparison of the error with other methods, and (7) conclusions of the obtained results.

2. Nonlinear Autoregressive Neural Networks

An Artificial Neural Network (ANN) is a type of neural network represented by a mathematical model inspired by the neural connections of the human brain. It is an intelligent system capable of recognizing time series patterns and nonlinear features.

Therefore, it is widely used to model nonlinear dynamic time series [28]. ANN incorporates artificial neurons to process information. It consists of single neurons connected to a network via weighted links. Each input is multiplied by a weight calculated by a mathematical function that determines the activation of the neurons. Another activation function calculates the output of the artificial neuron based on a certain threshold [29].

The output of a neuron can be written as Equation (1):

$$y = f\left(b + \sum_i w_i x_i\right) \quad (1)$$

where b is the bias of the neuron, the bias input to the neuron algorithm is an offset value that helps the signal exceed the threshold of the activation function, f is the activation function, w_i is the weight, x_i is the input, and y is the output.

Several types of ANNs have been presented in the literature, including Multilayer Perceptron (MLP), in which neurons are grouped into an input layer, one or more hidden layers, and an output layer. These also include RNNs such as Layer Recurrent Networks [30],

Time Delay Neural Networks (TDNN) [31], and NARNN [32]. In RNNs, the output of a dynamic system depends not only on the current inputs, but also on the history of inputs and states of the system. The NARNN is a recurrent dynamic network based on a linear autoregressive model with feedback connections, and consists of several network layers.

Humans do not start their thinking from scratch every second. As we read, we understand each word based on our understanding of the previous words. We never start thinking from scratch every time we do; our thoughts have permanence. A traditional ANN cannot do this, and it seems like a major shortcoming. For example, imagine that you want to classify what kind of event is happening at each point in a movie. It is not clear how a traditional ANN could use its reasoning about earlier events in the movie to inform later events, and RNN address this problem. They are networks with loops in them which allows information to persist.

An RNN is a type of artificial neural network that uses sequential or time series da-ta. These deep learning algorithms are commonly used for ordinal or temporal problems, such as language translation, natural language processing (NLP) [33,34], speech recognition, and image captioning [35]. They are distinguished by their "memory" because they take information from previous inputs to influence the current input and output. While traditional deep neural networks assume that inputs and outputs are independent of each other, the output of recurrent neural networks depends on previous elements within the sequence.

NARNNs are a type of RNN with memory and feedback capabilities. The output of each point is based on the result of the dynamic synthesis of the system before the current time. It has great advantages for modeling and simulating dynamic changes in time series [36]. Typical NARNNs mainly consist of an input layer, a hidden layer, an output layer and an input delay function, the basic structure of which is shown in Figure 1.

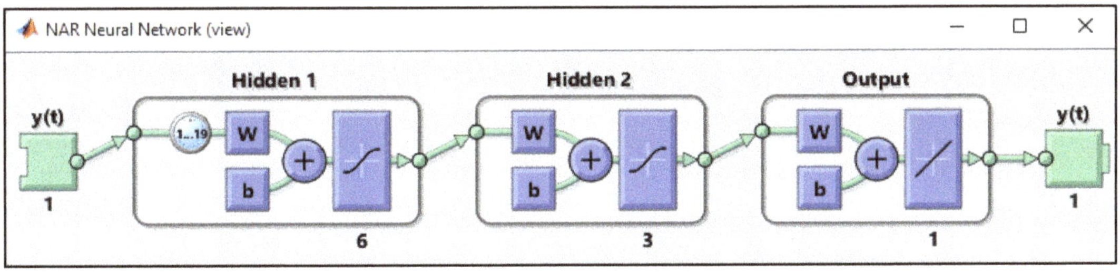

Figure 1. Standard NARNN schematic structure of the neural networks.

In Figure 1, *y(t)* is the output of the NARNN, 1..19 represents the delay order, w is the joint weight and b is the threshold of NARNNs. The model of NARNN networks can be expressed as in Equation (2), where d is the delay order and f are a nonlinear function, where the future values depend only on the previous values d of the output signal.

From the equation, it can be seen that the value of *y(t)* is determined by the values of $y(t-1), \ldots, y(t-d)$, which indicates that based on the continuity of data development, the model uses past values to estimate the current value [37,38].

The prediction method of the NARNN model adopts the recursive prediction method. The main purpose of this prediction method is to reproduce the predicted value one step ahead.

The future values of the time series *y(t)* are predicted only from the past values of this series. This type of prediction is called Nonlinear Autoregression (NAR) and can be written as Equation (2):

$$y(t) = f(y(t-1), \ldots y(t-d)) \quad (2)$$

This model can be used to predict financial instruments, but it does not use additional sequences [39].

Looking at Figure 2, NARNN represents the entire neural network. Figure 3 "Unrolled" represents the individual layers, or time steps, of the NARNN network. Each layer corresponds to a single piece of data [40,41].

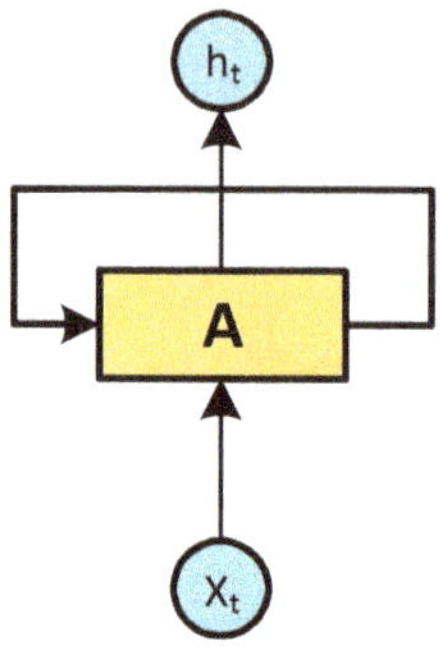

Figure 2. NARNN have loops.

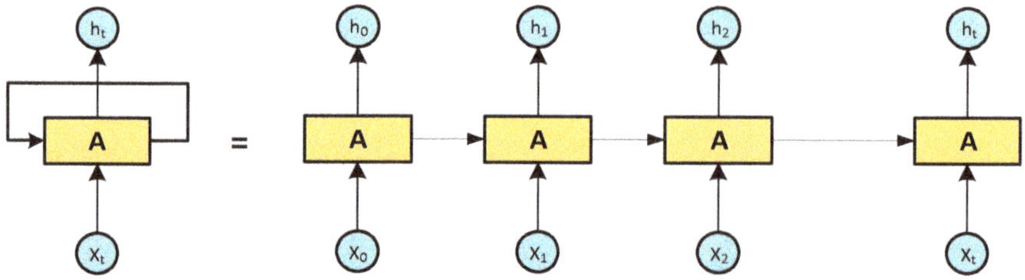

Figure 3. Unrolled NARNN.

Predicting a sequence of values in a time series is also known as multi-pass forecasting. Closed-loop networks can perform multi-step forecasting. When external feedback is missing, closed-loop networks can still make predictions using internal feedback. In NARNN prediction, the future values of a time series are predicted only from the past values of that series.

The current literature provides a history of very extensive research on the use of NARNNs in the following areas:

- The use of NARNN in medical devices such as continuous glucose monitors and drug delivery pumps that are often combined with closed-loop systems to treat chronic diseases, for error detection and correction due to their predictive capabilities [42].
- The use of NARNNs as Chinese e-commerce sales forecasting to develop purchasing and inventory strategies for EC companies [43], to support management decisions [44], the effects of air pollution on respiratory morbidity and mortality [45], the relationship between time series in the economy [46], to model and forecast the prevalence of COVID-19 in Egypt. [47], etc.

3. Discrete Mycorrhiza Optimization Algorithm

Most of the world's plant species are associated with mycorrhizal fungi in nature; this association involves the interaction of fungal hyphae on plant roots. Hyphae extend from the roots into the soil, where they absorb nutrients and transport them through the mycelium to the colonized roots [48]. Some hyphae connect host plants in what is known as a Mycorrhizal Network (MN). The MN is subway and is difficult to understand. As a result, plant and ecosystem ecologists have largely overlooked the role of MNs in plant community and ecosystem dynamics [49].

It is clear that most MN are present and provide nutrition to many plant species. This has important implications for plant competition for soil nutrients, seedling formation, plant succession and plant community and ecosystem dynamics [50].

Plant mycorrhizal associations have large-scale consequences throughout the ecosystem [51,52]. The origins of plant-fungal symbiosis are ancient and have been proposed as a mechanism to facilitate soil colonization by plants 400 Mya [53,54]. Mycorrhizal symbiosis is a many-to-many relationship: plants tend to form symbioses with a diverse set of fungal species and, similarly, fungal species tend to be able to colonize plants of different species [55].

In Figure 4 we can see that through the MN resources such as carbon (CO_2) from plants to fungi and water, phosphorus, nitrogen and other nutrients from fungi to plants are exchanged, in addition to an exchange of information through chemical signals when the habitat feels threatened by fire, floods, pests, or predators. It should be noted that this exchange of resources can be between plants of the same species or of different species. Figure 5 shows the symbiosis between plants and the fungal network and how the carbon in the form of sugars flows from the plants to the MN and how the MN fixes the nutrients in the roots of the plants.

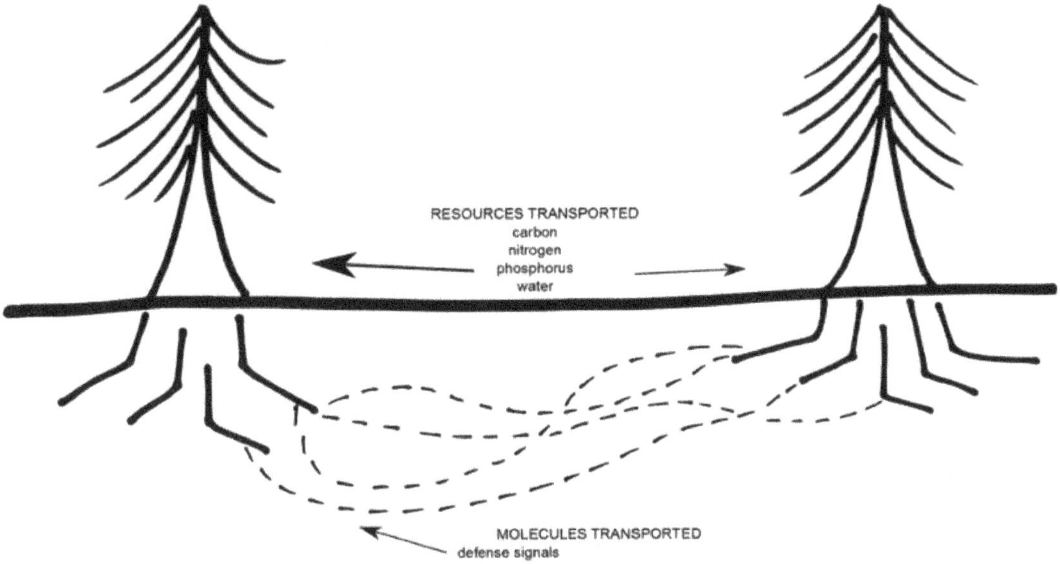

Figure 4. Signaling and resource exchange through MN.

The Nobel optimization algorithm DMOA is inspired by the nature of the Mycorrhiza Network (MN) and plant roots with this intimate interaction between these two organisms (plant roots and the network of MN fungi), a symbiosis is generated and it has been discovered that in this relationship [56–60]:

1. There is a communication between plants, which may or may not be of the same species, through a network of fungi (MN).
2. There is an exchange of resources between plants through the fungal network (MN).
3. There is a defensive behavior against predators that can be insects or animals, for the survival of the whole habitat (plants and fungi).
4. The colonization of a forest through a fungal network (MN) thrives much more than a forest where there is no exchange of information and resources.

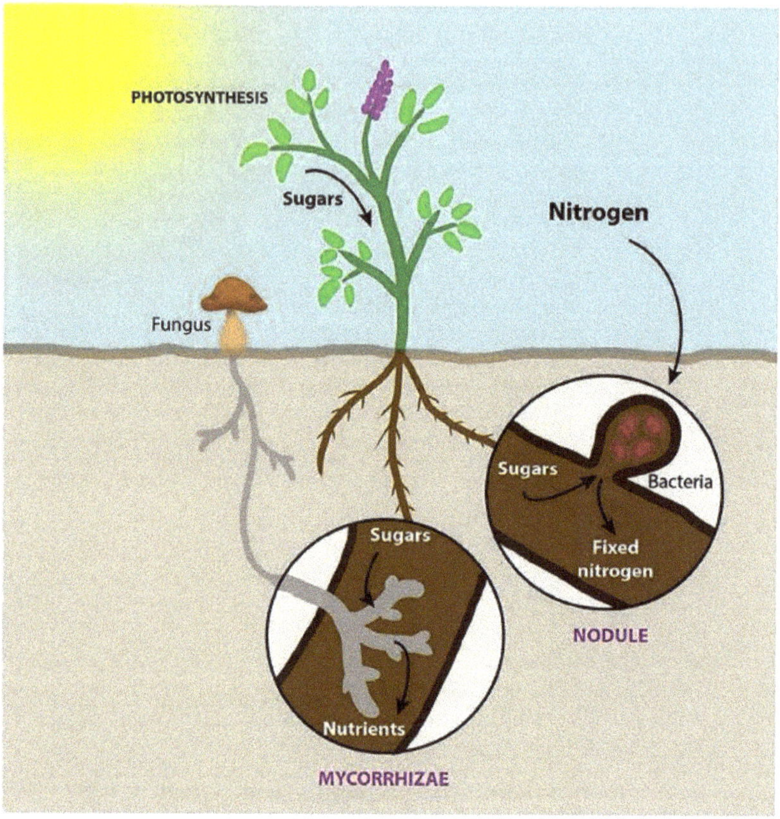

Figure 5. Symbiosis between plant roots and MN.

The launch and publication of the DMOA algorithm has just been carried out in 2022 [61].

Figure 6 describes the flowchart of the DMOA algorithm: we initialize the parameters such as dimensions, epochs, number of iterations, etc., and we also initialize the two populations of plants and mycorrhizae; with these populations we find the best fitness of plants and mycorrhizae, while with these results we use the biological operators. The first operator is represented by the Lotka-Volterra System of Discrete Equations (LVSDE) Cooperative Model [62], whose result has inference on the other two models represented by LVSDE, Defense and Competitive [63,64], and in this frequency we evaluate the fitness to determine if it is better than the previous one and we update the same as the populations, if not we continue with the next iteration and continue the calculation with the biological operators. If the stop condition is fulfilled we obtain the last solution before evaluation and the algorithm ends.

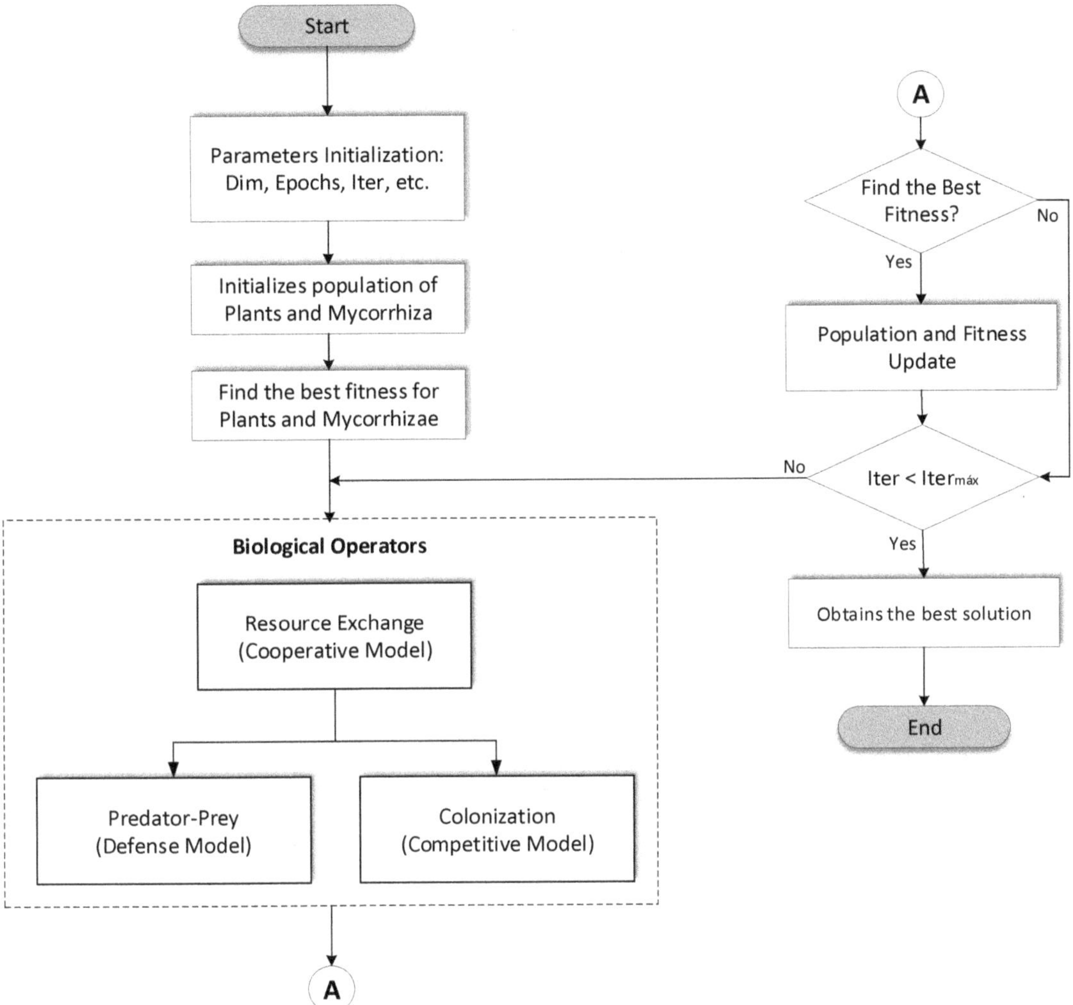

Figure 6. DMOA Flowchart.

4. Proposed Method

The proposed method is to use the Discrete Mycorrhiza Optimization Algorithm (DMOA) to optimize the architecture of the Nonlinear Autoregressive Neural Network (NARNN), and as input data we use the Mackey-Glass chaotic time series. In Figure 7 and Algorithm 1 we can find the DMOA-NARNN flowchart and DMOA-NARNN pseudocode, respectively. The DMOA algorithm is explained in Figure 6 in the previous section, in this flowchart we include the optimization of the NARNN, evaluating its results by means of the RMSE, until we manage to find the minimum error of that architecture through the iterations and the populations of the DMOA algorithm (Algorithm 1).

Algorithm 1 DMOA-NARNN Pseudocode. Discrete Mycorrhiza Optimization Algorithm (DMOA)

Objective min or max f(x), x = ($x_1, x_2, \ldots, x_d$)
Define parameters (a, b, c, d, e, f, x, y)
Initialize a population of n plants and mycorrhiza with random solutions
Find the best solution ***fit*** *in the initial population*
while *(t < maxIter)*
    for *i* = 1:*n (for n plants and Mycorrhiza population)*
        $X_p = abs(FitA)$
        $X_m = abs(FitB)$
end for
$a = minorX_p$
$d = minorX_m$
Apply (LV-Cooperative Model)
$x_i^{t+1} = \frac{(ax_i - bx_i y_i)}{(1 - gx_i)}$
$y_i^{t+1} = \frac{(dy_i + ex_i y_i)}{(1 + hy_i)}$
if $x_i < y_i$
    $x^t = x_i$
else
    $x^t = y_i$
end if
rand ([1 2])
  if *(rand = 1)*
      Apply (LV-Predator-Prey Model)
      $x_i^{t+1} = ax_i(1 - x_i) - bx_i y_i$
      $y_i^{t+1} = dx_i y_i - gy_i$
else
      Apply (LV- Competitive Model)
      $x_i^{t+1} = \frac{(ax_i - bx_i y_i)}{(1 + gx_i)}$
      $y_i^{t+1} = \frac{(dy_i - ex_i y_i)}{(1 + hy_i)}$
end if
Evaluate new solutions.
NARNN-Architecture
Evaluate Error
Error minor?
Update NARNN-Architecture.
Find the current best NARNN-Architecture solution.
end while

Difference equations often describe the evolution of a particular phenomenon over time. For example, if a given population has discrete generations, the size of $(n + 1)$ 1st generation $x(n + 1)$ is a function of the nth generation $x(n)$. This relationship is expressed by Equation (3):

$$x(n+1) = f(x(n)) \tag{3}$$

We can look at this issue from another perspective. You can generate a sequence from the point x_0, Equation (4):

$$x_0, f(x_0), f(f(x_0)), f(f(f(x_0))), \ldots \tag{4}$$

$f(x_0)$ is called the first iterate of x_0 under f.

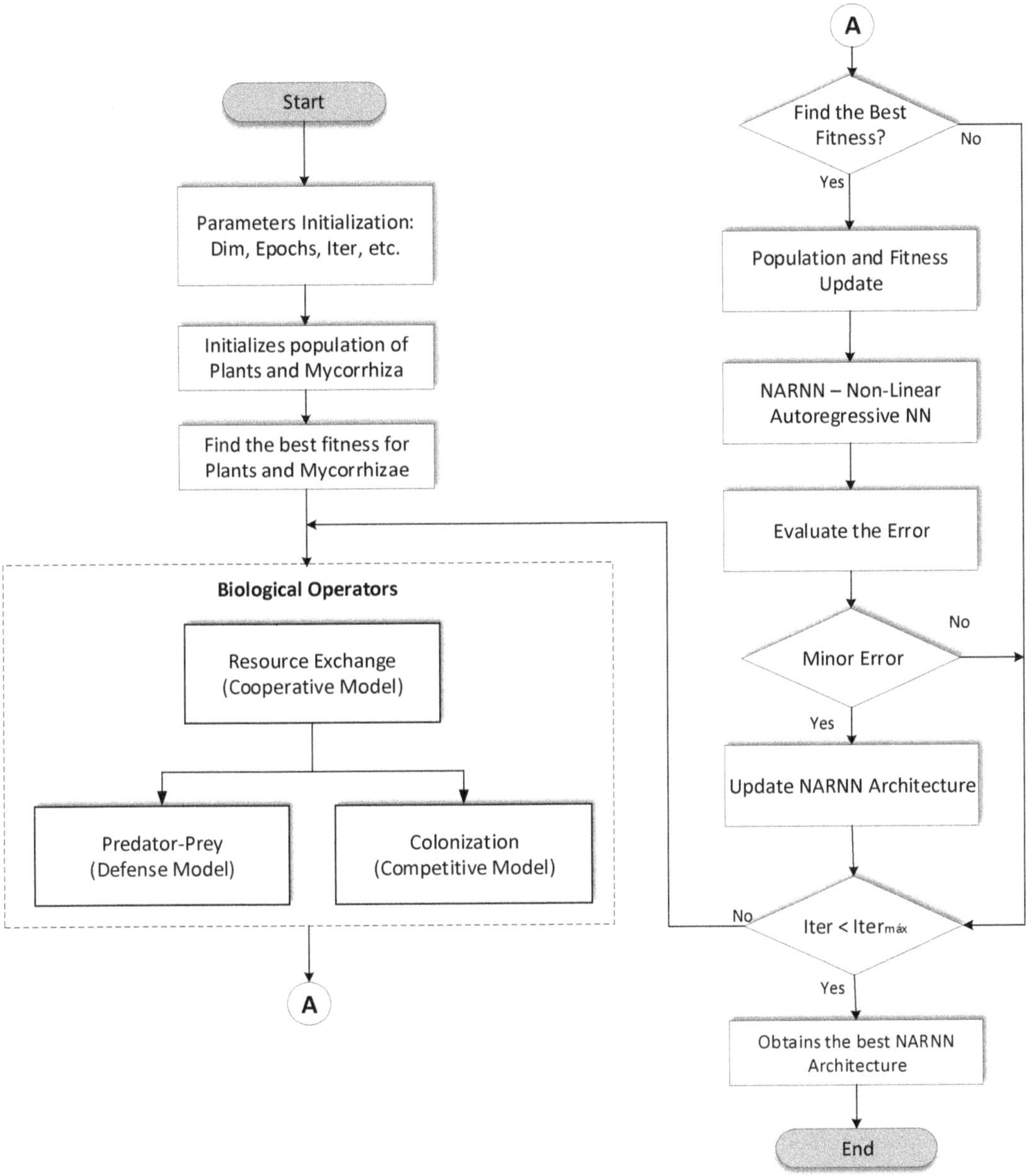

Figure 7. DMOA-NARNN Flowchart.

Discrete models driven by difference equations are more suitable than continuous models when reproductive generations last only one breeding season (no overlapping generations) [65,66].

An example would be a population that reproduces seasonally, that is, once a year. If we wanted to determine how the population size changes over many years, we could collect data to estimate the population size at the same time each year (say, shortly after the breeding season ends). We know that between the times at which we estimate population size, some individuals will die and that during the breeding season many new individuals will be born, but we ignore changes in population size from day to day, or week to week, and look only at how population size changes from year to year. Thus, when we build a mathematical model of this population, it is reasonable that the model only predicts the population size for each year shortly after the breeding season. In this case, the underlying variable, time, is represented in the mathematical model as increasing in discrete one-year increments.

The LVSDE Equations (5)–(10), have many uses in applied science. These models were first developed in mathematical biology, after which research spread to other fields [67–71].

Discrete Equations (5) and (6) Cooperative Model (Resource-Exchange), for both species, where parameters a, b, d, e, g, and h are positive constants, x_i and y_i represent the initial conditions of the population for both species and are positive real numbers [72].

The biological operators are represented by LVSDE, the mathematical description of the Discrete Equations (7) and (8) Defense Model (Predator-Prey), where the parameters a, b, d and g are positive constants, x_i and y_i represent the initial population conditions for both species and are positive real numbers [73,74].

Discrete Equations (9) and (10) Competitive Model (Colonization), for two species, where the parameters a, b, d, e, g, and h are positive constants, x_i and y_i are the populations for each of the species respectively and are positive real numbers. Each of the parameters of the above equations is described in Table 1, [74].

$$x_i^{t+1} = \frac{(ax_i - bx_iy_i)}{(1 - gx_i)} \tag{5}$$

$$y_i^{t+1} = \frac{(dy_i + ex_iy_i)}{(1 + hy_i)} \tag{6}$$

$$x_i^{t+1} = ax_i(1 - x_i) - bx_iy_i \tag{7}$$

$$y_i^{t+1} = dx_iy_i - gy_i \tag{8}$$

$$x_i^{t+1} = \frac{(ax_i - bx_iy_i)}{(1 + gx_i)} \tag{9}$$

$$y_i^{t+1} = \frac{(dy_i - ex_iy_i)}{(1 + hy_i)} \tag{10}$$

Table 1 contains the parameters used in all the experiments performed in this research, both those of the DMOA algorithm and those of the NARNN neural network.

The theory of Differential Equations, as well as that of Equations by Differences, can be found in Youssef N. Raffoul. *Qualitative Theory of Volterra Difference Equations* [75], Sigrun Bodine et al., *Asymptotic Integration of Differential and Difference Equations* [76], Takashi Honda et al., *Operator Theoretic Phenomena of the Markov Operators which are Induced by Stochastic Difference Equations* [77], Ronald E. Mickens, *Difference Equations Theory, Applications and Advanced Topics* [78], and Konrad Kitzing, et al., *A Hilbert Space Approach to Difference Equations* [79].

The metric for measuring error is RMSE (Root Mean Square Error) or root mean square deviation, which is one of the most commonly used measures for evaluating the quality of predictions. It shows how far predictions fall from measured true values using Euclidean distance Equation (11), where n is the number of data points, y_i is the *ith* measurement and $\hat{y}_i$ is the expected prediction [80,81].

$$RMSE = \sqrt{\sum_{i=1}^{n} (\hat{y}_i - y_i)^2} \quad (11)$$

Table 1. DMOA-NARNN Parameters.

Parameter	Description	Value
DMOA—Parameters:		
x_i^{t+1}	Population x at time t	
y_i^{t+1}	Population y at time t	
x_i	Grow rates of populations x at time t	
y_i	Grow rates of populations y at time t	
t	time	
a	Population growth rate x	0.01
b	Influence of population x on itself	0.02
g	Influence of population y on population x	0.06
d	Population growth rate y	0
e	Influence of population x on population y	1.7
h	Influence of population y on itself	0.09
x	Initial population in x	0.0002
y	Initial population in y	0.0006
	In the absence of population $x = 0$, In the absence of population $y = 0$	
	a, *b*, *c*, *d*, *e* and *f*—are positive constants	
Population	Population size	20
Populations	Number of populations	2
Dimensions	Dimensions size	30, 50, 100
Epochs	Number of epochs	30
Iterations	Iteration's size	30, 50, 100, 500
NARNN—Parameters:		
h	Hidden Layers	2, 3
n	Neurons	2–10
	Vector time delay	01:06:19

Mackey-Glass

Chaotic and random time series are both disordered and unpredictable. In extreme cases, the data are so mixed up that those consecutive values seem unrelated to each other. Such disorder would normally eliminate the ability to predict future values from past data.

The Mackey-Glass chaotic time series Equation (12) is a nonlinear differential equation of time delay, and this equation is widely used in the modeling of natural phenomena to make comparisons between different forecasting techniques and regression models [82–84], where $a = 0.1$, $b = 0.2$, and $\tau = 17$ are real numbers, t is the time, and with this setting the series produces chaotic behavior, and we can compare the forecasting performance of DMOA-NARNN with other models in the literature.

$$\dot{y}(t) = -by(t) + \frac{cy(t-\tau)}{1 + y^{10}(t-\tau)} \quad (12)$$

5. Results

This section shows the results of the experiments performed in the research involving the Non-Optimized and Optimized results of the method.

Table 2 presents 10 different non-optimized NARNN architectures using only the Mackey-Glass chaotic time series; in the table the columns are represented by: N—Experiment Number, Experiment Name, S—Sample size, T—Training, V—Validation, P—Prediction, HL—Hidden Layers of the NARNN, E—Number of experiment and RMSE (Root Mean Square Error), while the best architecture of the non-optimized NARNN is found in experiment number 4, with the RMSE of 0.1670.

Table 2. Results of non-optimized experiments.

N	Name	S	T	V	P	HL	E	RMSE
1	NARNN401	700	300	400	400	[1 1]	3000	0.2777
2	NARNN402	700	400	300	300	[1 1]	3000	0.1683
3	NARNN403	1000	400	600	600	[3 1]	3000	0.2307
4	NARNN404	1000	500	500	500	[9 2]	3000	0.1670
5	NARNN405	1000	600	400	400	[2 2]	3000	0.2488
6	NARNN406	1500	600	900	900	[4 1 9]	3000	0.2550
7	NARNN407	1500	700	800	800	[8 5 2]	3000	0.2158
8	NARNN408	1500	800	700	700	[1 2 1]	3000	0.4001
9	NARNN409	1500	900	600	600	[2 3 2]	3000	0.2810
10	NARNN410	1500	1000	500	500	[6 6 8]	3000	0.1712

In Figures 8–13, the y axes represent the input values (Validation-Training) and output values of the samples (Prediction-Error), the x axis represents the number of samples in time, Name is the name of the experiment, Samples is the total number of samples in the experiment, Training is the number of samples for training, Error is the minimum error obtained in the experiment, and *HL* represents the number of neurons in the hidden layers.

Figure 8 shows the behavior of the data for 1000 samples of the NARNN403, obtaining an RMSE of 0.2307, with the reference data at the top of the figure.

Figures 9 and 10 show the data behavior for 1000 samples of the NARNN404 and NARNN405, obtaining an RMSE of 0.167 and 0.2488, respectively, with the reference data at the top of each figure.

Table 3 shows the results of 39 NARNN architectures optimized with the DMOA algorithm using the Mackey-Glass chaotic time series, in the table the columns are represented by: N - Experiment Number, Experiment Name, S—Sample size, T—Training, V—Validation, P—Prediction, HL—Hidden Layers of the NARNN, I—Number of iterations, Tt—total time of the experiments in seconds, T—time in which the best result was found and RMSE (Root Mean Square Error). The best architecture of the non-optimized NARNN is found in experiment number 31, with the RMSE of 0.0023.

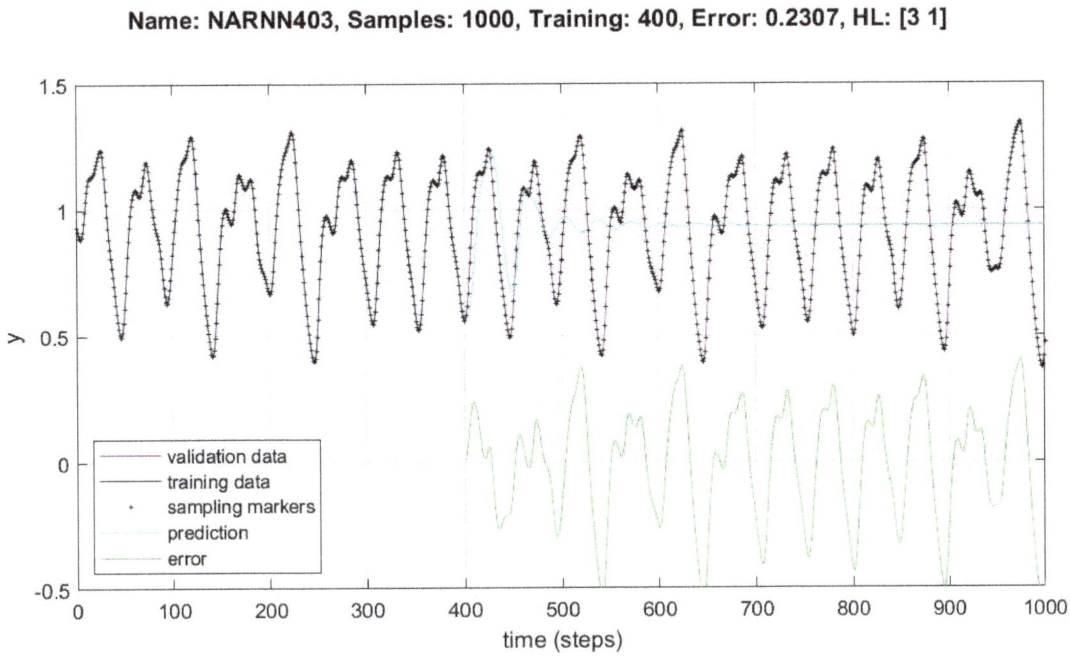

Figure 8. Performance of NARNN403 for 1000 samples.

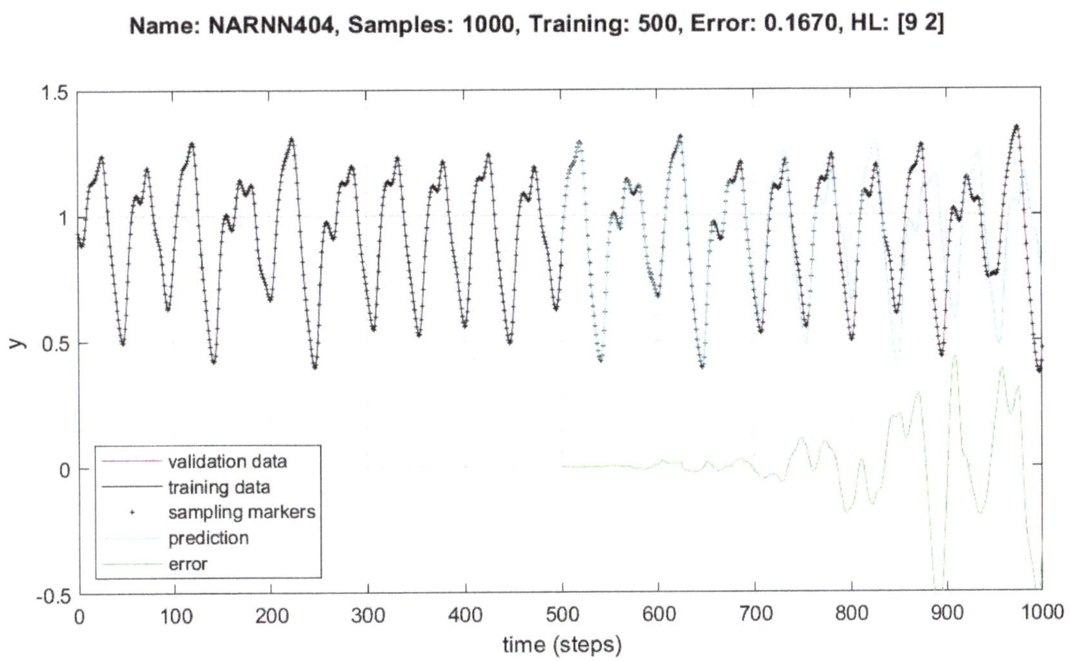

Figure 9. Performance of NARNN404 for 1000 samples.

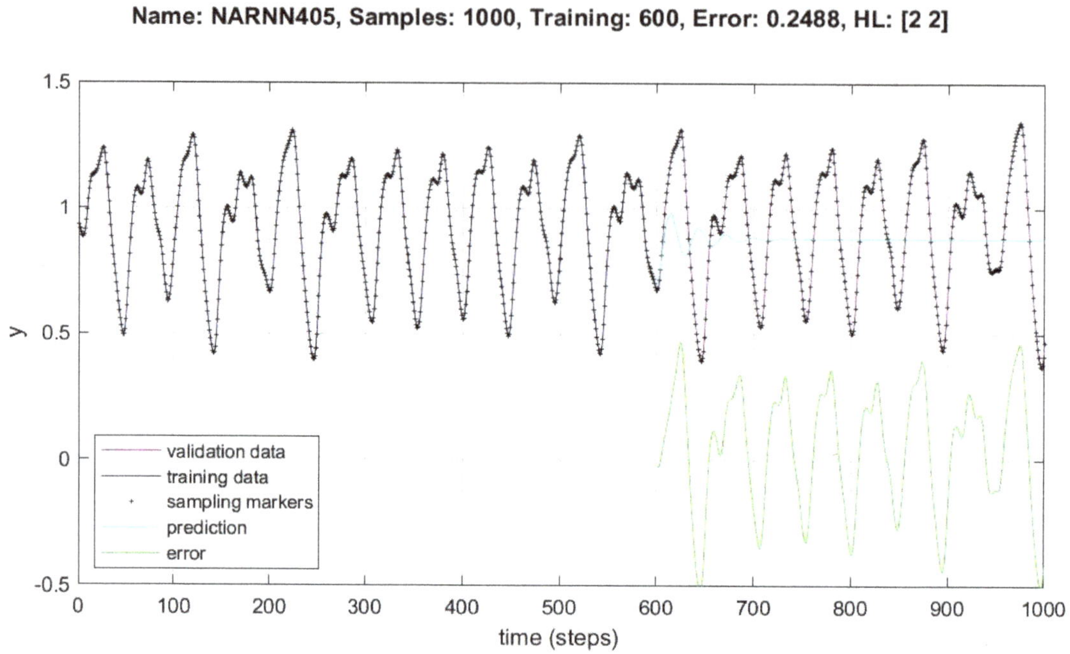

Figure 10. Performance of NARNN405 for 1000 samples.

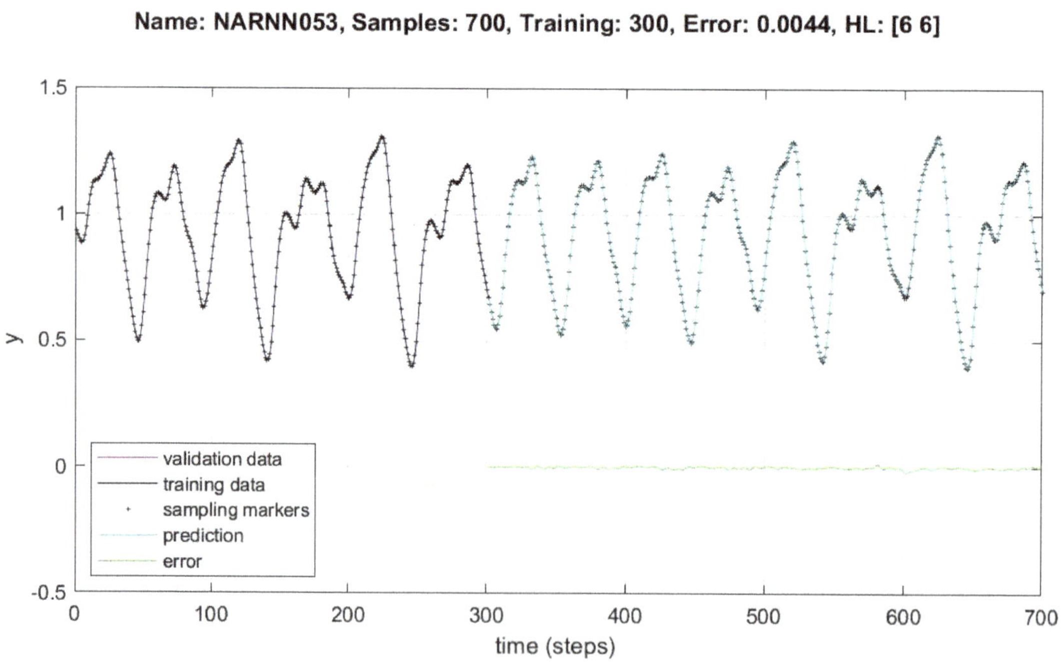

Figure 11. Performance of NARNN053 for 700 samples.

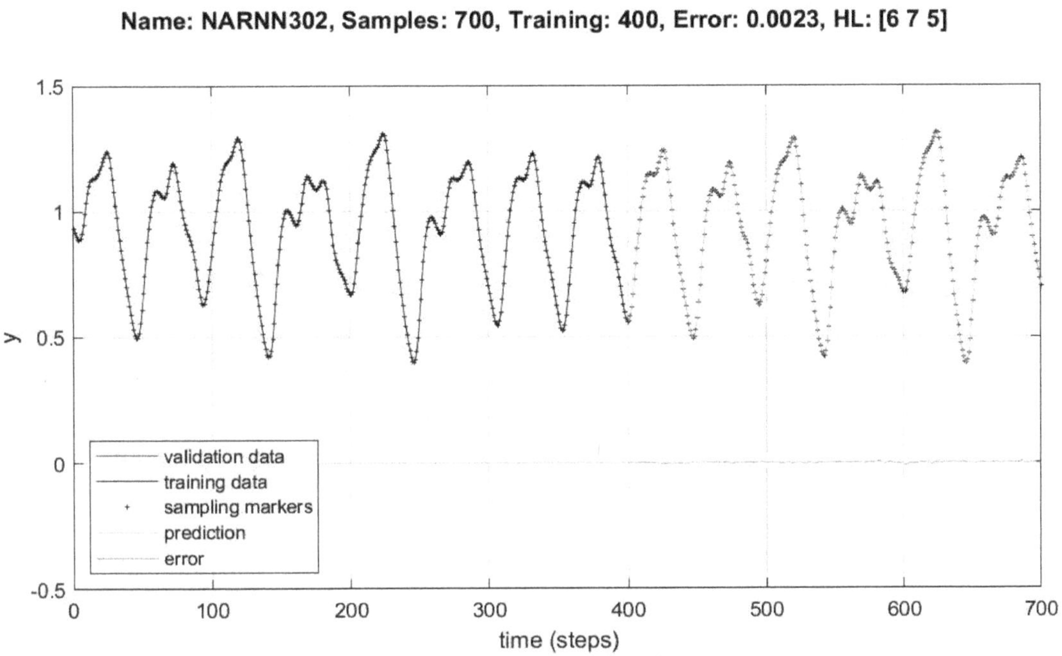

Figure 12. Performance of NARNN302 for 700 samples.

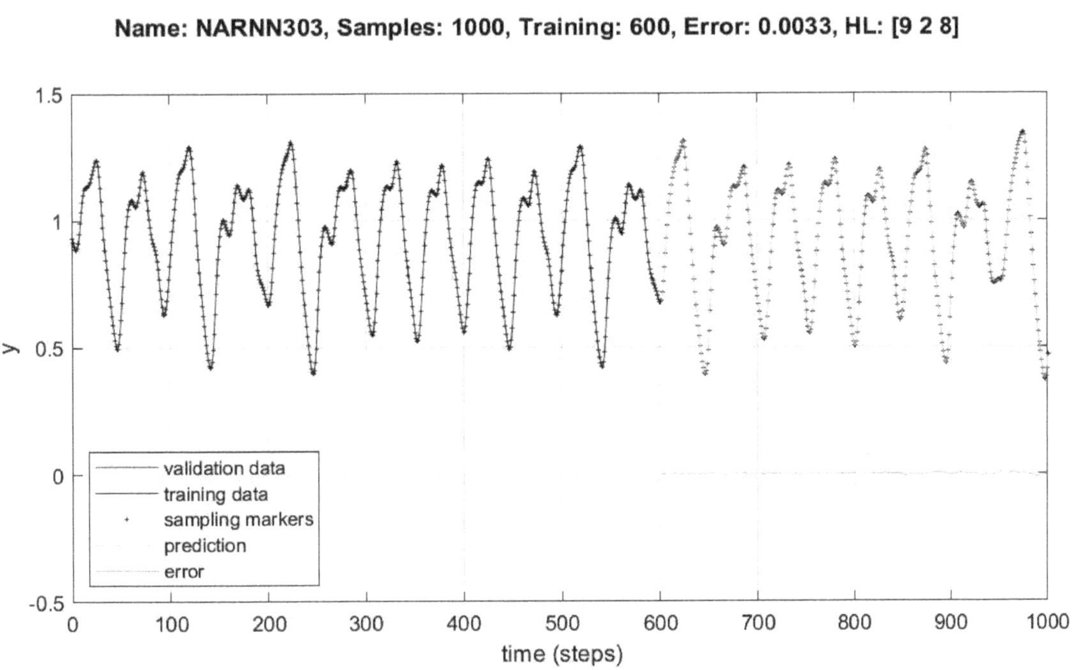

Figure 13. Performance of NARNN303 for 1000 samples.

Table 3. Results of optimized experiments.

N	Name	S	T	V	P	HL	I	Tt	T	RMSE
1	NARNN041	700	300	400	400	[5 6]	900			0.0114
2	NARNN042	700	300	400	400	[5 6]	1000			0.0054
3	NARNN043	700	300	400	400	[5 6]	1000			0.1012
4	NARNN053	700	300	400	400	[6 6]	500		405	0.0044
5	NARNN055	700	300	400	400	[6 1]	100		4	0.0202
6	NARNN056	700	300	400	400	[6 6]	500		486	0.0067
7	NARNN057	700	300	400	400	[4 3]	1000		807	0.0075
8	NARNN058	700	300	400	400	[8 7]	2500		154	0.0131
9	NARNN058r	700	300	400	400	[6 5]	2500		75	0.0202
10	NARNN059	700	300	400	400	[5 4]	1000		445	0.0061
11	NARNN060	700	300	400	400	[6 2]	1000		259	0.0081
12	NARNN061	700	300	400	400	[5 6]	1000		496	0.0044
13	NARNN062	700	400	300	300	[5 6]	5000		688	0.0024
14	NARNN201	700	400	300	300	[9 4]	5000	2948	1476	0.0024
15	NARNN202	1000	600	400	400	[7 5]	5000		1905	0.0035
16	NARNN203	1000	500	500	500	[4 7]	1000		681	0.0084
17	NARNN204	1000	400	600	600	[8 3]	1000		917	0.0144
18	NARNN205	1000	400	600	600	[7 8]	5000	2849	2741	0.0076
19	NARNN206	1000	500	500	500	[8 5]	5000	3262	2517	0.0059
20	NARNN207	1000	600	400	400	[6 2]	5000	3659	1822	0.0047
21	NARNN208	1500	600	900	900	[7 9]	5000	3666	1686	0.0187
22	NARNN209	1500	700	800	800	[7 7]	5000	4039	2665	0.0104
23	NARNN210p	1500	800	700	700	[7 3]	6000	5750	5226	0.0122
24	NARNN211	1500	900	600	600	[7 3]	3000	5439	1658	0.0136
25	NARNN212	1500	900	600	600	[5 6]	3000	6479	606	0.0055
26	NARNN213	1500	800	700	700	[5 6]	3000	12685	470	0.0080
27	NARNN214p	1500	700	800	800	[5 6]	5000	6183	4739	0.0157
28	NARNN215	1500	600	900	900	[5 6]	3000	2848	1910	0.0230
29	NARNN311	1500	1000	500	500	[6 1]	3000	6919	1917	0.0047
30	NARNN301	700	300	400	400	[7 8 3]	2000	1371	1313	0.0052
31	NARNN302	700	400	300	300	[6 7 5]	3000	5353	1235	0.0023
32	NARNN303	1000	600	400	400	[9 2 8]	3000	2348	2003	0.0033
33	NARNN304	1000	500	500	500	[4 5 4]	3000	5072	1334	0.0040
34	NARNN305	1000	400	600	600	[8 7 1]	3000	4435	2414	0.0070
35	NARNN306	1500	600	900	900	[5 5 1]	3000	5610	2309	0.0851
36	NARNN307	1500	700	800	800	[8 4 1]	3000	3151	1664	0.0245
37	NARNN308	1500	800	700	700	[8 1 1]	3000	3161	76	0.0127
38	NARNN309	1500	900	600	600	[8 2 7]	3000	3175	1502	0.0098
39	NARNN310	1500	1000	500	500	[8 3 8]	3000	3333	1050	0.0074

Figures 11–13 show the data behavior for 700, 700 and 1000 samples of the NARNN053, NARNN302 and NARNN303, obtaining an RMSE of 0.0044, 0.0023 and 0.0033, respectively, with the reference data at the top of each figure.

As for the complexity of the DMOA algorithm, it is a linear order algorithm that uses the discrete equations of Lotka-Volterra Equations (5)–(10), and in the search to find the global minimum it performs iterations and in each cycle it compares the best previous local minimum with the lowest current minimum and updates the value in the case that this is the case. As for the times, Table 3 shows the times Tt which represents the total time (seconds) of the experiment and T (seconds) the time in which the DMOA algorithm found the lowest local minimum; in terms of its efficiency the algorithm took 1235 s, about 21 min, to find the lowest minimum 0.0023, which seems to us a short time compared to the times used by the method [22] of up to 3 h and a half, the method [21], its experiments took up to 81 h to find the lowest minimum and as for the method [23] it does not provide the times of its experiments.

5.1. Statistical Data

Table 4 shows 30 experiments with eight non-optimized NARNNN architectures. Each column represents the total number of samples and the number of training samples used for each architecture (700 × 300), and at the end of the table we can find the results of the total sum, mean and standard deviation for each column.

Table 4. 30 Experiments with non-optimized NARNN.

No	Non Optimized							
	2 Hidden Layers					3 Hidden Layers		
	700 × 300	700 × 400	1000 × 400	1000 × 500	1000 × 600	1500 × 600	1500 × 700	1500 × 800
1	3.58×10^{-1}	3.17×10^{-1}	2.84×10^{-1}	2.45×10^{-1}	3.01×10^{-1}	9.88×10^{-1}	2.78×10^{-1}	7.27×10^{-1}
2	2.97×10^{-1}	2.39×10^{-1}	2.81×10^{-1}	2.32×10^{-1}	2.96×10^{-1}	3.41×10^{-1}	2.72×10^{-1}	5.39×10^{-1}
3	2.84×10^{-1}	2.35×10^{-1}	2.77×10^{-1}	2.26×10^{-1}	2.90×10^{-1}	3.39×10^{-1}	2.44×10^{-1}	4.67×10^{-1}
4	2.81×10^{-1}	2.23×10^{-1}	2.66×10^{-1}	2.25×10^{-1}	2.55×10^{-1}	3.38×10^{-1}	2.36×10^{-1}	4.35×10^{-1}
5	2.68×10^{-1}	2.22×10^{-1}	2.60×10^{-1}	2.22×10^{-1}	2.49×10^{-1}	3.27×10^{-1}	2.34×10^{-1}	4.27×10^{-1}
6	2.66×10^{-1}	2.22×10^{-1}	2.47×10^{-1}	2.06×10^{-1}	2.47×10^{-1}	3.19×10^{-1}	2.33×10^{-1}	4.23×10^{-1}
7	2.65×10^{-1}	2.21×10^{-1}	2.39×10^{-1}	2.03×10^{-1}	2.44×10^{-1}	3.16×10^{-1}	2.31×10^{-1}	4.18×10^{-1}
8	2.62×10^{-1}	2.19×10^{-1}	2.39×10^{-1}	1.83×10^{-1}	2.44×10^{-1}	3.07×10^{-1}	2.30×10^{-1}	4.18×10^{-1}
9	2.62×10^{-1}	2.16×10^{-1}	2.34×10^{-1}	1.75×10^{-1}	2.43×10^{-1}	3.00×10^{-1}	2.30×10^{-1}	4.08×10^{-1}
10	2.52×10^{-1}	2.13×10^{-1}	2.33×10^{-1}	1.74×10^{-1}	2.40×10^{-1}	2.92×10^{-1}	2.27×10^{-1}	4.02×10^{-1}
11	2.51×10^{-1}	2.12×10^{-1}	2.32×10^{-1}	1.73×10^{-1}	2.36×10^{-1}	2.73×10^{-1}	2.27×10^{-1}	3.98×10^{-1}
12	2.50×10^{-1}	2.11×10^{-1}	2.31×10^{-1}	1.72×10^{-1}	2.34×10^{-1}	2.73×10^{-1}	2.21×10^{-1}	3.94×10^{-1}
13	2.50×10^{-1}	2.10×10^{-1}	2.31×10^{-1}	1.71×10^{-1}	2.33×10^{-1}	2.69×10^{-1}	2.19×10^{-1}	3.91×10^{-1}
14	2.49×10^{-1}	2.09×10^{-1}	2.29×10^{-1}	1.69×10^{-1}	2.31×10^{-1}	2.63×10^{-1}	2.16×10^{-1}	3.89×10^{-1}
15	2.49×10^{-1}	2.08×10^{-1}	2.29×10^{-1}	1.69×10^{-1}	2.15×10^{-1}	2.60×10^{-1}	2.14×10^{-1}	2.53×10^{-1}
16	2.43×10^{-1}	2.07×10^{-1}	2.29×10^{-1}	1.68×10^{-1}	2.11×10^{-1}	2.47×10^{-1}	2.04×10^{-1}	2.51×10^{-1}
17	2.43×10^{-1}	2.00×10^{-1}	2.28×10^{-1}	1.67×10^{-1}	2.10×10^{-1}	2.33×10^{-1}	1.97×10^{-1}	2.42×10^{-1}
18	2.43×10^{-1}	1.95×10^{-1}	2.27×10^{-1}	1.66×10^{-1}	2.08×10^{-1}	2.29×10^{-1}	1.95×10^{-1}	2.38×10^{-1}
19	2.36×10^{-1}	1.93×10^{-1}	2.23×10^{-1}	1.66×10^{-1}	2.07×10^{-1}	2.29×10^{-1}	1.87×10^{-1}	2.33×10^{-1}
20	2.36×10^{-1}	1.89×10^{-1}	2.17×10^{-1}	1.66×10^{-1}	2.05×10^{-1}	2.27×10^{-1}	1.86×10^{-1}	2.27×10^{-1}
21	2.36×10^{-1}	1.89×10^{-1}	2.13×10^{-1}	1.54×10^{-1}	2.02×10^{-1}	2.12×10^{-1}	1.86×10^{-1}	2.25×10^{-1}
22	2.33×10^{-1}	1.89×10^{-1}	2.01×10^{-1}	1.52×10^{-1}	2.02×10^{-1}	2.06×10^{-1}	1.82×10^{-1}	2.25×10^{-1}
23	2.33×10^{-1}	1.88×10^{-1}	1.95×10^{-1}	1.35×10^{-1}	2.02×10^{-1}	2.04×10^{-1}	1.75×10^{-1}	2.24×10^{-1}
24	2.28×10^{-1}	1.82×10^{-1}	1.90×10^{-1}	1.30×10^{-1}	2.01×10^{-1}	2.03×10^{-1}	1.72×10^{-1}	2.23×10^{-1}
25	2.26×10^{-1}	1.78×10^{-1}	1.85×10^{-1}	1.19×10^{-1}	2.00×10^{-1}	1.98×10^{-1}	1.69×10^{-1}	2.22×10^{-1}
26	2.25×10^{-1}	1.78×10^{-1}	1.85×10^{-1}	1.19×10^{-1}	1.97×10^{-1}	1.96×10^{-1}	1.67×10^{-1}	2.19×10^{-1}
27	2.24×10^{-1}	1.77×10^{-1}	1.74×10^{-1}	1.14×10^{-1}	1.74×10^{-1}	1.96×10^{-1}	1.59×10^{-1}	2.16×10^{-1}
28	2.23×10^{-1}	1.70×10^{-1}	1.60×10^{-1}	1.09×10^{-1}	1.62×10^{-1}	1.95×10^{-1}	1.58×10^{-1}	2.15×10^{-1}
29	2.10×10^{-1}	1.70×10^{-1}	1.45×10^{-1}	1.01×10^{-1}	1.32×10^{-1}	1.92×10^{-1}	1.57×10^{-1}	2.14×10^{-1}
30	1.96×10^{-1}	1.65×10^{-1}	5.20×10^{-2}	6.12×10^{-2}	7.29×10^{-2}	1.61×10^{-1}	1.50×10^{-1}	1.94×10^{-1}
Sum:	$7.48 \times 10^{+0}$	$6.15 \times 10^{+0}$	$6.53 \times 10^{+0}$	$4.97 \times 10^{+0}$	$6.54 \times 10^{+0}$	$8.34 \times 10^{+0}$	$6.16 \times 10^{+0}$	$9.86 \times 10^{+0}$
Mean:	2.49×10^{-1}	2.05×10^{-1}	2.18×10^{-1}	1.66×10^{-1}	2.18×10^{-1}	2.78×10^{-1}	2.05×10^{-1}	3.29×10^{-1}
SD:	3.00×10^{-2}	2.91×10^{-2}	4.59×10^{-2}	4.25×10^{-2}	4.58×10^{-2}	1.44×10^{-1}	3.40×10^{-2}	1.27×10^{-1}

Table 5 shows 30 experiments with eight optimized NARNNN architectures; each column represents the total number of samples and the number of training samples used by each architecture (700 × 300), and at the end of the table we can find the results of the total sum, mean and standard deviation for each column.

Table 5. 30 Experiments with optimized NARNN.

No	Optimized							
	2 Hidden Layers					3 Hidden Layers		
	700 × 300	**700 × 400**	**1000 × 400**	**1000 × 500**	**1000 × 600**	**1500 × 600**	**1500 × 700**	**1500 × 800**
1	1.72×10^{-1}	4.62×10^{-2}	2.63×10^{-1}	1.97×10^{-1}	2.71×10^{-1}	3.88×10^{-1}	2.85×10^{-1}	2.93×10^{-1}
2	1.52×10^{-1}	4.44×10^{-2}	2.34×10^{-1}	1.97×10^{-1}	1.92×10^{-1}	3.07×10^{-1}	2.67×10^{-1}	2.67×10^{-1}
3	1.35×10^{-1}	4.21×10^{-2}	2.32×10^{-1}	1.92×10^{-1}	1.92×10^{-1}	3.06×10^{-1}	2.53×10^{-1}	2.66×10^{-1}
4	1.34×10^{-1}	4.07×10^{-2}	2.31×10^{-1}	1.91×10^{-1}	1.91×10^{-1}	2.97×10^{-1}	2.43×10^{-1}	2.36×10^{-1}
5	1.18×10^{-1}	3.98×10^{-2}	2.30×10^{-1}	1.87×10^{-1}	1.91×10^{-1}	2.94×10^{-1}	2.39×10^{-1}	2.29×10^{-1}
6	1.09×10^{-1}	3.73×10^{-2}	2.28×10^{-1}	1.86×10^{-1}	1.89×10^{-1}	2.90×10^{-1}	2.36×10^{-1}	2.28×10^{-1}
7	1.07×10^{-1}	3.59×10^{-2}	2.27×10^{-1}	1.86×10^{-1}	1.89×10^{-1}	2.86×10^{-1}	2.25×10^{-1}	2.27×10^{-1}
8	9.68×10^{-2}	3.40×10^{-2}	2.26×10^{-1}	1.83×10^{-1}	1.89×10^{-1}	2.74×10^{-1}	2.22×10^{-1}	2.24×10^{-1}
9	9.67×10^{-2}	3.26×10^{-2}	2.22×10^{-1}	1.80×10^{-1}	1.86×10^{-1}	2.74×10^{-1}	2.21×10^{-1}	2.23×10^{-1}
10	9.61×10^{-2}	3.08×10^{-2}	2.20×10^{-1}	1.75×10^{-1}	1.83×10^{-1}	2.72×10^{-1}	2.19×10^{-1}	2.17×10^{-1}
11	9.54×10^{-2}	3.06×10^{-2}	2.18×10^{-1}	1.74×10^{-1}	1.74×10^{-1}	2.70×10^{-1}	2.17×10^{-1}	2.13×10^{-1}
12	9.23×10^{-2}	2.19×10^{-2}	2.18×10^{-1}	1.73×10^{-1}	1.69×10^{-1}	2.66×10^{-1}	2.16×10^{-1}	2.09×10^{-1}
13	9.01×10^{-2}	2.18×10^{-2}	2.16×10^{-1}	1.72×10^{-1}	1.69×10^{-1}	2.62×10^{-1}	2.14×10^{-1}	2.08×10^{-1}
14	8.57×10^{-2}	2.14×10^{-2}	2.12×10^{-1}	1.72×10^{-1}	1.57×10^{-1}	2.53×10^{-1}	2.04×10^{-1}	2.07×10^{-1}
15	8.11×10^{-2}	1.91×10^{-2}	2.09×10^{-1}	1.72×10^{-1}	1.50×10^{-1}	2.53×10^{-1}	1.78×10^{-1}	2.05×10^{-1}
16	7.47×10^{-2}	1.91×10^{-2}	2.08×10^{-1}	1.71×10^{-1}	1.40×10^{-1}	2.51×10^{-1}	1.74×10^{-1}	2.02×10^{-1}
17	7.13×10^{-2}	1.79×10^{-2}	1.89×10^{-1}	1.69×10^{-1}	1.35×10^{-1}	2.44×10^{-1}	1.73×10^{-1}	1.99×10^{-1}
18	6.11×10^{-2}	1.77×10^{-2}	1.89×10^{-1}	1.69×10^{-1}	1.33×10^{-1}	2.37×10^{-1}	1.71×10^{-1}	1.98×10^{-1}
19	5.99×10^{-2}	1.31×10^{-2}	1.85×10^{-1}	1.67×10^{-1}	1.32×10^{-1}	2.33×10^{-1}	1.58×10^{-1}	1.98×10^{-1}
20	5.76×10^{-2}	1.25×10^{-2}	1.78×10^{-1}	1.66×10^{-1}	1.32×10^{-1}	2.28×10^{-1}	1.53×10^{-1}	1.90×10^{-1}
21	5.43×10^{-2}	1.12×10^{-2}	1.78×10^{-1}	1.66×10^{-1}	1.29×10^{-1}	2.27×10^{-1}	1.53×10^{-1}	1.78×10^{-1}
22	5.35×10^{-2}	1.02×10^{-2}	1.77×10^{-1}	1.64×10^{-1}	1.29×10^{-1}	2.24×10^{-1}	1.52×10^{-1}	1.66×10^{-1}
23	3.96×10^{-2}	9.31×10^{-3}	1.76×10^{-1}	1.45×10^{-1}	1.28×10^{-1}	2.21×10^{-1}	1.50×10^{-1}	1.65×10^{-1}
24	3.58×10^{-2}	9.11×10^{-3}	1.74×10^{-1}	1.42×10^{-1}	1.28×10^{-1}	2.12×10^{-1}	1.45×10^{-1}	1.61×10^{-1}
25	2.82×10^{-2}	6.99×10^{-3}	1.71×10^{-1}	1.37×10^{-1}	1.27×10^{-1}	1.98×10^{-1}	1.44×10^{-1}	1.58×10^{-1}
26	2.27×10^{-2}	5.98×10^{-3}	1.71×10^{-1}	1.30×10^{-1}	1.26×10^{-1}	1.98×10^{-1}	1.42×10^{-1}	1.58×10^{-1}
27	2.13×10^{-2}	5.63×10^{-3}	1.54×10^{-1}	1.16×10^{-1}	1.26×10^{-1}	1.92×10^{-1}	1.41×10^{-1}	9.36×10^{-2}
28	1.89×10^{-2}	4.96×10^{-3}	1.19×10^{-1}	1.15×10^{-1}	1.24×10^{-1}	1.91×10^{-1}	1.38×10^{-1}	6.64×10^{-2}
29	1.84×10^{-2}	4.21×10^{-3}	1.03×10^{-1}	8.25×10^{-2}	1.23×10^{-1}	1.88×10^{-1}	1.37×10^{-1}	5.36×10^{-2}
30	1.14×10^{-2}	3.52×10^{-3}	5.66×10^{-2}	1.51×10^{-2}	1.11×10^{-1}	1.75×10^{-1}	1.35×10^{-1}	4.84×10^{-2}
Sum:	$2.29 \times 10^{+0}$	6.50×10^{-1}	$5.81 \times 10^{+0}$	$4.79 \times 10^{+0}$	$4.71 \times 10^{+0}$	$7.51 \times 10^{+0}$	$5.70 \times 10^{+0}$	$5.69 \times 10^{+0}$
Mean:	7.63×10^{-2}	2.17×10^{-2}	1.94×10^{-1}	1.60×10^{-1}	1.57×10^{-1}	2.50×10^{-1}	1.90×10^{-1}	1.90×10^{-1}
SD:	4.21×10^{-2}	1.38×10^{-2}	4.35×10^{-2}	3.81×10^{-2}	3.52×10^{-2}	4.60×10^{-2}	4.47×10^{-2}	5.91×10^{-2}

5.2. Hypothesis Test

Equation (13) represents Hypothesis Testing, Null Hypothesis Equation (14) and Alternative Hypothesis Equation (15), with which comparisons were made between the non-optimized and optimized experiments of the method proposed here.

$$z = \frac{(\overline{x}_1 - \overline{x}_2) - D_0}{\sqrt{\frac{\sigma_1^2}{n_1} + \frac{\sigma_2^2}{n_2}}} \tag{13}$$

$$H_0 : \mu_1 \geq \mu_2 \tag{14}$$

$$H_a : \mu_1 < \mu_2 \; claim \tag{15}$$

where $\overline{x}_1$ is the Mean of sample 1, $\overline{x}_2$ Mean of sample 2, σ_1 Standard Deviation of sample 1, σ_2 Standard Deviation of sample 2, n_1 Number of sample data 1, n_2 Number of sample data 2, $\mu_1 - \mu_2 = D_0$ and $\mu_1 - \mu_2 = D_0$.

Significance Level α = 0.05, Confidence Level = 95%, Confidence Level = 1 − α; 1−0.05 = 0.95 o 95%, Since the p-value is less than 0.01, the null hypothesis is rejected.

Tables 6 and 7 show the results of the hypothesis testing done on the non-optimized and optimized methods shown above; of the eight different architectures, the test results show that in only six were the optimized NARNNs better, and the non-optimized NARNNs were better in two.

Table 6. Data for the eight non-optimized and optimized architectures.

	NARNN—Non-OPTIMIZED					NARNN—OPTIMIZED				
			HL and N					HL and N		
N	**Name**	**Error**	**1**	**2**	**3**	**Name**	**Error**	**1**	**2**	**3**
1	NARNN401	0.2777	1	1		NARNN053	0.0052	6	6	
2	NARNN402	0.1683	1	1		NARNN062	0.0023	5	6	
3	NARNN403	0.2307	3	1		NARNN205	0.0070	7	8	
4	NARNN404	0.1670	9	2		NARNN206	0.0040	8	5	
5	NARNN405	0.2488	2	2		NARNN207	0.0033	6	2	
6	NARNN406	0.2550	4	1	9	NARNN306	0.0851	5	5	1
7	NARNN407	0.2158	8	5	2	NARNN307	0.0245	8	4	1
8	NARNN408	0.4001	1	2	1	NARNN308	0.0127	8	1	1

Table 7. Hypothesis test results of the eight non-optimized and optimized NARNNs.

Samples				Results	
				Non-Optimized	**Optimized**
Total	**T**	**V**	**P**	***p*-Value**	
700	300	400	400		8.25×10^{-26}
700	400	300	300		6.00×10^{-38}
1000	400	600	600		4.23×10^{-2}
1000	500	500	500	5.59×10^{-1}	
1000	600	400	400		3.13×10^{-7}
1500	600	900	900	3.25×10^{-1}	
1500	700	800	800	1.46×10^{-1}	
1500	800	700	700		1.14×10^{-6}

In Table 6, N and Name represent the number and name of the experiment, respectively. Error is the minimum error found, HL are the Hidden Layers of neural network (1, 2, 3), and N is the number of neurons in each HL. In Table 7, the samples are represented by Total number of samples, T is the training samples, V is the validation samples, P represents the prediction, and *p*-value represents the results of the hypothesis test.

5.3. Comparisone with Other Methods

Table 8 shows the comparison with other methods that performed experimentation with the chaotic Mackey-Glass time series, and it can be seen from the table that the lowest error belongs to the optimized NARNN-302.

Table 8. Error comparison with three different methods and the non-optimized and optimized NARNN.

N	Experiment Description	Error	Serie	Ref
1	Genetic Algorithm—Ensemble ANFIS—T1FLS—IT2FLS	0.0219	MG	[21]
2	Ensemble Neural Network Architecture	0.008945	MSE	[22]
	2 Modules, 2 Hidden Layers, 2116 and 2128 Neurons respectively, PSO Optimized			
3	SNR(dB) ANFIS IT2FNN-0 IT2FNN-1 IT2FNN-2 IT2FNN-3	0.0028	MG	[23]
	30 0.0225 0.0106 0.0079 0.0045 0.0028			
4	NARNN-DMOA (No-Optimized)	0.167	MG	
5	NARNN-DMOA (Optimized)	0.0023	MG	

In Table 8, case number 1, the method is the Optimization of the Fuzzy Integrators in Ensembles of ANFIS Model for Time Series Prediction [21], where the authors use the Mackey-Glass chaotic time series, with genetic optimization of Type-1 Fuzzy Logic System (T1FLS) and Interval Type-2 Fuzzy Logic System (IT2FLS) integrators in Ensemble of ANFIS models and evaluate the results through Root Mean Square Error (RMSE). ANFIS is a hybrid model of a neural network implementation of a TSK (Takagi-Sugeno-Kang) fuzzy inference system. ANFIS applies a hybrid algorithm which integrates BP (Backpropagation) and LSE (least square estimation) algorithms, and thus it has a fast learning speed.

Case number 2 refers to the method using Particle Swarm Optimization of ensemble neural networks with fuzzy aggregation for time series prediction of the Mexican Stock Exchange [22]. In this case, the authors propose an ensemble neural network model with type-2 fuzzy logic for the integration of responses; in addition, the particle swarm optimization method determines the number of modules of the ensemble neural network, the number of layers and number of neurons per layer, and thus the best architecture of the ensemble neural network is obtained. Once this architecture is obtained, the results of the modules with type-1 and type-2 fuzzy logic systems are added, the inputs to the fuzzy system are the responses according to the number of modules of the network, and this is the number of inputs of the fuzzy system.

Case number 3 refers to the Application of Interval Type-2 Fuzzy Neural Networks (IT2FNN) in non-linear identification and time series prediction (MG) [23]. The authors propose IT2FNN models that combine the uncertainty management advantage of type-2 fuzzy sets with the learning capabilities of neural networks. One of the main ideas of this approach is that the proposed IT2FNN architectures can obtain similar or better outputs than type-2 interval fuzzy systems using the Karnik and Mendel (KM) algorithm, but with lower computational cost, which is one of the main disadvantages of KM mentioned in many papers in the literature. Cases 4 and 5 have already been explained earlier in this article.

By making a brief description of the techniques of the different methods above, we can observe the complexity of their designs using optimization algorithms such as PSO and GAs as optimizers, robust Ensemble Neural Networks, T1FLS and IT2FLS, in comparison with our method that uses the optimization algorithm DMOA and NARNNN, which are neural networks with short memory, and according to the results are made precisely for the prediction of time series. In a future work we plan to perform experiments with the RNN LSTM networks, which have short- and long-term memories.

6. Discussion of Results

The use of metaheuristics in the optimization of methods is a constant in all research work in artificial intelligence, and in this work the DMOA algorithm was used to optimize the architecture of the NARNN neural network using the MG chaotic series as input data. We also performed experiments without optimizing the NARNN network, while with the optimization we performed experiments with 39 different architectures, and without optimization we performed experiments with 10 different architectures. When we performed the optimization we found an extremely fast algorithm that found the right architecture with very satisfactory results. Of the 39 different optimized architectures, the one that gave us the best results was number 31 (narAll303) Table 3, a NARNN network with three hidden layers of 6, 7, and 5 neurons, respectively. With this architecture we performed 3000 experiments with a total time of 5353 s and in the second 1235 we obtained the best result of 0.0023 (error). Of the 10 experiments without optimization, with architecture number 4 (narAll404) Table 2, a NARNN network with two hidden layers of 9 and 2 neurons, respectively, we also performed 3000 experiments with this architecture and obtained the best result of 0.1670 (error). We performed eight hypothesis tests under equal conditions with these results and found that in five tests the NARNN architectures optimized with the DMOA algorithm were better and in three tests the non-optimized architectures were better, as shown in Table 6. We also performed error comparisons with three other different methods of which the DMOA-NARNN was better, as shown in Table 8.

7. Conclusions

A total of 49 different architectures were designed, of which 10 non-optimized and 39 were optimized by the DMOA algorithm, 30,000 experiments were performed with the non-optimized architectures, and approximately 110,000 experiments were performed with the optimized architectures. A total of 700, 1000 and 1500 samples were generated with the MG chaotic time series, of which between 300 and 1000 were used for training, between 300 and 900 were used for validation in different combinations, and between 300 and 900 points were generated as prediction points, as can be seen in Tables 2 and 3. The design of the NARNN architectures were two and three hidden layers, with neurons in the range of 2–9, and the graphs of the most representative results of the non-optimized and optimized NARNNs are presented in Figures 8–13.

The optimization of the NARNN network with the DMOA algorithm obtained good results, better than without optimizing the network, and better than the other methods with which it was compared, although not all of the optimized architectures were better in the hypothesis test (only five of them were), the results of the error were much better, as can be seen in Table 7. In the comparison with other methods, the results were also better, as demonstrated in Table 8. We were also able to verify that the DMOA optimization algorithm is fast and efficient, which was really the reason for this research. We wish to continue investigating the efficiency of the algorithm in the optimization of architectures with other types of neural networks, also in Fuzzy Logic Systems Type-1 and Type-2, and also to do the same with the optimization algorithm CMOA (Continuous Mycorrhiza Optimization Algorithm). In addition, the proposed algorithm can be applied to robots, microsystems, sensors, devices, MEMS, microfluidics, piezoelectricity, motors, biosensors, 3D printing, etc.

We also intend to conduct further research and experimentation with the DMOA method and other time series. We will also consider the DMOA and the LSTM (Long Short-Term Memory) Neural Regression Network for Mackey-Glass time series, weather and financial forecasting, and we are interested in hybridizing the method with Interval Type-2 Fuzzy Logic System (IT2FLS), and Generalized Type-2 Fuzzy Logic System (GT2FLS).

Author Contributions: Conceptualization, F.V. and O.C.; methodology, F.V. and P.M.; software, H.C.-O.; validation, H.C.-O. and F.V.; formal analysis, H.C.-O.; investigation, O.C. and P.M.; resources, O.C.; writing—original draft preparation, F.V. and O.C.; writing—review and editing, H.C.-O. and P.M.; visualization, F.V.; supervision, F.V. and O.C. All authors have read and agreed to the published version of the manuscript.

Funding: This research received no external funding.

Institutional Review Board Statement: Not Applicable.

Informed Consent Statement: Not Applicable.

Data Availability Statement: Not applicable.

Acknowledgments: We would like to thank Tecnologico Nacional de Mexico (TecNM) for the support and Consejo Nacional de Ciencia y Tecnologia for the Ph.D. scholarship provided to Hector Carreon-Ortiz.

Conflicts of Interest: The authors declare that they have no conflicts of interest.

References

1. Diwekar, U.M. *Introduction to Applied Optimization*; Springer Nature Switzerland AG: Cham, Switzerland, 2020; pp. 1–3.
2. Ghaemi, M.B.; Gharakhanlu, N.; Rassias, T.M.; Saadati, R. *Advances in Matrix Inequalities*; Springer Nature Switzerland AG: Cham, Switzerland, 2021; pp. 3–4.
3. Lange, K. *Optimization Second Edition*; Springer Text in Statistics; Biomathematics, Human Genetics, Statistics University of California; Springer: New York, NY, USA, 2013; pp. 1–3. [CrossRef]

4. Kochenderfer, M.J.; Wheeler, T.A. *Algorithms for Optimization*; The MIT Press Cambridge: London, UK, 2019; pp. 1–4.
5. Adam, S.P.; Alexandropoulos, S.N.; Pardalos, P.M.; Vrahatis, M.N. No Free Lunch Theorem: A Review. In *Approximation and Optimization*; Springer Optimization and Its Applications; Demetriou, I., Pardalos, P., Eds.; Springer: Cham, Switzerland, 2019; Volume 145.
6. Bianchi, F.M.; Maiorino, E.; Kampffmeyer, M.C.; Rizzi, A.; Jenssen, R. An overview and comparative analysis of Recurrent Neural Networks for Short Term Load Forecasting. *arXiv* **2017**, arXiv:1705.04378.
7. Schäfer, A.M.; Zimmermann, H.G. *Recurrent Neural Networks Are Universal Approximators*; Artificial Neural Networks—ICANN 2006. ICANN 2006. Lecture Notes in Computer Science; Kollias, S.D., Stafylopatis, A., Duch, W., Oja, E., Eds.; Springer: Berlin, Heidelberg, 2006; Volume 4131.
8. Brownlee, J. *Deep Learning for Time Series Forecasting Predict the Future with MLPs, CNNs and LSTMs in Python*; Machine Learning Mastery: San Juan, PR, USA, 2019; pp. 6–7.
9. Graves, A. Sequence transduction with recurrent neural networks. *arXiv* **2012**, arXiv:1211.3711.
10. Graves, A. Generating sequences with recurrent neural networks. *arXiv* **2013**, arXiv:1308.0850.
11. Pascanu, R.; Mikolov, T.; Bengio, Y. On the difficulty of training Recurrent Neural Networks. In Proceedings of the 30th International Conference on Machine Learning, ICML 2013, JMLR.org, Atlanta, GA, USA, 16–21 June 2013.
12. Mikolov, T. Statistical Language Models Based on Neural Networks. Ph.D. Thesis, Brno University of Technology, Brno-střed, Czechia, 2012.
13. Sutskever, I.; Martens, J.; Hinton, G. Generating Text with Recurrent Neural Networks. In Proceedings of the 28th International Conference on Machine Learning. ICML 2011, Bellevue, WA, USA, 28 June–2 July 2011.
14. Graves, A. Practical variational inference for neural networks. In *Advances in Neural Information Processing Systems*; Curran Associates Inc.: Red Hook, NY, USA, 2011; pp. 2348–2356.
15. Mikolov, T.; Sutskever, I.; Chen, K.; Corrado, G.S.; Dean, J. Distributed representations of words and phrases and their compositionality. In *Advances in Neural Information Processing Systems*; Curran Associates Inc.: Red Hook, NY, USA, 2013; pp. 3111–3119.
16. Oord, A.; Dieleman, S.; Zen, H.; Vinyals, K.S.O.; Graves, A.; Kalchbrenner, N.; Senior, A.; Kavukcuoglu, K. A generative model for raw audio. *arXiv* **2016**, arXiv:1609.03499.
17. Graves, A.; Schmidhuber, J. Offline handwriting recognition with multidimensional recurrent neural networks. In *Advances in Neural Information Processing Systems*; Curran Associates Inc.: Red Hook, NY, USA, 2009; pp. 545–552.
18. Graves, A.; Fernández, S.; Liwicki, M.; Bunke, H.; Schmidhuber, J. Unconstrained On-line Handwriting Recognition with Recurrent Neural Networks. In Proceedings of the Advances in Neural Information Processing Systems, Vancouver, BC, Canada, 3–6 December 2007; pp. 577–584.
19. Gregor, K.; Danihelka, I.; Graves, A.; Rezende, D.; Wierstra, D. DRAW: A recurrent neural network for image generation. In Proceedings of the 32nd International Conference on Machine Learning, PMLR, Lille, France, 6–11 July 2015.
20. Hochreiter, S.; Schmidhuber, J. Long short-term memory. *Neural Comput.* **1997**, *9*, 1735–1780. [CrossRef] [PubMed]
21. Soto, J.; Melin, P. Optimization of the Fuzzy Integrators in Ensembles of ANFIS Model for Time Series Prediction: The case of Mackey-Glass. In Proceedings of the 2015 Conference of the International Fuzzy Systems Association and the European Society for Fuzzy Logic and Technology (IFSA-EUSFLAT-15), Gijón, Spain, 30 June 2015; pp. 994–999.
22. Pulido, M.; Melin, P.; Castillo, O. Particle swarm optimization of ensemble neural networks with fuzzy aggregation for time series prediction of the Mexican Stock Exchange. *Inf. Sci.* **2014**, *280*, 188–204. [CrossRef]
23. Castillo, O.; Castro, J.R.; Melin, P.; Rodríguez-Díaz, A. Application of interval type-2 fuzzy neural networks in non-linear identification and time series prediction. *Soft Comput.* **2013**, *18*, 1213–1224. [CrossRef]
24. Amador-Angulo, L.; Castillo, O. Amador-Angulo, L.; Castillo, O. A Fuzzy Bee Colony Optimization Algorithm Using an Interval Type-2 Fuzzy Logic System for Trajectory Control of a Mobile Robot. In *Mexican International Conference on Artificial Intelligence*; Springer: Cham, Switzerland, 2015.
25. Zangeneh, M.; Aghajari, E.; Forouzanfar, M. A Review on Optimization of Fuzzy Controller Parameters in Robotic Applications. *IETE J. Res.* **2020**, 1–10. [CrossRef]
26. Peraza, C.; Ochoa, P.; Castillo, O.; Geem, Z.W. Interval-Type 3 Fuzzy Differential Evolution for Designing an Interval-Type 3 Fuzzy Controller of a Unicycle Mobile Robot. *Mathematics* **2022**, *10*, 3533. [CrossRef]
27. Jiang, Y.; Yin, S.; Dong, J.; Kaynak, O. A Review on Soft Sensors for Monitoring, Control and Optimization of Industrial Processes. *IEEE Sens. J.* **2020**, *21*, 12868–12881. [CrossRef]
28. Bradley, E.; Kantz, H. Nonlinear time-series analysis revisited. *Chaos Interdiscip. J. Nonlinear Sci.* **2015**, *25*, 097610. [CrossRef]
29. Benmouiza, K.; Cheknane, A. Forecasting hourly global solar radiation using hybrid k-means and nonlinear autoregressive neural network models. *Energy Convers. Manag.* **2013**, *75*, 561–569. [CrossRef]
30. Long, D.; Zhang, R.; Mao, Y. Recurrent Neural Networks With Finite Memory Length. *IEEE Access.* **2019**, *7*, 12511–12520. [CrossRef]
31. Ji, W.; Chan, C. Prediction of hourly solar radiation using a novel hybrid model of ARMA and TDNN. *Solar Energy* **2011**, *85*, 808–817. [CrossRef]
32. Taherdangkoo, R.; Tatomir, A.; Taherdangkoo, M.; Qiu, P.; Sauter, M. Nonlinear Autoregressive Neural Networks to Predict Hydraulic Fracturing Fluid Leakage into Shallow Groundwater. *Water* **2020**, *12*, 841. [CrossRef]

33. Kumar, A.; Irsoy, O.; Su, J.; Bradbury, J.; English, R.; Pierce, B.; Ondruska, P.; Gulrajani, I.; Socher, R. Ask Me Anything: Dynamic Memory Networks for Natural Language Processing. In Proceedings of the International conference on machine learning, Palo Alto, CA, USA, 6 July–11 July 2015; p. 97.
34. Young, T.; Hazarika, D.; Poria, S.; Cambria, E. Recent Trends in Deep Learning Based Natural Language Processing. *IEEE Comput. Intell. Mag.* **2018**, *13*, 55–75. [CrossRef]
35. Kalimuthu, M.; Mogadala, A.; Mosbach, M.; Klakow, D. Fusion Models for Improved Image Captioning. In *ICPR International Workshops and Challenges, ICPR 2021, Lecture Notes in Computer Science*; Springer: Cham, Switzerland, 2021; Volume 12666.
36. Yassin, I.M.; Zabidi, A.; Salleh, M.K.M.; Khalid, N.E.A. Malaysian tourism interest forecasting using nonlinear auto regressive (NAR) model. In Proceedings of the 3rd International Conference on System Engineering and Technology, New York, NY, USA, 19–20 August 2013; pp. 32–36.
37. Raturi, R.; Sargsyan, H. A Nonlinear Autoregressive Scheme for Time Series Prediction via Artificial Neural Networks. *J. Comput. Commun.* **2018**, *6*, 14–23. [CrossRef]
38. Ahmed, A.; Khalid, M. A Nonlinear Autoregressive Neural Network Model for Short-Term Wind Forecasting. In Proceedings of the 2017 9th IEEE-GCC Conference and Exhibition (GCCCE), Manama, Bahrain, 8–11 May 2017.
39. MATLAB 2022b. *Deep Learning Toolbox Reference*; The MathWorks, Inc.: Natick, MA, USA, 2022; pp. 2200–2203.
40. Padilla, C.; Hashemi, R.; Mahmood, N.; Latva-aho, M. A Nonlinear Autoregressive Neural Network for Interference Prediction and Resource Allocation in URLLC Scenarios. In Proceedings of the 2021 International Conference on Information and Communication Technology Convergence (ICTC), Jeju Island, Republic of Korea, 20–22 October 2021.
41. Adedeji, P.A.; Akinlabi, S.A.; Ajayi, O.O.; Madushele, N. Non-Linear Autoregressive Neural Network (NARNET) with SSA filtering for a university Campus Energy Consumption Forecast. *Procedia Manuf.* **2019**, *33*, 176–183. [CrossRef]
42. Olney, B.; Mahmud, S.; Karam, R. Efficient Nonlinear Autoregressive Neural Network Architecture for Real-Time Biomedical Applications. In Proceedings of the 2022 IEEE 4th International Conference on Artificial Intelligence Circuits and Systems (AICAS), Incheon, Republic of Korea, 13–15 June 2022; pp. 411–414.
43. Li, M.; Ji, S.; Liu, G. Forecasting of Chinese E-Commerce Sales: An Empirical Comparison of ARIMA, Nonlinear Autoregressive Neural Network, and a Combined ARIMA-NARNN Model. *Math. Probl. Eng. Vol.* **2018**, *2018*, 1–12. [CrossRef]
44. Kummong, R.; Supratid, S. Long-term forecasting system using wavelet – nonlinear autoregressive neural network conjunction model. *J. Model. Manag.* **2019**, *14*, 948–971. [CrossRef]
45. Davood, N.K.; Goudarzi, G.R.; Taghizadeh, R.; Asumadu-Sakyi, A.B.; Fehresti-Sani, M. Long-term effects of outdoor air pollution on mortality and morbidity–prediction using nonlinear autoregressive and artificial neural networks models. *Atmos. Pollut. Res.* **2021**, *12*, 46–56.
46. Domaschenko, D.; Nikulin, E. Forecasting time series of the market indicators based on a nonlinear autoregressive neural network. *Stat. Econ. Vol.* **2017**, *2017*, 4–9. [CrossRef]
47. Saba, A.I.; Elsheikh, A.H. Forecasting the prevalence of COVID-19 outbreak in Egypt using nonlinear autoregressive artificial neural networks. *Process. Saf. Environ. Prot.* **2020**, *141*, 1–8. [CrossRef] [PubMed]
48. Newman, E.I. Mycorrhizal links between plants: Their functioning and ecological significance. *Adv. Ecol. Res.* **1988**, *18*, 243–270.
49. Bahram, M.; Põlme, S.; Kõljalg, U.; Tedersoo, L. A single European aspen (Populus tremula) tree individual may potentially harbour dozens of Cenococcum geophilum ITS genotypes and hundreds of species of ectomycorrhizal fungi. *FEMS Microbiol. Ecol.* **2010**, *75*, 313–320. [CrossRef]
50. Schimel, J.P.; Bennett, J. Nitrogen mineralization: Challenges of a changing paradigm. *Ecology* **2004**, *85*, 591–602. [CrossRef]
51. Averill, C.; Turner, B.L.; Finzi, A.C. Mycorrhiza-mediated competition between plants and decomposers drives soil carbon storage. *Nature* **2014**, *505*, 543–545. [CrossRef] [PubMed]
52. Dickie, I.A.; Koele, N.; Blum, J.D.; Gleason, J.D.; McGlone, M.S. Mycorrhizas in changing ecosystems'. *Botany* **2014**, *92*, 149–160. [CrossRef]
53. Redecker, D.; Kodner, R.; Graham, L.E. Glomalean Fungi from the Ordovician. *Science* **2000**, *289*, 1920–1921. [CrossRef]
54. Humphreys, C.P.; Franks, P.J.; Rees, M.; Bidartondo, M.I.; Leake, J.R.; Beerling, D.J. Mutualistic mycorrhiza-like symbiosis in the most ancient group of land plants. *Nat. Commun.* **2010**, *1*, 103. [CrossRef]
55. Lang, C.; Seven, J.; Polle, A. Host preferences and differential contributions of deciduous tree species shape mycorrhizal species richness in a mixed Central European forest. *Mycorrhiza* **2010**, *21*, 297–308. [CrossRef]
56. Simard, S.W. Mycorrhizal Networks Facilitate Tree Communication, Learning, and Memory. In *Memory and Learning in Plants*; Signaling and Communication in Plants; Baluska, F., Gagliano, M., Witzany, G., Eds.; Springer: Cham, Switzerland, 2018.
57. Castro-Delgado, A.L.; Elizondo-Mesén, S.; Valladares-Cruz, Y.; Rivera-Méndez, W. Wood Wide Web: Communication through the mycorrhizal network. *Tecnol. Marcha J.* **2020**, *33*, 114–125.
58. Beiler, K.J.; Simard, S.W.; Durall, D.M. Topology of tree-mycorrhizal fungus interaction networks in xeric and mesic Douglas-fir forests. *J. Ecol.* **2015**, *103*, 616–628. [CrossRef]
59. Simard, S.W.; Asay, A.; Beiler, K.; Bingham, M.; Deslippe, J.; He, X.; Philip, L.; Song, Y.; Teste, F. Resource Transfer Between Plants Through Ectomycorrhizal Fungal Networks. In *Mycorrhizal Networks. Ecological Studies*; Horton, T., Ed.; Springer: Dordrecht, The Netherlands, 2015; Volume 224.
60. Gorzelak, M.A.; Asay, A.K.; Pickles, B.J.; Simard, S.W. Inter-plant communication through mycorrhizal networks mediates complex adaptive behaviour in plant communities. *AoB Plants* **2015**, *7*, plv050. [CrossRef] [PubMed]

61. Carreon, H.; Valdez, F.; Castillo, O. A New Discrete Mycorrhiza Optimization Nature-Inspired Algorithm. *Axioms* **2022**, *11*, 391. [CrossRef]
62. Liu, P.; Elaydi, S.N. Discrete Competitive and Cooperative Models of Lotka–Volterra Type. *J. Comput. Anal. Appl.* **2001**, *3*, 53–73.
63. Muhammadhaji, A.; Halik, A.; Li, H. Dynamics in a ratio-dependent Lotka–Volterra competitive-competitive-cooperative system with feedback controls and delays. *Adv. Differ. Equ.* **2021**, *230*, 1–14. [CrossRef]
64. Din, Q. Dynamics of a discrete Lotka-Volterra model. *Adv. Differ. Equ.* **2013**, *2013*, 95. [CrossRef]
65. Liu, X. A note on the existence of periodic solutions in discrete predator–prey models. *Appl. Math. Model.* **2010**, *34*, 2477–2483. [CrossRef]
66. Zhou, Z.; Zou, X. Stable periodic solutions in a discrete periodic logistic equation. *Appl. Math. Lett.* **2003**, *16*, 165–171. [CrossRef]
67. Krabs, W. A General Predator-Prey Model. *Math. Comput. Model. Dyn. Syst.* **2003**, *9*, 387–401. [CrossRef]
68. Allen, L.J.S. *An Introduction to Mathematical Biology*; Pearson Prentice Hall: New York, NY, USA, 2007; pp. 89–140.
69. Brauer, F.; Castillo-Chavez, C. *Mathematical Models in Population Biology and Epidemiology*, 2nd ed.; Springer: New York, NY, USA; Dordrecht, The Netherlands; Heidelberg, Germany, 2012; pp. 123–134.
70. Müller, J.; Kuttler, C. Methods and Models in Mathematical Biology, Deterministic and Stochastic Approaches. In *Lecture Notes on Mathematical Modelling in the Life Sciences*; Springer-Verlag: Berlin/Heidelberg, Germany, 2015.
71. Voroshilova, A.; Wafubwa, J. Discrete Competitive Lotka–Volterra Model with Controllable Phase Volume. *Systems* **2020**, *8*, 17. [CrossRef]
72. Saha, P.; Bairagi, N.; Biswas, M. On the Dynamics of a Discrete Predator-Prey Model. In *Trends in Biomathematics: Modeling, 337 Optimization and Computational Problems*; Mondaini, R., Ed.; Springer: Berlin, Germany, 2018; pp. 219–232.
73. Zhao, M.; Xuan, Z.; Li, C. Dynamics of a discrete-time predator-prey system. In *Advances in Difference Equations 2016*; Springer Open: Berlin/Heidelberg, Germany, 2016.
74. Chou, C.S.; Friedman, A. Introduction to Mathematical Biology, Modeling, Analysis, and Simulations. In *Springer Undergraduate Texts in Mathematics and Technology*; Springer International Publishing: Cham, Switzerland, 2016.
75. Raffoul, Y.N. *Qualitative Theory of Volterra Difference Equations*; Springer Nature Switzerland AG: Cham, Switzerland, 2018; pp. 55–89.
76. Bodine, S.; Lutz, D.A. *Asymptotic Integration of Differential and Difference Equations*; Springer International Publishing: Cham, Switzerland, 2015; pp. 237–282.
77. Honda, T.; Iwata, Y. Operator Theoretic Phenomena of the Markov Operators which are Induced by Stochastic Difference Equations. In *Advances in Difference Equations and Discrete Dynamical Systems. ICDEA 2016*; Springer Proceedings in Mathematics & Statistics; Elaydi, S., Hamaya, Y., Matsunaga, H., Pötzsche, C., Eds.; Springer: Singapore, 2017; Volume 212.
78. Mickens, R.E. *Difference Equations Theory, Applications and Advanced Topics*; CRC Press: Boca Raton, FL, USA, 2015; pp. 83–115.
79. Kitzing, K.; Picard, R.; Siegmund, S.; Trostorff, S.; Waurick, M. A Hilbert Space Approach to Difference Equations. In *Difference Equations, Discrete Dynamical Systems and Applications, ICDEA 2017*; Springer Proceedings in Mathematics & Statistics; Elaydi, S., Pötzsche, C., Sasu, A., Eds.; Springer: Cham, Switzerland, 2019; Volume 287, pp. 285–307.
80. Castro, J.R.; Castillo, O.; Melin, P.; Rodríguez-Díaz, A. Building Fuzzy Inference Systems with a New Interval Type-2 Fuzzy Logic Toolbox. In *Transactions on Computational Science I*; Springer: Berlin/Heidelberg, Germany, 2008.
81. Chai, T.; Draxler, R.R. Root mean square error (RMSE) or mean absolute error (MAE)?– Arguments against avoiding RMSE in the literature. *Geoscientific Model Development.* **2014**, *7*, 1247–1250. [CrossRef]
82. Saeed, W.; Ghazali, R. Chaotic Time Series Forecasting Using Higher Order Neural Networks. *Int. J. Adv. Sci. Eng. Inf. Technol.* **2016**, *6*, 624–629.
83. Martínez-García, J.A.; González-Zapata, A.M.; Rechy-Ramírez, E.J.; Tlelo-Cuautle, E. On the prediction of chaotic time series using neural networks. *Chaos Theory Appl.* **2022**, *4*, 94–103.
84. López-Caraballo, C.H.; Salfate, I.; A Lazzús, J.; Rojas, P.; Rivera, M.; Palma-Chilla, L. Mackey-Glass noisy chaotic time series prediction by a swarm-optimized neural network. *J. Physics: Conf. Ser.* **2016**, *720*, 012002. [CrossRef]

micromachines

MDPI

Article

Development of an Assessment Model for the Effect of the Replacement of Minimal Artificial Ossicles on Hearing in the Inner Ear

Junyi Liang [1], Jiakun Wang [2,3], Wenjuan Yao [2,3,*] and Mianzhi Wang [2,3]

1 Genomic Medicine Institute, Lerner Research Institute, Cleveland Clinic Foundation, Cleveland, OH 44106, USA
2 School of Mechanics and Engineering Science, Shanghai University, Shanghai 200072, China
3 Shanghai Institute of Applied Mathematics and Mechanics, Shanghai 200072, China
* Correspondence: wjyao@shu.edu.cn

Abstract: Due to ethical issues and the nature of the ear, it is difficult to directly perform experimental measurements on living body elements of the human ear. Therefore, a numerical model has been developed to effectively assess the effect of the replacement of artificial ossicles on hearing in the inner ear. A healthy volunteer's right ear was scanned to obtain CT data, which were digitalized through the use of a self-compiling program and coalescent Patran-Nastran software to establish a 3D numerical model of the whole ear, and a frequency response of a healthy human ear was analyzed. The vibration characteristics of the basilar membrane (BM) after total ossicular replacement prosthesis (TORP) implantation were then analyzed. The results show that although the sound conduction function of the middle ear was restored after replacement of the TORP, the sensory sound function of the inner ear was affected. In the low frequency and medium frequency range, hearing loss was 5.2~10.7%. Meanwhile, in the middle–high frequency range, the replacement of a middle ear TORP in response to high sound pressure produced a high acoustic stimulation effect in the inner ear, making the inner ear structures susceptible to fatigue and more prone to fatigue damage compared to the structures in healthy individuals. This developed model is able to assess the effects of surgical operation on the entire hearing system.

Keywords: hearing system; assessing model; a minimal artificial ossicles chain; total ossicular replacement prosthesis; the inner ear

Citation: Liang, J.; Wang, J.; Yao, W.; Wang, M. Development of an Assessment Model for the Effect of the Replacement of Minimal Artificial Ossicles on Hearing in the Inner Ear. *Micromachines* **2023**, *14*, 483. https://doi.org/10.3390/mi14020483

Academic Editor: Isabelle Huynen

Received: 29 January 2023
Revised: 14 February 2023
Accepted: 17 February 2023
Published: 19 February 2023

1. Introduction

Since the beginning of the 21st century, the two journals "*Science*" and "*Nature*" have respectively reported that "The problems of deafness are deeper and more complex, which are even more important than those of blindness" and "Restoration of auditory function remains a lofty, (challenge), but achievable goal for deaf patients and for scientists. Because hearing loss is a multifactorial problem arising from many possible causes, it is unlikely that any single strategy will be an effective cure-all" [1]. Thus, up to now, the investigation of hearing mechanisms in humans is still a medical challenge for mankind [2].

Middle ear lesions can lead to conductive deafness. At present, artificial ossicular replacement is an effective method for the treatment of conductive deafness. Some research has been conducted in this relevant area. Murphy [3] investigated and evaluated the degree of hearing recovery after total ossicular replacement prosthesis (TORP) implantation. Yao et al. studied the effect of artificial ossicular replacements with different materials on postoperative hearing recovery [4]. Kelly et al. [5] investigated the effect of different artificial ossicular quality on sound transmission. Marchese et al. researched the effect of stapes footplate (SF) perforation on hearing improvement in the human ear [6].

Kelly and Fisch [5,7] conducted a study on human ear hearing recovery after TORP implantation using different shapes of artificial ossicular implants. Kelly [5] found that the amplitude of vibrations was more similar to that of a healthy ear when implanted with a Kurz TORP; while Fisch's results showed that the L-shaped Fisch titanium total prosthesis implant was functionally superior to comparable small columned titanium prostheses [7]. Morris et al. [8] studied the effect of different lengths of artificial ossicles on sound transmission. In the ossicular reconstruction, three different lengths of partial ossicular replacement prostheses were used, and the best stapes vibration results in their models were achieved with shorter prostheses that resulted in reduced tension [8]. Vincent et al. used new technology to make the TORP implant in artificial ossicular connection more reliable and stable [9]. Liu et al. researched a new suspension middle ear hearing aid device, including the influence of a suspension oscillator on sound propagation [10]. Huttenbrink et al. investigated improvements in human hearing after TORP implantation based on experiments [11].

The function of the middle ear is sound conduction, which converts the sound signals received by the external ear into mechanical signals and conducts them to the inner ear, thus causing a pressure difference between the vestibular and tympanic steps that causes the basilar membrane (BM) in the inner ear to vibrate. Vibration of the basilar membrane in the inner ear is the key step in the perception of sound. Because of the complexity and delicacy of the inner ear, few scholars have carried out experimental measurements of the BM in the human inner ear. Gundersen et al. experimentally researched the vibration of the BM and the malleus head of the human ear [12]. Stenfelt et al. [13] investigated the vibration characteristics of the BM and the osseous spiral lamina in human cadavers under bone conduction and air conduction stimuli, respectively. Nakajima et al. found that when the artificial prosthesis has a certain contact area with the cochlear window, cochlear window excitation with the prosthesis can produce a pressure similar to normal conditions [14].

The finite element (FE) simulation model compensates for the difficulty of conducting experiments. Lim et al. [15] established a nonlinear active cochlear model to study the steady-state and transient-state response of BM movement. Kim et al. established a finite element (FE) model including the middle ear and the straightened inner ear to further understand the mechanism of bone conduction hearing [16]. Yao et al. developed an analytical model of the BM and a numerical simulation model of the cochlea to discover new vibration modes at low frequencies [17] and performed a time-domain analysis of the cochlear BM [18].

According to previous research reports, numerical simulation is an effective and feasible method for analyzing the biomechanical behavior of hearing in the human ear. It also provides an effective method for analyzing and predicting the outcome of hearing recovery in the human ear after TORP replacement surgery. So far, the study of the vibration characteristics of the spiral BM by simulating the TORP using numerical simulation methods has not been reported to the public. Therefore, based on the CT images obtained by scanning the right ear of healthy volunteers in Zhongshan Hospital, which is affiliated with Fudan University, a three-dimensional numerical model of the whole human ear was developed. In this paper, the influence of TORP replacement on the vibration characteristics of the spiral BM was analyzed in order to provide a numerical simulation platform for clinical surgical treatment.

2. Materials and Methods

2.1. Establishment of an FE Model of the Human Ear

2.1.1. Establishment of the Model

The geometric dimensions of the numerical model were based on images obtained from CT scanning of healthy human ears at Zhongshan Hospital (GE lightspeed VCT 64 slice spiral CT machine; scanning parameters: collimation 0.625 mm, tube rotation time 0.4 s, reconstruction layer thickness 0.625 mm, interval 0.5–0.625 mm, and a voxel size of $1.0 \times 1.0 \times 0.75$ mm^3). CT scanning images were digitized by a self-compiling program and then imported into the Patran software to reconstruct the three-dimensional model of the human ear. The mesh was then divided and the boundary conditions and material parameters were set. Finally, three-dimensional fluid–solid coupling dynamic analysis was carried out using the Nastran software. This study was approved by the Ethics Committee of the Zhongshan Hospital affiliated to Fudan University. Written informed consent was obtained from the patient.

The mesh of the whole human ear finite element (FE) numerical model was divided into the following elements: 1. The gas in the external ear canal: 7581 nodes, 600 six-node pentahedral (Wedge6) elements and 6600 eight-node hexahedral (Hex8) elements, and element properties defined as Fluid. 2. Tympanic membrane (TM): 361 nodes, 30 three-node triangular (Tria3) elements and 330 four-node quadrilateral (Quad4) elements, and element properties defined as 2D-Membrane. 3. TM and malleus connector: 264 nodes, 135 eight-node hexahedral (Hex8) elements, and element properties defined as Solid. 4. The ossicular chain–ligament–tendon: 6254 nodes, 60 eight-node hexahedral (Hex8) elements and 26,567 four-node tetrahedral (Tet4) elements, and element properties defined as Solid. 5. The perilymph of the inner ear: 5872 nodes, 3852 eight-node hexahedral (Hex8) elements, and element properties defined as Fluid. 6. Oval window (vestibular window): 45 nodes, 32 four-node quadrilateral (Quad4) elements, and element properties defined as 2D-Membrane. 7. Round window (fenestra cochleae): 25 nodes, 16 four-node quadrilateral (Quad4) elements, and element properties defined as 2D-Membrane. 8. BM: 505 nodes, 400 four-node quadrilateral (Quad4) elements, and element properties defined as 2D-Membrane. The mesh of the structure of the human ear is shown in Figure 1.

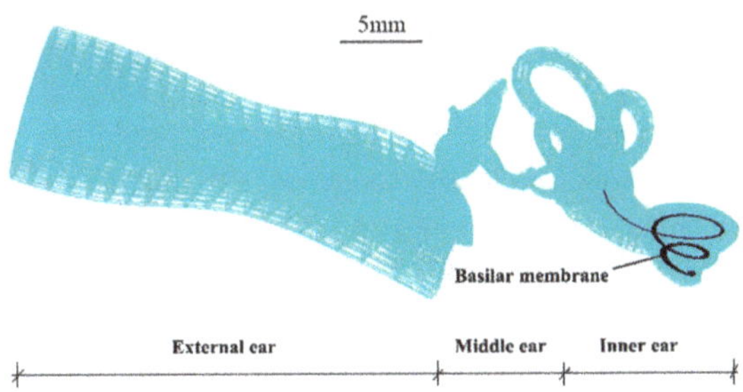

Figure 1. Numerical model of the whole human ear (the BM is shown).

2.1.2. Material Properties

The material properties of the ossicular chain numerical model are shown in Table 1 [19–24], and the material properties of the soft tissue finite element model (FEM) are shown in Table 2 [24,25]. The Poisson's ratio of each part of the middle ear structure is 0.3, the structural damping coefficient is 0.4, the viscosity of the fluid is 0.001 NS/m^2, and the damping coefficient β of the fluid is 0.0001 s.

Table 1. Material properties of the numerical model of the middle ear ossicular chain [19–24].

Structure	Density ($kg{\cdot}m^{-3}$)	Published Data	Young's Modulus (Pa)	Published Data
Malleus head	2.55×10^3	Kirikae et al.	1.41×10^{10}	Herrmann et al.
Malleus neck	4.53×10^3	Kirikae et al.	1.41×10^{10}	Herrmann et al.
Malleus handle	3.70×10^3	Kirikae et al.	1.41×10^{10}	Herrmann et al.
Incus body	2.36×10^3	Kirikae et al.	1.41×10^{10}	Herrmann et al.
Incus short process	5.08×10^3	Kirikae et al.	1.41×10^{10}	Herrmann et al.
Incus long process	2.26×10^3	Kirikae et al.	1.41×10^{10}	Herrmann et al.
Stapes	2.2×10^3	Kirikae et al.	1.41×10^{10}	Herrmann et al.
Incudomalleolar joint	3.2×10^3	Sun et al.	1.41×10^{10}	Sun et al.
Incudostapedial joint	1.2×10^3	Sun et al.	0.6×10^6	Wada et al.
TM (par tensa)	1.2×10^3	Wada et al.	3.5×10^7	2.0×10^7 Bekesy et al.
TM (par flaccida)	1.2×10^3	Wada et al.	1.0×10^7	4.0×10^7 Kirikae et al.

Table 2. Material properties of the numerical model of middle ear soft tissue [24,25].

Soft Tissue	Young's Modulus (MPa)	
	FEM	Published Data
Tympanic annulus ligament	0.6	0.6 (Wada et al.)
Anterior mallear ligament	10	2.1 (Gan et al.)
Lateral mallear ligament	6.7	6.7 (Gan et al.)
Superior mallear ligament	4.9	4.9 (Gan et al.)
Tensor tympani tendon	8.7	7 (Gan et al.), 2.6 (Wada et al.)
Superior incudal ligament	4.9	4.9 (Gan et al.)
Posterior incudal ligament	6.5	6.5 (Gan et al.)
Stapedial tendon	5.2	5.2 (Wada et al.)
Stapedial annulus ligament	0.2	0.2 (Wada et al.)

The material properties of the inner ear structure shown above were obtained from the relevant published references [26–28]. The material properties of each part of the inner ear in the numerical model in this paper are as follows: Oval window: the elastic modulus is E = 0.2 MPa, Poisson's ratio is $\mu = 0.3$, density is $\rho = 1200$ kg/m^3, and the damping coefficient is $\beta = 0.5 \times 10^{-4}$ s. Round window: the elastic modulus is E = 0.35 MPa, Poisson's ratio is $\mu = 0.3$, and the damping coefficient is $\beta = 0.5 \times 10^{-4}$ s. Lymphatic fluid (scala vestibuli, scala tympani, scala media, 3 semicircular canals, and lymphatic fluid in the vestibuli): density is $\rho = 1000$ kg/m^3, sound velocity is C = 1400 m/s, the damping coefficient is $\beta = 1.0 \times 10^{-4}$ s, and viscous damping is D = 0.001 NS/m. BM: As the length of the BM changes, the elastic modulus decreases linearly from 50 MPa at the base of the cochlea to 15 MPa at the middle and then decreases linearly to 3 MPa at the apex. The damping coefficient β varies linearly from 0.2×10^{-3} s at the base to 0.1×10^{-2} s at the apex, with a Poisson's ratio of 0.3.

2.1.3. Boundary Conditions of the Whole Ear FEM

Because the boundary conditions need to be provided for the numerical model's calculations according to the structural characteristics of the human ear and the connection relationship and related characteristics between the structures of the human ear, the boundaries of some tissues of the human ear are set appropriately based on the mechanical principle. The details are as follows:

(1) Application of 80 dB SPL (0.2 Pa), 90 dB SPL (0.632 Pa), and 105 dB SPL (3.56 Pa) surface pressure to the opening surface of the external ear canal or TM to simulate pure tone sound pressure stimulation (100–10,000 Hz);

(2) The positions of soft tissues (tensor tympani, superior mallear ligaments, anterior mallear ligaments, lateral mallear ligament, superior incudal ligament, posterior incudal ligament, stapedial tendon) associated with the temporal bone were defined as the fixed constraint (constrain all displacement and all rotation);

(3) The outer edge of the TM's annular ligament was defined as the hinged constraint (only constrains all displacement, not rotation);

(4) The outer edge of the SF annular ligament was defined as the fixed constraint (constrain all displacement and all rotation);

(5) The outer edge of the oval window and the round window were fixed constraints;

(6) The three edges of the BM (both sides and the base of the cochlea) were considered as hinged constraints (only constrains all displacement, not rotation);

(7) The external ear canal wall and the inner ear bony labyrinth wall were set as the rigid wall;

(8) The TM, SF, and annular ligament were set up as a fluid–solid coupling interface.

The FE numerical model of the human ear with boundary constraints is shown in Figure 2.

Figure 2. Boundary constraints of the whole ear in the FEM.

2.2. Establishment of the TORP Model

In this study, an artificial ossicular TORP (produced by the Medtronic Xomd Company in the USA) was applied for FE analysis. The size of the TORP (363) was as follows: the length was 6.7 mm in total and the diameter of its columella was 0.9 mm. In addition, the undersurface diameter of the round terminal disc was 3.2 mm, and the end face of the columella was a plane. The model and mesh of the TORP were established using the Patran FE software. The mesh was divided into 539 nodes (100 eight-node hexahedral (Hex8) elements and 325 six-node pentahedral (Wedge6) elements) and the element property was defined as Solid. The round terminal disc of the TORP was connected to the center of the TM and its columella's end face was connected to the SF or oval window membrane. Additionally, a 0.5 mm cartilage slice was set between the TM and the undersurface of the round terminal disc of the TORP. The material properties of the titanium in the TORP were as follows [5]: the elastic modulus was 116 GPa, density was 4500 kg·m^{-3}, and the Poisson's ratio was 0.33. The mesh division of the model is shown in Figure 3.

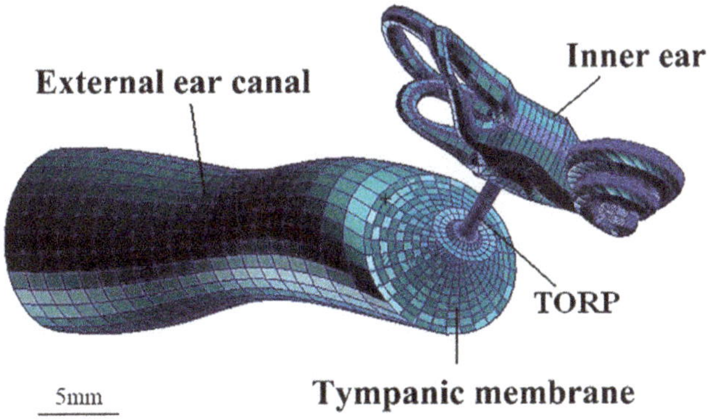

Figure 3. Whole ear FEM after TORP replacement.

3. Results and Discussion

3.1. Verification of the FEM of the Human Ear

3.1.1. Load Acting on the TM

Frequency response analysis was performed by applying a sound pressure of 90 dB SPL (0.632 Pa) and 105 dB SPL (3.56 Pa) to the TM without considering the effect of the external ear canal.

The frequency response curves of the umbo and SF were obtained through numerical simulation analysis and compared to the experimental data of Gan et al. [29,30], as shown in Figures 4 and 5.

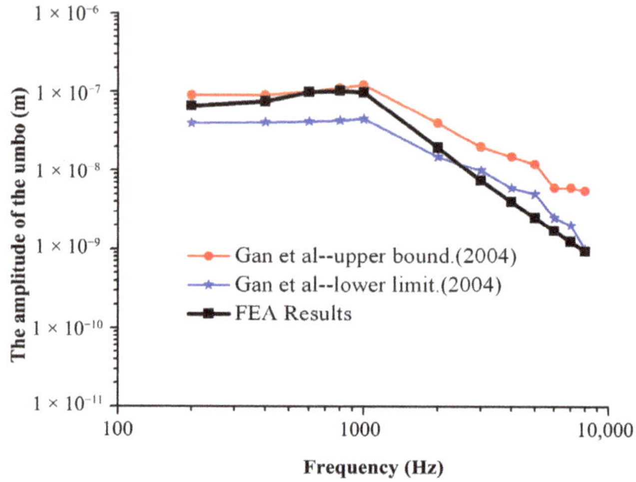

Figure 4. Comparison of umbo's amplitude between FEA and the experiment (90 dB).

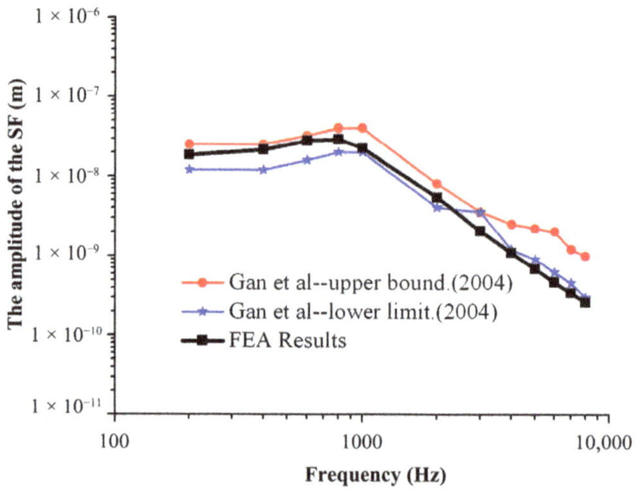

Figure 5. Comparison of the SF's amplitude between FEA and the experiment (90 dB).

As can be seen from Figure 4 in the frequency range of 200–2000 Hz, the amplitude–frequency response curve of the umbo obtained from the numerical simulation of this model is within the upper and lower limits of the curve obtained from the experiment of Gan et al., and the numerical simulation results are consistent with the experimental data. In the frequency range of 2000–10,000 Hz, the numerical simulation results are slightly lower than the experimental results of Gan et al., and the average error is about 18.7%.

It can be seen from Figure 5 that the amplitude–frequency response curve for the SF obtained by numerical simulation with this model is within the upper and lower limits of the curve obtained from the experiment of Gan et al. in the frequency range of 200 Hz to 2000 Hz, and the numerical simulation results are consistent with the experimental data. In the frequency range of 2000 Hz to 10,000 Hz, the numerical simulation results are slightly lower than the experimental data of Gan et al., and the average error is about 16.5%.

3.1.2. Verification of the FEM of the Spiral BM

Since the vibration of external sound through the TM and the ossicular chain drives the vibration of the SF, it is the vibration of the SF that provokes the response of the BM through cochlear lymphatic fluid, and the vibration of the stapes has a direct relationship with the response of the BM. The ratio of the amplitude of the BM to the amplitude of the SF represents the response of the BM caused by the vibration of the SF per unit, which clearly reflects the amplification characteristics and frequency selection characteristics of the BM.

Considering the influence of the spiral BM in the cochlea, a sound pressure of 90 dB SPL was applied to the opening surface of the external ear canal for frequency response analysis, as shown in Figure 2.

The BM data derived from the spiral inner ear numerically simulated by this model, and the experimental data at 12 mm from the base of the BM obtained by Gundersen et al. and Stenfelt et al. [12,13], were compared to some extent, and the BM to SF amplitude ratio–frequency response curves were constructed, as shown in Figure 6.

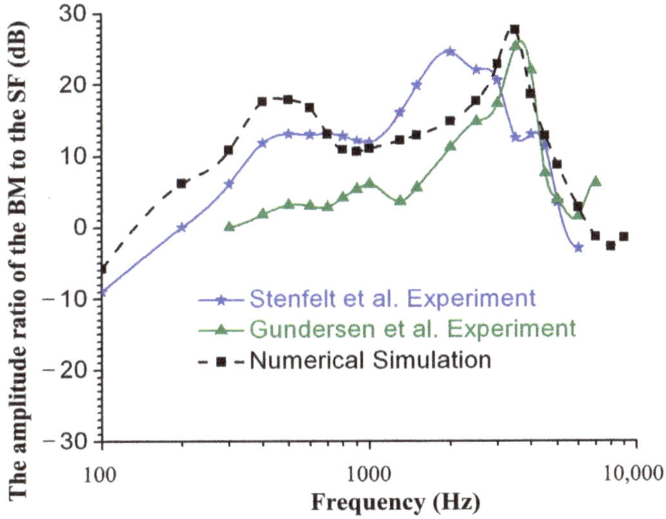

Figure 6. Comparison of the numerical simulation of this model with experimental data (90 dB).

As shown in Figure 6, in the frequency range of 100–1000 Hz, the amplitude ratio–frequency response curve of the BM and SF obtained by numerical simulation with this model is slightly higher than that of Gundersen et al., Stenfelt et al., and other experimental curves, and the average error is about 12%. The peak value of this frequency range appears at 400 Hz, which is between the Gundersen et al. experimental peak and the Stenfelt et al. experimental peak. In the frequency range of 1000–3000 Hz, the amplitude ratio–frequency response curve of the BM to the SF obtained by numerical simulation is between the Gundersen et al. experimental curve and the Stenfelt et al. experimental curve. In the frequency range of 3000–8000 Hz, the amplitude ratio–frequency response curve of the BM to the SF obtained by numerical simulation is close to the experimental curve of Gundersen et al., and the curve trend is consistent with the peak value of this frequency range. The peak value appears at about 3500 Hz, and the average error in this interval is about 6%.

In Figure 6, comparing these curves, the amplitude ratio–frequency response curves of the BM to the SF measured by Gundersen et al. and Stenfelt et al. have two peaks, and those derived from numerical simulations of the spiral cochlea in the model also have two peaks over the entire frequency range (100–8000 Hz), one at approximately 400 Hz and the other at 3500 Hz, which occur in the low and mid-frequency ranges, respectively. This shows that the vibration characteristics of the BM obtained from FE method simulation of the spiral inner ear (including the spiral BM) are closer to those obtained from inner ear experiments, that is, FE simulation of the spiral inner ear may also accurately reflect the inner ear characteristics of the actual human inner ear, thus providing good theoretical guidance and help for clinical research in this area. Figure 7 more clearly shows the error in the results of the model calculation in this paper and experimental results in terms of frequency. In the frequency range of 1000–3000 Hz, the average error was minimal (only 1.6%). In the frequency range of 100–1000 Hz, the average error was 12%. In the frequency range of 1000–10,000 Hz, the average error was 6%. The overall average error was 8%.

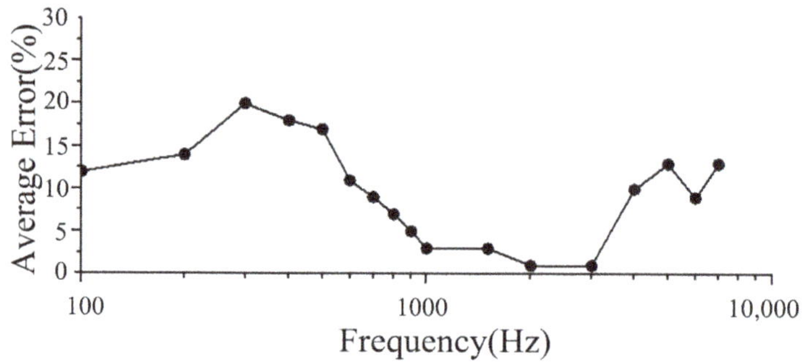

Figure 7. The curve of error between calculations in this model and the experiment.

In summary, the SF and umbo data obtained from the human ear FEM simulations in this paper are close to the experimental data in terms of amplitude and trend, respectively, thus verifying the correctness of this model. In addition, the simulation of nodal vibrations on the spiral BM and a comparison with experimental data were used to verify the correctness of the spiral cochlea (contains the spiral BM).

3.2. Numerical Analysis of the Vibration Characteristics of the BM after Replacement with a TORP

In this paper, the frequency response analysis of the FEM of the whole hearing system after replacement with a titanium TORP in the frequency range of 100–10,000 Hz was performed by applying a sound pressure of 90 dB SPL over the opening surface of the external ear canal while considering the effect of the spiral BM built into the cochlea, as shown in Figure 8.

Figure 8. FEM of the whole ear after replacement with a TORP (showing the spiral BM).

Since the vibration of external sound through the TM and the ossicular chain drives the vibration of the SF, it is the vibration of the SF that provokes the response of the BM through cochlear lymphatic fluid, and the vibration of the stapes has a direct relationship with the response of the BM. In order to clearly reflect the amplification characteristics and frequency selection characteristics of the spiral BM after replacement with a titanium TORP, the ratio of BM amplitude to the amplitude of the SF is used in this paper, which indicates the BM response caused by the vibration of the SF.

The amplitude–frequency response curves of the nodes at the center of the SF and at 12 mm from the base of the cochlea on the BM were obtained by simulation analysis, and the relationship between the ratio of the BM amplitude and the stapes amplitude at 12 mm from the base of the cochlea on the BM and frequency was calculated. The BM data were also compared to those of the healthy human ear, as shown in Figure 9.

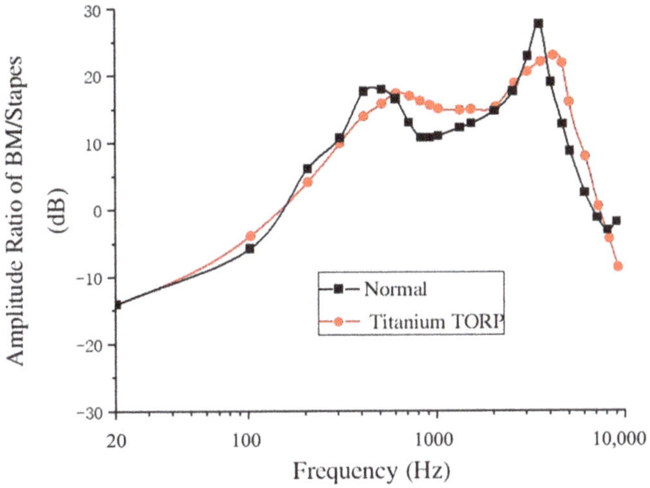

Figure 9. Comparison of the characteristics of the BM after TORP replacement with those of normal conditions (90 dB).

From Figure 9, it can be seen that, in the frequency range of 20–10,000 Hz, the amplitude ratio–frequency response curve of the BM and the SF after replacement with a TORP is closer to the vibration characteristics of the BM of the healthy ear in terms of amplitude and overall trend. There are also two peaks at 600 Hz and 4000 Hz, and the peak of the amplitude ratio–frequency response curve of the BM to the SF after replacement with a TORP is somewhat more lagged than in healthy persons, where the two peaks occur at approximately 400 Hz and 3500 Hz.

In the frequency range of 100–600 Hz, the amplitude ratio of the BM to the SF increased gradually with frequency after the TORP replacement, and the curve was slightly lower than that of healthy persons. In this frequency range, the difference between the amplitude ratio of the BM and the SF was about 1.03 dB to 3.72 dB, with an average relative error of about 5.2% compared with that of healthy persons.

In the frequency range of 600–2000 Hz, the BM to SF amplitude ratio after TORP replacement tended to decrease slowly with frequency, and the curve was somewhat higher than that of healthy persons. In this frequency range, the difference between the BM and SF amplitude ratios compared to healthy persons was about 0.67 dB to 5.33 dB, with an average relative error of about 8.3%.

In the frequency range of 2000–4000 Hz, the BM to SF amplitude ratio after TORP replacement tended to increase slowly with frequency, and the curve was somewhat lower than that of healthy persons. In this frequency range, the difference between the amplitude ratio of the BM and the SF was about 2.01 dB to 5.54 dB compared with that of healthy persons, with an average relative error of about 10.7%.

In the frequency range of 4000–7000 Hz, the amplitude ratio of the BM to the SF after TORP replacement decreased linearly with frequency, and the curve was higher than that of healthy persons. In this frequency range, compared with the amplitude ratio of the BM to the SF in healthy persons, the difference was about 1.89–9.22 dB, and the average relative error was about 9.1%.

In the frequency range of 7000–10,000 Hz, the amplitude ratio of the BM to the SF after TORP replacement was lower than that of healthy persons. At this frequency range, the amplitude ratio of the BM to the SF was about 1.32–6.43 dB different from that of healthy persons.

A fluid–solid coupling cloud diagram of the whole hearing system (ossicular chain, a displacement diagram of the BM, and a pressure diagram of the external ear canal air and inner ear perilymph) is shown in Figures 10 and 11.

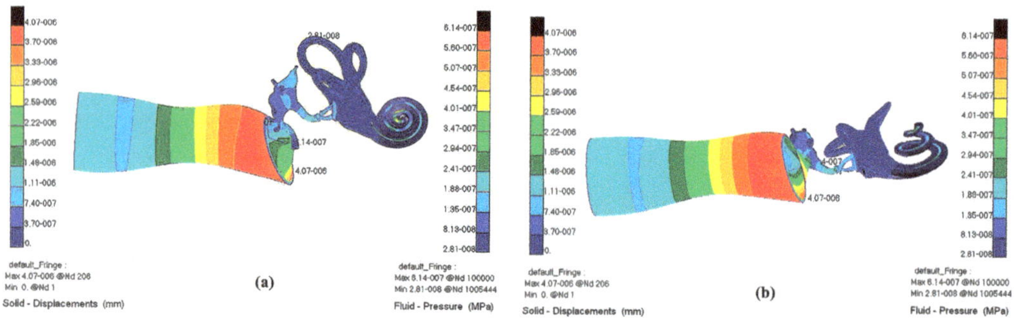

Figure 10. Fluid–solid coupling cloud diagram of the whole hearing system. (**a**) Visual angle 1; (**b**) Visual angle 2.

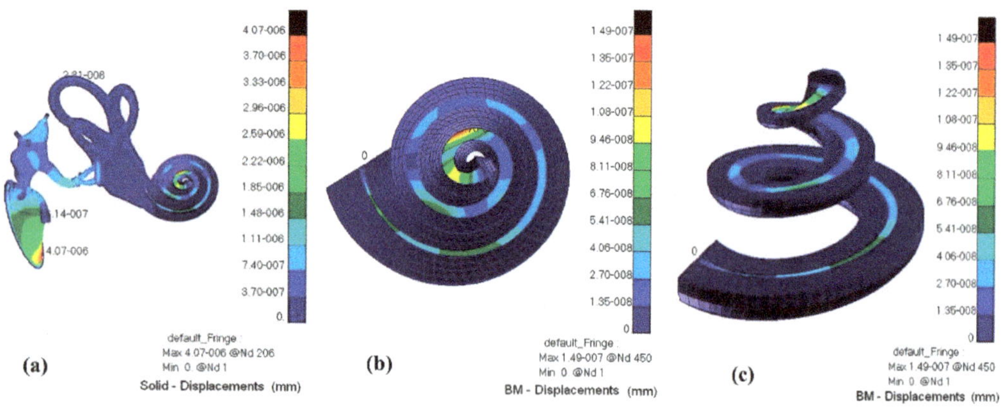

Figure 11. Displacement cloud diagram of the middle ear ossicular chain and BM. (**a**) Ossicular chain in middle ear; (**b**) BM's displacement diagram (visual angle 1); (**c**) BM's displacement diagram (visual angle 2).

Through numerical simulation with this model, the displacement cloud diagram of the BM after TORP replacement at different frequencies (500 Hz, 1000 Hz, 4000 Hz) is compared with that of healthy human ears, as shown in Figures 12 and 13 below.

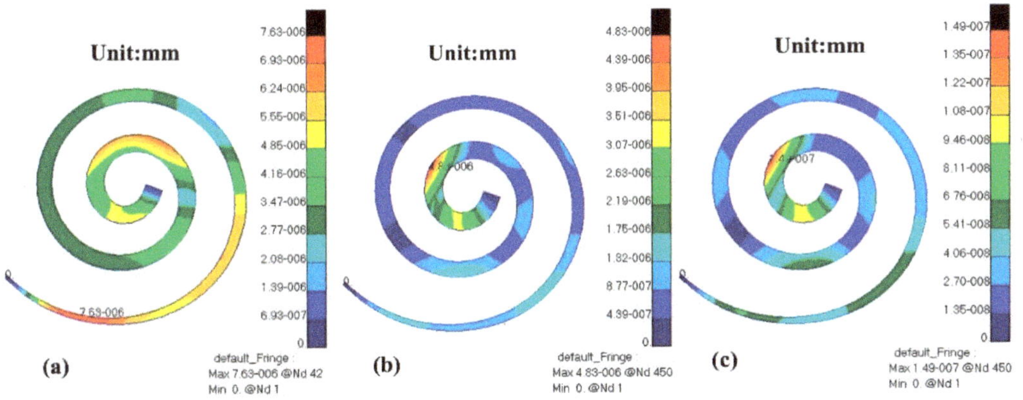

Figure 12. BM displacement cloud diagram of a healthy ear at different frequencies. (**a**) 500 Hz; (**b**) 1000 Hz; (**c**) 4000 Hz.

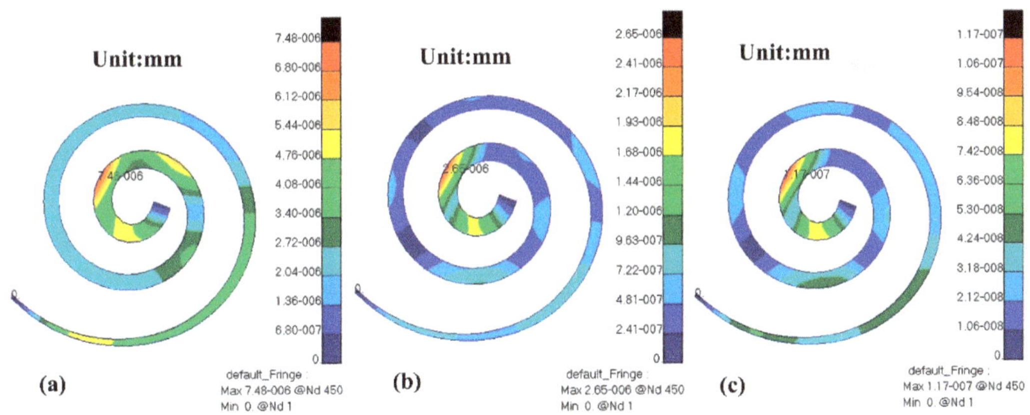

Figure 13. Displacement cloud diagram of a human ear BM at different frequencies after TORP replacement. (**a**) 500 Hz; (**b**) 1000 Hz; (**c**) 4000 Hz.

From the comparison of Figures 12 and 13, the following can be determined:

At 500 Hz, the displacement of the BM in the healthy human ear ranged from 0 to 7.63×10^{-6} mm, and the maximum value of this displacement was observed near the base of the cochlea. The displacement of the BM in the human ear after TORP replacement ranged from 0 to 7.48×10^{-6} mm, and the maximum displacement value was observed near the apex of the cochlea. BM displacement after TORP was smaller than in the healthy human ear, with the maximum displacement occurring at different positions; however, the values of the displacements were in the same order of magnitude.

At 1000 Hz, the displacement of the BM in the healthy human ear ranged from 0 to 4.83×10^{-6} mm, and the maximum value of displacement at this time was close to the apex of the cochlea. Displacement of the BM in the human ear after TORP replacement ranged from 0 to 2.65×10^{-6} mm, and the maximum displacement value was observed close to the apex of the cochlea. The position of the maximum displacement at this frequency is closer to that of the BM in the healthy human ear, though the values are somewhat different. Compared to the healthy human ear, BM displacement after TORP replacement decreased.

At 4000 Hz, the displacement of the BM in the healthy human ear ranged from 0 to 1.49×10^{-7} mm, and the maximum displacement value was observed in the first lap of the upper cochlea. The displacement of the BM in the human ear after TORP replacement ranged from 0 to 1.17×10^{-7} mm, and the position of maximum displacement at this frequency was basically the same as it was in the healthy human ear. BM displacement after TORP replacement was smaller than it was in the healthy human ear, though the values of the displacements were in the same order of magnitude.

4. Conclusions

After the TORP replacement based on the whole hearing system, the vibration characteristics of the spiral BM changed at low and high frequencies. Compared to the healthy human ear, the position of the peak amplitude value in the BM after TORP replacement was somewhat more lagged than in healthy persons based on the amplitude–frequency response curve. In the low frequency range, BM amplitude after TORP replacement was lower than in the healthy human ear, and the decrease increased as frequency increased, with an average relative difference between the two of 5.2%. In the middle frequency range, BM amplitude after TORP replacement increased slowly with increasing frequency, amplitude was lower than in the healthy human ear, and the average relative difference between the two was 10.7%. In the middle–high frequency range, BM amplitude after TORP replacement is higher than in the healthy human ear, and the average relative difference between the two was 9.1%. In the high frequency range, BM amplitude after TORP replacement was lower than in the healthy human ear, and the average relative difference between the two was 8.6%.

In conclusion, TORP replacement in the middle ear also has some infaust effects on the acoustic sensing function of the inner ear, especially in the low frequency and medium frequency range. Meanwhile, in the middle–high frequency range, TORP replacement in the middle ear in response to high sound pressure produces a high acoustic stimulation effect on the inner ear, making the inner ear structures susceptible to fatigue and more prone to fatigue damage.

The numerical model in this paper can not only study the level of sound conduction recovery in the human middle ear, but also predict the effect of clinical surgery involving TORP replacement on the whole hearing system.

Author Contributions: Conceptualization, J.L. and W.Y.; methodology, J.L.; software, J.W.; validation, J.L., J.W. and W.Y.; formal analysis, M.W.; writing—original draft preparation, J.W.; writing—review and editing, J.L.; funding acquisition, W.Y. All authors have read and agreed to the published version of the manuscript.

Funding: This research was funded by the Key projects of the National Natural Science Foundation of China, grant number 11932010.

Data Availability Statement: Data are contained within the article.

Acknowledgments: We would like to thank Zhongshan Hospital, which is affiliated with Fudan University, for providing the fresh right ear specimen and the Shanghai Synchrotron Radiation Facility for CT scanning.

Conflicts of Interest: The authors declare no conflict of interest.

References

1. Géléoc, G.S.; Holt, J.R. Sound strategies for hearing restoration. *Science* **2014**, *344*, 1241062. [CrossRef] [PubMed]
2. Jia, S.; He, D.Z. Motility-associated hair-bundle motion in mammalian outer hair cells. *Nat. Neurosci.* **2005**, *8*, 1028–1034. [CrossRef] [PubMed]
3. Murphy, T.P. Hearing Results in Pediatric Patients with Chronic Otitis Media After Ossicular Reconstruction with Partial Ossicular Replacement Prostheses and Total Ossicular Replacement Prostheses. *Laryngoscope* **2000**, *110*, 536–544. [CrossRef] [PubMed]
4. Yao, W.; Guo, C.; Luo, X. Study on effects of partial ossicular replacement prostheses with different materials on hearing restoration. *J. Mater. Sci. Mater. Med.* **2013**, *24*, 515–522. [CrossRef]
5. Kelly, D.J.; Prendergast, P.J.; Blayney, A.W. The Effect of Prosthesis Design on Vibration of the Reconstructed Ossicular Chain: A Comparative Finite Element Analysis of Four Prosthese. *Otol. Neurotol.* **2003**, *24*, 11–19. [CrossRef]

6. Marchese, M.R.; Cianfrone, F.; Passali, G.C.; Paludetti, G. Hearing results after stapedotomy: Role of the prosthesis diameter. *Audiol. Neurootol.* **2007**, *12*, 221–225. [CrossRef]
7. Fisch, U.; May, J.; Linder, T.; Naumann, L.C. A New L-Shaped Titanium Prosthesis for Total Reconstruction of the Ossicular Chain. *Otol. Neurotol.* **2004**, *25*, 891–902. [CrossRef] [PubMed]
8. Morris, D.P.; Bance, M.; Vanwijhe, R.G.; Kiefte, M.; Smith, R. Optimum Tension for Partial Ossicular Replacement Prosthesis Reconstruction in the Human Middle Ear. *Laryngoscope* **2004**, *114*, 305–308. [CrossRef] [PubMed]
9. Vincent, R.; Sperling, N.M.; Oates, J.; Osborne, J. Ossiculoplasty with Intact Stapes and Absent Malleus: The Silastic Banding Technique. *Otol. Neurotol.* **2005**, *26*, 846–852. [CrossRef]
10. Liu, H.; Xue, L.; Yang, J. Modeling the effect of cochlear windows activity on reverse stimulation under the role of physiological third windows. *Appl. Acoust.* **2020**, *169*, 107473. [CrossRef]
11. Huttenbrink, K.B.; Beutner, D.; Zahnert, T. Clinical Results with an Active Middle Ear Implant in the Oval Window. *Adv. Otorhinolaryngol.* **2010**, *69*, 27–31. [PubMed]
12. Gundersen, T.; Skarstein, O.; Sikkeland, T. A study of the vibration of the basilar membrane in human temporal bone preparations by the use of the mossbauer effect. *Acta Otolaryngol.* **1978**, *86*, 225–232. [CrossRef] [PubMed]
13. Stenfelt, S.; Puria, S.; Hato, N.; Goode, R.L. Basilar membrane and osseous spiral lamina motion in human cadavers with air and bone conduction stimuli. *Hear. Res.* **2003**, *181*, 131–143. [CrossRef]
14. Nakajima, H.H.; Merchant, S.N.; Rosowski, J.J. Performance considerations of prosthetic actuators for round-window stimulation. *Hear. Res.* **2010**, *263*, 114–119. [CrossRef] [PubMed]
15. Lim, K.M.; Steele, C.R. Response suppression and transient behavior in a nonlinear active cochlear model with feed-forward. *Int. J. Solids Struct.* **2003**, *40*, 5097–5107. [CrossRef]
16. Kim, N.; Homma, K.; Puria, S. Inertial Bone Conduction: Symmetric and Anti-Symmetric Components. *J. Assoc. Res. Otolaryngol.* **2011**, *12*, 261–279. [CrossRef] [PubMed]
17. Yao, W.; Liang, J.; Ren, L.; Ma, J.; Zhao, Z.; Wang, J.; Xie, Y.; Dai, P.; Zhang, T. Revealing the actions of the human cochlear basilar membrane at low frequency. *Commun. Nonlinear. Sci. Numer. Simul.* **2021**, *104*, 106043. [CrossRef]
18. Yao, W.; Zhao, Z.; Wang, J.; Duan, M. Time-domain analysis of a three-dimensional numerical model of the human spiral cochlea at medium intensity. *Comput. Biol. Med.* **2021**, *136*, 104756. [CrossRef]
19. Wada, H.; Metoki, T.; Kobayashi, T. Analysis of dynamic behavior of human middle ear using a finite-dement method. *J. Acoust. Soc. Am.* **1992**, *92*, 3157–3168. [CrossRef]
20. Herrmann, G.; Liebowitz, H. Mechanics of bone fractures. In *Fracture: An Advanced Treatise*; Academic Press: New York, NY, USA, 1972; pp. 772–840.
21. Kirikae, I. *The Structure and Function of the Middle Ear*; University of Tokyo Press: Tokyo, Japan, 1960.
22. Békésy, G.V. *Experiments in Hearing*; McGraw-Hill: New York, NY, USA, 1960.
23. Sun, Q.; Gan, R.Z.; Chang, H.K.; Dormer, K.J. Computer integrated finite element modeling of human middle ear. *Biomech. Model. Mechanobiol.* **2002**, *1*, 109–122. [CrossRef]
24. Wada, H.; Koike, T.; Kobayashi, T. Three-dimensional finite element method (FEM) analysis of the human middle ear. In *Research and Otosurgery: Proceedings of the International Workshop on Middle Ear Mechanics in Research and Otosurgery*; Dresden University Press: Dresden, Germany, 1997; pp. 76–80.
25. Gan, R.Z.; Sun, Q.; Dyer, R.K.; Chang, K.H.; Dormer, K.J. Three dimensional modeling of middle ear biomechanics and its application. *Otol. Neurotol.* **2002**, *23*, 271–280. [CrossRef] [PubMed]
26. Gan, R.Z.; Reeves, B.P.; Wang, X. Modeling of Sound Transmission from Ear Canal to Cochlea. *Ann. Biomed. Eng.* **2007**, *35*, 2180–2195. [CrossRef] [PubMed]
27. Zhang, X.; Gan, R.Z. A Comprehensive Model of Human Ear for Analysis of Implantable Hearing Devices. *IEEE Trans. Biomed. Eng.* **2011**, *58*, 3024–3027. [CrossRef] [PubMed]
28. Naidu, R.C.; Mountain, D.C. Longitudinal coupling within the basilar membrane, recticular liminae. In *Recent Developments in Auditory Mechanics*; World Scientific: Teaneck, NJ, USA, 2000; pp. 123–129.
29. Gan, R.Z.; Feng, B.; Sun, Q. Three-Dimensional Finite Element Modeling of Human Ear for Sound Transmission. *Ann. Biomed. Eng.* **2004**, *32*, 847–859. [CrossRef]
30. Gan, R.Z.; Wood, M.W.; Dormer, K.J. Human middle ear transfer function measured by double laser interferometry system. *Otol. Neurotol.* **2004**, *25*, 423–435. [CrossRef]

micromachines

MDPI

Article

Dental Lesion Segmentation Using an Improved ICNet Network with Attention

Tian Ma [1], Xinlei Zhou [1,*], Jiayi Yang [1], Boyang Meng [1], Jiali Qian [1], Jiehui Zhang [1] and Gang Ge [2,*]

[1] College of Computer Science and Technology, Xi'an University of Science and Technology, Xi'an 710054, China
[2] Department of Electrical and Computer Engineering, National University of Singapore, Singapore 117583, Singapore
* Correspondence: zxl10261@163.com (X.Z.); ggeup@nus.edu.sg (G.G.)

Abstract: Precise segmentation of tooth lesions is critical to creation of an intelligent tooth lesion detection system. As a solution to the problem that tooth lesions are similar to normal tooth tissues and difficult to segment, an improved segmentation method of the image cascade network (ICNet) network is proposed to segment various lesion types, such as calculus, gingivitis, and tartar. First, the ICNet network model is used to achieve real-time segmentation of lesions. Second, the Convolutional Block Attention Module (CBAM) is integrated into the ICNet network structure, and large-size convolutions in the spatial attention module are replaced with layered dilated convolutions to enhance the relevant features while suppressing useless features and solve the problem of inaccurate lesion segmentations. Finally, part of the convolution in the network model is replaced with an asymmetric convolution to reduce the calculations added by the attention module. Experimental results show that compared with Fully Convolutional Networks (FCN), U-Net, SegNet, and other segmentation algorithms, our method has a significant improvement in the segmentation effect, and the image processing frequency is higher, which satisfies the real-time requirements of tooth lesion segmentation accuracy.

Keywords: attention model; asymmetric convolution; ICNet; tooth lesions

Citation: Ma, T.; Zhou, X.; Yang, J.; Meng, B.; Qian, J.; Zhang, J.; Ge, G. Dental Lesion Segmentation Using an Improved ICNet Network with Attention. *Micromachines* **2022**, *13*, 1920. https://doi.org/10.3390/mi13111920

Academic Editor: Arman Roohi

Received: 16 October 2022
Accepted: 4 November 2022
Published: 7 November 2022

1. Introduction

Continuous advancements in the field of computer vision have propelled online intelligent diagnosis and treatment system research. As human living conditions have improved, the interest in dental lesions has increased, and the impact of the epidemic has made hospital diagnoses inconvenient. The following problems occur in clinical settings: (1) Some lesions are similar to the tooth structure, and doctors are prone to miss or misdiagnose the problem. (2) With an increased number of consultations, reviewing numerous films has increased the workload for doctors, resulting in slower review times and a lack of timely feedback to patients. Because of an unequal distribution of medical resources, patients in remote areas lack access to in-depth treatment options. As a result, the online clinic industry continues to emerge to meet the additional needs. The dental lesion identification system can play a role in pre-diagnosis and auxiliary diagnoses so that patients with less severe problems can save time in seeing a doctor, and patients with severe diseases can be diagnosed thoroughly and can obtain their dental lesion diagnosis results anytime and anywhere. This reduces the number of on-site diagnoses necessary during the epidemic and prevents spreading of the COVID-19 virus. The study of real-time segmentation algorithms for dental lesions has become the key to developing intelligent dental lesion detection systems.

Dental diagnosis technology primarily enhances or segments X-ray films and optical coherence tomography (OCT) images to assist doctors in their diagnoses. Lee et al. [1] proposed vertical intensity transform function (VIFT) to solve the problem of lessening the illumination-based differences in tooth grayscales, then used the K-means algorithm and

Markov random field to specify the detection range and finally segmented the candidate block of dental calculus. However, the accuracy of these traditional methods for specific dental diagnosis tasks still has room for improvement, and for a single image with multiple lesions, traditional methods cannot identify and segment multiple categories.

At present, many scholars are also trying to use deep learning methods for dental diagnoses. Kreis [2] used Convolutional Neural Network (CNN) to detect periodontal bone loss (PBL) on panoramic dental X-rays. Casalegno et al. [3] used near-infrared light transmission (NILT) images for caries segmentation. Jae-Hong Lee [4] evaluated the effectiveness of the deep CNN algorithm in detecting and diagnosing caries on X-rays of the periapical period. Yu [5] and others also evaluated the performance of the CNN in the classification of bones by lateral head measurement. Recently, Wen et al. [6] used deep learning methods to detect tooth lesions. They built a multitask network structure, and the model was composed of three subnets: FNet (feature extraction subnet), LNet (location subnet), and CNet (classification subnet). This work mainly detected dental calculus, gingivitis, and soft deposits. They identified calculus and gingivitis with different color candidate frames, while for soft deposits, only an image-level classification was performed. The problem with this work is that the shape of the tooth lesion is irregular. The rectangular labeling method learns the characteristics of the surrounding normal tissues by using the network, and the detection result range is larger than the actual pixel range of the lesion. In addition, the LNet subnet in the network model is a two-stage positioning network, so the real-time nature of the model is a problem. Different from previous work, we used an intraoral camera to collect RGB images of dental lesions and used deep learning methods to segment multiple lesions at the pixel level to more accurately detect the scope of the lesions.

Many diagnostic methods are based on currently popular 3D imaging techniques such as CT, 3DMD, and others [7]. Therefore, 3D point cloud data segmentation is also one of the hot research topics. Karatas et al. introduce basic 3D image segmentation methods and summarize the current status of 3D imaging techniques and evaluate their application in orthodontics. To address the sparsity of point cloud data, Graham et al. proposed new sparse convolutional operations SSCNs [8] to handle sparse data more efficiently. Liu et al. proposed a convolutional BEACon network [9] with embedded attentional boundaries for point cloud instance segmentation, which combines geometry and color into attentional weights based on how humans perceive geometry and color for object recognition motives. SMU-Net [10] uses saliency mapping to guide the primary and secondary networks to learn foreground saliency and background saliency representations, respectively, to obtain good segmentation results, but its effectiveness for edge segmentation of fuzzy lesions on small-scale datasets needs to be improved.

Deep learning uses end-to-end training to predict complex models and is able to accomplish lesion segmentation in complex scenarios. Long et al. [11] proposed a fully convolutional network based on CNN, which, for the first time, achieved a pixel-level classification. They cleverly used convolutional layers to replace fully connected layers that contained complex calculations and used deconvolutions to restore the original sizes after convolutions. Subsequently, encoder–decoder was widely used in semantic segmentation. The proposed SegNet [12] network is similar to FCN, with the difference that the location of the maximum value is recorded during the maximum pooling operation in the decoding operation, and then a nonlinear upsampling is achieved by the corresponding pooling index during decoding. A sparse feature map is obtained after upsampling, and then a dense feature map is obtained by ordinary convolution, and the upsampling is repeated. This reduces the amount of computation in the encoding phase. The U-Net [13] proposed by Ronneberger et al. is an improvement on the encoder–decoder architecture, which by connecting the corresponding layers of the encoder and decoder, the low-level and high-level features are merged to bridge the gap.

In the actual segmentation scenario, we have to pursue real-time performance while ensuring accuracy. ENet [14] believes that the decoding structure is only used for the output of upsampling coding and is only used for fine-tuning the edge details, so it does not need to be particularly deep. In addition, the full convolution process is very time-consuming, so ENet only uses one layer of full convolution and makes use of fewer parameters and obtains a faster speed. However, ENet guarantees real-time performance while giving up a certain accuracy rate, resulting in lower segmentation accuracy. ICNet [15] uses the pyramid pooling module of PSPNet [16] to fuse multiscale context information and divide the network structure into three branches: low resolution, medium resolution and high resolution. It uses low resolution to complete the semantic segmentation, and a high-resolution strategy to refine the segmentation results improves the model's segmentation accuracy. In addition, its use of cascading labels to guide the training of each branch speeds up the model's convergence and prediction and improves real-time performance.

The above segmentation network has good performance in the segmentation of remote sensing images [17], street view images [18], and lesion images [19], but it does not have an application in dental lesions thus far. We found that there is oversegmentation and under-segmentation when testing with segmentation algorithms such as FCN, SegNet, ENet, etc. The analysis believes that tooth lesions and normal tooth tissues are similar in texture and color and have similar features, which makes it difficult to segment tooth lesions and their edges correctly. In addition, the segmentation of dental lesions needs to ensure accuracy in a reasonable time, thus meeting the needs of the diagnostic equipment and massive amounts of data. In this article, we propose to integrate the attention mechanism into the ICNet network to solve the above two problems. The salient points and contributions of this paper are as follows:

1. The self-built dental lesion dataset included four types of lesions: calculus, gingivitis, tartar, and worn surfaces and was preprocessed with the ACE color equalization algorithm for overexposed images caused by light sources.
2. The lightweight Convolutional Block Attention Module (CBAM) attention module is integrated into the low and middle branches of the image cascade network (ICNet) network so that the high-resolution branches can better guide the features of the low and middle branches, and the large-size convolution of spatial attention uses stacked hollow volumes for product replacement.
3. The regular convolution in the low- and medium-resolution branches are replaced with asymmetric convolutions to reduce the computational effort.

2. Related Work

In this section, we introduce the semantic segmentation architecture and model related to our method. These architectures and models are widely used in image segmentation tasks.

2.1. Encoder–Decoder Network

Our segmentation network is based on the encoder-decoder network structure. In 2015, the FCN [11] proposed by Long et al. used a convolution layer to replace the complete convolution layer of the entire network and used the deconvolutional layer for upsampling to restore the segmentation results. This fully convolutional network is called an encoder–decoder network [20–22]. Since then, most image segmentation networks have adopted the form of codec networks. Based on FCN, U-Net [13] built a more complex decoder, adding compensation at the corresponding level to compensate for local information. SegNet [12] is a further extension of U-Net, which implements the maximum pixel pool operation in the encoder model and reduces the amount of calculation in the decoding stage. ICNet [15] first changes the input image size to one-half and one-quarter of the original image and combines the original image to form the input image, which is input into the low-, medium-, and high-resolution branches. There are many network layers at the low and medium resolutions, but the image resolution is low, which saves calculation time. Although the high-resolution branch has a large image, the number of input network layers is small, resulting in a relatively small time overhead, thus achieving the real-time goal. The feature maps extracted from each branch are fused through the CFF [15] module, and finally, the segmentation results are obtained through decoding.

2.2. Attention Mechanism

A limitation of CNN is that it is difficult to effectively learn the global information, and the segmentation of the details of an image is not perfect. In recent years, researchers have integrated feature fusion and attention mechanisms [22–25] into network models to increase the network's ability to learn global features to better segment the details. PspNet is proposed through the pyramid pooling module, which aggregates contextual information based on different regions and extends pixel-level features to a specially designed global pyramid pooling. Local and global clues work together to make the final prediction more reliable. In recent years, a large number of studies have proven that introducing the attention module into the network model can effectively improve performance. SA-UNet [26] introduced a spatial attention module SE [27], which can infer attention maps along the spatial dimension and multiply the attention maps with the input feature maps for adaptive feature refinement. The effect of this work on retinal vessel segmentation exceeds the original U-Net of Ronneberger et al. [24] proposed an attention gate structure. The AG module is connected to the end of each hop connection, and the attention mechanism is implemented for the extracted features. In our proposed model, we use the convolution block attention module CBAM, which takes the input image and applies the attention sequence to the channel and then applies it to the spatial dimension. The result of CBAM is a weighted feature map that takes into account the channels and spatial regions of the input image.

2.3. Multiple Forms of Convolution

Many studies have proposed various forms of convolutions to improve the performance of the segmentation network. Szegedy et al. [28] first proposed the concept of a 1×1 convolution, and its main function was to reduce the dimensionality and save calculation costs. The 1×1 convolution is not different from a conventional convolution in terms of the convolution method; the main difference lies in the application scenario and function. They also proposed an asymmetric convolution; that is, the convolution of n*n can be replaced by the convolution of $1 \times n$ followed by $n \times 1$ so that the effect obtained is the same as that of a conventional convolution and the calculation amounts are reduced. A depth separable convolution [29] is an innovation based on a 1×1 convolution, and it includes two parts, a deep convolution and 1×1 convolution. The purpose of a convolution is to convolve each of the inputs separately using a convolution kernel to convolve it; that is, the channels are separated and then combined. Koltun et al. [30] first used a dilated convolution for image segmentation. A dilated convolution is the process of expanding the convolution kernel by adding some spaces between the elements of the convolution kernel, which increases the receptive field without increasing the number of parameters. The DeepLabV2 algorithm proposed by Papandreou et al. [31] uses a dilated convolution to extract features. Many studies have used a dilated convolution [31,32] to replace a conventional convolution.

3. Methods

3.1. Adding the CBAM Attention Module to Low- and Medium-Resolution Branches

The network framework in this paper is mainly an ICNet network, which consists of three network branches: low-, medium-, and high-resolution branches, as shown in Figure 1. In our designed network, the low-resolution network is designed according to the first 11 convolutional blocks of Resnet50, using dilated convolution to enhance the feature perception field and asymmetric convolution to reduce the computational effort. The medium-resolution branch is designed according to the first 5 convolution blocks of Resnet50, using asymmetric convolution in the last two convolutions, and both branch network heads include a 7×7 convolution and maximum average pooling. The high-resolution branch is designed as a lightweight network, which consists of three convolution blocks, including 3 × 3 convolution, Relu activation function, BN layer, 1 × 1 convolution, and Relu activation function, in that order.

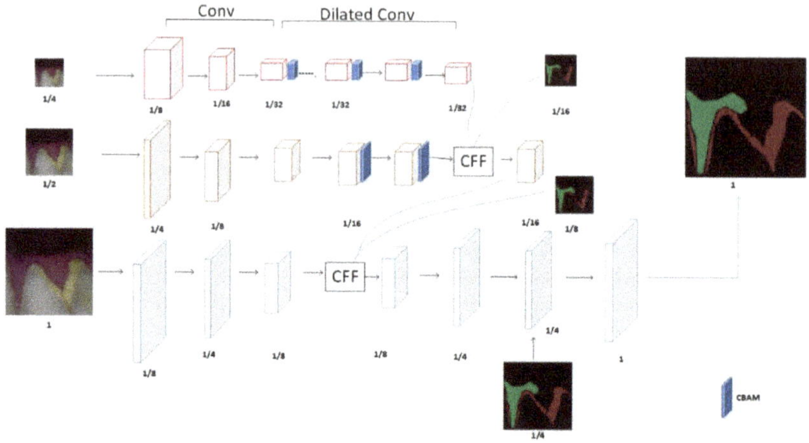

Figure 1. Improved ICNet network structure.

There is a morphological similarity between dental lesions and normal teeth, and the edges are difficult to distinguish. In addition, although ICNet retains most of the semantic information when performing feature extraction in low-resolution branches, there is a loss of detail and edge information, which results in an unsatisfactory segmentation accuracy. In this paper, the CBAM module is added to the low-resolution branch, as shown in Figure 1. Through the serial combination of channel and spatial attention, important features can be enhanced, and unimportant features can be suppressed, thereby improving the performance of the network and improving the model's ability to learn details.

CBAM applies attention to both channel and spatial dimensions, as shown in Figure 2. CBAM, like the SE module, can be embedded in most current mainstream networks. It can improve the feature extraction ability of the network model without significantly increasing the number of calculations and parameters.

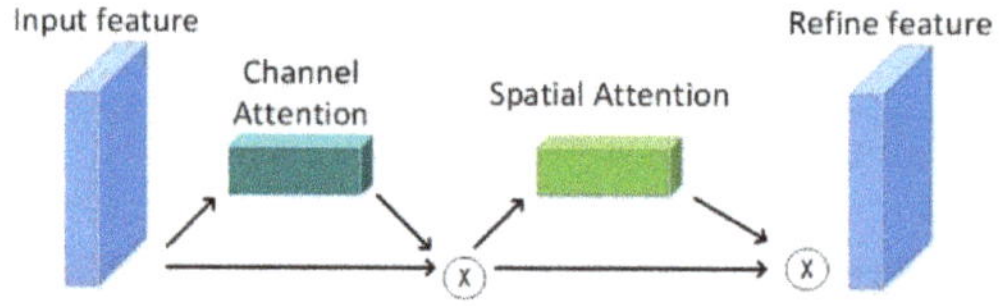

Figure 2. The serial structure of channel attention and spatial attention.

CBAM includes two pieces of content, the channel attention module and spatial attention module; that is, the channel attention CAM and spatial attention SAM.

The network embedded with the CBAM mechanism first performs global pooling and maximum pooling on the constitutionally generated feature maps F using channel attention mapping. The pooling result is connected to the multilayer perceptron for the addition operation, and the channel weight coefficient Mc is generated through the sigmoid activation function. Finally, this weighting coefficient is multiplied by the original feature map F to obtain the feature map F′ after the channel weighting adjustment as shown in Figure 3. The channel attention mapping process is shown in Equation (1).

$$M_c = \delta(MLP(AvgPool(F)) + MLP(MaxPool(F))) \quad (1)$$

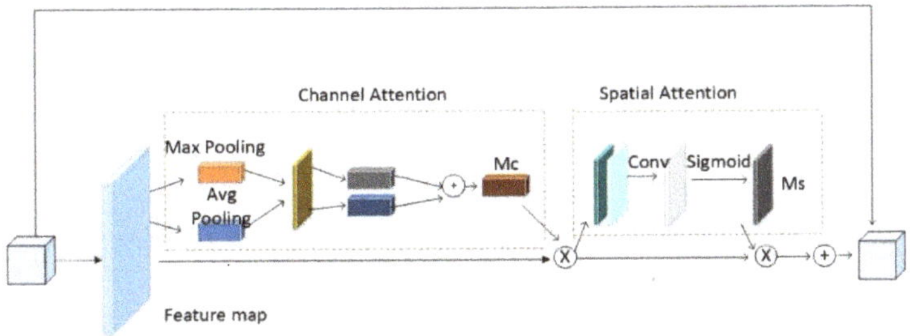

Figure 3. Embedding CBAM outside the convolution block.

In Equation (1), MLP stands for the multilayer perceptron; δ is the activation function.

After that, the feature map generated by the channel attention is sent to the spatial attention for processing. Spatial attention mapping performs a serial connection of global maximum pooling and average pooling on the weighted feature map F', uses convolution to reduce the dimensionality into a single-channel feature map, and uses the sigmoid function to activate the spatial feature matrix Ms. The weight matrix and the feature map F are subjected to the dot multiplication operation to obtain the final required spatial feature map F'', and the spatial attention mapping process is shown in Equation (2):

$$M_s(F) = \delta\left(f^{7\times 7}\{AvgPool(F); MaxPool(F)\}\right) \tag{2}$$

In the formula, $f^{7\times 7}$ represents that the convolution kernel is a 7×7 convolution layer; δ is the activation function; and; represents the serial connection.

In this paper, we replace the 7×7 large convolution kernel in the spatial attention module with two 3×3 dilated convolutions with a dilated rate of 2, reducing the number of parameters in the spatial attention module. The modified CBAM model is shown in Figure 4, where X and Y represent the input and output feature matrices, respectively, and C, W, and H and C^1, W^1, and H^1 represent the three-dimensional information of X and Y, respectively.

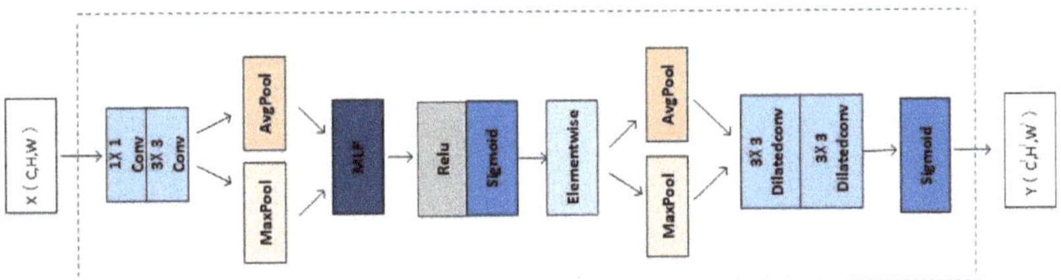

Figure 4. Use stacked dilated convolution to replace the spatial attention structure diagram of large-size convolution.

To suppress the influence of the useless features on the model, the CBAM attention model is connected to the outside convolution of the last three convolution blocks in the one-quarter resolution branch and the last two convolution blocks in the one-half resolution to improve the segmentation accuracy. The convolution block in front of the low- and medium-resolution branches mainly extracts image features, and then the attention layer is used to enhance the feature extraction so that the obtained features are more accurate, and the high-resolution branch can better guide us in low resolution and medium resolution to better achieve the segmentation effect.

Finally, the feature maps generated by the three branches are fused through the CFF module, as show in Figure 5.

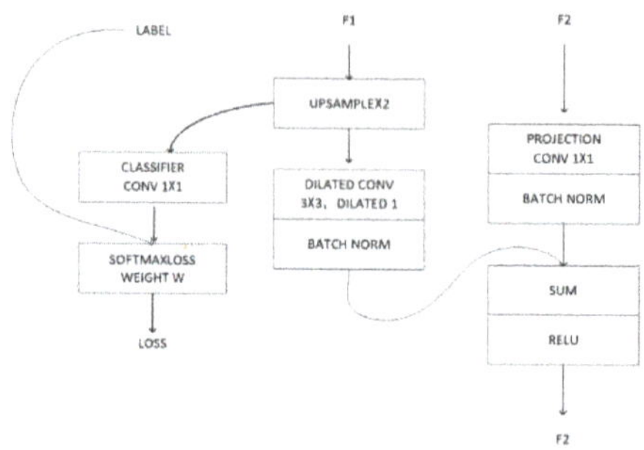

Figure 5. Feature maps F1 and F2 are fused through CFF.

The CFF module has three inputs, the feature maps F1, F2, and label. For F1, it is upsampled twice to make it the same size as F2, and then the features of F1 are refined by convolution with a hole of size 3 × 3 and a dilation rate of 2. For F2, it is convolved by 1 × 1 to make it the same number of channels as F1, and then the features are normalized using the BN layer. This is then added to the F1 features obtained above to obtain F2′. To enhance the learning of feature F1, the upsampled features of 1 are guided using an auxiliary label, optimizing the loss. Where, for the first CFF module, F1 and F2 are the features obtained from the low- and medium-resolution branches, respectively.

3.2. Asymmetric Convolution Replaces Regular Convolution

A large convolution kernel can create a larger receptive field, but it also means there are more parameters. An asymmetric convolution can greatly reduce the calculation amounts in the convolution stage without reducing the accuracy, which reduces the size of the model and improves the real-time segmentation of the model. This article replaces part of the regular convolution with an asymmetric convolution to further reduce the quantity of ICNet calculations.

Adding the CBAM attention module to the ICNet network results in a small increase in the calculations of the model. Compared with conventional convolution, asymmetric convolution can greatly reduce the amount of calculation in the convolution stage without losing accuracy. First, an n × 1 convolution is performed, then a 1 × n convolution is performed, as shown in Figure 6, which is consistent with the result of directly performing an n × n convolution, but the scale of the multiplication operation changes from n × n to 2 × n; so, the larger the n, the more obvious the effect of asymmetric convolution in reducing the calculation amounts. In this paper, the one-half and one-quarter resolution branches replace the 3 × 3 convolution with a 3 × 1 convolution, and then a 1 × 3 convolution is joined, as shown in Figure 6, thereby reducing the calculation amounts.

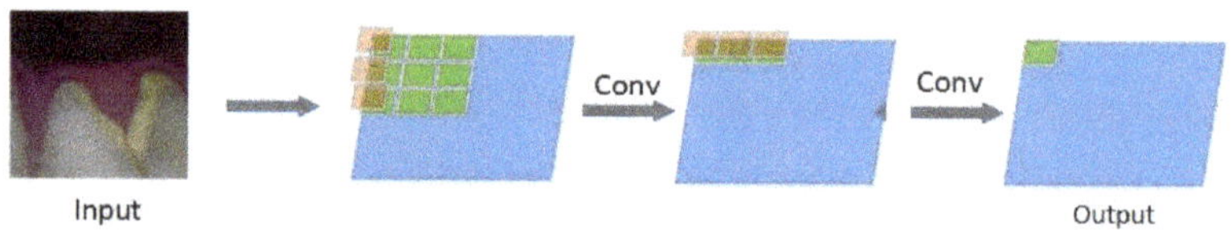

Figure 6. Asymmetric convolution process.

4. Experiment

4.1. Datasets

4.1.1. Data Collection

An intraoral camera was used to collect images of dental calculus, gingivitis, tartar, and worn surfaces. Two hundred images of dental lesions were obtained by shooting from three angles: the exterior, interior, and top views. The number of occurrences of various lesions in the collected images are shown in Table 1.

Table 1. The number of times that various lesions appear on the collected images.

Calculus	Gingivitis	Tartar	Worn Surfaces
209	215	33	100

4.1.2. Data Augmentation

To expand the difference between the samples and ensure the generalization ability of the later model training, randomly crop the collected dental lesion pictures and restore the original size and flip it over, then adjust the contrast, brightness, saturation, etc. Then, the original tooth lesion image is scaled to 512 × 512 pixels according to the principle of proportional invariance, the distorted image is removed by manual screening, and 400 dental lesion images are selected as the original dataset. Figure 7 shows that dataset example and Figure 8 shows that the number of various lesions before and after data augmentation. Finally, the Label Me labeling tool was used to imitate the PASCAL VOC2012 dataset format to manually label the four categories of calculus, gingivitis, tartar, and wear surface in the lesion image and the dataset was divided into a training set, validation set, and test set at a ratio of 7:2:1.

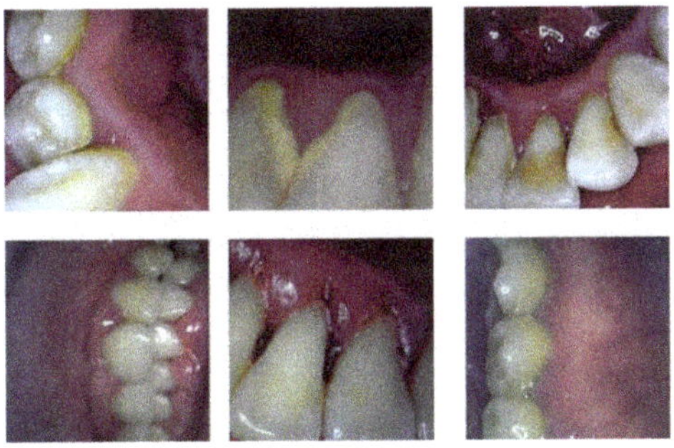

Figure 7. Example of combined dataset.

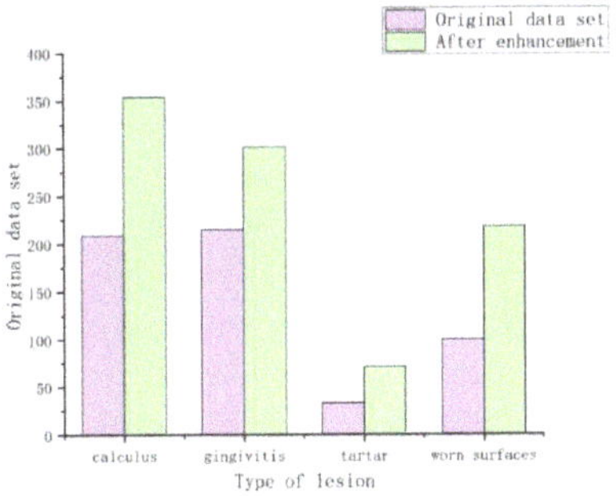

Figure 8. Comparison of the number of occurrences of each lesion on the original dataset and the combined dataset.

The tartar data is small, but it has more prominent features such as color and texture. To cope with the small amount of tartar data, we incorporate some training techniques in the training process, such as pre-training the tartar dataset to initialize the weights and then using the pre-trained model to train our multi-class lesion segmentation model. We also oversample the tartar data during training by random replication, so that the prediction is more accurate for small amounts of tartar, and we add dropout to the network to prevent overfitting of the tartar data.

In addition, the tooth lesion dataset was overexposed, resulting in unclear edges and unclear details in the lesion image. The histogram of the normal tooth image is more balanced, while the overexposed tooth image has too many pixels with high brightness, which causes the histogram to shift to the right, as show in Figure 9.

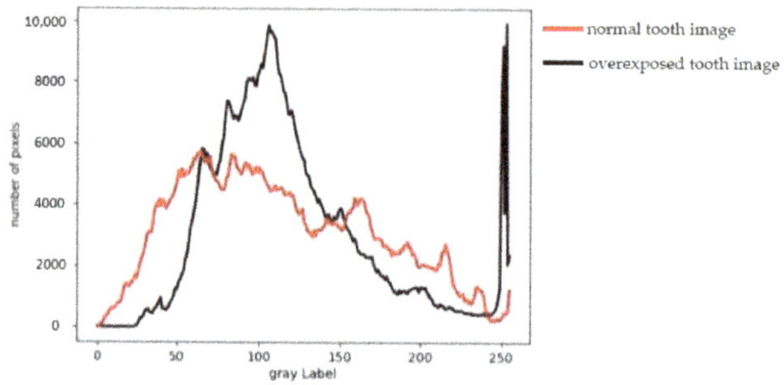

Figure 9. Comparison of gray histogram between a normal tooth image and an overexposed tooth image.

This article uses the ACE [33] automatic color equalization algorithm to color balance the unclear images in the dataset. The algorithm considers the spatial positional relationship between color and brightness in the image, performs adaptive filtering of local characteristics, makes image brightness and color adjustments and contrast adjustments with local and nonlinear characteristics, and satisfies the gray world theory hypothesis and white speckle hypothesis. The ACE algorithm consists of two steps. The first step is to adjust the color/spatial domain of the image, complete the chromatic aberration correction of the image and obtain a spatially reconstructed image, as shown in Equation (3).

$$R_c(p) = \sum_{j=subset} \frac{r(I_c(p) - I_c(j))}{d(p,j)} \quad (3)$$

In the formula, R_c is the intermediate result, $I_c(p) - I_c(j)$ is the brightness difference between two different points, d(p, j) is the distance degree function, and r (*) is the degree performance function, which must be an odd function. This step can adapt to the local image contrast, r (*) can amplify small differences, enrich large differences and expand or compress the dynamic range according to the local content. It is generally agreed that r (*) is:

$$r(n) = \begin{cases} 1, & X < T \\ X/T, & -T < X < T \\ -1, & X > T \end{cases} \quad (4)$$

4.2. Metrics

In order to compare the performance of our method, we use the training model to segment the test set and compare the segmented image with the masked label. In terms of segmentation accuracy, our metrics mainly include pixel accuracy (Acc), average interaction ratio (mIoU), and F1 score. In terms of real-time performance, we mainly compare the amount of calculation and the reasoning time of a single picture. We perform semantic segmentation on tooth lesions, which is a pixel-level segmentation. In the field of deep learning image segmentation in computer vision, the mIoU value is an important metric to measure the accuracy of image segmentation. Assuming there are $k + 1$ classes, P_{ij} represents the number of pixels whose actual class is class i but whose predicted result is class j. The calculation formula of mIoU is shown in Equation (5).

$$\text{MIou} = \frac{1}{K+1}\sum_{i=0}^{k}\frac{P_{ii}}{\sum\limits_{i=0}^{k} p_{ij} + \sum\limits_{i=0}^{k} p_{ji} - p_{ii}} \tag{5}$$

4.3. Loss Function

ICNet adds a loss weight to each branch training and optimizes the weighted SoftMax cross-entropy, and its loss function L can be expressed as:

$$\text{L} = \text{W}_1\text{L}_1 + \text{W}_2\text{L}_2 + \text{W}_3\text{L}_3 \tag{6}$$

where L_1, L_2 and L_3 are the loss of the low-, medium-, and high-resolution branches, respectively, and W_1, W_2 and W_3 are the weights of the loss function of the low-, medium-, and high-resolution branches, respectively. Normally, if the high-resolution branch weight W_1 is set to 1, the weights W_2 and W_3 of the medium-resolution and low-resolution branches are 0.4 and 0.16, respectively.

4.4. Experimental Details

We evaluated our method on a self-built dental lesion dataset. During the training process, we use Resnet50 as our backbone network and we use the SGD optimization method for training when the loss of the validation set does not decrease in 20 epochs. The data transformation of loss and acc in the training process is shown in Figure 10. Our method has a faster convergence rate. We replace the large-size convolution in the spatial attention module in CBAM and perform ablation experiments. Finally, we enhanced and reduced the brightness of the original image and tested the segmentation results of our model under different brightness levels.

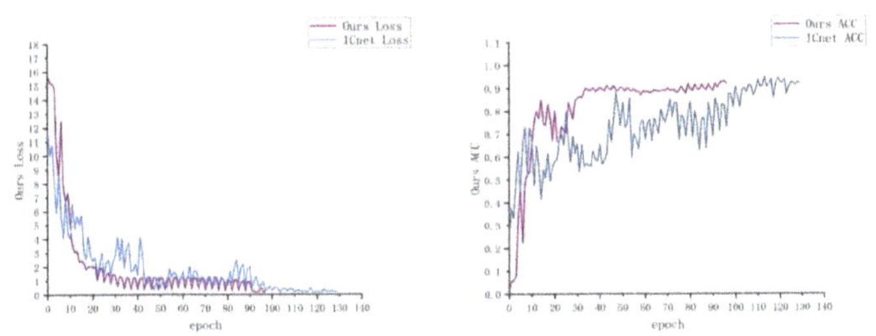

Figure 10. Comparison of loss and ACC curve.

The experiments in this article are all based on the TensorFlow deep learning framework, completed on the Bit hub cloud server, and the graphics card information is gtx1080.

5. Results

5.1. Contrast Test with Other Segmentation Algorithms

At present, with the development of convolutional neural networks, an increasing number of deep learning methods are used for semantic image segmentation. However, the segmentation performance of different tasks and methods is significantly different. To further test the pros and cons of this method for real-time semantic segmentation of dental lesion images, the training parameters of the above four models are the same as those of the improved ICNet. They are all trained based on the strategy of automatically saving the optimal model and then tested on the verification set. The test indicators mainly start from the two aspects of segmentation accuracy and time performance and include Acc, mIoU, the F1 score, and the reasoning time of a single picture. It can be seen from Table 2 that our method has the highest Acc, mIoU, and F1 score, with scores of 0.8897, 78.67%, and 0.8890, respectively. Compared with ICNet, the improved ICNet segmentation accuracy, mIoU, and F1 scores are 0.0384, 3.91%, and 0.0397 higher, respectively.

Table 2. Comparison of various segmentation indices of different algorithms.

Model	Acc	mIoU	F1_Score	Times (ms)
FCN8	0.8215	68.17	0.8045	833
ENet	0.8160	62.30	0.7749	696
U-Net	0.8875	78.61	0.8838	739
SegNet	0.8284	69.24	0.8162	805
ICNet	0.8513	74.76	0.8493	**307**
Ours	**0.8897**	**78.67**	**0.8890**	395

In addition, from the results of the visualization in Figure 11, U-Net and SegNet have oversegmentations. That is, the no lesion part is segmented, while ENet and ICNet have undersegmentations, which makes it difficult to correctly identify normal tooth tissues and lesions. The improved ICNet greatly improves the oversegmentation, and the result is closer to the label.

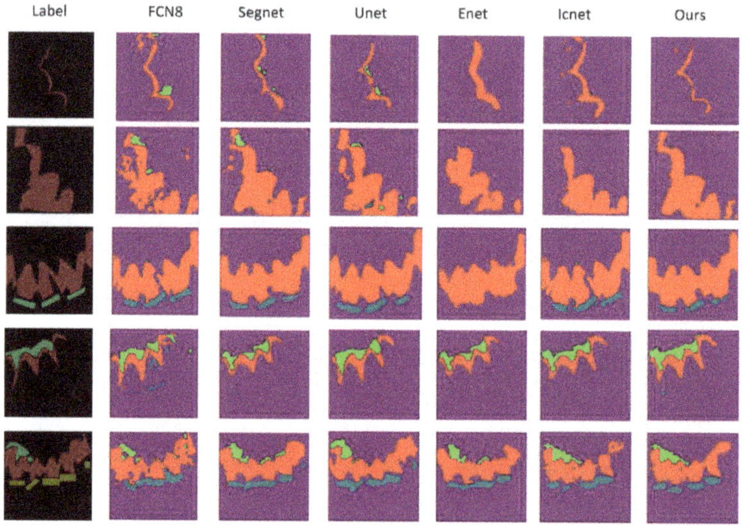

Figure 11. Visual comparison chart of segmentation results of different algorithms.

In terms of real-time segmentation, ENet, UNet, SegNet, ICNet, and the improved ICNet took 833 ms, 696 ms, 739 ms, 805 ms, 307 ms, and 395 ms for a single image, respectively. Compared with ENet, U-Net, and SegNet, the improved ICNet shortens the time by 34.63%, 38.43%, and 43.47%, respectively, and shortens the time by nearly half compared with FCN and increases less time compared with ICNet. From Table 3, we can see that the improved ICNet is slightly more computationally expensive than ICNet.

Table 3. Comparison of calculations before and after ICNet improvement.

Model	FLOPs
ICNet	13,524,726
Ours	15,608,042

From the above evaluation indicators, it can be seen that our method has the best segmentation effect, and when the segmentation time is close to ICNet, the three indicators of our method have improved to varying degrees. The indicators have improved to varying degrees. Among the visualization results of all methods, our segmentation effect is the closest to the real label. The abscissa of Figure 12 is the frequency of division, the ordinate is MIOU, and the upper-right corner of the figure is the optimal method. Figure 12 shows that our method is optimal in terms of accuracy and time.

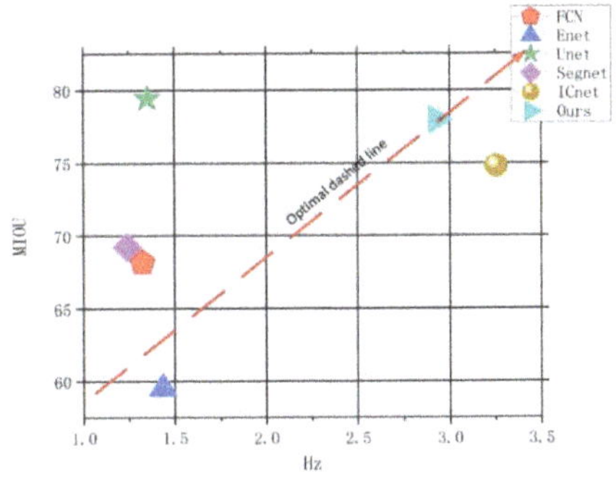

Figure 12. Comparison of accuracy and efficiency of different segmentation algorithms.

Among the four types of lesions, dental calculus and wear surface are the most similar to normal tooth tissue characteristics. Figure 13 shows the segmentation visualization results of dental calculus and wear surface by our method and other algorithms. FCN, U-Net, and SegNet oversegment for calculus. ENet and ICNet have certain undersegmentation problems. The visualization effect of our method is closest to the label. Among the visualization results of wear surface segmentation, ENet segmentation has the worst effect, the segmentation results of the other algorithms are not much different, and the results are relatively close.

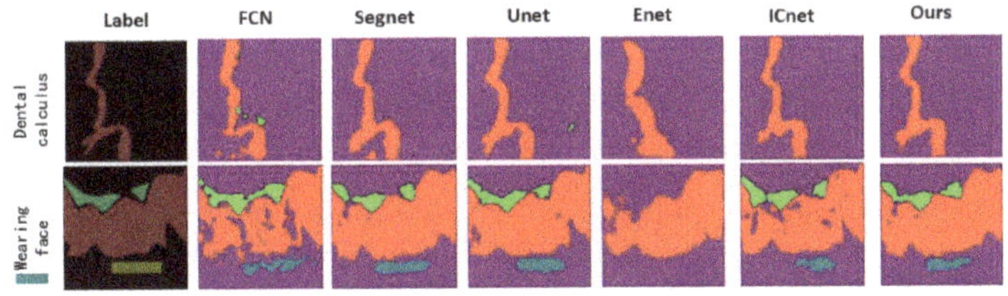

Figure 13. Visual comparison of segmentation results of dental calculus and worn surface.

5.2. Segmentation under Different Brightness

When an oral image is collected, the brightness of each image is difficult to maintain, and the brightness becomes one of the important factors that affects the quality of the segmentation. Therefore, this article compares the segmentation and dental lesion images under different brightness levels. We set the original image brightness to P, reduce the image brightness to 0.7 times that of the original image, and increase it to 1.3 times that of the original image for image segmentation.

It can be seen from Table 4 that when the brightness is adjusted to 0.7 P or 1.3 P, the segmentation accuracy is reduced. The analysis believes that it is more difficult to identify the characteristics of the lesion in the image that is too bright and too dark. The ICNet segmentation accuracy dropped by 2.33% and 4.5%, respectively, and our method's segmentation accuracy dropped by 2.68% and 5.34%, respectively. According to the visualization results, in Figure 14, ICNet has an undersegmentation at 0.7 P and an oversegmentation at 1.3 P. Our method is closer to the real label at 0.7 and 1.3 P.

Table 4. mIoU comparison of our method with the ICNet method at different luminance.

	mIoU		
Model	0.7P	P	0.8P
ICNet	58.50	60.83	56.38
Ours	58.65	61.33	55.99

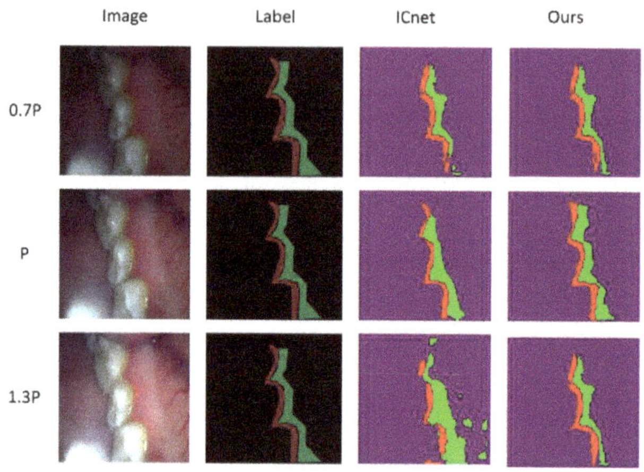

Figure 14. Visualization comparison of segmentation results under different brightness.

5.3. Ablation Experiment of CBAM

We replace the 7 × 7 large-size convolution kernel in the spatial attention in CBAM with a convolution kernel with a size of 3 × 3 and a dilated rate of 2. As shown in Table 4, the standard convolution with a convolution kernel size of 7 × 7 is represented by ICnet + $CBAM_{7 \times 7}$; the convolution kernel size is 3 × 3, and the void convolution with the void ratio is 2, using ICNet + $CBAM_{3 \times 3}$ means. We use a dilated convolution with a convolution kernel size of 3 × 3 and a dilated rate of 2 twice, denoted by $CBAM_{2 \times 3 \times 3}$. Compared with the parameter quantity of ICNet + $CBAM_{3 \times 3}$ as the benchmark, we set the parameter increment of ICNet + $CBAM_{3 \times 3}$ to 0.

From Table 5, we can see that after replacing the 7 × 7 convolution in the spatial attention with two 3 × 3 convolutions, the mIoU increased by 0.55% when the number of parameters was reduced. Experiments show that a 3 × 3 dilated convolution with a dilated rate of 2 has the same perceptual field as a standard convolution of 7 × 7, and the effect of both is approximate. In addition, comparing the results of the third group of experiments, it is found that stacking multiple convolutional layers with dilated layers can also improve the feature expression abilities of the spatial attention module.

Table 5. Replace the 7 × 7 convolution in the spatial attention in CBAM with a different convolution.

Model	MIOU	Add Param
ICNet + $CBAM_{3 \times 3}$	76.84	0
ICNet + $CBAM_{7 \times 7}$	77.42	240
ICNet + $CBAM_{2 \times 3 \times 3}$	77.97	180

6. Conclusions

In this paper, for a small sample of dental lesion datasets, the data are augmented by random cropping and flipping. The ACE automatic color equalization algorithm addresses the problems of blurred images of the lesion caused by the light source. Combining the augmented data and processed image with the original dataset into a new dataset makes the resulting model more generalizable.

The edge of the tooth lesion is highly similar to the normal tooth, and the edge is difficult to subdivide. Moreover, although ICNet obtains most of the semantic information when performing feature extraction in low- and medium-resolution branches, the details are easily lost, resulting in inaccurate lesion segmentations. Therefore, this article adds a lightweight CBAM module to the feature extraction stage, which can better obtain the semantic information of the image so that the high-resolution branch can better guide the low-resolution-generated feature map, thereby improving the segmentation accuracy.

We replace the 3×3 convolution with an asymmetric convolution in the low- and medium-resolution convolution stages, which further reduces the computational complexity of the model. We also ensure that the time it takes to improve the accuracy of the model does not increase significantly.

Although our method has significantly improved the segmentation of a variety of dental lesions, the light intensity has a large impact on the segmentation effect. In addition, the labeling of lesion images requires a great deal of effort. In future research, we will establish a highly generalized, weakly supervised network to solve the problems of difficult labeling and the large impact of the light.

Author Contributions: Conceptualization, X.Z. and T.M.; methodology, X.Z.; software, X.Z. and B.M.; validation, J.Q., J.Z. and G.G.; formal analysis, G.G. and J.Y.; investigation, B.M.; resources, G.G.; data curation, T.M. and J.Z.; writing—original draft preparation, X.Z.; writing—review and editing, X.Z.; supervision, T.M.; supervision, J.Y. All authors have read and agreed to the published version of the manuscript.

Funding: This research was supported by the National Natural Science Foundation of China (Grant No. 62101432 and 62102309) and, in part, by Shaanxi Natural Science Fundamental Research Program Project (No. 2022JM-508).

Data Availability Statement: Not applicable.

Conflicts of Interest: The authors declare no conflict of interest.

References

1. Lee, C.Y.; Chuang, C.C.; Chen, G.J.; Huang, C.C.; Lee, S.Y.; Lin, Y.H. Automated segmentation of dental calculus in optical coherence tomography images. *Sens. Mater.* **2018**, *30*, 2517–2529. [CrossRef]
2. Krois, J.; Ekert, T.; Meinhold, L.; Golla, T.; Kharbot, B.; Wittemeier, A.; Dörfer, C.; Schwendicke, F. Deep learning for the radio graphic detection of periodontal bone loss. *Sci. Rep.* **2019**, *9*, 84–95. [CrossRef]
3. Casalegno, F.; Newton, T.; Daher, R.; Abdelaziz, M.; Lodi-Rizzini, A.; Schürmann, F.; Krejci, I.; Markram, H. Caries detection with near-infrared trans illumination using deep learning. *J. Dent. Res.* **2019**, *98*, 1227–1233. [CrossRef]
4. Lee, J.-H.; Kim, D.-H.; Jeong, S.-N.; Choi, S.-H. Diagnosis and prediction of periodontally compromised teeth using a deep learning-based convolutional neural network algorithm. *J. Periodontal Implant Sci.* **2018**, *48*, 114–123. [CrossRef]
5. Yu, H.; Cho, S.; Kim, M.; Kim, W.; Kim, J.; Choi, J. Automated skeletal classification with lateral cephalometry based on artificial intelligence. *J. Dent. Res.* **2020**, *99*, 249–256. [CrossRef]
6. Li, W.; Liang, Y.; Zhang, X.; Liu, C.; He, L.; Miao, L.; Sun, W. A deep learning approach to automatic gingivitis screening based on classification and localization in RGB photos. *Sci. Rep.* **2021**, *11*, 16831. [CrossRef]
7. Karatas, O.H.; Toy, E. Three-dimensional imaging techniques: A literature review. *Eur. J. Dent.* **2014**, *8*, 132–140. [CrossRef]
8. Graham, B.; Engelcke, M.; Van Der Maaten, L. 3D semantic segmentation with submanifold sparse convolutional networks. *Proc. IEEE Conf. Comput. Vis. Pattern Recognit.* **2018**, *102*, 9224–9232.
9. Liu, T.; Cai, Y.; Zheng, J.; Thalmann, N.M. BEACon: A boundary embedded attentional convolution network for point cloud instance segmentation. *Vis. Comput.* **2022**, *38*, 2303–2313. [CrossRef]
10. Ning, Z.; Zhong, S.; Feng, Q.; Chen, W.; Zhang, Y. SMU-Net: Saliency-Guided Morphology-Aware U-Net for Breast Lesion Segmentation in Ultrasound Image. *IEEE Trans. Med. Imaging* **2021**, *41*, 476–490. [CrossRef]
11. Long, J.; Shelhamer, E.; Darrell, T. Fully Convolutional Networks for Semantic Segmentation. *IEEE Trans. Pattern Anal. Mach. Intell.* **2017**, *39*, 640–651.
12. Badrinarayanan Vijay and Kendall Alex and Cipolla Roberto. SegNet: A Deep Convolutional Encoder-Decoder Architecture for Image Segmentation. *IEEE Trans. Pattern Anal. Mach. Intell.* **2017**, *39*, 2481–2495. [CrossRef]
13. Ronneberger, O.; Fischer, P.; Brox, T. *U-Net: Convolutional Networks for Biomedical Image Segmentation*; Springer: Cham, Switzerland, 2015.

14. Paszke, A.; Chaurasia, A.; Kim, S.; Culurciello, E. ENet: A Deep Neural Network Architecture for Real-Time Semantic Segmentation. 2016. *arXiv* **2016**, arXiv:1606.02147.
15. Zhao, H.; Qi, X.; Shen, X.; Shi, J.; Jia, J. *ICNet for Real-Time Semantic Segmentation on High-Resolution Images*; Springer: Cham, Switzerland, 2018.
16. Zhao, H.; Shi, J.; Qi, X.; Wang, X.; Jia, J. Pyramid Scene Parsing Network. In Proceedings of the 2017 IEEE Conference on Computer Vision and Pattern Recognition (CVPR), Honolulu, HI, USA, 21–26 July 2017; pp. 6230–6239. [CrossRef]
17. Liu, S.; Ye, H.; Jin, K.; Cheng, H. CT-UNet: Context-Transfer-UNet for Building Segmentation in Remote Sensing Images. *Neural Process. Lett.* **2021**, *53*, 4257–4277. [CrossRef]
18. Tang, Q.; Liu, F.; Jiang, J.; Zhang, Y. EPRNet: Efficient Pyramid Representation Network for Real-Time Street Scene Segmentation. *IEEE Trans. Intell. Transp. Syst.* **2021**, *23*, 7008–7016. [CrossRef]
19. Garcia-Garcia, A.; Orts-Escolano, S.; Oprea, S.; Villena-Martinez, V.; Garcia-Rodriguez, J. A Review on Deep Learning Techniques Applied to Semantic Segmentation. *arXiv* **2017**.
20. Hu, S.; Ning, Q.; Chen, B.; Lei, Y.; Zhou, X.; Yan, H.; Zhao, C.; Tang, T.; Hu, R. Segmentation of aerial image with multi-scale feature and attention model. In *Artificial Intelligence in China*; Springer: Singapore, 2020; pp. 58–66.
21. Yi, Y.; Zhang, Z.; Zhang, W.; Zhang, C.; Li, W.; Zhao, T. Semantic segmentation of urban buildings from VHR remote sensing imagery using a deep convolutional neural network. *Remote Sens.* **2019**, *11*, 1774. [CrossRef]
22. Zhang, Z.; Liu, Q.; Wang, Y. Road extraction by deep residual u-net. *IEEE Geosci. Remote Sens. Lett.* **2018**, *15*, 749–753. [CrossRef]
23. Jie, H.; Li, S.; Gang, S. Squeeze and Excitation Networks. *IEEE Trans. Pattern Anal. Mach. Intell.* **2017**, *42*, 99.
24. Oktay, O.; Schlemper, J.; Folgoc, L.L.; Lee, M.; Heinrich, M.; Misawa, K.; Mori, K.; McDonagh, S.; Hammeria, N.Y.; Kainz, B.; et al. Attention U-Net: Learning where to look for the pancreas. *arXiv* **2018**, arXiv:1804.03999.
25. Woo, S.; Park, J.; Lee, J.Y.; Kweon, I.S. CBAM: Convolutional Block Attention Module. In *European Conference on Computer Vision*; Springer: Cham, Switzerland, 2018.
26. Guo, C.; Szemenyei, M.; Yi, Y.; Wang, W.; Chen, B.; Fan, C. SAUNet: Spatial Attention U-Net for Retinal Vessel Segmentation. In Proceedings of the 2020 25th International Conference on Pattern Recognition (ICPR), Milan, Italy, 10–15 January 2021; pp. 1236–1242. [CrossRef]
27. Wang, X.; Girshick, R.; Gupta, A.; He, K. Non-local Neural Networks. In Proceedings of the IEEE/CVF Conference on Computer Vision and Pattern Recognition, Salt Lake City, UT, USA, 18–23 June 2018; pp. 7794–7803. [CrossRef]
28. Szegedy, C.; Vanhoucke, V.; Ioffe, S.; Shlens, J.; Wojna, Z. Rethinking the Inception Architecture for Computer Vision. In Proceedings of the 2016 IEEE Conference on Computer Vision and Pattern Recognition (CVPR), Las Vegas, NV, USA, 27–30 June 2016; pp. 2818–2826.
29. Chollet, F. Xception: Deep Learning with Depth wise Separable Convolutions. In Proceedings of the 2017 IEEE Conference on Computer Vision and Pattern Recognition (CVPR), Honolulu, HI, USA, 21–26 July 2017.
30. Yu, F.; Koltun, V. Multi-Scale Context Aggregation by Dilated Convolutions. In Proceedings of the 4th International Conference on Learning Representations, ICLR 2016, San Juan, Puerto Rico, 2–4 May 2016.
31. Chen, L.C.; Papandreou, G.; Kokkinos, I.; Murphy, K.; Yuille, A.L. DeepLab: Semantic Image Segmentation with Deep Convolutional Nets, Atrous Convolution, and Fully Connected CRFs. *IEEE Trans. Pattern Anal. Mach. Intell.* **2018**, *40*, 834–848. [CrossRef]
32. Zhang, K.; Zuo, W.; Gu, S.; Zhang, L. Learning deep CNN denoiser prior for image restoration. *Proc. IEEE Conf. Comput. Vis. Pattern Recognit.* **2017**, *2808*, 3929–3938.
33. Xie, Z.; Huang, Y.; Zhu, Y.; Jin, L.; Liu, Y.; Xie, L. Aggregation Cross-Entropy for Sequence Recognition. In Proceedings of the 2019 IEEE/CVF Conference on Computer Vision and Pattern Recognition (CVPR), Long Beach, CA, USA, 15–20 June 2019.

micromachines

Article

Human Motion Pattern Recognition and Feature Extraction: An Approach Using Multi-Information Fusion

Xin Li [1,*], Jinkang Liu [1], Yijing Huang [1], Donghao Wang [1] and Yang Miao [2,3]

1 School of Mechanical and Materials Engineering, North China University of Technology, Beijing 100144, China; ncut_liujinkang@126.com (J.L.); ncut_huangyijing@163.com (Y.H.); ncut_wangdonghao@126.com (D.W.)
2 Faculty of Materials and Manufacturing, Beijing University of Technology, Beijing 100124, China; miaoyang@vip.126.com
3 Beijing Key Laboratory of Advanced Manufacturing Technology, Beijing University of Technology, Beijing 100124, China
* Correspondence: lixin2020@ncut.edu.cn

Abstract: An exoskeleton is a kind of intelligent wearable device with bioelectronics and biomechanics. To realize its effective assistance to the human body, an exoskeleton needs to recognize the real time movement pattern of the human body in order to make corresponding movements at the right time. However, it is of great difficulty for an exoskeleton to fully identify human motion patterns, which are mainly manifested as incomplete acquisition of lower limb motion information, poor feature extraction ability, and complicated steps. Aiming at the above consideration, the motion mechanisms of human lower limbs have been analyzed in this paper, and a set of wearable bioelectronics devices are introduced based on an electromyography (EMG) sensor and inertial measurement unit (IMU), which help to obtain biological and kinematic information of the lower limb. Then, the Dual Stream convolutional neural network (CNN)-ReliefF was presented to extract features from the fusion sensors' data, which were input into four different classifiers to obtain the recognition accuracy of human motion patterns. Compared with a single sensor (EMG or IMU) and single stream CNN or manual designed feature extraction methods, the feature extraction based on Dual Stream CNN-ReliefF shows better performance in terms of visualization performance and recognition accuracy. This method was used to extract features from EMG and IMU data of six subjects and input these features into four different classifiers. The motion pattern recognition accuracy of each subject under the four classifiers is above 97%, with the highest average recognition accuracy reaching 99.12%. It can be concluded that the wearable bioelectronics device and Dual Stream CNN-ReliefF feature extraction method proposed in this paper enhanced an exoskeleton's ability to capture human movement patterns, thus providing optimal assistance to the human body at the appropriate time. Therefore, it can provide a novel approach for improving the human-machine interaction of exoskeletons.

Keywords: feature extraction; artificial intelligence; motion pattern recognition; wearable sensors; EMG; IMU

Citation: Li, X.; Liu, J.; Huang, Y.; Wang, D.; Miao, Y. Human Motion Pattern Recognition and Feature Extraction: An Approach Using Multi-Information Fusion. *Micromachines* **2022**, *13*, 1205. https://doi.org/10.3390/mi13081205

Academic Editors: Xiao Xiao and Jose Luis Sanchez-Rojas

Received: 13 June 2022
Accepted: 26 July 2022
Published: 29 July 2022

1. Introduction

Exoskeleton is an intelligent wearable device, which can assist the movement of human beings' upper or lower limbs. In recent years, it has become a research hotspot in the field of robotics. By wearing an exoskeleton, the wearer's motor abilities and muscular endurance are enhanced, and the wearer can perform tasks that would otherwise be impossible [1]. Based on the above advantages, exoskeletons have a broad range of application prospects in medical rehabilitation, logistics, and military fields [2–6]. Exoskeletons rely on human-machine interaction to function efficiently. They need to recognize the current motion mode of a human body in real time so as to make corresponding motion at the appropriate time and determine the necessary assistance to the human body. Although the technology of exoskeletons has been greatly improved, there are still some deficiencies in human motion pattern recognition, wherein exoskeletons lack the ability to fully recognize the actions and intentions of human wearers. Therefore, exoskeletons cannot help the wearer at the right time, a shortcoming also highlighted in reference [7]. It is therefore necessary to study human motion pattern recognition. In this paper, the recognition of human lower limb motion pattern will be deeply studied.

In the whole process of lower limb movement pattern recognition, firstly, the human lower limb movement information should be obtained based on data acquisition equipment. Then, the obtained movement information should be extracted with features. Finally, the features are input into the classifier to identify various human movement patterns. If the motion information acquisition and feature extraction methods are not appropriate, the accuracy of human movement mode recognition will be affected. At present, a lot of work has been carried out on motion pattern recognition. Song et al. [8] used EMG sensor to collect data of five movement patterns (up the stairs, down the stairs, sit, stand and walk), and extracted frequency domain features and time domain features, including the mean frequency, median frequency, mean absolute value, wave length, slope sign change, variance, integrated, zero-crossing, and Willison amplitude. The random forest (RF) algorithm optimized by grid search method identified five motion patterns with an average recognition rate of 97.5%. Lopez-delis et al. [9] characterized knee motion patterns from EMG signals of erector spine muscles, extracted three features of mean absolute value, waveform length, and auto-regressive model and used linear discriminant analysis (LDA), K-nearest neighbor (KNN) and support vector machine (SVM) classification algorithms to identify eight human motion patterns, respectively. The recognition accuracy was greater than 95%. Xi et al. [10] extracted fifteen features (including time domain, frequency domain, time-frequency domain and entropy) and five classification algorithms for recognizing seven activities of daily life from the obtained surface EMG signals, and finally determined the optimal surface EMG features and classifiers. Peng et al. [11] used the designed plantar pressure sensing shoes to collect the plantar pressure of human body under different movement modes, and extracted five characteristics of average value, standard deviation, maximum value, minimum value, and difference deviation. The proposed KPCA-SVM classifier can recognize different motion patterns with an accuracy of 91.1%. Zhang et al. [12] collected data under dynamic (walking) and static (sitting, standing and lying) activities of the elderly by using IMU. Mean absolute value, zero-crossing, slope sign change, and waveform length were obtained, and the SVM algorithm was used to classify the above activities. A better classification effect has been achieved. Dhindsa et al. [13] obtained the EMG data of subjects during the process from sitting on a chair to standing up, extracted 15 features, and input them into the classifier after dimensionality reduction, and evaluated the performance of different classifiers. The SVMQ classifier performed best, and its recognition accuracy can reach about 92.2%. Gupta et al. [14] used acceleration sensors tied to the waist to obtain acceleration signals and extracted seven features including mean trend, windowed mean difference, detrended fluctuation analysis (DFA) coefficient, variation trend, windowed variance difference, energy uncorrelated, and maximum acceleration difference. They used Naive Bayes and KNN classifier to identify six different daily activities. The results showed that each individual activity was more than 95%. However, there are two main problems with the above work. Firstly, since the movement of human lower limbs is an extremely complex process, it needs to involve many types of data before it can be characterized comprehensively. It is difficult to fully obtain the motion information of human lower limbs by relying on a single type of sensor data. In addition, if the single sensor is affected by external uncertainty, it may lead to wrong recognition results (that is, it is difficult to meet the requirements of reliability by using a single sensor). Secondly, in the aspect of feature extraction, the time domain and frequency domain features mentioned in the above work are predefined and are manually designed based on previous research results and experience. When people use these methods to extract features, they often need to have some professional knowledge of mathematics and signal processing to select suitable features for extraction. Features extracted in this way are easily affected by the combination and number of different features, so it takes a lot of time to select the best combination and number of features, and features selected by manual experience may not be able to obtain the best classification accuracy.

In view of the above problems, this paper is based on the advantages that EMG sensor can collect human biological information and IMU can intuitively reflect the three-dimensional kinematics information of the human body [15]. These two sensors are used to obtain more comprehensive information of human lower limb movement so as to improve the accuracy of human motion pattern recognition. Then, a feature extraction method based on Dual Stream CNN-ReliefF is proposed. It can automatically extract well differentiated fusion features from EMG and IMU data, and the extracted features have a high accuracy of motion pattern recognition. Thus, the major contributions of this paper are as follows:

Firstly, this paper presents a bioelectronic motion information acquisition device based on the fusion of EMG and IMU sensors. Compared with the information based on a single sensor, our device can collect more human lower limb motion information, which mainly consist of bioelectric information (EMG signal) and biological kinematics information (acceleration, angular velocity and angle of joint motion). It is suggested that the combination of the two kinds of information will enhance the accuracy and reliability of motion pattern recognition.

Secondly, a feature extraction method based on Dual Stream CNN-ReliefF is proposed. This method is improved on the basis of the Dual Stream CNN recognition model. The improved Dual Stream CNN-ReliefF method can automatically extract well differentiated fusion features from EMG and IMU data. The Dual Stream CNN-ReliefF feature extraction method proposed in this paper avoids the problems that the traditional manual design feature extraction involves, including a need to rely on manual experience, tedious steps, and low accuracy of feature recognition. Compared with single sensor (EMG or IMU) and Single Stream CNN or manual designed feature extraction methods, the feature extraction based on Dual Stream CNN-ReliefF shows better performance in terms of visualization performance and recognition accuracy. This method was used to extract features from EMG and IMU data of six subjects and input these features into four different classifiers. The motion pattern recognition accuracy of each subject under the four classifiers is above 97%, with the highest average recognition accuracy reaching 99.12%.

The structure of the paper is as follows: Section 2 introduces the mechanism of human lower limb movement, the wearable bioelectronics device of human lower limb movement information fusion, and the specific experimental scheme, respectively, and also introduces the feature extraction methods based on manual designs, Single Stream CNN and Dual Stream CNN-ReliefF. Section 3 presents the visual analysis of the features extracted by the three feature extraction algorithms and the accuracy analysis of motion pattern recognition under different classifiers. Section 4 presents the conclusions of our research.

2. Materials and Methods

2.1. Analysis of Human Lower Limb Movement Mechanism

Human lower limb movement is mainly completed by three parts: bone, bone connection, and skeletal muscle. Bones are the basic framework of human body; they cannot move itself and can be regarded as a rigid connecting rod. The connection between bones (that is, joints) can be regarded as motion pairs. Under the action of the nervous system, bone connection provides an external force for the joint through the contraction and relaxation of skeletal muscles in order to realize joint movement. Therefore, human lower limb movement relies on the synergistic effect between multiple joints and multiple muscle groups [16]. In order to comprehensively obtain the movement information of human lower limbs, it is necessary to obtain the movement information of each joint and muscle group of lower limbs during the movement of human body. Comprehensive movement information is helpful to improve the accuracy of human movement pattern recognition. All of the joints of the human body can only carry out rotating motion but not translational motion, which is decided by the particularity of human skeleton structure. Usually, kinematics concepts are used to represent joint motion information, such as angular velocity and angle and acceleration of joint rotation. The activity status of human muscle groups belongs to biological information. EMG signals are the superposition of action potential sequences of motor units generated during muscle contraction, which can reflect the intensity of muscle contraction. Therefore, EMG signals have been widely used in the recognition and prediction of human motion intentions [17].

2.2. Wearable Bioelectronics Device of Human Lower Limb Movement Information Fusion

From the above analysis of movement mechanism, it can be found that in order to comprehensively obtain the movement information of human lower limbs, a device is required to collect not only human biological information but also human kinematics information. In this paper, we proposed a set of wearable bioelectronics with human lower limb motion information acquisition device based on the EMG sensor and IMU. The device uses non-invasive EMG sensors to acquire the surface EMG signals generated by the muscles that play major roles in the movement of human lower limbs. For the collection of human kinematics information, IMU, a wearable inertial sensor, can usually be directly or indirectly placed on the human body, and can generate acceleration and rotation signals corresponding to human actions [18] so as to intuitively reflect the three-dimensional kinematics information of human lower limbs. Therefore, IMU was used to measure the angular velocity, angle and acceleration of joint rotation of human lower limbs, and these obtained data were used to represent the kinematics information. The entire wearable bioelectronics device is shown in Figure 1.

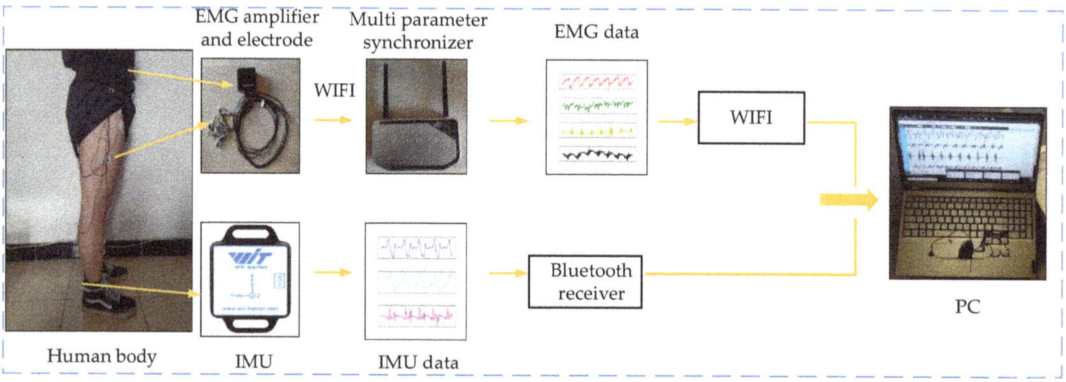

Figure 1. Wearable bioelectronics device of human lower limb movement information fusion.

The EMG sensor (Neu Sen WM, 2000 Hz) is developed by China Changzhou Neuracle Technology Co., Ltd. (Changzhou, China). The product can simultaneously collect four channels of human surface EMG signals, and has the advantages of signal stability and high shielding. IMU (BWT901CL, 200 Hz) adopts micro-electro-mechanical system (MEMS) technology based high performance three-dimensional motion posture measurement system developed by China Shenzhen Wit-Motion Technology Co., Ltd. (Shenzhen, China). It adopts advanced filtering technology, which can effectively reduce measurement noise and improve measurement accuracy. It has a built-in three-axis accelerometer and three-axis gyroscope, which can collect the acceleration, angular velocity, and angle of object rotation, respectively. The basic principle of the wearable bioelectronics device can be described as follows: three-dimensional motion information of human lower limbs is obtained through IMU, and IMU returns acceleration, angular velocity and angle signals of three axes at a sampling rate of 200 Hz, thus forming original IMU data. Then, the original IMU data collected by the Bluetooth receiver is transmitted to the PC for storage. The EMG sensor returns four channels of human lower limb EMG signals at a sampling rate of 2000 Hz and transmits them to the amplifier. Since the human EMG signals are weak, the amplifier needs to amplify the signals. Then, the signal is sent wirelessly to the multi parameter synchronizer to form the initial raw EMG signal. Finally, the data will be transmitted to the PC for storage through a wireless mode.

2.3. Installation of the EMG and IMU Sensors

To obtain a human EMG signal by an EMG sensor, first of all it is necessary to determine the attachment point of the electrode in the EMG sensor in the human lower limb muscles; a reasonable electrode attachment point can better extract a human EMG signal. By consulting the biomedical experts from Aerospace Central Hospital (Asch, Beijing, China), and referring to relevant references [19,20], we learned that some muscles of the human lower limbs play a major role in the process of exercise, as shown in the following Table 1:

Table 1. Muscles and functions that play a major role in lower limb movement.

Muscle	Motor Function
rectus femoris (RF)	stretch the calf, bend the thigh
vastus lateralis (VL)	stretch the calf
gastrocnemius (GA)	bend the calf, lift the heel, fix the knee, balance the body
musculus peroneus longus (ML)	foot valgus, plantarflexion
tibialis anterior (TA)	dorsiflexion, varus, adduction
biceps femoris (BF)	bend the calf, stretch the thigh, rotate the calf outward
soleus (SL)	bend the calf, lift the heel, fix the knee, balance the body
semitendinosus (SD)	stretch the thigh, bend the calf, rotate the thigh inside

In the course of our experiment, we found that some muscles in lower limbs are not convenient for the posting of EMG electrodes. Meanwhile, the EMG signals of some muscles are relatively weak, which is not conducive to the analysis of human motion patterns. Based on our experimental findings and the suggestions of biomedical experts, we finally selected four muscles (rectus femoris (RF), tibialis anterior (TA), biceps femoris (BF) and gastrocnemius (GA)) as the attachment points of EMG electrodes.

According to the IMU installation, we choose to place the IMU on the thigh, calf, and heel to obtain the kinematic information of human lower limbs. Then, we found that although all IMUs have been fixed with bandages. The IMU installed on the thigh and calf move up and down to a certain extent during human movement, which will affect the stability and accuracy of data acquisition, while the IMU installed on the heel shows better stability. Based on the IMU data collected in the thigh, calf, and heel during the experiment, it is found that there exist many burrs and drifts in the kinematic image of IMU installed on the thigh and calf. However, the kinematic information obtained by IMU installed on the heel shows stability, regularity, and smoothness. To sum up, we choose to install it on the heel to measure the acceleration, angular velocity, and angle of the ankle joint rotation around the Y-axis of IMU (the X-axis is perpendicular to the ground upward, Y-axis parallel to the ground and faces to the right, and Z-axis opposite to the forward direction of human body). Since the motion amplitude of the human body in the X and Z axes of IMU is small and the motion of the exoskeleton in these two directions is generally passive, the motion in the X and Z axes will be ignored in this paper.

Before the motion information collection test, in order to reduce the influence of environmental conditions (such as humidity, electro-magnetic interference, human subject's conditions, etc.) on EMG and IMU information, we took the following series of measurements:

(1) The experiment of human motion information collection is carried out in the laboratory, and the indoor temperature and humidity remain relatively constant.
(2) We removed the hair from the tested muscle, and wiped the skin with alcohol. Then, we used conductive paste to reduce the interference of skin on EMG signals and improve the conductivity and stability of EMG electrodes.
(3) The backing material of the EMG electrode used is non-woven fabric, which has better air permeability. Therefore, it is suitable for long-term skin use and can reduce the interference of sweat and other factors on EMG signals.
(4) The selected IMU and EMG sensors are transmitted wirelessly, which avoids the interference to the normal movement of the human body during data acquisition. The IMU uses the self-developed and improved Kalman filter fusion algorithm to solve the triaxial acceleration, angular velocity, and angle data, which avoids the situation that IMU is vulnerable to electro-magnetic interference and drift under dynamic conditions. The accuracy of acceleration, angular velocity and angle data output under dynamic conditions can reach 0.01 g, 0.05°/s and 0.03°, respectively. The EMG sensor has ultra-low input noise (<0.7 uVpp) and ultra-high input impedance (>1 GOhm), which can ensure high-quality EMG signals.
(5) Based on the IMU and EMG data obtained under the above conditions, we also performed filtering (see Section 2.4 for details) to eliminate noise and interference. Through the above methods, we ensure the reliability and authenticity of the collected data.

After the placement of the electrodes and IMU, the subjects will perform some simple lower limb movements to test whether the EMG and IMU data are clear or whether there is obvious noise. The attachment points of EMG electrodes on human lower limb muscles and IMU placement are shown in Figure 2.

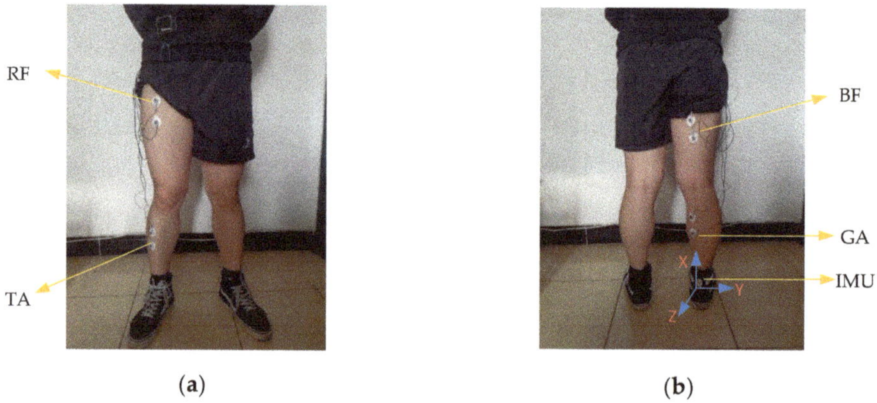

Figure 2. Electrode attachment point and IMU position. (**a**) Frontal side of human lower limbs; (**b**) back side of human lower limbs.

2.4. *Lower Limb Movement Information Collection Scheme*

This paper will consider four daily activities of the human body, including running, level ground walking, stair ascent, and ramp ascent. These daily activities can be easily carried out in the laboratory and no additional equipment is required. The four motion modes are illustrated in Figure 3.

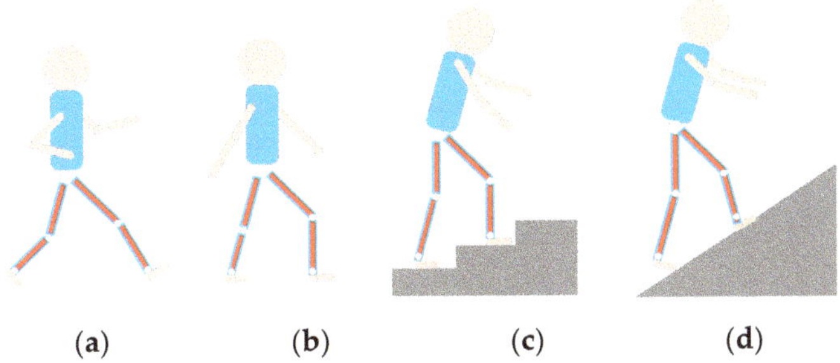

Figure 3. Four daily patterns of human movement. (**a**–**d**) represent running, level ground walking, stair ascent and ramp ascent.

The subjects collected the motion data of running, level ground walking, and ramp ascent on the treadmill (STAR TRAC, 10-TRx, Core health and fitness group in Vancouver, Washington, DC, USA) and stair ascent on the climbing machine (RISING, LMX-1100, Nantong, China). When we start walking or stop walking, we should stay still for 10 s so that we can better distinguish the two different states of static and walking when analyzing data. Four groups of data were collected for each motion pattern, and the duration of each group was 5 min. In order to minimize the impact of muscle fatigue on data collection, 50 min of rest is required after each motion mode. The participants included six healthy subjects, all of whom had given informed consent before participating in this study. After data collection, we use cubic spline interpolation to increase the sampling frequency of IMU from 200 Hz to 2000 Hz, which ensure that the sampling frequencies of EMG (2000 Hz) and IMU are synchronous and accurate. In addition, noise will be mixed in the process of data collection, and too much noise will adversely affect the accuracy of recognition. Therefore, denoising is necessary. Compared with other denoising methods, Butterworth can maintain good characteristics in its passband and stopband, which helps to retain useful information in the process of EMG and IMU signal denoising to obtain less signal distortion and noise. The noise of EMG signal is mainly caused by poor contact between electrode and body surface. Since the spectrum of EMG signals is mainly distributed in the range of 20 to 500 Hz [21], we use a Butterworth filter to filter out all noises outside this range. In addition, we use Butterworth low-pass filter to filter IMU data at a cut-off frequency of 10 Hz [22]. Some filtered data are shown in Figure 4a–d.

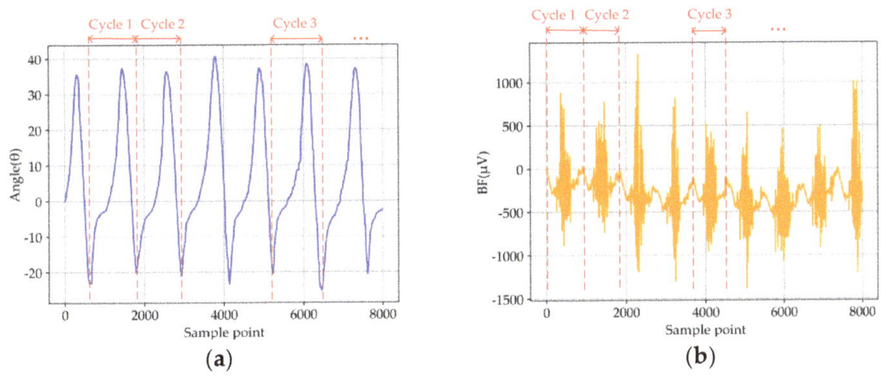

Figure 4. *Cont.*

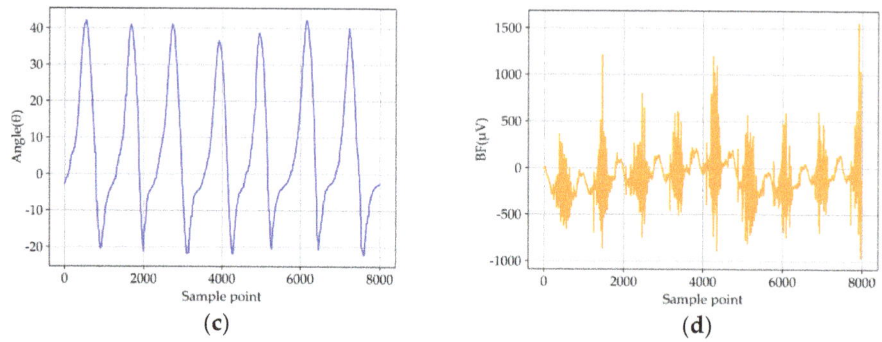

Figure 4. Parts of the test data obtained from the first and second groups of running mode tests. (**a**) Angle signal by IMU in the first group test; (**b**) EMG signal of BF in the first group test; (**c**) angle signal by IMU in the second group test; (**d**) EMG signal of BF in the second group test.

Although the quality of the measurements and the filtering algorithm have been implemented, the repeatability and reproducibility of the collected data should be verified. According to the four activity patterns of the lower limbs, including running, level ground walking, stair ascent and ramp ascent, we choose the running pattern as the analytic target, which impact the sensors data more than the others due to the large and rapid motion. Figure 4 showed parts of the IMU and EMG data from the first and second groups of running mode tests. The detailed testing method and results are as follows:

(1) According to the data repeatability, the IMU and EMG sensors were fixed on the lower limbs under the same conditions in one test. Then, the data illustrate the periodical variations (partial data of IMU and EMG sensors are shown in Figure 4a,b). Therefore, the similarity coefficients between the five random cycles test data of each IMU signal were calculated by using the Pearson correlation, which were higher than 0.87. Next, the fluctuation of the average value of iEMG (integrated EMG) showed a slight oscillation, which the average value and standard deviation of RF, TA, BF, and GA are 188.609 ± 3.109, 127.313 ± 2.050, 262.253 ± 3.948 and 459.762 ± 3.274, respectively. To sum up, the results indicate that the data of five random cycles with each signal presented preferable similarity and repeatability.

(2) According to the reproducibility of the data, another three group tests with the running motion mode were carried out under the same position, same sensor, and same nominal program (partial data of IMU and EMG sensors are shown in Figure 4c,d). Then, a similar approach was also adopted to validate the reproducibility with the data under the same sample periods in the four groups. Specifically, the similarity coefficients were calculated for IMU data, and the average value of iEMG were calculated for EMG data intercepted by each group of tests. The conclusions showed the similarity coefficients among each group of IMU signal data intercepted were higher than 0.85, and the average value and standard deviation of RF, TA, BF, and GA are 188.609 ± 3.109, 127.313 ± 2.050, 262.253 ± 3.948 and 459.762 ± 3.274, respectively. To sum up, it is indicated that the reproducibility among each group of data under the different tests can be demonstrated.

In summary, the collected IMU and EMG data showed available repeatability and reproducibility based on the above analysis, which can be input into the feature extraction.

2.5. Feature Extraction

2.5.1. Extracting Features Based on Traditional Manual Design

Feature extraction plays an important part in pattern recognition, and the extraction of appropriate features will have a higher resolution, which will have an impact on the recognition accuracy [23]. The original human motion data extracted through the experiment can not be directly applied to the classification algorithm due to the large amount of data and poor performance. Therefore, it is necessary to extract the features of the original data. The processed data is more expressive and more conducive to the identification of the algorithm. Traditional feature extraction methods based on manual design tend to rely too much on manual experience and the combination and number of different features will affect the accuracy of classification. We will calculate the time and frequency domain features commonly used by IMU and EMG signals in motion pattern recognition [24–26], in order to compare with the feature extraction method based on Dual Stream CNN-ReliefF proposed in this paper. The specific feature expressions of IMU and EMG signals are shown in Tables 2 and 3:

For IMU data, we calculate four time domain features (MAV, AVR, RMS, ZC) and two frequency domain features (MDF, MNP). For EMG data, we calculate six time domain features (MAV, AVR, RMS, WAMP, ZC, WL) and two frequency domain features (MDF, MNP).

2.5.2. Feature Extraction Based on Dual Stream CNN-ReliefF

In recent years, CNN, an artificial intelligence algorithm, has become a hot research topic in many fields, especially in image classification. Traditional CNN classification is generally based on the structure of single stream CNN, whose model structure mainly includes the convolution layer (Conv), pooling layer (Pool), full connection layer (FC) and output layer, as shown in Figure 5. The convolution layer performs convolution operation on the input image, extracts the image features, and then outputs the convolution results by forming feature mapping through activation function. The Pooling layer is also called the down-sampling layer, whose purpose is to compress the feature graph obtained by Convolution layer, so as to optimize the computational efficiency of the model. The full connection layer maps the distributed feature representation obtained through the convolution layer and pooling layer to the sample tag space. Finally, the softmax function is applied in the output layer to export the classification results, and the error between the results and the real classification results is calculated and the weight and paranoid value are updated by back propagation.

Table 2. Time and frequency domain features expression of IMU.

Feature	Mathematical Definition
Mean Absolute Value (MAV)	$\frac{1}{N}\sum_{i=1}^{N}\lvert x_i\rvert$
Variance (VAR)	$\frac{1}{N-1}\sum_{i=1}^{N}x_i^2$
Root Mean Square (RMS)	$\mathrm{RMS}=\sqrt{\frac{1}{N}\sum_{i=1}^{N}x_i^2}$
Zero Crossing (ZC)	$\begin{cases} \mathrm{ZC}=\sum_{i=1}^{N}z_i\times zc_i \\ zc_i=\mathrm{sgn}(-x_ix_{i+1}) \\ z_i=\begin{cases}1,\lvert x_i-x_{i+1}\rvert>\delta_z \\ 0,else\end{cases}\end{cases}$
Median Frequency (MDF)	$\sum_{i=f_{min}}^{\mathrm{MDF}}p_i=\sum_{i=\mathrm{MDF}}^{f_{max}}p_i$
Mean Power (MNP)	$\mathrm{MNP}=\sum_{j=1}^{M}p_j/M$

Table 3. Time and frequency domain features expression of EMG.

Feature	Mathematical Definition
Mean Absolute Value (MAV)	$\frac{1}{N}\sum_{i=1}^{N}\lvert x_i\rvert$
Variance (VAR)	$\frac{1}{N-1}\sum_{i=1}^{N}x_i^2$
Root Mean Square (RMS)	$\mathrm{RMS}=\sqrt{\frac{1}{N}\sum_{i=1}^{N}x_i^2}$
Wilson Amplitude(WAMP)	$\mathrm{WAMP}=\sum_{i=1}^{N-1}[f(\lvert x_n-x_{n+1}\rvert)]$ $f(x)=\begin{cases}1, if\ x\geq threshold\\ 0, \quad otherwise\end{cases}$
Zero Crossing (ZC)	$\mathrm{ZC}=\sum_{i=1}^{N}z_i\times zc_i$ $zc_i=\mathrm{sgn}(-x_ix_{i+1})$ $z_i=\begin{cases}1, \lvert x_i-x_{i+1}\rvert>\delta_z\\ 0, else\end{cases}$
Waveform Length (WL)	$\sum_{i=1}^{N-1}\lvert x_{i+1}-x_i\rvert$
Median Frequency (MDF)	$\sum_{i=f_{min}}^{\mathrm{MDF}}p_i=\sum_{i=\mathrm{MDF}}^{f_{max}}p_i$
Mean power (MNP)	$\mathrm{MNP}=\sum_{j=1}^{M}p_j/M$

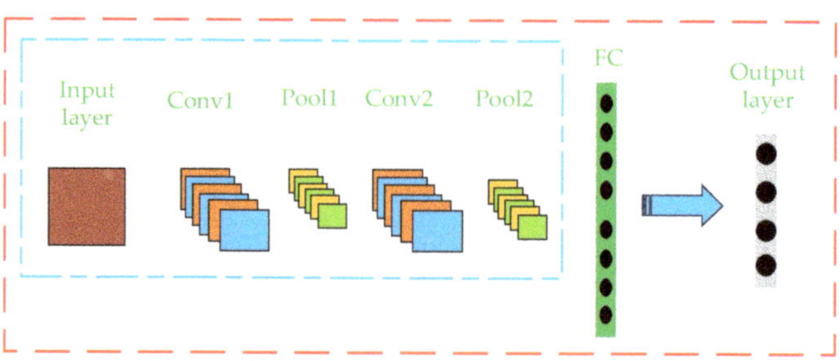

Figure 5. Single Stream CNN algorithm structure diagram.

The dual Stream CNN recognition model is improved based on Single Stream CNN and is widely used in video action recognition [27]. The main idea is to send RGB images (representing spatial information) and optical flow images (representing temporal information) to two neural networks for feature extraction, fuse the extracted features, and finally output the classification results. Based on the above ideas, the authors designed a Dual Stream CNN feature extractor based on multiple timing information fusion. The basic principle of the extractor can be specifically described as follows: First, a complete Dual Stream CNN recognition model is built (whose structure is shown in the blue-line box in Figure 6). IMU and EMG signals were input into the model to carry out convolution and pooling operations, respectively, and two eigenvalues were obtained. Then, the two eigenvalues are fused at the convergence layer and calculated at the full connection layer. Finally, the fusion features after operation are transferred to the output layer to export the classification results. In this process, the parameters of each layer in the model are determined and retained through continuous iteration and updating of weights and deviations by back propagation algorithm. According to the loss rate and accuracy rate, the training state of the current model is judged and the excess number is adjusted in time. Finally, the batch size of the training model was set as 64, the learning rate was 0.001, and the number of iterations was 200. Use ReLu function as activation function of each convolution layer. The cross-entropy loss function is used to measure the gap between the predicted result and the actual result. The above process is the training process of Dual Stream CNN recognition model. Then, the output layer in the trained Dual Stream CNN recognition model is removed, and the remaining network structure is used to extract data features (that is, a feature extractor only used for feature extraction is formed; its structure is shown in the yellow-line box in Figure 6). In this way, the feature extractor can be used to extract fusion features from the data set composed of IMU and EMG. In this process, the feature extractor can perform in-depth feature re-mining for the two different time series information and, finally, extract well differentiated fusion features. However, the features extracted by Dual Stream CNN model still have too high data dimensions. For motion pattern recognition, when the feature dimension of the data sample is high, two aspects will be affected. First, the feature dimension is high, which will cause redundancy among features, affecting the recognition accuracy. Second, the increase of feature dimension will greatly prolong the time of model training. Therefore, it is necessary to reduce the dimension of extracted features. ReliefF is a heuristic search filter method generated by Kononenko [28] based on Kira's work, which can process incomplete data and solve noisy multi-classification and regression problems [29,30]. At the same time, ReliefF provides feature selection with high search accuracy and efficiency. Therefore, the ReliefF feature selection method was introduced to eliminate redundant features in the initial feature set extracted by Dual Stream CNN so as to reduce the dimension of features, thus improving the classification accuracy of the model and reducing the training time of the model. The processing flow of ReliefF algorithm for feature selection is as follows:

(1) First, sample S is randomly selected from the initial feature set.

(2) From the samples with the same label as the sample, samples k are determined according to the nearest neighbor principle, constituting the sample subset H.

(3) According to the nearest neighbor principle, samples k with different labels from the sample are successively determined to form the sample subset M.

(4) Calculate the weight $W(N)$ of each feature N.

$$W(N) = W(N) - \frac{1}{mk}\sum_{i=1}^{k} diff(N,S,H_i) + \frac{1}{mk}\sum_{C\notin class(C)}\left[\frac{P(C)}{1-P(class(S))}\sum_{j=1}^{k} diff(N,S,M_j(C))\right] \quad (1)$$

where $P(C)$ represents the proportion of samples with category C in the training set, and $P(class(S))$ represents the proportion of samples with the same category as the sample S. if N is continuous,

$$diff(N,S_1,S_2) = \frac{|S_1(N) - S_2(N)|}{\max(N) - \min(N)} \quad (2)$$

if N is discrete,

$$diff(N,S_1,S_2) = \begin{cases} 0, S_1(N) = S_2(N) \\ 1, S_1(N) \neq S_2(N) \end{cases} \quad (3)$$

where S_1 and S_2 represent the value of feature N on sample S_1 and S_2, respectively.

(5) The above process is iterated m times in order to select m random samples. Finally, filter the features whose weight corresponding to the calculated features is less than the set threshold.

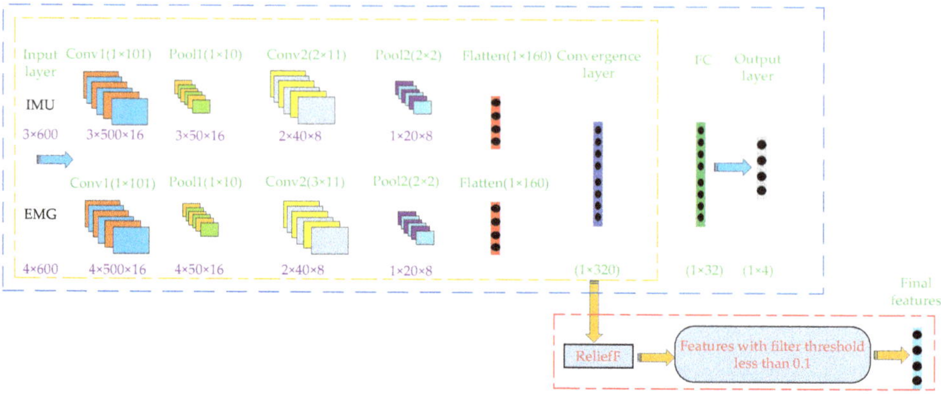

Figure 6. The whole process of feature extraction based on Dual Stream CNN-ReliefF.

The above is the working principle of the Dual Stream CNN-ReliefF feature extraction model proposed in this paper. The whole process of the Dual Stream CNN-ReliefF feature extraction method is shown in Figure 6.

The specific feature extraction and fusion process is as follows:

(1) The data set from IMU sensors is divided into the training set, verification set and test set according to the ratio of 7:1:2, and the data set of EMG sensors is also divided according to the above method. In order to facilitate the input of IMU and EMG data into the model, the input size of IMU data is set as 3×600, where 3 is the number of channels (acceleration, angular velocity, angle), and 600 is the length of input data. Then, set the input size of EMG data as 4×600, where 4 is the number of channels (RF, TA, BF, GA), and 600 is the length of input data.

(2) Build a complete Dual Stream CNN recognition model. The purpose of inputting the training sets of IMU and EMG into the model is to train the model. In the whole process of model training, the convolution kernel size of EMG data twice convolution operation is set to 1×101 and 3×11. The convolution kernel depth is 16 and 8, respectively. The pool kernel size of the two pooling operations is set to 1×10 and 2×2, respectively. Compared with EMG data, IMU data is only different from EMG data in the convolution kernel setting in the second convolution process, which size is set to 2×11, and the other parameters are the same setting. After two convolution and pooling operations, the two types of training sets are expanded into 1×160 eigenvectors, respectively. Then, the two feature vectors are fused and operated in the convergence layer and the full connection layer to generate 1×320 and 1×32 feature vectors, respectively. Finally, the feature vector is input into the Output layer to export the classification results. After the above steps, the final required features can be formed, and the formed features will be transmitted to the classifier for motion pattern recognition. After the above steps, the complete Dual Stream CNN recognition model training is completed.

(3) The training set and test set of IMU and EMG are input into the trained Dual Stream CNN recognition model, and the features before the full connection layer are reserved, which are the fusion features of the training set and test set respectively. Finally, ReliefF algorithm is used to filter the above fusion features with weights lower than 0.1.

After completing the above steps, the fusion features of training set and test set extracted by the Dual Stream CNN-ReliefF model can be obtained, respectively, and the extracted features will be input into different classifiers for motion pattern recognition.

2.6. Classifier

After feature extraction, another key point is to select the classifier. Since when different feature vectors and classifiers are combined, the accuracy of recognition will be different [31,32]. In this paper, we choose the commonly used classifiers in the field of motion pattern recognition, which are SVM, KNN, Decision Tree (DT) and RF [33–35]. In Section 3.2, we input the features extracted based on Dual Stream CNN-ReliefF, Single Stream CNN and manual design methods into the four classifiers, respectively, for comparative analysis of recognition accuracy.

3. Results and Discussions

3.1. Visual Analysis of Extracted Features Based on Different Methods and Different Types of Sensor Data

In order to verify the feature extraction capability of the Dual Stream CNN-ReliefF method proposed in this paper, it was used for feature extraction of fused EMG and IMU data, while Single Stream CNN was used for feature extraction of single EMG, single IMU, and fusion data, respectively. The extracted features are visualized in 3D images. The feature visualization images extracted based on Dual Stream CNN-ReliefF are shown in Figure 7a, and the feature visualization images extracted based on Single Stream CNN are shown in Figure 7b–d. Then, time and frequency domain features were extracted from single EMG, single IMU, and their fusion data by manual design method. Some features with better effects were visualized as shown in Figure 7e,f, respectively.

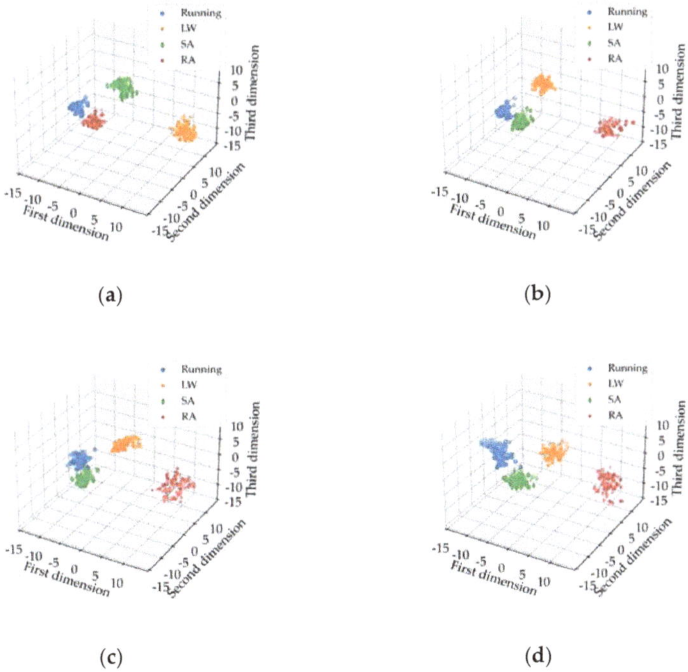

Figure 7. *Cont.*

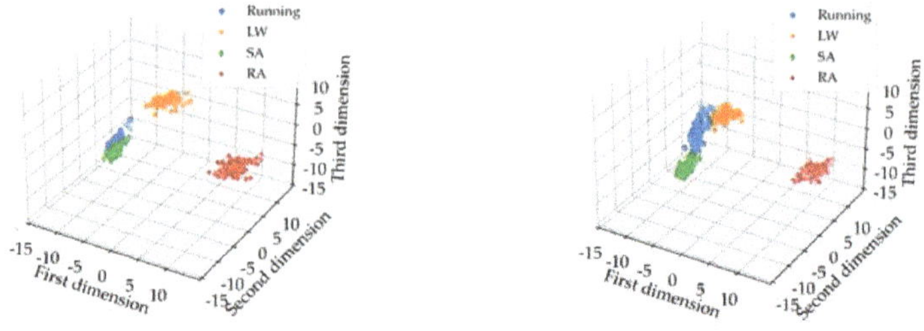

(e) (f)

Figure 7. Visualization of feature extraction based on different methods and different types of sensor data. LW (level ground walking), SA (stair ascent), RA (ramp ascent). (**a**) represent the features extracted by using Dual Stream CNN-ReliefF based on the fusion of EMG and IMU data; (**b–d**) represent the features extracted by using Single Stream CNN based on a single EMG, a single IMU and the fusion of the two data, respectively; (**e**,**f**) represent the features with better visual effects extracted by manual design based on a single EMG, a single IMU and the fusion of the two data, respectively.

As can be seen from Figure 7, the feature aggregation of the same motion mode and the feature differentiation of different motion modes extracted by Dual Stream CNN-ReliefF method are obviously better than those extracted by Single Stream CNN and manual design. The features extracted by Single Stream CNN and manual design have partial aliasing and poor feature aggregation. In addition, it can be seen that the features of the motion modes extracted by Dual Stream CNN-ReliefF and Single Stream CNN based on the data fusion are better than the features extracted by single EMG and single IMU data. In conclusion, we can verify that the proposed Dual Stream CNN-ReliefF method can effectively extract the features of each motion mode from the original signals. Furthermore, it can effectively characterize the four motion modes at the feature visualization performance. The visual performance of the extracted features based on the fusion of EMG and IMU data is better than that of the extracted features based on EMG and IMU data alone.

3.2. Recognition Accuracy Analysis of Extracted Features Based on Different Methods and Different Types of Sensor Data

In order to compare the difference in feature extraction ability among Single Stream CNN, manual design and Dual Stream CNN-ReliefF. The Dual Stream CNN-ReliefF method is used to extract features from the fused EMG and IMU data. Then these features are input into SVM, KNN, DT, and RF classifiers based on five-fold cross-validation to obtain the accuracy of motion pattern recognition. Meanwhile, the Single Stream CNN and manual design are used to extract features from a single EMG signal, a single IMU, and the fusion of the two data. Then, these features are also input into SVM, KNN, DT and RF classifiers based on five-fold cross-validation to obtain the accuracy of motion pattern recognition.

The recognition accuracy of feature extraction based on Dual Stream CNN-ReliefF and Single Stream CNN is shown in Figure 8a–d, and the recognition accuracy of feature extraction based on manual design is shown in Figure 8e–h. In addition, in the abscissa of Figure 8a–d, ISL, ESL, and IESL represent feature extraction of single IMU data, EMG data and fusion of two kinds of data based on Single Stream CNN algorithm, respectively. IEDL means feature extraction of fused EMG and IMU data based on Dual Stream CNN-ReliefF. In Figure 8e–h abscissa, ITD, IFD and ITF represent the time and frequency features extracted based on single IMU data, as well as the combination of time domain features and frequency domain features, respectively. ETD, EFD, and ETF represent the time and frequency features extracted based on single EMG data, as well as the combination of time domain features and frequency domain features, respectively. "+" indicates the fusion of time domain or frequency domain features extracted from EMG or IMU data.

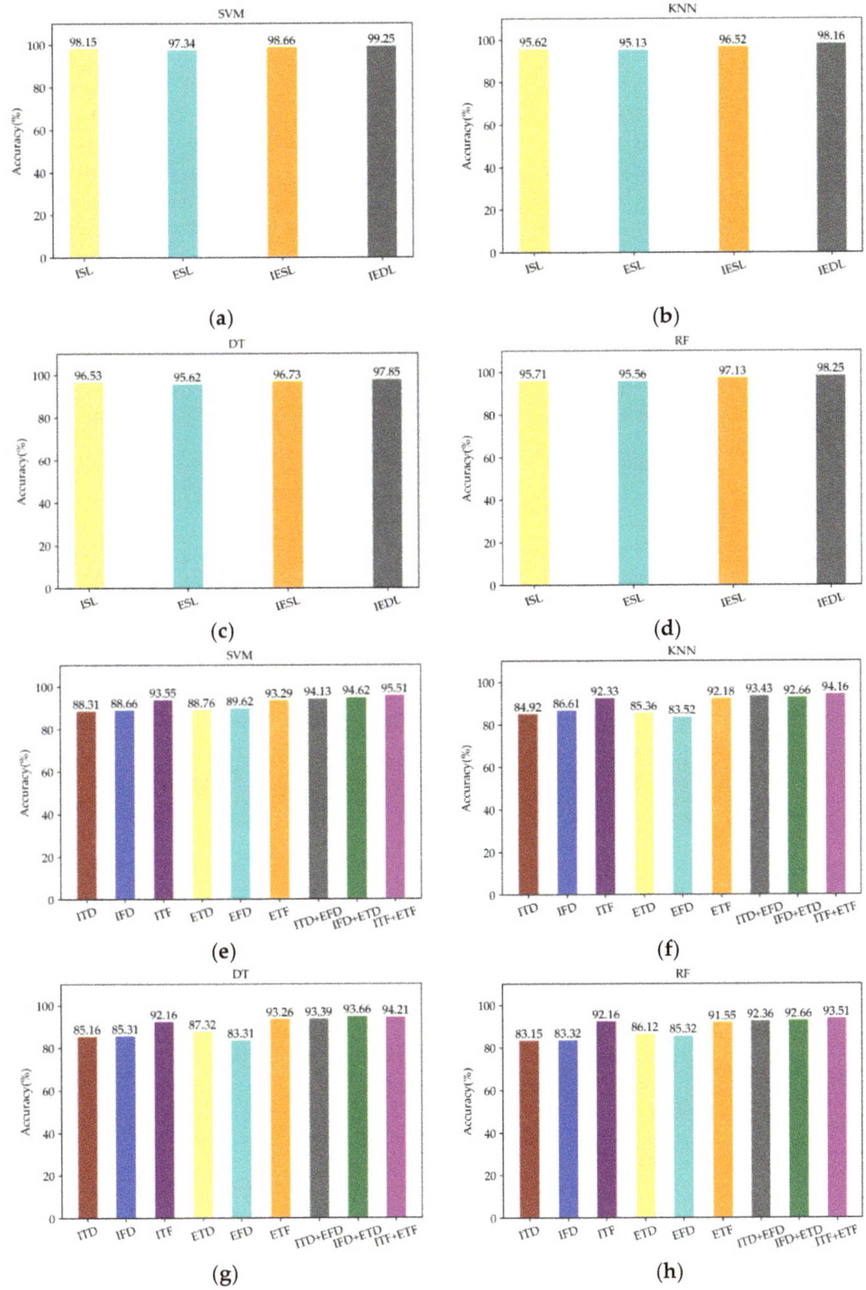

Figure 8. Recognition accuracy analysis of extracted features based on different methods and different types of sensor data. (**a**–**d**) represent the recognition accuracy of features extracted based on Dual Stream CNN-ReliefF and Single Stream CNN under SVM, KNN, DT and RF classifiers, respectively; (**e**–**h**) represent the recognition accuracy of features extracted based on manual design under SVM, KNN, DT and RF classifiers, respectively.

As can be seen from Figure 8, the recognition accuracy of the features extracted by Dual Stream CNN-ReliefF is higher than that extracted by Single Stream CNN and manual design methods in the four classification algorithms, furthermore, the recognition accuracy of the four classification algorithms reaches more than 97%, and the best recognition accuracy can reach 99.25%. The feature extraction method based on manual design not only has tedious steps, but also the combination of different features will affect the recognition accuracy. Therefore, researchers need to rely on manual experience to pick out feature combinations with high recognition accuracy. However, relying only on manual experience is not always able to select the best combination of features, so the best recognition accuracy may not be obtained. In this paper, the best recognition accuracy is only 95.51% by selecting different feature combinations. In addition, as can be seen from Figure 8, the data based on EMG and IMU signals fusion in four kinds of classification algorithm accuracy are higher than single sensor (EMG or IMU) data. This indicates that the biological information and kinematic information brought by EMG and IMU signals can better represent the motion status of human lower limbs, thus improving the accuracy of human motion pattern recognition, indicating that the data fusion of the two sensors is effective.

3.3. Discussion

In order to further verify the generalization ability of the features extracted by the Dual Stream CNN-ReliefF method for motion pattern recognition of different subjects. We used the Dual Stream CNN-ReliefF method for feature extraction of EMG and IMU data from six healthy subjects. The features of each subject are input into SVM, KNN, DT, and RF classifiers based on five-fold cross-validation to obtain the accuracy of motion pattern recognition. The specific results are shown in Table 4.

Table 4. The accuracy of motion pattern recognition of different subjects under five-fold cross-validation.

Subject	Age	Height [cm]	SVM	KNN	DT	RF	Average Accuracy
S1	23	167	99.25%	98.16%	97.85%	98.25%	98.37%
S2	24	170	99.46%	99.15%	98.36%	98.63%	98.90%
S3	26	165	99.69%	97.68%	99.13%	97.66%	98.54%
S4	28	172	99.13%	99.38%	97.46%	98.17%	98.53%
S5	29	176	99.73%	99.34%	98.25%	99.16%	99.12%
S6	30	173	99.67%	98.61%	98.74%	99.31%	99.08%

As can be seen from Table 4, the motion pattern recognition accuracy of each subject under the four classifiers is above 97%, with the highest average recognition accuracy reaching 99.12%.

In addition, we compare the wearable bioelectronics device and the Dual Stream CNN-ReliefF feature extraction method proposed in this paper with other references. The specific comparison is shown in Table 5.

Table 5. Comparison of the methods and experimental results.

Research	Sensor	Feature	Motion Pattern	Accuracy
Song [8]	EMG	9 time-frequency features	5 motion patterns	97.5%
Peng [11]	Plantar pressure	5 time-frequency features	5 motion patterns	91.1%
Dhindsa [13]	EMG	15 time-frequency features	2 motion patterns	92.2%
Gupta [14]	Acceleration	7 time-frequency features	6 motion patterns	>95%
This work	EMG + IMU	Dual Stream CNN-ReliefF	4 motion patterns	99.12%

Since the movement of human lower limbs is complex, which involves different kinds of information type. Compared with Re [8,11,13,14], the information obtained by a single sensor is insufficient to characterize the motion of human lower limbs, which will affect the recognition accuracy of human motion patterns. The wearable bioelectronics device proposed in this paper can collect more information of human lower limb motion, namely bioelectric information (EMG) and biological kinematics information (acceleration, angular velocity and angle of joint motion). It is suggested that the combination of these two kinds of information is more helpful to improve the accuracy of motion pattern recognition. In addition, Refs. [8,11,13,14] adopts the traditional manual design method to extract features. The steps of extracting features based on this method are not only tedious, but also the features selected by relying on human experience may not obtain best recognition accuracy when they are input into the classifier. In contrast, the Dual Stream CNN-ReliefF method proposed in this paper can automatically extract well differentiated fusion features from EMG and IMU data. The proposed method was used to extract features from six subjects, and these features were input into four different classifiers (SVM, KNN, DT, RF). The motion pattern average recognition accuracy obtained through five-fold cross-validation was up to 99.12%. However, compared with other references, the inadequacy of this paper is the lack of more research on the types of human lower motion pattern, such as jumping, squatting and so on.

Therefore, the wearable bioelectronics device based on EMG and IMU with the Dual Stream CNN-ReliefF feature extraction method proposed in this paper can improve the accuracy of human motion pattern recognition, and have better generalization ability for the motion pattern recognition of different subjects.

4. Conclusions

This paper firstly analyzed the motion mechanism of human lower limbs, and designed a set of wearable bioelectronics device based on EMG and IMU sensors, which can obtain biological information and kinematics information of human lower limbs. Based on the information fusion, we collected the lower limb movement information of four common movement modes of human body. Then, a feature extraction method based on Dual Stream CNN-ReliefF is proposed, which can automatically extract well differentiated fusion features from EMG and IMU data. Finally, the Dual Stream CNN-ReliefF was adopted for feature extraction of the fusion data of EMG and IMU. Meanwhile, the Single Stream CNN and manual design were used for feature extraction of single EMG, single IMU, and their fusion data, respectively. The extracted features based on the above three methods are analyzed and compared in feature visualization and recognition accuracy. We concluded that the wearable bioelectronics device information fusion and Dual Stream CNN-ReliefF feature extraction method proposed in this paper enhanced an exoskeleton's ability to capture human movement patterns, thus providing optimal assistance to the human body at the appropriate time. Therefore, it can provide a novel approach for improving the human-machine interaction of exoskeletons. In the future, on the one hand, we will study the recognition of more motion patterns of human lower limbs; on the other hand, we will conduct indepth research on the motion pattern recognition of human upper limbs, so as to provide certain direction for the motion pattern recognition of exoskeletons in human upper and lower limbs.

Author Contributions: Conceptualization, Y.M. and X.L.; resources, X.L.; data curation, D.W. and J.L.; validation, J.L.; writing—original draft preparation, X.L. and J.L.; writing—review and editing, X.L. and Y.H. All authors have read and agreed to the published version of the manuscript.

Funding: This research was funded by the Natural Science Foundation of China under Grant 62103007 and Beijing Natural Science Foundation under Grant L202020.

Institutional Review Board Statement: Ethical review and approval were waived for this study, due to no enforced nor uncomfortable restriction to the human wearer during the study period.

Data Availability Statement: The data presented in this study are available on request from the corresponding author on reasonable request.

Conflicts of Interest: The authors declare no conflict of interest.

References

1. Zheng, Y.; Song, Q.; Liu, J.; Song, Q.; Yue, Q. Research on motion pattern recognition of exoskeleton robot based on multimodal machine learning model. *Neural. Comput. Appl.* **2020**, *32*, 1869–1877. [CrossRef]
2. Kalita, B.; Narayan, J.; Dwivedy, S.K. Development of Active Lower Limb Robotic-Based Orthosis and Exoskeleton Devices: A Systematic Review. *Int. J. Soc. Robot* **2021**, *13*, 775–793. [CrossRef]
3. Pérez Vidal, A.F.; Rumbo Morales, J.Y.; Ortiz Torres, G.; Sorcia Vázquez, F.D.J.; Cruz Rojas, A.; Brizuela Mendoza, J.A.; Rodríguez Cerda, J.C. Soft Exoskeletons: Development, Requirements, and Challenges of the Last Decade. *Actuators* **2021**, *10*, 166. [CrossRef]
4. Islam, M.R.; Assad-Uz-Zaman, M.; Brahmi, B.; Bouteraa, Y.; Wang, I.; Rahman, M.H. Design and Development of an Upper Limb Rehabilitative Robot with Dual Functionality. *Micromachines* **2021**, *12*, 870. [CrossRef]
5. Wang, X.; Feng, Y.; Zhang, J.; Li, Y.; Niu, J.; Yang, Y.; Wang, H. Design and Analysis of a Lower Limb Rehabilitation Training Component for Bedridden Stroke Patients. *Machines* **2021**, *9*, 224. [CrossRef]
6. Shi, Y.; Guo, M.; Zhong, H.; Ji, X.; Xia, D.; Luo, X.; Yang, Y. Kinetic Walking Energy Harvester Design for a Wearable Bowden Cable-Actuated Exoskeleton Robot. *Micromachines* **2022**, *13*, 571. [CrossRef]
7. Zhang, X.; Tao, S. Research on Pattern Recognition of Lower Limb Motion Based on Convolutional Neural Network. *Wirel. Commun. Mob. Comput.* **2022**, *2022*, 4717413. [CrossRef]
8. Song, G.; Wang, Y.; Wang, M.; Li, Y. Lower Limb Movement Intent Recognition Based on Grid Search Random Forest Algorithm. In Proceedings of the 3rd International Conference on Robotics, Control and Automation, Chengdu, China, 11–13 August 2018; pp. 225–229.
9. Lopez-Delis, A.; Delisle-Rodriguez, D.; Villa-Parra, A.C.; Bastos-Filho, T. Knee Motion Pattern Classification from Trunk Muscle Based on sEMG Signals. In Proceedings of the 2015 37th Annual International Conference of the IEEE Engineering in Medicine and Biology Society (EMBC), Milan, Italy, 25–29 August 2015; pp. 2604–2607.
10. Xi, X.; Tang, M.; Miran, S.M.; Luo, Z. Evaluation of Feature Extraction and Recognition for Activity Monitoring and Fall Detection Based on Wearable sEMG Sensors. *Sensors* **2017**, *17*, 1229. [CrossRef]
11. Peng, Z.; Cao, C.; Liu, Q.; Pan, W. Human Walking Pattern Recognition Based on KPCA and SVM with Ground Reflex Pressure Signal. *Math. Probl. Eng.* **2013**, *2013*, 143435. [CrossRef]
12. Zhang, J.; Soangra, R.; Lockhart, T.E. Automatic Detection of Dynamic and Static Activities of the Older Adults Using a Wearable Sensor and Support Vector Machines. *Science* **2020**, *2*, 62. [CrossRef]
13. Dhindsa, I.S.; Agarwal, R.; Ryait, H.S. Performance evaluation of various classifiers for predicting knee angle from electromyography signals. *Expert Syst.* **2019**, *36*, e12381. [CrossRef]
14. Gupta, P.; Dallas, T. Feature Selection and Activity Recognition System Using a Single Triaxial Accelerometer. *IEEE Trans. Bio-Med. Eng.* **2014**, *61*, 1780–1786. [CrossRef] [PubMed]
15. Fang, C.; He, B.; Wang, Y.; Cao, J.; Gao, S. EMG-Centered Multisensory Based Technologies for Pattern Recognition in Rehabilitation: State of the Art and Challenges. *Biosensors* **2020**, *10*, 85. [CrossRef]
16. Li, J.; Li, G.; Zhang, L.; Yang, D.; Wang, H. Advances and Key Techniques of Soft Wearable Lower Limb Power-assisted Robots. *Acta Autom. Sin.* **2020**, *46*, 427–438.
17. Chen, W.; Li, J.; Zhu, S.; Zhang, X.; Men, Y.; Wu, H. Gait Recognition for Lower Limb Exoskeletons Based on Interactive Information Fusion. *Appl. Bionics Biomech.* **2022**, *2022*, 9933018. [CrossRef] [PubMed]
18. Chen, C.; Jafari, R.; Kehtarnavaz, N. A survey of depth and inertial sensor fusion for human action recognition. *Multimed. Tools Appl.* **2017**, *76*, 4405–4425. [CrossRef]
19. Wang, J.; Wang, L.; Xi, X.; Miran, S.M.; Xue, A. Estimation and Correlation Analysis of Lower Limb Joint Angles Based on Surface Electromyography. *Electronics* **2020**, *9*, 556. [CrossRef]
20. Gao, F.; Tian, T.; Yao, T.; Zhang, Q. Human Gait Recognition Based on Multiple Feature Combination and Parameter Optimization Algorithms. *Comput. Intell. Neurosci.* **2021**, *2021*, 6693206. [CrossRef]
21. Novak, D.; Riener, R. A survey of sensor fusion methods in wearable robotics. *Robot. Auton. Syst.* **2015**, *73*, 155–170. [CrossRef]
22. Yin, Z.; Zheng, J.; Huang, L.; Gao, Y.; Peng, H.; Yin, L. SA-SVM-Based Locomotion Pattern Recognition for Exoskeleton Robot. *Appl. Sci.* **2021**, *11*, 5573. [CrossRef]
23. Bunderson, N.E.; Kuiken, T.A. Quantification of Feature Space Changes with Experience during Electromyogram Pattern Recognition Control. *IEEE Trans. Neural Syst. Rehabil. Eng.* **2012**, *20*, 239–246. [CrossRef]
24. Phinyomark, A.; Phukpattaranont, P.; Limsakul, C. Feature reduction and selection for EMG signal classification. *Expert Syst. Appl.* **2012**, *39*, 7420–7431. [CrossRef]
25. Afzal, T.; White, G.; Wright, A.B.; Iqbal, K. Locomotion Mode Identification for Lower Limbs using Neuromuscular and Joint Kinematic Signals. In Proceedings of the 2014 36th Annual International Conference of the IEEE Engineering in Medicine and Biology Society, Chicago, IL, USA, 26–30 August 2014; pp. 4071–4074.

26. Chen, J.; Sun, Y.; Sun, S. Improving Human Activity Recognition Performance by Data Fusion and Feature Engineering. *Sensors* **2021**, *21*, 692. [CrossRef] [PubMed]
27. Yang, T.; Chen, Z.; Yue, W. Spatio-temporal two-stream human action recognition model based on video deep learning. *J. Comput. Appl.* **2018**, *38*, 895.
28. Kononenko, I. Estimating attributes: Analysis and Extensions of Relief, European Conference on Machine Learning. In Proceedings of the European Conference on Machine Learning (ECML), Catania, Italy, 6–8 April 1994; Springer: Berlin/Heidelberg, Germany, 1994; pp. 171–182.
29. Padovan, B.; Sackmann, S.; Eymann, T.; Pippow, I. A Prototype for an Agent-Based Secure Electronic Marketplace Including Reputation-Tracking Mechanisms. *Int. J. Electron. Commer.* **2002**, *6*, 93–113. [CrossRef]
30. Azzedin, F.; Maheswaran, M. Integrating Trust into Grid Resource Management Systems. In Proceedings of the International Conference on Parallel Processing, Vancouver, BC, Canada, 21–21 August 2002; pp. 47–54.
31. Huang, H.; Kuiken, T.A.; Lipschutz, R.D. A Strategy for Identifying Locomotion Modes Using Surface Electromyography. *IEEE Trans. Bio-Med. Eng.* **2008**, *56*, 65–73. [CrossRef]
32. Scheme, E.; Englehart, K. Electromyogram pattern recognition for control of powered upper-limb prostheses: State of the art and challenges for clinical use. *J. Rehabil. Res. Dev.* **2011**, *48*, 643–659. [CrossRef]
33. Shi, K.; Huang, R.; Mu, F.; Peng, Z.; Yin, J.; Cheng, H. Channel Synergy-based Human-Robot Interface for a Lower Limb Walking Assistance Exoskeleton. In Proceedings of the 2021 43rd Annual International Conference of the IEEE Engineering in Medicine and Biology Society (EMBC), Electronic Network, Mexico, 1–5 November 2021; pp. 1076–1081.
34. Wu, Y.; Hu, X.; Wang, Z.; Wen, J.; Kan, J.; Li, W. Exploration of Feature Extraction Methods and Dimension for sEMG Signal Classification. *Appl. Sci.* **2019**, *9*, 5343. [CrossRef]
35. Olmo, M.D.; Domingo, R. EMG Characterization and Processing in Production Engineering. *Materials* **2020**, *13*, 5815. [CrossRef]

micromachines

Article

Screw Analysis, Modeling and Experiment on the Mechanics of Tibia Orthopedic with the Ilizarov External Fixator

Peng Su [1], Sikai Wang [1], Yuliang Lai [1], Qinran Zhang [1] and Leiyu Zhang [2,*]

[1] School of Electromechanical Engineering, Beijing Information Science and Technology University, Beijing 100124, China; supeng@bistu.edu.cn (P.S.); skywang0306@foxmail.com (S.W.); laiyuliang666@163.com (Y.L.); zhangqinran@bistu.edu.cn (Q.Z.)

[2] Beijing Key Laboratory of Advanced Manufacturing Technology, Beijing University of Technology, Beijing 100124, China

* Correspondence: zhangleiyu@bjut.edu.cn; Tel.: +86-010-15210986251

Abstract: The Ilizarov external fixator plays an important role in the correction of complex malformed limbs. Our purpose in this work was to reveal the transmission of adjustable forces between the external fixator and the broken bone, and express the stress distribution at the end of the broken bone during the orthopedic treatment. Firstly, the screw model of the fixator was established and the theoretical relationship between the adjustable force and the stress was obtained. A sheep tibia was taken as a representative research object and its ediTable 3D entity was obtained by CT scanning. Then the mechanical model of the fixator and tibia was built using the ABAQUS software. Correction experiments were performed on the sheep tibia to measure the adjustable/support forces and tensions of the tibia. The measured results were imported to the screw and mechanical model, and the theoretical and simulation values were calculated. The theoretical tensions calculated by the screw model had a similar shape and doubled the value compared with that of the measured results. The transfer efficiency between the two results was improved and kept at about 50% after the initial 2~3 periods. The maximum stress occurring at the surface of the broken bone end was near the Kirschner wire pinhole. The simulation results for the tensions from the mechanical model showed a similar change trend, and the value was slightly higher. A biomechanical model of the Ilizarov external fixator was derived and verified through calculations, simulations and experiments. The change law of the adjustable forces and the tensions existing in the broken sheep tibias is presented herein, and offers a helpful contribution to orthopedic treatment.

Keywords: Ilizarov external fixator; screw theory; orthopedic treatment; tension of broken bone; stress distribution

check for updates

Citation: Su, P.; Wang, S.; Lai, Y.; Zhang, Q.; Zhang, L. Screw Analysis, Modeling and Experiment on the Mechanics of Tibia Orthopedic with the Ilizarov External Fixator. *Micromachines* **2022**, *13*, 932. https://doi.org/10.3390/mi13060932

Academic Editors: Xiao Xiao and Emmanuel Stratakis

Received: 11 April 2022
Accepted: 9 June 2022
Published: 11 June 2022

1. Introduction

As a surgical instrument, the Ilizarov external fixator plays an important role in the correction of complex malformed limbs. It can adjust the configuration of the bone external fixator according to the treatment progress, and may affect the results of surgery in the orthopedic process. Watson proposed that the engineering structure of the external fixator is related to bone healing [1]. However, this orthopedic treatment is mostly based on doctors' experience, and its orthopedic mechanism, including the tension at the broken bone and the transfer efficiency, has not been well studied. A mechanical study of the external fixator can reveal the mechanical relationship between the adjustable force and the broken bone end, which is helpful for doctors in adjusting external fixators [2].

As a practical and simple theoretical calculation method, screw theory has attracted the attention of many researchers and has been widely used in kinematics and static analysis of the mechanism. Based on the screw theory, Niu J, Wang H, Shi H, et al. analyzed the constraints and basic motion of the leg mechanism and solved the kinematics of the leg mechanism [3]. Researchers from Beihang university proposed an analytical algorithm for measuring three-dimensional forces and three-dimensional torques, which can achieve high precision measurement of interactive forces on the Stewart platform [4]. Cunfeng Kang and Zixiao Liu et al. proposed a position transformation method that can generate circular trajectory more accurately [5]. To investigate the motion around the libration points, Yingjing Qian, Zixiao Liu et al. transformed the origin of the coordinate system from the center of the system to translational points through coordinate transformation [6]. Further study of screw theory reveals that it also provides various feasible schemes in the configuration of parallel mechanism, such as 3CRU translational parallel mechanism, 2-UR-RRU parallel robot, etc. [7,8], which provide ideas for theoretical analysis and experimental study of the Ilizarov external fixator.

Due to the special environment of bone injury in the body, it is impossible to reproduce the true intervention process of orthopedic devices on broken bones. Therefore, many researchers have studied the relationship between various orthopedic devices and fracture healing through finite element analysis. The axial tension and compression of the tibia are real-time detected by simulating the actual motion state of the bone external fixator [9]. Karunratanakul K used a finite element model to predict the stiffness of the fixer [10]. Malayan researchers used polyethylene tubes to represent the tibia bone and applied the above methods to study the mechanical behaviors of different configurations of bone external fixators [11]. Although the various configurations and biological environment of the fixator have been discussed above, few of them have studied the stress and strain of the fixator and the tibia's broken end during the entire orthopedic process.

The mechanical model of the Ilizarov external fixator is the basis of numerical simulation, which can provide better prediction of mechanical properties and comparative analysis of the correction process. This study intended to establish an accurate mechanical model of the fixator and simulate the force transmission process in the normal fracture healing process using the screw theory. A tibia fracture correction and recovery experiment was performed on a special experimental bench. Referring to Cunfeng Kang and Zixiao Liu et al., the finite element analysis of the welding pipeline was carried out and compared with the experiment of [12], the measured data were imported into the mechanical model and the loading process was simulated using ABAQUS software. The simulated and actual tensions and stresses of the tibia were compared to improve understanding of the force transmission relationship between the Ilizarov external fixator and the broken tibia end.

2. Materials and Methods

2.1. Screw Model of Ilizarov External Fixator

The Ilizarov external fixator has various configurations in clinical applications. The specifications of the hole rings and the diameters and angles of the Kirschner wires effect the healing of broken bone tissue [9,13] and are determined according to the actual situations of patients. The corresponding classic parameters are shown in Table 1.

Table 1. Classic parameters of Ilizarov external fixator.

Parameter	Value	Parameter	Value
Hole ring	2	Diameter of hole ring	190 mm
One-way hinge	4 pairs	Diameter of Kirschner wire	2.5 mm
Broken bone space	10 mm	Pulling speed	1 mm/d
Tibia varus angle	6.85°	Number of adjustments	10 times

In orthopedic treatment, the mechanical analysis between the adjustable rods and the tibial broken end is insufficient, and it is difficult to achieve quantitative orthopedic and precise assessment, which greatly hinders the corresponding applications. The general kinematics analysis method is the D-H parameter method, but this method has a fundamental defect: all the motion is in the x-axis and z-axis, while it cannot represent the motion in the y axis. Therefore, it is necessary to use another method to establish the kinematics model: the screw theory. Using the screw to describe the kinematics of the rigid body has two advantages: first, it only needs two coordinate systems—the basic coordinate system and the tool coordinate system—to describe the motion of the rigid body from the whole, so as to avoid the singularity caused by using a local coordinate system in the D-H parameter method; secondly, the screw method describes the geometric meaning of rigid body motion clearly, which avoids the disadvantages of abstract mathematical symbols, thus greatly simplifying the analysis of the mechanism [14]. The screw theory can describe the rigid motion of each joint as a rotation around the joint axis and a translational motion along this axis. In the existing external fixator, only two of the screw coordinate systems of the broken bone end and the adjustable/support rod with respect to the base platform need to be established to avoid the singularities of the traditional D-H parametric method and simplify the kinematics and dynamics analyses [15]. Meanwhile, the rigid motion in the orthopedic treatment can be clearly described.

In the kinematics analysis, the motion screw and wrench represent an instantaneous motion of a rigid body and the force/moment acting on a rigid body, respectively, where there exists a dual relationship between them [16,17]. Hence, a set of dual vectors in space can represent the angular/linear velocities in the kinematics and the force/moment in rigid-body mechanics. The wrench can be compounded by a force along an axis and a torque around the axis. The generalized force acting on the rigid body consists of a moving component f (pure force) and a rotating component τ (pure torque), which can be expressed in a six-dimensional vector in the inertial coordinate system [18]:

$$F = \begin{pmatrix} f \\ \tau \end{pmatrix} = \begin{pmatrix} s \\ r \times s + hs \end{pmatrix}, \quad h\ is\ finite, \tag{1}$$

$$F = \begin{pmatrix} 0 \\ \tau \end{pmatrix} = \begin{pmatrix} 0 \\ s \end{pmatrix}, \ h = \infty, \tag{2}$$

where f, $\tau \in R^3$, h is called the pitch and denotes the ratio of moment to force, $r \times s$ is the distance of the axis of rotation from the origin of the inertial coordinate system $\{o_1\}$, and F denotes the wrench along the screw motion s.

The adjustable and support rods are defined as two kinematic chains by the screw theory. Since the adjustable and support rods are symmetrical and parallel to each other, they possess the same kinematic features, and only one of them needs to be analyzed in detail. During the orthopedic treatment, the movements of the kinematic chain can be decomposed into a translation motion along the rod and a rotational motion around the hinge. The adjustable forces can be transmitted to the moving platform and achieve predetermined orthopedic movements of the severed tibia. The inertial coordinate system $\{o_1\}$ of the adjustable rod and the coordinate system $\{o_2\}$ of the broken tibia end are established on the initial posture of this fixator, as shown in Figure 1.

During the orthopedic period, the adjustable rod only has a translational motion along the axis o_1z_1, and the spinor ξ_1 of the rod can be expressed as:

$$w_1 = \begin{pmatrix} 0 \\ 0 \\ 0 \end{pmatrix}, v_1 = \begin{pmatrix} 0 \\ 0 \\ 1 \end{pmatrix}, \xi_1 = \begin{pmatrix} 0 \\ v_1 \end{pmatrix} = \begin{pmatrix} 0 \\ 0 \\ 0 \\ 0 \\ 0 \\ 1 \end{pmatrix},$$

where w_1 and v_1 are defined as the angular/linear velocity of the adjustable rod. However, the adjustable rod moves only along the axis o_1z_1 in the initial stage of orthosis. The spinor ξ_1 only contains the velocity v_1, which is straight up, and the angular velocity w_1 is equal to zero.

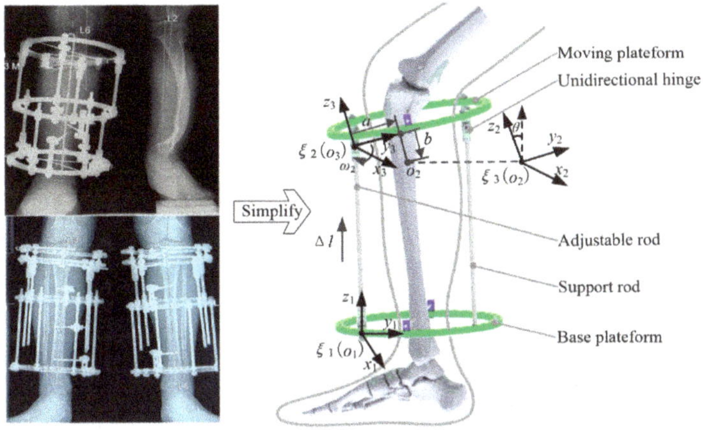

Figure 1. Kinematic model of Ilizarov external fixator.

The upper end of the adjustable rod is hinged to a unidirectional hinge. The unidirectional hinge has a rotational motion around the axis o_2x_2, where the corresponding spinor ξ_2 is:

$$w_2 = \begin{pmatrix} 1 \\ 0 \\ 0 \end{pmatrix}, r_2 = \begin{pmatrix} 0 \\ 0 \\ l \end{pmatrix}, \xi_2 = \begin{pmatrix} w_2 \\ r_2 \times w_2 \end{pmatrix} = \begin{pmatrix} 1 \\ 0 \\ 0 \\ 0 \\ l \\ 0 \end{pmatrix},$$

where w_2 and v_2 are defined as the angular/linear velocity of the unidirectional hinge, and r_2 denotes the corresponding position in the coordinate system $\{o_1\}$.

The Lie algebra $\hat{\xi}_i$ ($i = 1, 2$) attitude change matrix $e^{\theta_i \hat{\xi}_i}$ of the spinor ξ_i can be calculated:

$$\hat{\xi}_i = \begin{bmatrix} \hat{\omega}_i & v_i \\ 0 & 0 \end{bmatrix}, \quad (3)$$

$$e^{\theta_i \hat{\xi}_i} = \begin{bmatrix} e^{\theta_i \hat{w}_i} & (I - e^{\theta_i \hat{w}_i})(w_i \times v_i) + \theta_i \cdot w_i \cdot w_i^T \cdot v_i \\ 0 & 1 \end{bmatrix}, \quad (4)$$

Substituting Equation (4) into Equation (5), the function of $e^{\theta_i \hat{\xi}_i}$ with respect to Δl can be solved:

$$e^{\theta_1 \hat{\xi}_1} = \begin{pmatrix} 1 & 0 & 0 & 0 \\ 0 & 1 & 0 & 0 \\ 0 & 0 & 1 & \Delta l \\ 0 & 0 & 0 & 1 \end{pmatrix}, \quad (5)$$

$$e^{\theta_2 \hat{\xi}_2} = \begin{pmatrix} 1 & 0 & 0 & 0 \\ 0 & \cos\theta & -\sin\theta & l\sin\theta \\ 0 & \sin\theta & \cos\theta & l(1-\cos\theta) \\ 0 & 0 & 0 & 1 \end{pmatrix}, \quad (6)$$

where θ_i denotes the rotation angle of the unidirectional hinge and l and Δl are the length and the change of the adjustable rod, respectively.

Substituting Equations (6) and (7) into the exponential product formula, the posture of the fixator can be obtained:

$$p(\theta) = e^{\theta_1 \hat{\xi}_1} e^{\theta_2 \hat{\xi}_2} p(0) = \begin{pmatrix} 1 & 0 & 0 & 0 \\ 0 & \cos\theta & -\sin\theta & l\sin\theta \\ 0 & \sin\theta & \cos\theta & \Delta l + l(1-\cos\theta) \\ 0 & 0 & 0 & 1 \end{pmatrix}, \tag{7}$$

where $p(\theta)$ is the attitude matrix of the unidirectional hinge and $p(0)$ is the attitude matrix of the initial posture in the coordinate system $\{o_1\}$:

$$p(0) = \begin{bmatrix} I & \begin{pmatrix} 0 \\ l \\ 0 \end{pmatrix} \\ 0 & 1 \end{bmatrix}.$$

The upper broken tibia is installed on the moving platform through the Kirschner wires. The moving platform drives the upper broken tibia to achieve the correction function with help of the hinge. The coordinate system $\{o_3\}$ is established at the center of the hinge. The coordinate system $\{o_2\}$ of the broken tibia end is obtained through shifting the coordinate system $\{o_3\}$ by the displacements a and b along the axes o_3y_3 and o_3z_3. Similarly, according to the transformation formula of the motion screw, the relationship between the wrenches $F(f_i\ \tau_i)$ (i = 1, 2) and $F(f_{o_2}\ \tau_{o_2})$ of the adjustable rod and the broken tibia end can be established.

$$F(f_{o_2}\ \tau_{o_2}) = R_x F(f_i\ \tau_i) = \begin{pmatrix} 1 & 0 & 0 \\ 0 & \cos\theta & -\sin\theta \\ 0 & \sin\theta & \cos\theta \end{pmatrix} F(f_i\ \tau_i), \tag{8}$$

where f_i and τ_i denote the force and the wrench in the coordinate system $\{o_1\}$.

2.2. Mechanical Model of Ilizarov External Fixator

A finite element model of the Ilizarov external fixator needs to be established corresponding to the screw model. In order to reduce the error caused by the clearance, the connection part is simplified, and the external fixator is divided into seven main parts: base platform, moving platform, adjustable rod 1, adjustable rod 2 and support rod3, support rod4, and tibia, as shown in Figure 2. At the same time, the operation is simplified and the force is applied directly on the adjustable rod, so that the measured value obtained is more accurate. The proximal and distal hole rings are the base and moving platforms. Two pieces of the broken tibia are fixed by Kirschner wires, which are installed on the hole ring through retaining clips. The adjustable and support rods are inserted in the proximal hole ring and locked during the orthopedic treatment. Similarly, the rods are connected to the distal hole ring through the unidirectional hinges. The hinges are inserted in the distal hole ring and fixed by nuts. The hinges, nuts and distal hole rings are taken as a part, then the rods and the proximal hole ring are taken as another part. The moving platform can rotate around the axes of the two hinges connected to the support rods.

The original 3D model of the sheep tibia is taken as the subject and obtained by CT scanning. The ediTable 3D entity can be generated by the softwares Mimics and Geomagics. The tibia entity is imported into the ABAQUS software to analyze its mechanical properties. The Ilizarov external fixator is made of stainless steel, which possesses the properties of the linear elasticity and isotropy. The tibia is composed of cancellous and compact bones, and osteoporosis affects their function, distribution areas and load capacity [19]. Hence, the tibia can be simplified as an isotropic material composed of cancellous and compact bones. The proper values of the Poisson ratio λ and elastic modulus E are assigned to the tibia model.

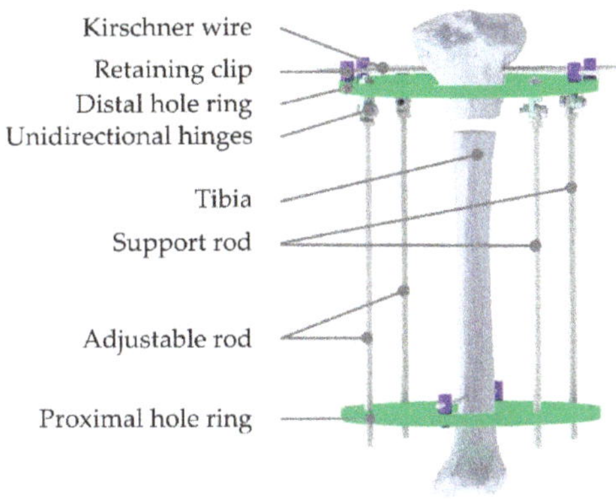

Figure 2. Schematic diagram of simulation model.

CT images can completely and accurately reflect the relationship between material characteristics and gray values. The CT images are imported into Mimics software and the values of the parameters are assigned to the mesh model. There is a linear relationship between the apparent density ρ of human bones and the gray value HU [20,21]:

$$\rho = \rho_{\max} HU / HU_{\max}, \tag{9}$$

where $\rho_{\max}$ and $HU_{\max}$ are the maximum of the parameters ρ and HU. Further, the relationship between the apparent density ρ and the elastic modulus E_i is as follows [22]:

$$E_1/\mathrm{MPa} = 2065 * \left[\rho/\left(\mathrm{g}\cdot\mathrm{cm}^{-3}\right)\right]^{3.09}, \tag{10}$$

$$E_2/\mathrm{MPa} = 1904 * \left[\rho/\left(\mathrm{g}\cdot\mathrm{cm}^{-3}\right)\right]^{1.64}, \tag{11}$$

where E_1, E_2 are the elastic moduli of compact and cancellous bones. The values of Poisson ratio λ of dense and cancellous bones are equal to 0.3.

The binding constraints are adopted in the bolts and nuts of the unidirectional hinges to simulate the behavior of bolt fastening. Finite element model parameters of the Kirschner wire, steel wire holder, and tibia are shown in Table 2, and the finite element model is shown in Figure 3.

Table 2. Finite element model parameters of fixator.

The Typical Sites	Number of Elements	Number of Nodes	Type of Mesh
Kirschner wire, steel wire holder	90,635	504,474	hexahedral mesh
tibia	67,973	127,614	tetrahedral mesh

The frictional coefficient between stainless steels was defined as 0.05. The tibia and Kirschner wires were defined as the common surface-to-surface contact and the frictional coefficient was equal to 0.1. During the adjustment of the fixator, the base platform was fixed and the osteotomy section was kept stationary relative to the base platform. The base platform and osteotomy section were completely constrained to simulate the fixation of the tibia to the moving platform.

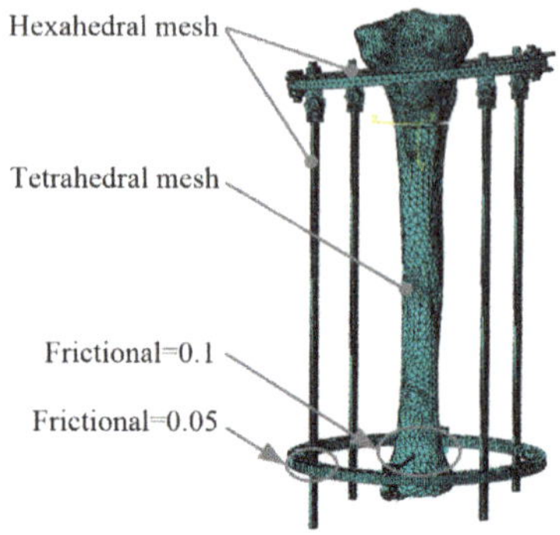

Figure 3. Finite element model of tibia orthopedic.

2.3. Measurement of Orthopedic Forces

The sheep tibia was taken as the research object in the orthopedic experiments. Compared with human tibia, the sheep tibia is of smaller diameter and higher hardness, matching the tensile strength of animal experiments [23,24]. Meanwhile, the sheep tibia can withstand greater stress over long periods of time. The tibia was taken from a sheep without obvious anatomical abnormalities that had been slaughtered within 3 h previously. The tibia was at an initial deviation angle of 6.85°, and was cut into two pieces in the osteotomy. The two pieces were installed to the fixator by the Kirschner wires. A six-dimensional force sensor was installed at the osteotomy site to accurately measure the tensions along three directions, as shown in Figure 4. A uniaxial force sensor was serially installed in each rod to measure the adjustable or support force. During the installation of the sensors, the tibia should be kept in the middle position and vertical posture.

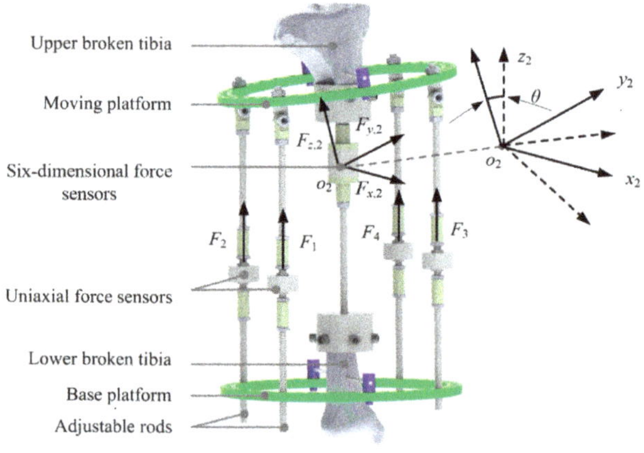

Figure 4. Installation of force sensors.

The process of fracture healing was simulated by changing the lengths of the two adjustable rods. According to the pulling speed in Table 1, the osteotomy surface of the upper broken tibia and the moving platform were rectified to the position parallel to the base platform by the standard limb lengthening methods [25].

The adjustable forces F_1 and F_2 were measured by the installed sensors and the support forces F_3 and F_4 were obtained. The tensions $F^{\mathrm{m}}_{x,2}$, $F^{\mathrm{m}}_{y,2}$, $F^{\mathrm{m}}_{z,2}$ between the upper and lower broken tibias along the three axes of $\{o_2\}$ were measured. The pre-tightening forces of the fixator and the osteotomy site were measured at the beginning of the orthopedic process. During the orthopedic experiments, the adjustable/support forces F_i and the tensions were measured and recorded in real time, as shown in Figure 5. Ten orthopedic experiments were completed and the measurement data had similar variation laws and consistency.

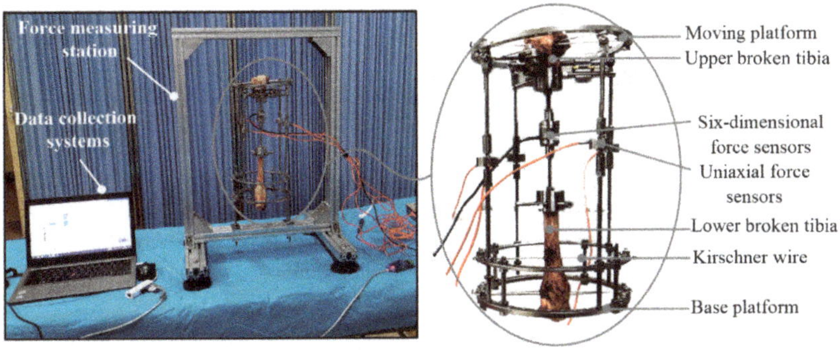

Figure 5. Measurement platform of the sheep tibia.

3. Results

3.1. Orthopedic Force Analysis

The adjustable rods were taken as the actuators to simulate the orthopedic treatment. The adjustable and support forces F_i in the rods had significant effects on the orthopedic treatment of the tibia. The tension $F^{m}_{z,2}$ at the broken tibia is an important factor to stimulate the process of the bone regeneration which should be continuous and periodic. Besides, the frequency of adjusting the rods and magnitude of the tension also should be appropriated [26,27]. Excessive tension will lead to bone disorders in the extension area of the broken bone. Otherwise, the lack of tension will lead to the healing of the broken tibia ahead of schedule [28].

The adjustable and support forces F_i during the orthopedic process are shown in Figure 6. The entire treatment is divided into 10 periods, each lasting one day. A period consists of correction and stay stages. The adjustable nut is rotated a single revolution for each correction stage, and the length of the rod is lengthened or shortened by 1 mm. As patients experience pain when the nut is rotated, a stay stage is added between two correction stages. In the stay stage, the adjustable and support forces remain almost unchanged in order to relieve the pain of patients. The clinical treatment of bone correction lasts 10 days, but it is difficult to work with fresh sheep tibias for such a long time. Hence, each period in the simulated experiment was compressed to 24 s.

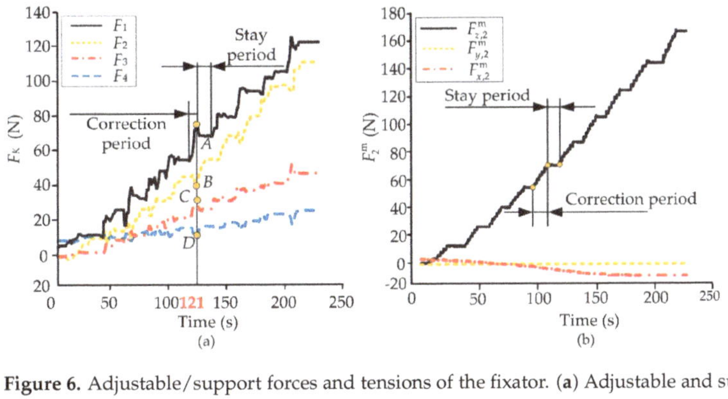

Figure 6. Adjustable/support forces and tensions of the fixator. (**a**) Adjustable and support forces on each rod. (**b**) The measured value of pull tension.

The changes in the forces emerging in the rods and tibia are shown in Figure 6. The adjustable rods used as actuators aggravated the generation of internal stress in the fixator. The forces F_1 and F_2 first increased and then decreased to a stable level during the correction stages. The support rods as passive components shared the loads of the fixator. The forces F_1 and F_2 were determined by the load distribution during the whole treatment. Since the load distribution was closely related to the initial pre-tightening state, and there were losses between the adjustable force and the support force, the support forces were always less than the adjustable forces. Meanwhile, during the correction stages, the changing trend of the support forces was opposite to that of the adjustable forces. All the four forces F_i remained unchanged during the stay stage. The sixth correction period was selected, and a dotted line was added at the time of 121 s (Figure 6a). The cross points of the dotted line and the force curves denoted by A, B, C, and D were extreme points of the four curves in this period. It can be seen that the four rods with motion consistency restricted each other. The tensions between the two broken pieces of tibia are depicted in Figure 6b. The tension $F^{m}_{z,2}$ along the axis of the broken end, which was important to the correction area, had similar changes compared with that of the forces F_1 or F_2. The tension $F^{m}_{z,2}$ increased gradually in the correction stage and remained unchanged in the stay stage. The other horizontal components, $F^{m}_{x,2}$ and $F^{m}_{y,2}$, which were harmful to the orthopedic treatment, remained at small values.

The load applied to the fixator through the adjustable rods and support rods was about 300 N, and the average stress transmitted to the broken bone end was 140 N. Compared with Gessmann's results [29] of indirectly pressurizing the four rings Ilizarov fixator with 300 N and the load on the broken bone end with 110–120 N, the results were slightly different, due to certain differences in structure and pressurization method. However, the similar trend proves the correctness of the experimental data.

3.2. Orthopedic Force Theoretical Calculate

The forces F_i and the rotation angle θ were substituted into Equation (8) where the rotation angle θ varied uniformly from 6.85° to 0°. The theoretical tensions at the extension area of the broken bone were calculated by the software Matlab, as shown in Figure 7.

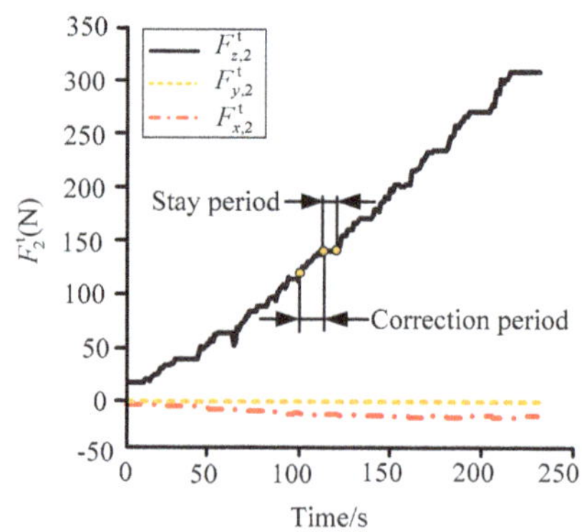

Figure 7. Theoretical tensions at the osteotomy.

The influences of pre-tightening, frictions in the fixator and the motion synchronization of the adjustable rods were disregarded in the theoretical calculation. The theoretical tensions $F^{t}_{z,2}$ at the correction and stay periods were consistent with the measured tension $F^{m}_{z,2}$. In addition, the other two tensions $F^{t}_{x,2}$ and $F^{t}_{y,2}$ all had a similar shape to that of $F^{m}_{x,2}$ and $F^{m}_{y,2}$ in the experiments. Hence, the screw model established in Section 2.1 is suitable for the actual movements of the fixator and has good simulation accuracy. However, the maximum of $F^{t}_{z,2}$ was nearly twice that of $F^{m}_{z,2}$. This was because the influences mentioned above led to the loss of the adjustable and support forces.

The transfer efficiency η_z between the measured and theoretical forces was defined to evaluate the efficiency of converting the adjustable and support forces into the tensions of broken tibia in orthopedic treatment, where η_z could be obtained by:

$$\eta_z = \frac{F^{t}_{z,2}}{F^{m}_{z,2}}. \tag{12}$$

The index η_z was calculated based on the obtained data in Figures 6b and 7, as shown in Figure 8. In the first 50 s of the experiment, the adjustable and support forces F_i were used for the pre-tightening of the fixator and rods. Few tensions were transmitted to the extension area of the broken bone. After 2~3 periods, the transfer efficiency η_z was improved and kept at about 50%.

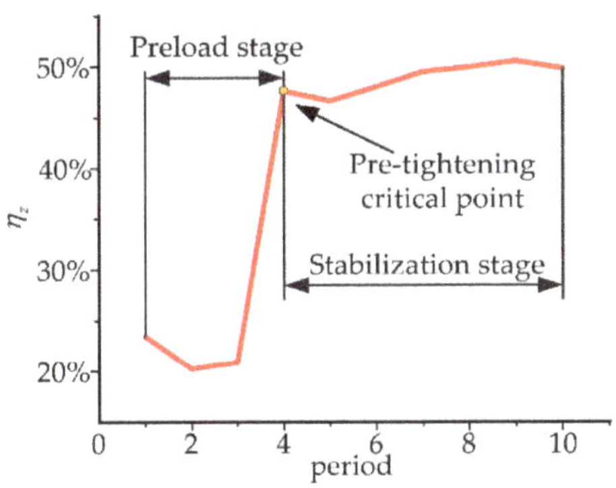

Figure 8. Transfer efficiency η_z.

3.3. Stress Distribution of Ilizarov External Fixator and Tibia

In order to analyze the stress distribution of the fixator and tibia, a finite element model of this mechanical system was established to simulate the correction periods under the software of ABAQUS, as shown in Figure 9.

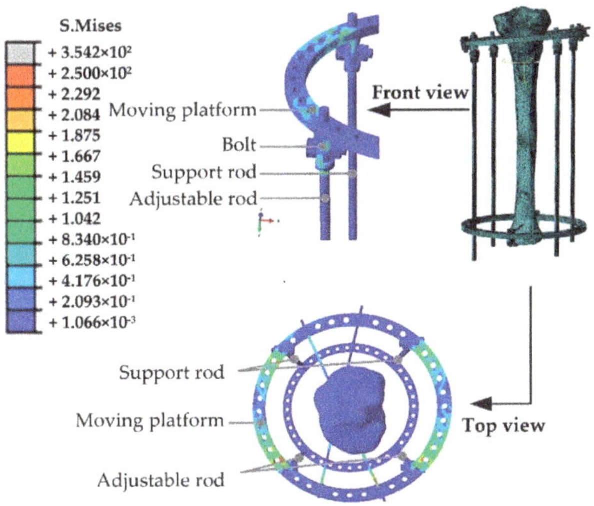

Figure 9. Stress distribution of Ilizarov external fixator at the tenth correction period.

The mean value of the two adjustable forces F_1 and F_2 in each correction period was applied to the end of the adjustable rod, as shown in Table 3. The tension $F^s_{z,2}$ in the tenth correction period reached the peak value, which had the most obvious stretch effect on the tibia end. Hence, this correction period was selected for further analysis, as shown in Figure 10. The adjustable forces were transferred from the adjustable rods to the Kirschner wires, and the upper broken tibia end was stretched. The stress concentration appeared in the unidirectional hinges connecting the moving platform and the adjustable/support rods.

Table 3. Average orthotic force in each treatment period.

Correction Period	F_1/N	F_2/N	$F^m_{z,2}$/N
1	0.56	8.56	5.26
2	12.18	20.78	11.03
3	16.89	30.31	15.58
4	27.28	40.03	48.11
5	35.81	53.04	59.75
6	43.17	63.83	72.83
7	58.65	77.15	93.52
8	70.51	91.92	111.37
9	90.77	99.39	130.25
10	99.62	117.54	145.39

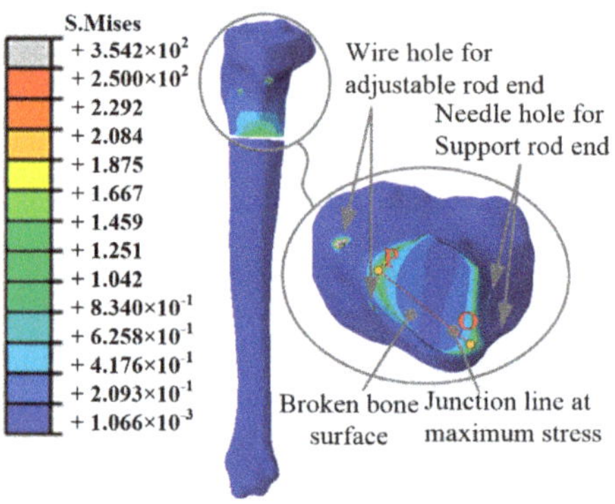

Figure 10. Mises stress state diagram of tibia.

During the simulation, the average of the stresses on the section of the broken bone surface along the axis o_2z_2 is multiplied by the total area is defined as the tensile force simulation value $F^s_{z,2}$, as shown in Table 4.

Table 4. Tensions $F^s_{z,2}$ of the broken bone surface.

Period/Time	1	2	3	4	5	6	7	8	9	10
Tension/N	3.667	22.32	32.71	50.89	68.53	81.28	120.00	130.60	151.31	155.0

On the outer surface of the broken bone, the maximum stress occurred at the Kirschner pinhole. Hence, the stresses on the surface were symmetrically distributed, and it was found that the maximum stress occurred at the broken bone end surface near the Kirschner wire pinhole, where points O and P were peak points close to the support and adjustable rods, respectively, as shown in Figure 10.

The tensions $F^s_{z,2}$ affecting the broken bone segment lead to changes in the biological environment/skeleton morphology. Figure 11 shows a line OP connecting the points O and P, with the stresses distributed on the line OP. The stresses decrease from the peak point O and increase to another peak point P. The stress of point P is higher than that of point O. The difference indicated that the tibia was significantly affected by the tensions during the treatment process. The orthopedic treatment of the broken tibia could be completed with the help of the tensions.

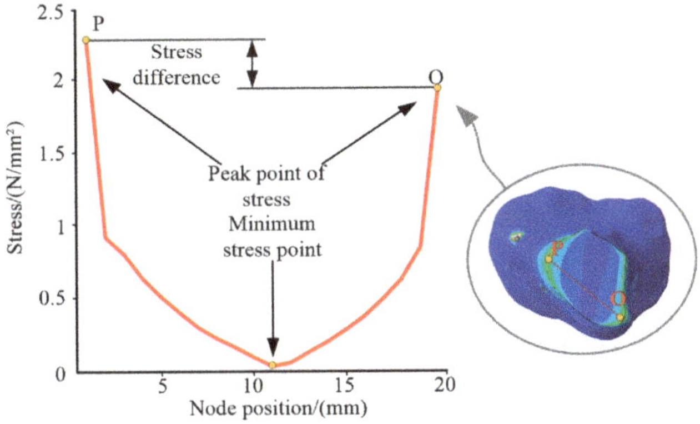

Figure 11. Stress distribution on the line OP.

3.4. Comparison of Theoretical, Simulation and Experimental Results

In order to better compare the accuracy of the three methods, the corresponding results of the tensions are shown in Figure 12. Point B is the pre-tightening point of the unidirectional hinges during the experiment. Before pre-tightening, the hinges and rods were still in a loose state, which led to gaps between various parts. Hence, the measured tensions $F^{m}_{z,2}$ were less than the simulation tensions $F^{s}_{z,2}$. After point B, the adjustable forces reached the critical point of pre-tightening and the gaps among the components disappeared. The measured tensions $F^{m}_{z,2}$ increased faster. However, the curves of the tensions $F^{m}_{z,2}$ and $F^{s}_{z,2}$ had similar shapes. During the orthopedic experiments, various factors influenced the force transmission efficiency of the adjustable forces and led to the difference between the results $F^{m}_{z,2}$ and $F^{s}_{z,2}$.

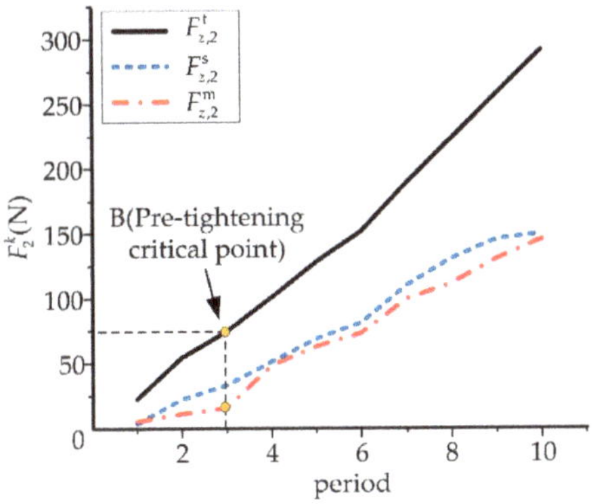

Figure 12. Comparison of the tensile force of the broken bone end.

The theoretical/measured/simulation results increased synchronously according to the ratio of force conduction efficiency, since the theoretical data did not consider friction and there was no gap problem. Hence, there was no pre-tightening critical point in the theoretical analysis, in which the growth rate was twice that of the simulated value, and the growth rate was maintained at a stable rate.

Through comparing the tensions obtained by the three methods, it was found that the occurrence of the pre-tension critical point was related to the installation of the fixator, the adjustable forces and the synchronization of the physician. In the stage of orthosis, these problems will lead to unsatisfactory correction results and should be eliminated as much as possible in the correction periods to improve orthopedic treatments.

4. Discussion

The Ilizarov external fixator has attracted the attention of researchers due to its versatile configuration and good clinical results, and because it mainly changes the skeletal force line by fixing or limiting the movement of the limb to achieve the function of orthopedic limb bone, which has good clinical prospects for the treatment of knee osteoarthritis and limb orthopedics. Many researchers have achieved results in the study of external fixators. Thiart G, Herbert C, Sivarasu S, et al. studied the influence of different connecting rod structures on the stability of the Ilizarov external fixator [30]. However, the clinical application relies on physician experience, and the orthopedic features are still not clearly explained. This study analyzed the mechanical characteristics of the Ilizarov external fixator by combining theoretical analysis, clinical experiment and finite element analysis, which provide a reliable basis for the application of external fixators. The results are beneficial for the development of bone external fixation techniques.

First of all, based on the screw theory, the screw model of the Ilizarov external fixator was deduced innovatively and the geometric explanation of the fixator was given. Afterwards, in contrast with the study of Karunratanakul K, who used resin rodlets to conduct simulation experiments [10], orthopedic experiments were conducted using sheep tibia innovatively, and the theoretical tensions based on geometric explanation were compared with the experimental tensions obtained by orthopedic experiments on sheep tibia. It was found that there are interactions and constraints between the adjustable rod and the support rod due to the structural configuration of the external fixator. The transfer efficiency between the fixator and the broken bone end was calculated and analyzed. The analysis of the experimental results can provide a direction and theoretical basis for the innovative design of orthopedic parameters.

On this basis, a finite element model was established to accurately describe the stress and strain of the fixator and tibia during the process of orthopedic treatment. In contrast with the study of Donaldson F E, Pankaj P et al., who used pipe for simulation analysis [31], we used CT scans of tibia for modeling to obtain a more accurate model, and the model was able to respond to some phenomena that were difficult to obtain in the experiments. Similarly to the research results of Donaldson F E, the maximum stress appeared in the pinhole of Kirschner wire [32], and it can be seen from the results that the stress value of the pinhole near the adjustable rods was greater than that of the support rods. These results can provide a theoretical basis for the precise regulation of the external fixator and assist physicians in formulating special treatment plans. The finite element model shows some adaptability and can be applied to the optimization of the Ilizarov external fixator and other orthopedic simulations.

Ganesharajah, Ganadhiepan, Saeed, et al. proposed that callus healing was related to stress between broken bone ends [33]. Based on the established biomechanical model, biomechanical analysis and prediction can be performed to evaluate the degree of callus healing and bone alignment. However, it is difficult for in vitro animal experiments to simulate the muscle system. The influences of cartilage and ligaments on the orthopedic process have not been considered in theoretical calculations and experiments and simulations. In the future, the osteomuscular system of the biomechanical model will continue to be improved to obtain more accurate research results.

Author Contributions: All authors made substantial contributions to this paper. Specifically, the experimental design was carried out by P.S., the experimental evaluation was performed by Q.Z., Y.L. was responsible for data collection, S.W. drafted the article and L.Z. critically revised the academic content. All authors have read and agreed to the published version of the manuscript.

Funding: This study was funded by the Natural Science Foundation of Beijing Municipality [Grant No. 19L2018 and 3202003], the National Natural Science Foundation of China [Grant No. 52005045], and the National Key R&D Program of China [Grant No. 2019YFC0119200].

Institutional Review Board Statement: Not applicable.

Informed Consent Statement: Not applicable.

Data Availability Statement: Not applicable.

Acknowledgments: Thanks to all those who contributed to this paper, including the experiment subjects and the people who provided technical support. They had writing assistance for this paper, and the entity that paid for this writing assistance was the Beijing Information Science and Technology University.

Conflicts of Interest: The authors declare no conflict of interest.

References

1. Watson, J.T.; Ripple, S.; Hoshaw, S.J.; Fyhrie, D. Hybrid External Fixation for Tibial Plateau Fractures. *Orthop. Clin. N. Am.* **2002**, *33*, 199–209. [CrossRef]
2. Chen, G.; Xiao, X.; Zhao, X.; Tat, T.; Bick, M.; Chen, J. Electronic Textiles for Wearable Point-of-Care Systems. *Chem. Rev.* **2021**, *122*, 3259–3291. [CrossRef] [PubMed]
3. Niu, J.; Wang, H.; Shi, H.; Pop, N.; Li, D.; Li, S.; Wu, S. Study on structural modeling and kinematics analysis of a novel wheel-legged rescue robot. *Int. J. Adv. Robot. Syst.* **2018**, *15*, 172988141775275. [CrossRef]
4. Ke, W.; Fuzhou, D.; Xianzhi, Z. Algorithm and experiments of six-dimensional force/torque dynamic measurements based on a Stewart platform. *Chin. J. Aeronaut.* **2016**, *29*, 1840–1851.
5. Kang, C.; Liu, Z.; Chen, S.; Jiang, X. Circular trajectory weaving welding control algorithm based on space transformation principle. *J. Manuf. Process.* **2019**, *46*, 328–336. [CrossRef]
6. Qian, Y.J.; Liu, Z.X.; Yang, X.D.; Hwang, I.; Zhang, W. Novel Subharmonic Resonance Periodic Orbits of a Solar Sail in Earth-Moon System. *J. Guid. Control. Dyn.* **2019**, *42*, 2532–2540. [CrossRef]
7. Wang, Z.; Zhang, N.; Chai, X.; Li, Q. Kinematic/dynamic analysis and optimization of a 2-URR-RRU parallel manipulator. *Nonlinear Dyn.* **2016**, *88*, 503–519. [CrossRef]
8. Li, B.; Li, Y.M.; Zhao, X.H.; Ge, W.M. Kinematic analysis of a novel 3-CRU translational parallel mechanism. *Mech. Sci.* **2015**, *6*, 57–64. [CrossRef]
9. Mitousoudis, A.S.; Magnissalis, E.A.; Kourkoulis, S.K. A biomechanical analysis of the Ilizarov external fixator. *EPJ Web Conf.* **2010**, *6*, 21002. [CrossRef]
10. Karunratanakul, K.; Schrooten, J.; Oosterwyck, H.V. Finite element modelling of a unilateral fixator for bone reconstruction: Importance of contact settings. *Med. Eng. Phys.* **2010**, *32*, 461–467. [CrossRef]
11. Tan, B.; Shanmugam, R.; Gunalan, R.; Chua, Y.; Hossain, G.; Saw, A. A biomechanical comparison between Taylor's Spatial Frame and Ilizarov external fixator. *Malays. Orthop. J.* **2014**, *8*, 35. [CrossRef] [PubMed]
12. Kang, C.; Shi, C.; Liu, Z.; Liu, Z.; Jiang, X.; Chen, S.; Ma, C. Research on the optimization of welding parameters in high-frequency induction welding pipeline. *J. Manuf. Process.* **2020**, *59*, 772–790. [CrossRef]
13. Solomin, L.N.; Green, S.A. *The Basic Principles of External Skeletal Fixation Using the Ilizarov and other Devices*; Springer: Milan, Italy, 2012.
14. Lu, H.Q. *Kinematics and Dynamics Research of Robot Manipulators and Its Application Based on Screw Theory*; Nanjing University: Nanjing, China, 2007.
15. Davidson, J.K.; Hunt, K.H.; Pennock, G.R. Robots and screw theory: Applications of kinematics and statics to robotics. *J. Mech. Des.* **2004**, *126*, 763–764. [CrossRef]
16. Guo, S.; Fang, Y.; Qu, H. Type synthesis of 4-DOF nonoverconstrained parallel mechanisms based on screw theory. *Robotica* **2012**, *30*, 31–37. [CrossRef]
17. Kong, X.; Gosselin, C.M. Type synthesis of 4-DOF SP-equivalent parallel manipulators: A virtual chain approach. *Mech. Mach. Theory* **2006**, *41*, 1306–1319. [CrossRef]
18. Kong, X.; Gosselin, C.M. Type synthesis of 3T1R 4-DOF parallel manipulators based on screw theory. *IEEE Trans. Robot. Autom.* **2004**, *20*, 181–190. [CrossRef]
19. Peng, L.; Bai, J.; Zeng, X.; Zhou, Y. Comparison of isotropic and orthotropic material property assignments on femoral finite element models under two loading conditions. *Med. Eng. Phys.* **2006**, *28*, 227–233. [CrossRef]
20. Baca, V.; Horak, Z.; Mikulenka, P.; Dzupa, V. Comparison of an inhomogeneous orthotropic and isotropic material models used for FE analyses. *Med. Eng. Phys.* **2008**, *30*, 924–930. [CrossRef]
21. Dempsey, I.J.; Southworth, T.M.; Huddleston, H.P.; Yanke, A.; Farr, J., II. Tibial tubercle osteotomy: Anterior, medial, and distal correction. *Oper. Tech. Sports Med.* **2019**, *27*, 150686. [CrossRef]

22. Younger, A.S.; Morrison, J.; MacKenzie, W.G. Biomechanics of external fixation and limb lengthening. *Foot Ankle Clin. N. Am.* **2004**, *9*, 433–448. [CrossRef]
23. Claes, L.; Augat, P.; Schorlemmer, S.; Konrads, C.; Ignatius, A.; Ehrnthaller, C. Temporary distraction and compression of a diaphyseal osteotomy accelerates bone healing. *J. Orthop. Res.* **2008**, *26*, 772–777. [CrossRef] [PubMed]
24. Spiegelberg, B.; Parratt, T.; Dheerendra, S.K.; Khan, W.S.; Jennings, R.; Marsh, D.R. Ilizarov principles of deformity correction. *Ann. R. Coll. Surg. Engl.* **2010**, *92*, 101. [CrossRef] [PubMed]
25. Gundes, H.; Buluc, L.; Sahin, M.; Alici, T. Deformity correction by Ilizarov distraction osteogenesis after distal radius physeal arrest. *Acta Orthop. Traumatol. Turc.* **2011**, *45*, 406–411. [CrossRef]
26. Kawoosa, A.A.; Wani, I.H.; Dar, F.A.; Sultan, A.; Qazi, M.; Halwai, M.A. Deformity correction about knee with Ilizarov technique: Accuracy of correction and effectiveness of gradual distraction after conventional straight cut osteotomy. *Ortop. Traumatol. Rehabil.* **2015**, *17*, 587. [CrossRef] [PubMed]
27. Gubin, A.V.; Borzunov, D.Y.; Malkova, T.A. The Ilizarov paradigm: Thirty years with the Ilizarov method, current concerns and future research. *Int. Orthop.* **2013**, *37*, 1533–1539. [CrossRef] [PubMed]
28. Ilizarov, G. The tension-stress effect on the genesis and growth of tissues: Part, I. The influence of stability of fixation and soft-tissue preservation. *Clin. Orthop.* **1989**, *238*, 249–281. [CrossRef]
29. Gessmann, J.; Citak, M.; Jettkant, B.; Schildhauer, T.A.; Seybold, D. The influence of a weight-bearing platform on the mechanical behavior of two Ilizarov ring fixators: Tensioned wires vs. half-pins. *J. Orthop. Surg. Res.* **2011**, *6*, 61. [CrossRef]
30. Thiart, G.; Herbert, C.; Sivarasu, S.; Gasant, S.; Laubscher, M. Influence of Different Connecting Rod Configurations on the Stability of the Ilizarov/TSF Frame: A Biomechanical Study. *Strateg. Trauma Limb Reconstr.* **2020**, *15*, 23. [CrossRef]
31. Watson, M.; Mathias, K.J.; Maffulli, N.; Hukins DW, L.; Shepherd, D.E.T. Finite element modelling of the Ilizarov external fixation system. *Proc. Inst. Mech. Eng. Part H* **2007**, *221*, 863–871. [CrossRef]
32. Donaldson, F.E.; Pankaj, P.; Simpson, A. Bone properties affect loosening of half-pin external fixators at the pin–bone interface. *Inj. Int. J. Care Inj.* **2012**, *43*, 1764–1770. [CrossRef]
33. Ganadhiepan, G.; Miramini, S.; Patel, M.; Mendis, P.; Zhang, L. Bone fracture healing under Ilizarov fixator: Influence of fixator configuration, fracture geometry, and loading. *Int. J. Numer. Methods Biomed. Eng.* **2019**, *35*, e3199. [CrossRef] [PubMed]

micromachines

MDPI

Article

Ergonomic Design and Performance Evaluation of H-Suit for Human Walking

Leiyu Zhang [1], Zhenxing Jiao [1], Yandong He [1] and Peng Su [2,*]

[1] Beijing Key Laboratory of Advanced Manufacturing Technology, Beijing University of Technology, Beijing 100124, China; zhangleiyu@bjut.edu.cn (L.Z.); jiaozhenxing0109@163.com (Z.J.); heyandong13@163.com (Y.H.)
[2] School of Electromechanical Engineering, Beijing Information Science and Technology University, Beijing 100192, China
* Correspondence: supeng@bistu.edu.cn; Tel.: +86-010-189-1102-7599

Abstract: A soft exoskeleton for the hip flexion, named H-Suit, is developed to improve the walking endurance of lower limbs, delay muscle fatigue and reduce the activation level of hip flexors. Based on the kinematics and biomechanics of the hip joints, the ergonomic design of the H-Suit system is clearly presented and the prototype was developed. The profile of the auxiliary forces is planned in the auxiliary range where the forces start at the minimum hip angle, reach the maximum (120 N) and end at 90% of each gait cycle. The desired displacements of the traction unit which consist of the natural and elastic displacements of the steel cables are obtained by the experimental method. An assistance strategy is proposed to track the profile of the auxiliary forces by dynamically adjusting the compensation displacement L_c and the hold time Δt. The influences of the variables L_c and Δt on the natural gaits and auxiliary forces have been revealed and analyzed. The real profile of the auxiliary forces can be obtained and is consistent with the theoretical one by the proposed assistance strategy. The H-Suit without the drive unit has little effect on the EMG signal of the lower limbs. In the powered condition, the H-Suit can delay the muscle fatigue of the lower limbs. The average rectified value (ARV) slope decreases and the median frequency (MNF) slope increases significantly. Wearing the H-Suit resulted in a significant reduction of the vastus lateralis effort, averaged over subjects and walking speeds, of 13.3 ± 2.1% ($p = 2 \times 10^{-5}$).

Keywords: soft robotic suit; hip assistance; gait prediction; performance evaluation; EMG signal

Citation: Zhang, L.; Jiao, Z.; He, Y.; Su, P. Ergonomic Design and Performance Evaluation of H-Suit for Human Walking. *Micromachines* **2022**, *13*, 825. https://doi.org/10.3390/mi13060825

Academic Editor: Xiao Xiao

Received: 11 April 2022
Accepted: 23 May 2022
Published: 25 May 2022

1. Introduction

The expanding elderly and disabled population poses considerable challenges to the current healthcare system [1]. A lower-limb exoskeleton has been developed to reduce the metabolic consumption of human walking or running [2–5] and is widely used in rehabilitation and strength enhancement [6]. According to the stiffness of the robot's main components, the lower-limb exoskeleton can be divided into two types, rigid and soft exoskeletons. The traditional rigid exoskeleton robots, which are worn by lower limbs and moved with the human body, imitate the biological structure of human lower extremities. The rigid linkages or components are taken as the main mechanical backbone [7,8]. The rigid exoskeleton can provide gait correction for patients with dyskinesia and walking assistance for healthy persons (such as BLEEX [9], HAL System [10], EKSO [11], and ReWalk [12]). However, the additional mass of lower extremities and the axis misalignment of the human–machine system after wearing can easily lead to the increase of additional torques, a disorder of natural gait, and dislocation of auxiliary forces [13–16]. The energy consumption of the human body will rise substantially and it is difficult to be applied to humans fast walking or running. Currently, assistive devices focus more on comfort and personalization [17], in order to overcome the shortcomings of the rigid exoskeleton mentioned above, many scholars have studied soft exoskeletons with better wearing comfort

(SEU-EXO [14], Exosuit [18], and Myosuit [19]), namely the soft power-assisted system, which can assist the lower limbs in the state of walking or running. The flexible driving units, such as the Bowden cable, pneumatic muscle and airbag, are applied to transmit the auxiliary forces/moments to the hip, knee or ankle joints [20].

The soft power-assisted system supported by the lower limb skeleton transfers the auxiliary forces through the flexible elements (nylon belt, flexible cable, pneumatic muscles, etc.) distributed along the skin surface to the single or multiple joints [21], and reduces the additional loads and metabolic consumption of the corresponding flexor/extensor muscles. Ding et al. [22] developed a functional coat called Exosuit which was worn on the skin surface of lower limbs. The multi-joint assistance was provided for the hip extension and ankle plantar flexion through Bowden cables and an external traction unit. Kim and Ding et al. [18,23] designed an Exosuit with a waist belt and tight trousers as the substrate and developed a portable drive device worn on the waist. The Exosuit system only provided auxiliary forces for hip extension and was suitable for walking and running at a specific speed. Aiming at the gravity intensive activities (sitting, standing, squatting, etc.) of the knee joint, Schmidt et al. [19] developed Myosuit, which consisted of corsetry, ligament layer and active layer, to achieve the function of the external bionic muscle by controlling the contraction and relaxation of the active layer. The pneumatic muscle was applied in a Power Assist Wear [24] to improve the independent walking ability of the elderly and the tension force was transmitted to the hip and knee joints by nylon belts. Tian et al. [25] developed a unilateral power-assisted exoskeleton based on the pneumatic muscle, which could assist the flexion and extension of the hip/knee joints and reduce the muscle activation by 14.6%. A flexible ankle-assisted exoskeleton developed by John et al. [26] proposed a cross-line power suit with four Bowden cables. The cables were distributed along the front and rear sides of the thigh and could be controlled independently to provide auxiliary force/moment for the hip joint. Based on the above studies, a soft assistance suit system, which consists of the suit body, Bowden cables and traction unit, is proposed for the hip flexion.

Based on the kinematics and biomechanics of the hip joint, a new soft power-assisted system named H-Suit is proposed to provide a suitable auxiliary force for the forward flexion of the hip joint. An assistance strategy is established to produce a desired profile of the auxiliary force. The fatigue and activation of hip flexors (iliopsoas muscle, rectus femoris and vastus lateralis, etc.) are analyzed comprehensively under different experimental conditions and walking speeds. Understanding how the H-Suit affects the hip flexors of human bodies is fundamental for quantifying its benefits and drawbacks, guiding a continuous data-driven design refinement. Furthermore, the H-Suit system can be worn for a long time for healthy persons, especially soldiers who need to walk a long distance with a heavy load. Besides, the additional mass and gait interference of lower limbs can be reduced as much as possible with the help of the ergonomic design method.

2. Ergonomic Design of the H-Suit

2.1. Kinematics and Biomechanics of Hip Joint

The lower extremities of humans are mainly composed of the hip, knee and ankle joints. The main activities of the lower limbs (walking or running) are mainly composed of flexion/extension (FL/EX) of the three joints. Hence, the muscle groups that dominate hip FL/EX are analyzed in this paper. During human walking, the flexor and extensor muscles work together to control FL/EX movements. Hip flexors mainly consist of the iliopsoas muscle, rectus femoris and vastus lateralis [27]. The main action interval of hip flexors and extensors is complementary in a single gait cycle T [28]. The activation times of the flexors are concentrated in the middle stage of gait cycle T, namely the late support phase and the prometaphase of the swing phase. The corresponding activation of extensors is distributed at the beginning and end stage of a gait cycle. The corresponding activation intensity can be easily detected and measured [29,30]. Besides, the leg lift that occurred in the flexion movements should overcome the gravity of lower limbs rapidly and have greater energy expenditures. Hence, the flexion movements are selected as the assistance object in this study. The kinematics and biomechanics of the hip joint in a single gait cycle T are shown in Figure 1 where θ_h and T_h denote the rotation angle and biological torque of the hip joint, respectively.

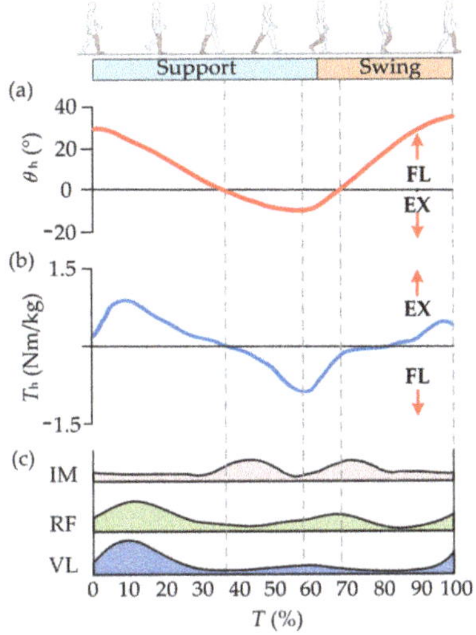

Figure 1. Kinematics and biomechanics of the hip joint. (**a**) When the hip joint moves to the absolute minimum angle (maximum extension), the thigh begins to flex forward and enter the early swing phase. (**b**) The flexion torque gradually increases with the forward flexion until the later stage of the swing phase. (**c**) The activations of the flexors are concentrated in the middle stage of the gait cycle [27].

2.2. Ergonomic Design of H-Suit

The H-Suit consists of a soft exosuit, external traction units and Bowden cables. The soft exosuit is composed of waist belts, thigh belts, inertial measurement units (IMUs) and load cells, as shown in Figure 2a. The waist and thigh belts are embraced around the waist and thighs separately. The anchors used to fix Bowden cables are sutured on the belt surfaces. In order to improve the rated output of the traction unit and the smoothness of auxiliary torques, the optimal positions of anchors (Figure 2b) were selected in our previous study [31]. Three IMUs are respectively stuck on the front sides of the waist and thigh to detect the kinematics information of the hip joint. Besides, the load cells are connected in series between the steel cable and the thigh anchor to measure the auxiliary forces in real-time. The power of the external traction units is transmitted to the soft exosuit through Bowden cables which consist of sheaths and steel cables. The H-Suit has two identical traction units for the right and left legs. One end of the sheath is fixed at the waist anchor and the other at the frame of the traction units. Similarly, one end of the steel cable is installed at the thigh anchor and the other at the slider of traction units. Based on the gait information, an expected profile of auxiliary forces can be applied to the H-Suit and assist flexions movements by controlling the traction displacements.

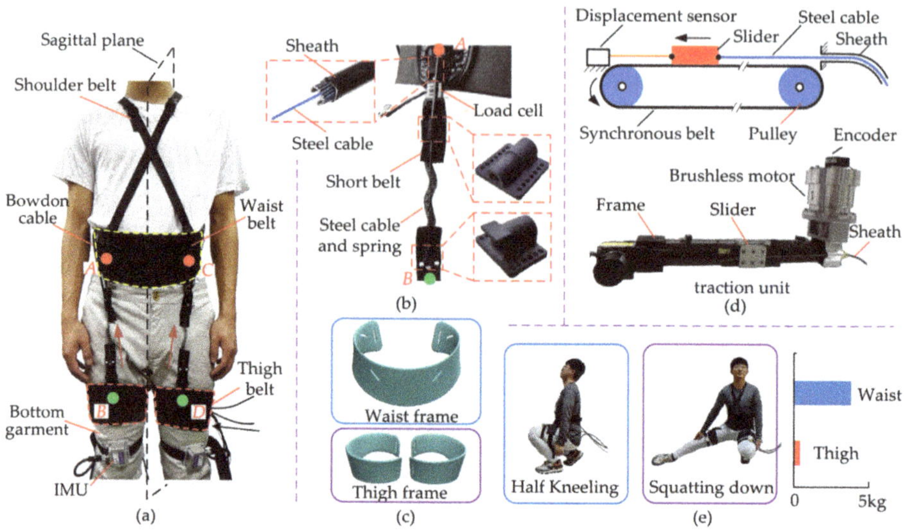

Figure 2. Development of the H-Suit system. (**a**) The H-Suit consists of a soft exosuit, external traction units and Bowden cables where *A*, *B*, *C* and *D* are anchor points. (**b**) The sheath and steel cable of the Bowden cable are respectively fixed in the upper and lower components. (**c**) The open frames of the waist and thigh are designed and fitted with the physiological structures of the human body. (**d**) The synchronous belt transmission is adopted in the traction unit and the steel cable is reciprocally dragged by the slider. (**e**) The H-Suit has few constraints on the dexterity of low limbs and daily activities.

According to the average body sizes of our research group (height 175 cm, weight 75 kg), four bottom garments are selected and bought. The waist and thigh belts consist of open frames and Velcro. The open frames are designed and fitted with the physiological structures of the human body to improve the corresponding twisting resistance (Figure 2c) [32]. Four short nylon belts (width 25 mm) are sewn on the anchor locations denoted by *A*, *B*, *C* and *D* (Figure 2a). Two types of anchor components with butterfly shapes are fixed on the short belts. The sheath is inserted and bonded in the hole of the upper component. The steel cable with a plug is mounted in the lower component, passed through the upper one and enters the sheath. In order to increase the cable resilience during the no-assistance period, a compression spring is added between the upper and lower components. Two load cells are installed at the upper short belts in series to measure the auxiliary forces. Furthermore, shoulder straps are connected at the front and back of the waist belt to avoid the dislocation of the H-Suit. In order to collect the gait information of lower limbs, two IMUs are fixed on the left and right thigh belts, respectively, and a third one at the waist as reference. IMUs only collect the motion angle in the human sagittal plane.

There are two external traction units for the right and left lower extremities. The synchronous belt with advantages of smooth motion and small backlash is adopted in the traction unit. A slider is reciprocally pulled by a synchronous belt driven by a brushless motor (TBM7646, Kollmorgen Inc., New York, NY, USA) and the steel cable is connected to this slider (Figure 2d). The tension and release of the cable are accomplished by those reciprocal motions. The traction displacements of the cable are measured by a draw-wire displacement sensor (WFS1000; Fiaye Inc., Shanghai, China).

The total mass except for the traction unit is 0.995 kg where 0.8 kg is distributed around the waist and 0.195 kg around the thigh. The H-Suit has few constraints on the dexterity of low limbs and daily activities. When the subjects wear the H-Suit, they can easily complete the actions of half-kneeling, squatting down and bending over, etc. (Figure 2e). The weight of each traction unit is 1.5 kg. The traction units are installed on a cart and used to test the assistance strategy. The cart can move with the subjects when evaluating the performance of the H-Suit.

2.3. Control Unit

The control unit of H-Suit is mainly composed of a PC, an interactive interface, and the control board of MCU. The gait detection/prediction algorithm and the control algorithm are integrated into the software. PC can communicate with MCU through a serial port to realize data receiving and sending, and all data of the gait, auxiliary forces and traction displacement are displayed constantly in the interactive interface. MCU receives the control instructions from PC and controls the brushless motors through the CAN bus to realize the assistance of hip flexion. The data of IMUs, load cells and displacement sensors are collected by MCU and sent to the PC in real-time. Besides, a DC power supply, electric relays and limit switches are necessary to guarantee the stable work of the control unit. In order to serve the performance evaluation of H-Suit expediently, a cart with four floors is designed and manufactured. The two traction units connected to the exosuit body through Bowden cables are fixed on the second floor and the control unit on the third floor. The exosuit is folded up and placed on the bottom layer when not in use. The subjects can wear the H-Suit and walk on the treadmill or indoors where the cart is put aside.

3. Assistance Strategy of H-Suit

It is necessary to analyze the activations of hip flexors and the characteristics of the lower limb's gait and optimize the assistance strategy. According to the parameters of lower limbs and the stiffness of the exosuit body, the desired auxiliary forces are transformed into the trajectory of the traction unit. Then, the assistance strategy of H-Suit is established to ensure smooth assistance for lower limbs.

3.1. Auxiliary Force

The assistance performance is directly affected by the starting/ending time and the peak value of the auxiliary forces. The hip flexion which dominates 40% of the whole gait cycle only can occur when going from the maximum extension (minimum θ_{min} of hip angle θ) to the maximum flexion (maximum θ_{max}). Further, the hip flexion is divided into two stages: ① The first stage is from θ_{min} to the zero-point θ_{zero}, ② The second one is from θ_{zero} to θ_{max}, as shown in Figure 3a. Compared with the first stage, the second one processes a longer duration. Besides, the zero-point θ_{zero}, the beginning of the second stage, can be easier predicted through the prediction algorithm of "Newton + triple exponential smoothing" proposed in the previous research [31]. When the lower limb reaches the maximum extension position (Figure 3a), the flexors are activated and start isometric contraction to produce biological forces for the hip flexion. When the hip angle θ is about zero in the region of flexion, the activation of the flexor muscles reaches the maximum and produces the maximum biological forces for hip flexion [33]. The activation decreases to a stable level when the lower limb reaches 90% of a gait cycle T. Therefore, the profile of the auxiliary force F_t begins at $t = t_{\theta,min}$, reaches the peak value $F_{t,max}$ at $t = t_{\theta,zero}$, and ends at $t = t_{90\%}$, as shown in Figure 3b.

Based on the biological force of the flexor muscles [27], the 15% of the biological peak force is taken as $F_{t,\,max}$ which is equal to 120 N. The profile of F_t is designed and the corresponding expression of F_t is obtained as follows:

$$F_t = 109.5 \sin\left[(1103P - 1.298) \times \frac{\pi}{180}\right] + 21.13 \cos\left[(2826P + 10.83) \times \frac{\pi}{180}\right] \quad (1)$$

where P is the percentage of the gait cycle T (60% < P < 90%). The time $t_{\theta,min}$ and $t_{\theta,zero}$ can be substituted by the parameter P which stays in dynamic changes according to the time

$t_{\theta,\text{min}}$ and gait cycle T. When the angle θ reaches θ_{min}, the percentage P corresponding to the time $t_{\theta,\text{min}}$ is obtained.

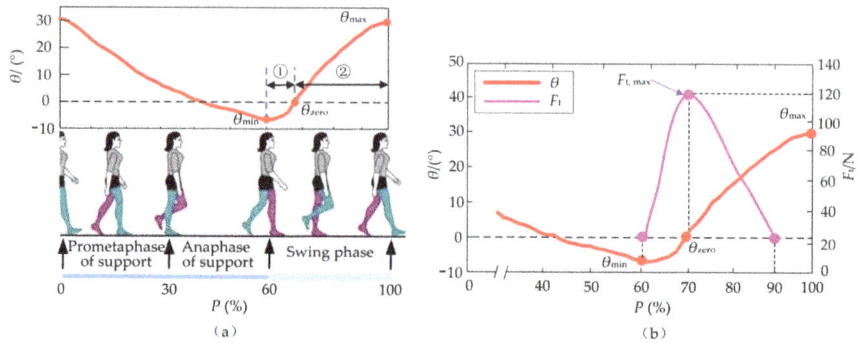

Figure 3. Profile of auxiliary forces F_t. (**a**) The hip flexion is divided into two stages: ① The first stage is from θ_{min} to the zero-point θ_{zero}, ② The second one is from θ_{zero} to θ_{max}. (**b**) The profile of the auxiliary force F_t begins at $t = t_{\theta,\text{min}}$, reaches the peak value $F_{t,\text{max}}$ at $t = t_{\theta,\text{zero}}$, and ends at $t = t_{90\%}$.

3.2. Desired Displacement of the Traction Unit

Based on the profile of forces F_t, the desired displacement of the traction unit can be calculated which is closely related to the hip angle θ and the total stiffness k_{total} of the H-Suit. A coordinate system xoy is built at the center o of the hip joint and the axis ox is horizontal, as shown in Figure 4a. The vectors $\boldsymbol{oA}$ and $\boldsymbol{oB}$ are the position vector of the waist and thigh anchors. According to the optimal position of A and B, the length of $\boldsymbol{oA}$ and $\boldsymbol{oB}$ denoted l_{oA} and l_{oB} can be calculated. The corresponding included angle α indicates the dynamic position of the thigh during the walking. The changes in the length l_{AB} of $\boldsymbol{AB}$, without the auxiliary forces, are the natural traction displacement L_n to adapt the positions of the thigh. The initial length l_0 of $\boldsymbol{AB}$ at the standing posture can be obtained by the cosine theorem:

$$l_0 = \sqrt{l_{oA}^2 + l_{oB}^2 - 2l_{oA} \cdot l_{oB} \cdot \cos\alpha_0} \tag{2}$$

where α_0 is the angle α in the standing posture.

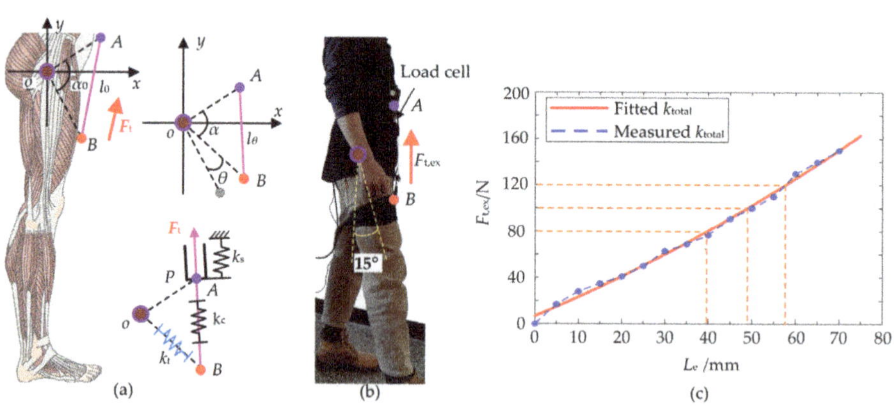

Figure 4. Measurement of the total stiffness. (**a**) The desired displacement consists of the natural displacement of Bowden cables and the elastic deformations of the human–machine system. (**b**) The total stiffness k_{total} was measured when the hip flexed forwards 15°. (**c**) The external force $F_{t,ex}$ at the equal intervals of displacements are presented and k_{total} can be fitted.

Assuming that the angle of the anchor A relative to the hip center is constant during the human walking, the anchor B is rotated around the center o and the angle α is obtained:

$$\alpha = \alpha_0 - \theta \tag{3}$$

Hence, the length l_{AB} of $\boldsymbol{AB}$ can be calculated:

$$l_{AB} = \sqrt{l_{oA}^2 + l_{oB}^2 - 2l_{oA} \cdot l_{oB} \cdot \cos\alpha} \tag{4}$$

The natural displacement L_n of the traction unit can be acquired:

$$L_n = l_{AB} - l_0 \tag{5}$$

The auxiliary force F_t is applied to the anchors along the vector $\boldsymbol{AB}$. There exist elastic deformations in the exosuit, Bowden cables and local tissues. Due to the intercoupling of the H-Suit and local tissues, it is difficult to measure and calculate the stiffness of three parts denoted by k_s, k_c and k_t, respectively. However, the total stiffness k_{total} can be measured by the experimental method [22,23]. A subject wore H-Suit and the right lower limb flexed forwards 15°, as shown in Figure 4b. A traction force $F_{t,ex}$ was applied to the anchors from 0 to 120 N with the help of the traction unit. The subject kept still to complete the experiments. The displacements of the traction unit were measured to indicate the elastic deformations L_e by the displacement sensor. The external force $F_{t,ex}$ at the equal intervals of displacements are presented in Figure 4c.

According to the measured data, the expression of k_{total} is fitted and calculated where k_{total} = 2100 N/m. The elastic deformations L_e of the human–machine system are obtained:

$$L_e = \frac{F_t}{k_{total}} \tag{6}$$

Hence, by combining Equations (5) and (6), the output displacement L of the traction unit can be acquired:

$$L = l_{AB} - l_0 + \frac{F_t}{k_{total}} \tag{7}$$

According to the profiles of F_t and θ, the desired displacement L can be calculated as shown in Figure 5. The displacement L reaches the maximum $L'_{max} = 106$ mm at $t = t_{90\%}$ and descends quickly to the initial position.

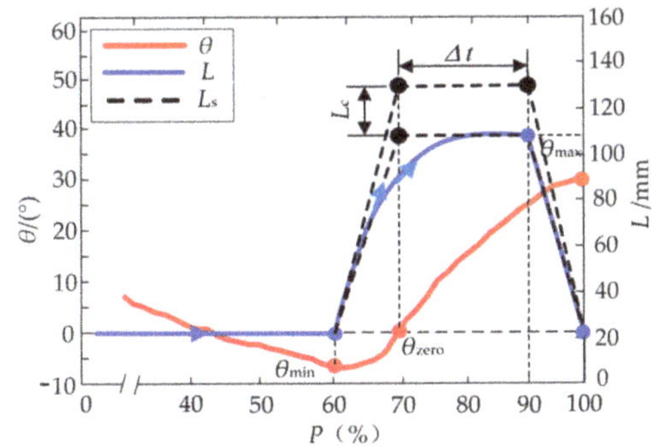

Figure 5. Desired displacement L.

3.3. Assistance Strategy

When the hip angle is about to reach the minimum, the traction unit will be moved along the profile of the displacement L to produce the corresponding auxiliary force F_t.

Then the traction unit returns to the initial position immediately and waits for the next gait cycle. The drivers need to detect the current positions of the servo motors and consume the majority of the internal resource of the MCU. Besides, the algorithms of gait detection and prediction also occupy part of the internal resource. The motion lags appear in the movements of the traction unit and seriously restrict the performance of the H-Suit. Hence, the theoretical profile of the displacement L is simplified to a trapezoidal profile L_s with three segments, as shown in Figure 5. The traction unit moves from the initial position to the maximum with a constant speed, stays for a hold time Δt and returns to the initial position immediately.

According to the activation times of the iliopsoas muscle and the rectus femoris, the traction unit reaches the maximum position at $\theta = \theta_{zero}$. The theoretical maximum L'_{max} of the trapezoidal profile L_s can be calculated:

$$L'_{max} = L'_{n} + L'_{e} \tag{8}$$

Where L'_n is the natural displacement L_n at $\theta = \theta_{max}/2$ (θ_{max} denotes the maximum of hip angle θ) and L'_e is the elastic deformations L_e at $F_{t,max}$ = 120 N. The deformation L'_e is acquired:

$$L'_{e} = \frac{F_{t,max}}{k_{total,120}} \tag{9}$$

where $k_{total,120}$ is the total stiffness k_{total} at $F_{t,ex}$ = 120 N.

Due to the initial wearing deviations and the disturbances of the external force F_t, the position deviations of the anchors are inevitable. A compensation displacement L_c is added to the theoretical maximum L'_{max} for compensating the deviations. Hence, the maximum L_{max} of the trapezoidal profile L_s during the assistance process can be obtained:

$$L_{max} = L'_{max} + L_c \tag{10}$$

The compensation displacement L_c and the hold time Δt mainly affect the magnitude and duration of the peak force $F_{t,\,max}$. With the help of the load cells, the auxiliary forces F_t during the gait cycle T_i can be measured and the actual maximum $F^i_{t,max}$ are selected. The displacement L_c can be obtained based on the forces $F^i_{t,max}$ of the last n gait cycles.

$$L_c = \frac{\sum_{i=1}^{n} (F^i_{t,max} - F_{t,max})}{n \cdot k_{total,120}} \tag{11}$$

The hold time Δt can improve the auxiliary profile and reduce the impacts of the forces F_t. Under the constant peak forces, a long hold time Δt can enhance the assistance performance, especially the heartbeat index. There is a linear relationship between heart rate and oxygen consumption [34]. However, if the hold time Δt is too long, it will hinder the backward extension of lower limbs and aggravate the metabolic consumption. The traction and release times of the steel cable denoted by $\Delta t_{traction}$ and $\Delta t_{release}$ can be measured by the encoder of the control unit. The auxiliary time ΔT of the forces F_t can be calculated:

$$\Delta T = \Delta t_{traction} + \Delta t_{release} + \Delta t \tag{12}$$

The auxiliary time ΔT is continuously changed for different walking speeds. Hence, during the assistance performance, the hold time Δt and the compensation displacement L_c are dynamically adjusted within the time ΔT according to the heartbeat index. Based on the control strategy, an algorithm is added to the control unit.

4. Performance Evaluation

4.1. Experimental Platform

There are three experiments for the performance evaluation: (I) Turn on H-Suit when walking with it (Powered); (II) Turn off H-Suit (Unpowered); (III) Not wear H-Suit (No-Suit). A subject performs each experiment by walking on the treadmill for 10 min and resting for 20 min, as shown in Figure 6. Eight subjects (average age 26, height 175 cm, weight 75 kg) wear the H-Suit and walk on the treadmill at the speed v_1 = 0.75 m/s, as shown in Figure 7. When the curve of the hip angle θ is stable, the established prediction algorithm will be executed and the angle θ is predicted in real-time. Then, the traction units are opened and moved along the profile of the output displacements. The data of the angle θ and force F_t are measured by load cells and IMUs and recorded in the upper computer to analyze the assistance strategy.

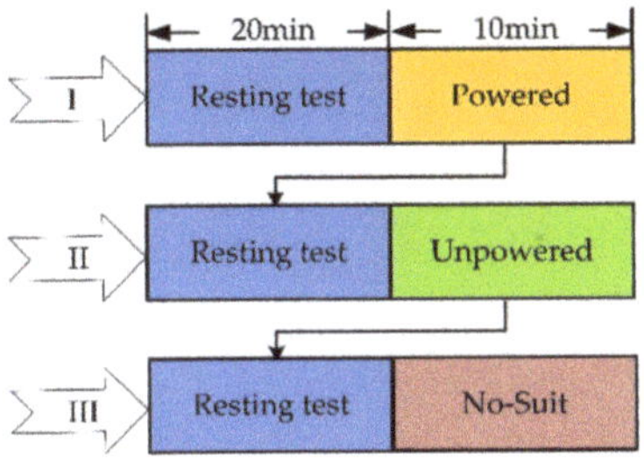

Figure 6. Scheme of assistance experiments. The subjects rested for 20 min before the test and walked in the state of Powered, Unpowered and No-Suit for 10 min.

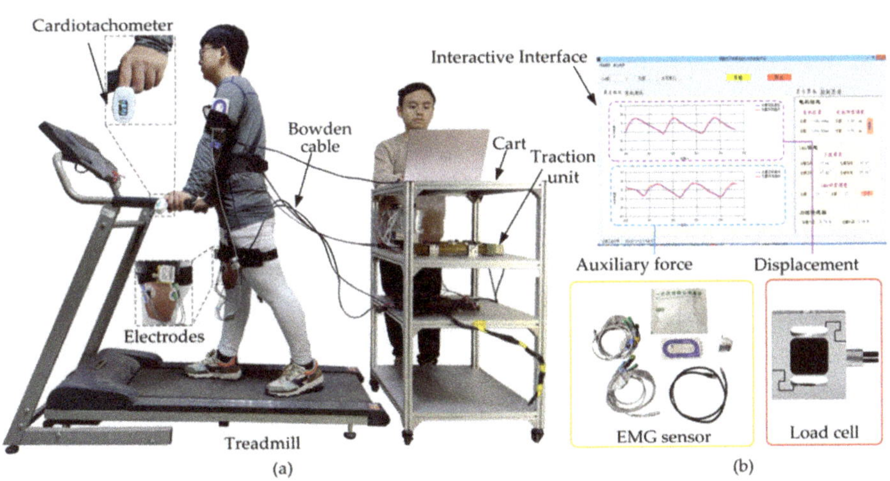

Figure 7. Experimental platform of the H-Suit. (**a**) The subjects walked on the treadmill at three different speeds. (**b**) The data of EMG, auxiliary forces and displacements are measured and depicted in the visual interface.

4.2. Evaluation of Auxiliary Force

During the assistance experiments, the initial values of compensation displacement L_c and the hold time Δt are 17 mm and 300 ms, respectively. When a subject walked on the treadmill at the speed v_1 = 0.75 m/s, the dynamic data of three gait cycles during the assistance process are shown in Figure 8. Compared with the real curve of θ, the prediction curve has a similar shape and advances at a time interval of 30 ms (Figure 8a). Meanwhile, the hip angle θ (such as the maximum, zero, minimum, etc.) can be forecasted early by a time of 30 ms. The predicted time of the minimum angle θ_{min}, which is the start time of the traction unit, will be sent to the control unit. The steel cables are pulled quickly to the maximum position (L_{max} = 123 mm) which is calculated by Equation (10), stayed for 300 ms and returned to the initial position immediately (Figure 8b). The traction of steel cables can produce the useful forces during the assistance phase where $F^i_{t,max}$ = 135 N. Since the compensation displacement L_c is larger than the appropriate value, the actual peak forces of F_t are bigger than the ideal force $F_{t,max}$ and has a slight impact on the hip joint. Similarly, the hold time Δt is also too long, the traction unit still stays at the maximum position even though the hip joint begins to stretch backward. When the H-Suit works without correctly adjusting the controller, the auxiliary forces F_t at the resistance phase hinder the extension movements at the later stage (Figure 8c). Hence, the improper values of L_c and Δt will interfere with the natural movements of lower limbs.

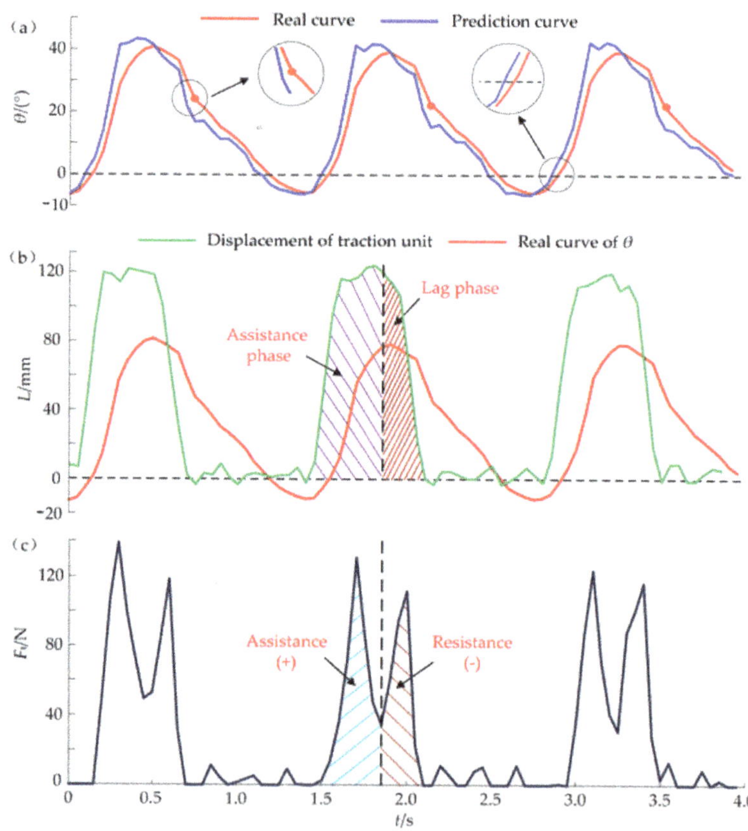

Figure 8. Dynamic data during the assistance process (v_1 = 0.75 m/s, L_c = 17 mm, Δt = 300 ms). (**a**) The prediction curve of the hip angle θ has a similar shape and advances a time interval compared with the real curve. (**b**) The output displacement L is determined by the parameters L_c and Δt. (**c**) The auxiliary forces F_t hinder the extension movements at the later stage due to the improper values of L_c and Δt.

According to the effects of the two parameters, the established algorithm of the control strategy begins to work during performance experiments. Before the experiments, the steel cables should keep a tension state ($F_t \leq 5$ N). A subject wears the H-Suit and walks on the treadmill at v_1 = 0.75 m/s. Based on the peak forces $F^i_{t,max}$ and the average heart rates measured by a cardiotachometer, the optimum values of L_c and Δt are obtained through the control algorithm after 10 gait cycles where $L_c = 7$ mm and $\Delta t = 200$ ms. The changes in the average heart rates were analyzed and discussed in our previous work [31] where the heart rates have decreased remarkably. The corresponding dynamic data are shown in Figure 9. It can be found that the minimum θ_{min} is predicted 50 ms in advance and the traction unit starts to move along the profile of the displacements L. The traction unit can release the steel cable immediately after the hold time Δt and not hinder the deceleration of the flexion movement. Then the right leg completes the extension movements. The switching time from the flexion to extension is shorter and the real curve of θ is smoother compared with that of Figure 8a. Hence, the added mass and the traction movement of steel cables have few influences on the natural gait of lower limbs. The assistance phase dominates the whole activation interval of the H-Suit and the lag phase almost disappears. Although the traction unit tries to move along the ideal displacement curve with high precision, the real curve of L still has a few deviations at the peak stage. The real profile of the auxiliary forces F_t is in good agreement with the theoretical profile (Figure 9c) where $F^i_{t,max}$ is approximately equal to 120 N. The subject feels it is easier to walk under the assistance condition.

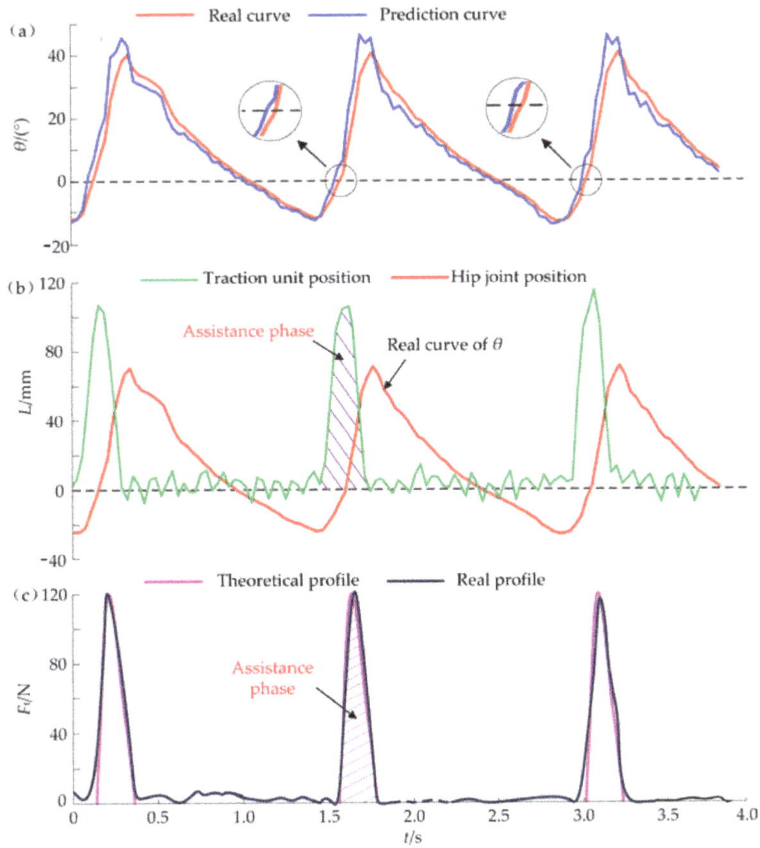

Figure 9. Experimental data of the effective assistance ($v_1 = 0.75$ m/s, $L_c = 7$ mm, $\Delta t = 200$ ms) (**a**) The switching time from the flexion to extension is shorter and the real curve of θ is smoother compared with that of Figure 8a. (**b**) The assistance phase dominates the whole activation interval of the H-Suit and the lag phase almost disappears. (**c**) The real profile of the auxiliary forces F_t is in good agreement with the theoretical profile.

4.3. EMG Signals of Rectus Femoris and Vastus Lateralis

In order to evaluate the performance of H-Suit, the EMG signals of the rectus femoris and vastus lateralis of the subjects were measured through pasting electrodes according to SENIAM standards [35]. The corresponding EMG data were obtained at a normal speed (v_2 = 1.25 m/s) and in three conditions (No-Suit, Unpowered, Powered), as shown in Figure 10. Figure 10a shows the EMG signals of the rectus femoris and vastus lateralis of 15 gait cycles under the condition of No-Suit. It can be found that the EMG signals of the two muscles have similar trends. While the periodic of the vastus lateralis is more obvious than the rectus femoris, the vastus lateralis was selected for further analysis. Figure 10b shows the EMG signals of the vastus lateralis of 5 gait cycles under the three conditions. The vastus lateralis has obvious periodic characteristics. Combined with the average rectified value (ARV) and median frequency (MNF) slopes in Figure 10c, it can be found that the change trends of the two indexes in the No-Suit and Unpowered are basically identical, while ARV decreases and MNF increases significantly at the powered condition. It shows that the H-Suit without the drive unit has little effect on the EMG signal of the lower limbs. In the powered condition, H-Suit can delay the muscle fatigue of the lower limbs where a steeper positive slope for ARV and a steeper negative one for MNF indicate a faster onset of fatigue [36]. The ARV slope during the powered condition was decreased by 0.5178 ± 0.1110/min and the MNF slope was increased by 0.2280 ± 0.0714/min.

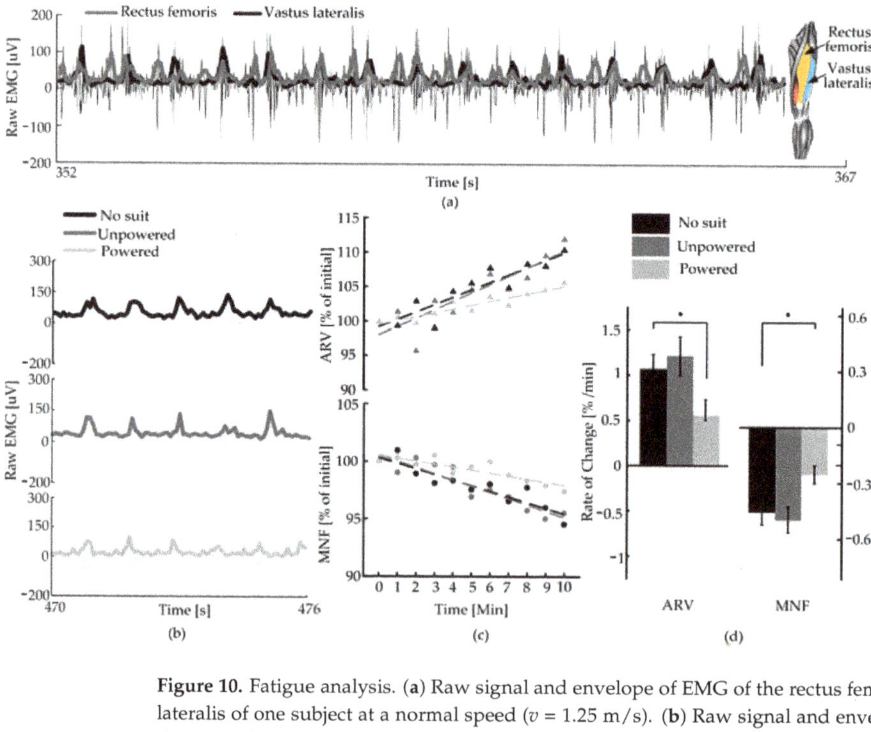

Figure 10. Fatigue analysis. (**a**) Raw signal and envelope of EMG of the rectus femoris and vastus lateralis of one subject at a normal speed (v = 1.25 m/s). (**b**) Raw signal and envelope of EMG of the vastus lateralis in the conditions of No-Suit, Unpowered and Powered. (**c**) Trend of the average rectified value (ARV) and median frequency (MNF) of the EMG signal of one subject in the 10 min of tests. Indexes are expressed in percentage of their initial value. (**d**) Slope of ARV and MNF. Both indexes confirm that wearing the H-Suit significantly reduces the onset of fatigue (p = 0.03 for ARV and p = 0.01 for the MNF). Error bars show the standard error of the mean.

In order to further evaluate the performance of the H-Suit, eight healthy subjects with a 10 kg load walked on the treadmill at three speeds (0.75 m/s, 1.25 m/s and 1.75 m/s) [37]. The electromyography (EMG) signals of the vastus lateralis were measured and analyzed, as shown in Figure 11. As the H-Suit provided the assistance for the flexion movements, it reduced the amount of effort that flexor muscle needed to exert. Figure 11a shows a representative case of the activation (raw EMG and its envelope) of the vastus lateralis during the three walking speeds, in both powered and unpowered conditions. The EMG's values under the powered condition decreased markedly. Besides, the corresponding excitation was delayed and the action time was shortened. The net change in the vastus lateralis effort (Figure 11b), evaluated as the difference in the root mean square (RMS) of the EMG signals between the powered and unpowered cases, was significantly smaller than 0 for all velocities ($p = 2 \times 10^{-3}$, $p = 7.0 \times 10^{-3}$, $p = 4 \times 10^{-3}$ at $v = 0.75$ m/s, 1.25 m/s, 1.75 m/s). Figure 11c shows the change in the activation of the valtus lateralis as a percentage of its activation in the unpowered condition. The performance of the H-Suit degraded for higher speeds. Wearing the H-Suit resulted in a significant reduction of the vastus lateralis effort, averaged over subjects and walking speeds, of 13.3 ± 2.1% ($p = 2 \times 10^{-5}$).

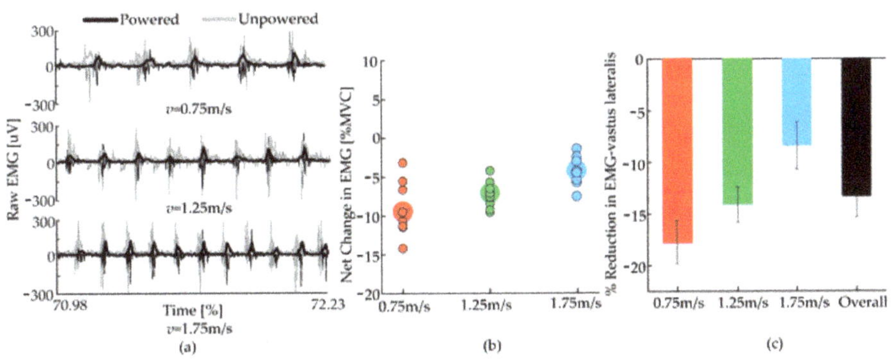

Figure 11. Changes in muscular activation. (**a**) Raw signal and envelope of the vastus lateralis' EMG. (**b**) Net change (Powered—Unpowered) of RMS of the EMG signals of the vastus lateralis, for the three walking speeds. Opaque contoured circles are the values for each individual subject, bigger circles indicate the mean over subjects. (**c**) Changes in the activation of the vastus lateralis, expressed as a percentage of its activation in the unpowered condition (net change/Unpowered). Error bars show the standard error of the mean.

5. Discussion

The ergonomic design of the H-Suit system is clearly presented, such as optimization of anchor points, comfort evaluation, analysis of biomechanical characteristics of lower limbs, etc. Compared with the development of the existed soft assistance systems [14,18,19], the design process is presented in more detail. The H-Suit processes the advantages of being lightweight, having high wearing comfort and has minimal impact on the natural gaits.

The profile of the auxiliary forces is planned in the auxiliary range where the forces start at the minimum of hip angle, reach the maximum (120 N) when the hip angle is equal to zero and end at 90% of each gait cycle. The beginning of the profile is dynamically changed according to the time of the minimum hip angle and the gait cycle. Compared with the profile proposed by Ding [22,23], our designed profile uses fewer memory resources and allows for faster processing speed. An assistance strategy is proposed to track the profile of the auxiliary forces by dynamically adjusting the compensation displacement L_c and the hold time Δt. The force-based position control [38] and admittance-based force tracking [39,40] are usually applied to track and compensate for the profile of auxiliary force. However, the assistance strategy proposed in this paper can achieve approximate performance and provide a new idea for the robotic suit. Furthermore, the influences of the variables L_c and Δt on the natural gaits and auxiliary forces have been analyzed to reveal the interactions between lower limbs and the H-Suit.

The changes in human metabolic consumption are essentially caused by the activation degree of muscles which can be evaluated by the slopes of the average rectified value and the median frequency [39]. During the performance evaluations, our H-Suit delayed the onset of fatigue and reduced the muscular activation of hip flexors. Wearing the H-Suit resulted in a significant reduction of the vastus lateralis effort, averaged over subjects and walking speeds, of 13.3 $\pm$ 2.1% ($p = 2 \times 10^{-5}$). A similar finding is described in [18], where a cable-driven suit for the hip extension is shown to reduce activation in muscles that do not cross the assisted joints (such as vastus lateralis). Unfortunately, a quantitative comparison here is not possible because of the different metrics used to assess fatigue in [18].

6. Conclusions

The advantages of a svelte and portable exosuit for the lower limbs, able to intuitively assist its wearer and reduce the effort required to walk, make it a good candidate for both industrial and clinical applications. Our results showed that the H-Suit can improve the walking endurance of lower limbs, delay muscle fatigue and reduce the activation level of hip flexors.

7. Limitations of the Study

The traction unit of the H-Suit is an external device and cannot be carried by the subjects. This study mainly analyzes the influences of the assistance strategy on the fatigue and activation of hip flexors. More flexors, especially the iliopsoas muscle, should be measured to evaluate metabolic consumption. Meanwhile, the experimental conditions (Powered, Unpowered and No-Suit) and walking speeds are not randomly distributed in the experiments of performance evaluation. In future research, a portable traction unit will be developed and subjects can complete all experiments when wearing the H-Suit system. The scheme of performance experiments will be improved by considering the randomization of experimental conditions.

Supplementary Materials: The following supporting information can be downloaded at: https://www.mdpi.com/article/10.3390/mi13060825/s1. Table S1: Parameters of the lower limb. All data generated or analyzed during this study are included in this published article. The mathematical model of the anchor locations is depicted in Appendix A.

Author Contributions: All authors developed the H-Suit body. L.Z., Z.J. and Y.H. developed the assistance strategy. All authors designed the experiment. L.Z. and Z.J. performed the experiment. L.Z., Y.H. and P.S. analyzed and interpreted the data. All authors prepared the manuscript. All authors provided critical feedback on the manuscript. All authors have read and agreed to the published version of the manuscript.

Funding: This research is partially supported by the projects of the National Natural Science Foundation of Beijing (No. 3202003) and the National Natural Science Foundation of China (No. 52005045).

Conflicts of Interest: Patents have been filed with the China Patent Office by the Beijing University of Technology, which describes the exosuit components documented in this manuscript. L.Z. is the author of those patents.

Appendix A

The coordinate system xoy is established where the origin o is located on the rotation center of the hip joint, as shown in Figure A1. The waist and thigh anchors are denoted by P and Q. The direction of the auxiliary force F is parallel to the direction of the vector $\boldsymbol{QP}$. The point Q_x is the projection point of the thigh anchor Q on the axis ox when the volunteer is upright. The direction of the vector $\boldsymbol{oQ_x}$ is parallel to the positive direction of the axis ox, and its modulus depends on the diameter of the human thigh, as shown in

Figure A1a. Since the upper body keeps upright during walking, the vertical displacement of the hip joint is small, and the rotation angle θ of the hip joint can be represented by the direction change of the vector$\boldsymbol{Q_xQ}$. Meanwhile, the angle θ of the hip joint extension stage is defined as negative, and the forward flexion stage as positive, as shown in Figure A1b,c. The vector $\boldsymbol{oR}$ is perpendicular to the vector $\boldsymbol{QP}$, and the modulus of $\boldsymbol{oR}$ represented as l_{OR} is the auxiliary force arm.

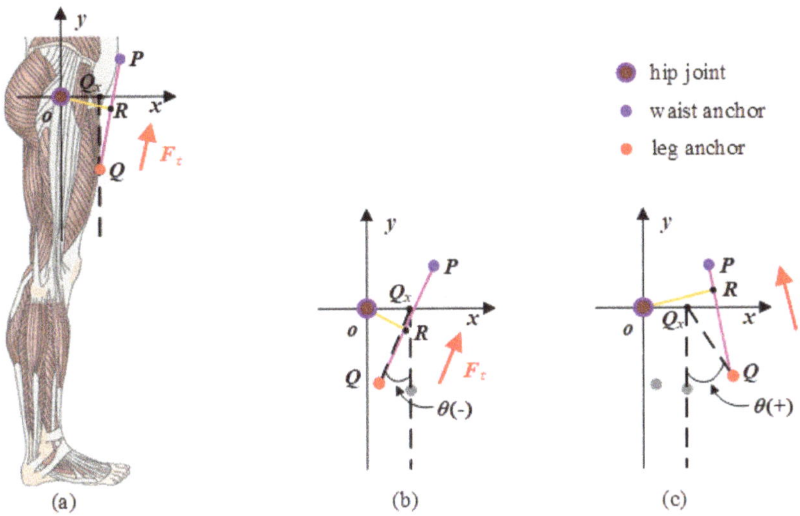

Figure A1. Simplified model of human lower limbs. (**a**) Standing phase (θ = 0). (**b**) Stretching stage (θ < 0). (**c**) Buckling stage (θ > 0).

In the process of walking, assuming that the upper body of the human remains upright, the position vector $\boldsymbol{oP}$ of the waist anchor P depends on the vertical distance between the waist anchor P and hip joint, and the waist thickness T_w of the human body. Since the waist anchor P is fixed on the waist belt, the variation range of P is small. Through comparing and measuring the walking posture of eight healthy subjects, the waist anchor P is fixed at 50 mm above the hip joint. In addition, the thigh anchor Q is fixed on the thigh belt. According to the above analysis, the direction of the vector $\boldsymbol{Q_xQ}$ is consistent with the rotation angle θ of the hip joint. The modulus of $\boldsymbol{Q_xQ}$ is equal to the vertical component of the distance between the thigh anchor Q and the hip joint, which is named as l_{ah}, therefore, the position vector $\boldsymbol{oQ}$ and $\boldsymbol{QP}$ of the thigh anchor Q can be represented as:

$$\boldsymbol{oQ} = \boldsymbol{oQ_x} + \boldsymbol{Q_xQ} \tag{A1}$$

$$\boldsymbol{QP} = \boldsymbol{oP} - (\boldsymbol{oQ_x} + \boldsymbol{Q_xQ}) \tag{A2}$$

Assuming the auxiliary force produced by the traction unit is nearly constant, the changes in the auxiliary torque can be indirectly reflected by calculating the length l_{oR} of the torque arm. Combined with the structure of the lower limbs shown in Figure A1, the length l_{oR} is obtained:

$$l_{oR} = l_{oQ} \cdot \sqrt{1 - \left(\frac{\boldsymbol{oQ} \cdot \boldsymbol{QP}}{l_{oQ} \cdot l_{QP}}\right)^2} \tag{A3}$$

where l_{oQ} is the modulus of the vector $\boldsymbol{oQ}$.

According to the measurement of eight subjects, the average body sizes of the subjects (male, 26 years, height 175 cm, weight 75 kg) during the walking are obtained, as shown in Table S1.

Substitute the above parameters into Equations (A2) and (A3), and the torque arm l_{oR} corresponding to different positions of the thigh anchor are calculated by the software MATLAB, as shown in Figure A2.

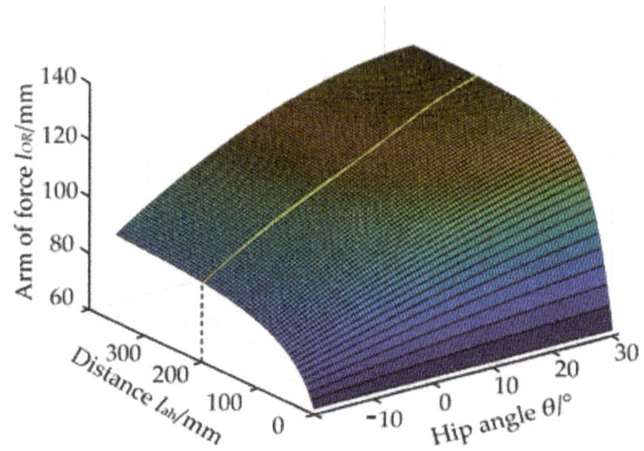

Figure A2. Auxiliary force arm in different states.

When the distance l_{ah} between the thigh anchor Q and the hip joint is more than 200 mm, the torque arm l_{oR} corresponding to the same hip joint angle remains approximately unchanged, and the curve of l_{oR} arm is changed smoothly. Moreover, in order to improve the servo performance of the wire rope during the assisting process, the length of $l_{a\mathrm{h}}$ should not be too large and the thigh anchor Q is fixed 200 mm below the hip joint. Hence, the waist anchor P is fixed at 50 mm above the hip joint and the thigh anchor Q at 200 mm below the hip joint.

References

1. Xiao, X.; Fang, Y.; Xiao, X.; Xu, J.; Chen, J. Machine-Learning-Aided Self-Powered Assistive Physical Therapy Devices. *ACS Nano* **2021**, *15*, 18633–18646. [CrossRef] [PubMed]
2. Dollar, A.M.; Herr, H. Lower Extremity Exoskeletons and Active Orthoses: Challenges and State-of-the-Art. *IEEE Trans. Robot.* **2008**, *24*, 144–158. [CrossRef]
3. Herr, H.M. Exoskeletons and Orthoses: Classification, Design Challenges and Future Directions. *J. Neuro Eng. Rehabil.* **2009**, *6*, 21–29. [CrossRef] [PubMed]
4. Collins, S.H.; Wiggin, M.B.; Sawicki, G.S. Reducing the Energy Cost of Human Walking Using an Unpowered Exoskeleton. *Nature* **2015**, *522*, 212–215. [CrossRef] [PubMed]
5. Mooney, L.M.; Rouse, E.J.; Herr, H. Autonomous Exoskeleton Reduces Metabolic Cost of Human Walking during Load Carriage. *Neuroeng. Rehabil.* **2014**, *11*, 80–91. [CrossRef] [PubMed]
6. Kazerooni, H. Exoskeletons for Human Power Augmentation. In Proceedings of the 2005 IEEE/RSJ International Conference on Intelligent Robots and Systems, Edmonton, AB, Canada, 2–6 August 2005; pp. 3459–3464.
7. Ohta, Y.; Yano, H.; Suzuki, R.; Yoshida, M.; Kawashima, N.; Nakazawa, K. A Two-degree-of-freedom Motor-powered Gait Orthosis for Spinal Cord Injury Patients. *Proc. Inst. Mech. Eng. Part H J. Eng. Med.* **2007**, *221*, 629–639. [CrossRef]
8. Shorter, K.A.; Kogler Géza, F.; Loth, E.; Durfee, W.K.; Hsiao-Wecksler, E.T. A Portable Powered Ankle-foot Orthosis for Rehabilitation. *J. Rehabil. Res. Dev.* **2011**, *48*, 459–472. [CrossRef]
9. Zoss, A.B.; Kazerooni, H.; Chu, A. Biomechanical Design of the Berkeley Lower Extremity Exoskeleton. *IEEE ASME Trans. Mechatron.* **2006**, *11*, 128–138. [CrossRef]
10. Yoko, I.; Michiko, I. *Global Rulemaking Strategy for Implementing Emerging Innovation: Case of Medical/Healthcare Robot, HAL by Cyberdyne (Japanese)*; Policy Discussion Papers; Research Institute of Economy, Trade and Industry (RIETI): Tokyo, Japan, 2019; pp. 1–27.
11. Stearns-Yoder, K.A.; Brenner, L.A. Novel Psychological Outcomes with Ekso Bionics Technology. *Arch. Phys. Med. Rehabil.* **2018**, *99*, e70–e71. [CrossRef]
12. Zeilig, G.; Weingarden, H.; Zwecker, M.; Dudkiewicz, I.; Bloch, A.; Esquenazi, A. Safety and Tolerance of the ReWalk Exoskeleton Suit for Ambulation by People with Complete Spinal Cord Injury: A Pilot Study. *J. Am. Paraplegia Soc.* **2012**, *35*, 96–101. [CrossRef]

13. Xiloyannis, M.; Anna-Maria, G.; Haufe, F.L.; Wolf, P.; Masia, L.; Riener, R. Soft Robotic Suits: State of the Art, Core Technologies, and Open Challenges. *IEEE Trans. Robot.* **2021**, 1–20. [CrossRef]
14. Park, J.; Park, H.; Kim, J. Performance Estimation of the Lower Limb Exoskeleton for Plantarflexion Using Surface Electromyography (sEMG) Signals. *J. Biomech. Sci. Eng.* **2017**, *12*, 1–9. [CrossRef]
15. Walsh, C.J.; Endo, K.; Herr, H.M. A Quasi-passive Leg Exoskeleton for Load-carrying Augmentation. *Int. J. Hum. Robot.* **2007**, *4*, 487–506. [CrossRef]
16. Schiele, A.; Van, D.H.; Frans, C.T. Influence of Attachment Pressure and Kinematic Configuration on Phri with Wearable Robots. *Appl. Bionics Biomech.* **2009**, *6*, 157–173. [CrossRef]
17. Xiao, X.; Xiao, X.; Zhou, Y.; Zhao, X.; Chen, G.; Liu, Z.; Wang, Z.; Lu, C.; Hu, M. An Ultrathin Rechargeable Solid-state Zinc Ion Fiber Battery for Electronic Textiles. *Sci. Adv.* **2021**, *7*, eabl3742. [CrossRef]
18. Kim, J.; Lee, G.; Heimgratner, R.; Arumukhom, R.D.; Karavas, N.; Nathanson, D.; Galiana, I.; Eckert-Erdheim, A.; Murphy, P.; Perry, D.; et al. Reducing the Metabolic Rate of Walking and Running with a Versatile, Portable Exosuit. *Science* **2019**, *365*, 668–672. [CrossRef]
19. Schmidt, K.; Duarte, J.E.; Grimmer, M. The Myosuit: Bi-articular Anti-gravity Exosuit that Reduces Hip Extensor Activity in Sitting Transfers. *Front. Neurorobotics* **2017**, *11*, 1–16. [CrossRef]
20. Li, J.; Li, G.; Zhang, L.; Yang, D.; Wang, H. Advances and Key Techniques of Soft Wearable Lower Limb Power-Assisted Robots. *Acta Autom. Sin.* **2020**, *46*, 427–438.
21. Yang, B.; Huang, J.; Chen, X.; Xiong, C.; Hasegawa, Y. Supernumerary Robotic Limbs: A Review and Future Outlook. *IEEE Trans. Med. Robot. Bionics* **2021**, *3*, 623–639. [CrossRef]
22. Ding, Y.; Galiana, I.; Asbeck, A.T.; De, R.S.M.; Bae, J.; Santos, T.R.; Araujo, V.L.; Lee, S.; Holt, K.G.; Walsh, C. Biomechanical and Physiological Evaluation of Multi-joint Assistance with Soft Exosuits. *IEEE Trans. Neural Syst. Rehabil. Eng.* **2017**, *25*, 119–130. [CrossRef]
23. Ding, Y.; Kim, M.; Kuindersma, S.; Walsh, C.J. Human-in-the-loop Optimization of Hip Assistance with a Soft Exosuit during Walking. *Sci. Robot.* **2018**, *3*, eaar5438. [CrossRef] [PubMed]
24. Hashimoto, Y.; Nakanishi, Y.; Saga, N.; Nagase, J.; Satoh, T. Development of Gait Assistive Device Using Pneumatic Artificial Muscle. In Proceedings of the 2016 Joint 8th International Conference on Soft Computing and Intelligent Systems (SCIS) and 17th International Symposium on Advanced Intelligent Systems (ISIS), Sapporo, Japan, 25–28 August 2016; pp. 710–713.
25. Tian, M.; Wang, X.; Wang, J.; Gan, Z. Design of a Lower Limb Exoskeleton Driven by Tendon-sheath Artificial Muscle. In Proceedings of the 2019 IEEE International Conference on Robotics and Biomimetics (ROBIO), Dali, China, 6–8 December 2019; pp. 91–96.
26. John, S.W.; Murakami, K.; Komatsu, M.; Adachi, S. Cross-wire Assist Suit Concept for Mobile and Lightweight Multiple Degree of Freedom Hip Assistance. In Proceedings of the 2017 International Conference on Rehabilitation Robotics (ICORR), London, UK, 17–20 July 2017; pp. 387–393.
27. Neumann Donald, A. *Kinesiology of the Musculoskeletal System: Foundations for Rehabilitation*; Elsevier: Amsterdam, The Netherlands, 2010; pp. 650–652.
28. Mundt, M.; Thomsen, W.; Witter, T.; Koeppe, A.; David, S.; Bamer, F.; Potthast, W.; Markert, B. Prediction of Lower Limb Joint Angles and Moments during Gait Using Artificial Neural Networks. *Med. Biol. Eng. Comput.* **2020**, *58*, 211–225. [CrossRef] [PubMed]
29. Kyeong, S.; Feng, J.; Ryu, J.K.; Park, J.J.; Lee, K.H.; Kim, J. Surface Electromyography Characteristics for Motion Intention Recognition and Implementation Issues in Lower-limb Exoskeletons. *Int. J. Control. Autom. Syst.* **2022**, *20*, 1018–1028. [CrossRef]
30. Chenyang, G.; Chunhua, R.; Meilin, Z. A Novel Method to Process Surface Electromyography Signal for Pedestrian Lower Limb Motion Pattern Recognition. *Trans. Inst. Meas. Control.* **2020**, *42*, 2492–2498.
31. Zhang, L.; He, Y.; Li, J.; Su, P.; Tao, C.; Ji, Y.; Dong, M. Ergonomic Design of Flexible Lower Limb Assist Exosuit and Gait Prediction. *J. Cent. South Univ. (Sci. Technol.)* **2021**, *52*, 1171–1184.
32. Meng, Q.; Zeng, Q.; Xie, Q.; Fei, C.; Kong, B.; Lu, X.; Wang, H.; Yu, H. Flexible Lower Limb Exoskeleton Systems: A review. *NeuroRehabilitation* **2022**, 1–24. [CrossRef]
33. Young, A.J.; Hannah, G.; Ferris, D.P. A Biomechanical Comparison of Proportional Electromyography Control to Biological Torque Control Using a Powered Hip Exoskeleton. *Front. Bioeng. Biotechnol.* **2017**, *5*, 37. [CrossRef]
34. Bot, S.D.; Hollander, A.P. The Relationship between Heart Rate and Oxygen Uptake during Non-steady State Exercise. *Ergonomics* **2000**, *43*, 1578–1592. [CrossRef]
35. Hermens, H.J.; Freriks, B.; Disselhorst-Klug, C.; Rau, G. Development of Recommendations for SEMG Sensors and Sensor Placement Procedures. *Electromyogr. Kinesiol.* **2000**, *10*, 361–374. [CrossRef]
36. Merletti, R.; Parker, P. *Electromyography: Physiology, Engineering, and Noninvasive Applications*; John Wiley & Sons: Hoboken, NJ, USA, 2004; pp. 238–246.
37. Au, S.K.; Weber, J.; Herr, H. Powered Ankle—Foot Prosthesis Improves Walking Metabolic Economy. *IEEE Trans. Robot.* **2009**, *25*, 51–66. [CrossRef]
38. Quinlivan, B.T.; Lee, S.; Malcolm, P.; Rossi, D.M.; Grimmer, M.; Siviy, C.; Karavas, N.; Wagner, D.; Asbeck, A.; Galiana, I.; et al. Assistance Magnitude Versus Metabolic Cost Reductions for A Tethered Multiarticular Soft Exosuit. *Sci. Robot.* **2017**, 1–17. [CrossRef] [PubMed]

39. Xiloyannis, M.; Chiaradia, D.; Frisoli, A.; Masia, L. Physiological and Kinematic Effects of A Soft Exosuit on Arm Movements. *Neuro Eng. Rehabil.* **2019**, *16*, 29–44. [CrossRef] [PubMed]
40. Lee, G.; Ding, Y.; Bujanda, I.G.; Karavas, N.; Zhou, Y.M.; Walsh, C.J. Improved assistive profile tracking of soft exosuits for walking and jogging with off-board actuation. In Proceedings of the 2017 IEEE/RSJ International Conference on Intelligent Robots and Systems (IROS), Vancouver, BC, Canada, 24–28 September 2017; pp. 1699–1706.

micromachines

MDPI

Article

Multifunctional Nanoplatform Based on Sunitinib for Synergistic Phototherapy and Molecular Targeted Therapy of Hepatocellular Carcinoma

Wenjing Xu [1,†], Meng Yang [2,†], Xuanlong Du [1], Hao Peng [1], Yue Yang [1], Jitao Wang [1] and Yewei Zhang [3,*]

1 School of Medicine, Southeast University, Nanjing 210009, China
2 Department of Ultrasound, State Key Laboratory of Complex Severe and Rare Diseases, Peking Union Medical College Hospital, Chinese Academy of Medical Sciences, Beijing 100005, China
3 Hepatopancreatobiliary Center, The Second Affiliated Hospital of Nanjing Medical University, Nanjing 210011, China
* Correspondence: zhangyewei@njmu.edu.cn
† These authors contributed equally to this work.

Abstract: Hepatocellular carcinoma (HCC) is a tumor that poses a serious threat to human health, with an extremely low five-year survival rate due to its difficulty in early diagnosis and insensitivity to radiotherapy and chemotherapy. To improve the therapeutic efficiency of HCC, we developed a novel multifunctional nanoplatform (SCF NPs) with an amphiphilic polymer (Ce6-PEG2000-FA) and a multitarget tyrosine kinase inhibitor sunitinib. SCF NPs showed superior therapeutical efficiency for HCC due to the synergetic effect of molecular targeted therapy and phototherapy. The Ce6-PEG2000-FA not only serves as a nanocarrier with excellent biocompatibility but also can act as a therapeutic reagent for photothermal therapy (PTT) and photodynamic therapy (PDT). Furthermore, the folic acid group of Ce6-PEG2000-FA enhanced the active targeting performance of SCF NPs. As a multitargeted tyrosine kinase inhibitor, sunitinib in SCF NPs can play a role in molecular targeted therapies, including tumor growth inhibition and anti-angiogenesis. In vivo experiments, SCF NPs showed multimode imaging capabilities, which can be used for tumorous diagnosis and intraoperative navigation. Meanwhile, SCF NPs showed outstanding synergetic tumor inhibition ability. Tumors of SCF NPs group with laser radiation were eradicated without any recrudescence after 14 days of treatment. Such theranostic nanoparticles offer a novel therapeutic tactic for HCC.

Keywords: multifunctional nanoplatform; sunitinib; photodynamic therapy; photothermal therapy; hepatocellular carcinoma

Citation: Xu, W.; Yang, M.; Du, X.; Peng, H.; Yang, Y.; Wang, J.; Zhang, Y. Multifunctional Nanoplatform Based on Sunitinib for Synergistic Phototherapy and Molecular Targeted Therapy of Hepatocellular Carcinoma. *Micromachines* **2023**, *14*, 613. https://doi.org/10.3390/mi14030613

Academic Editors: Nam-Trung Nguyen and Songkil Kim

Received: 17 January 2023
Revised: 23 February 2023
Accepted: 27 February 2023
Published: 7 March 2023

1. Introduction

Hepatocellular carcinoma (HCC), accounting for 90% of primary liver cancers, is the sixth most common malignancy worldwide and the second leading cause of cancer deaths in China [1,2]. It is mainly caused by chronic hepatitis B and C virus infection [3]. Currently, the potential therapeutic options for early-stage patients with HCC include liver transplantation, surgical resection, and local ablation [4]. However, only 30% of patients present with resectable tumors, and the poor liver reserve capacities and high postoperative recurrence hinder tolerance to surgical treatment [5,6]. HCC is inherently resistant to radiotherapy and chemotherapy due to the expression of multiple drug-resistance genes [7]. Furthermore, HCC is often asymptomatic in the early stage, missing the optimal period for treatment [8]. Therefore, novel diagnostic and therapeutic approaches are essential to be explored for treating this lethal disease.

Sunitinib, a novel small-molecule multitargeted tyrosine kinase inhibitor, has been reported to inhibit HCC growth [9–11]. In a recent study, synergizing sunitinib and radiofrequency ablation as a novel therapeutic strategy significantly suppressed HCC growth by triggering the antitumor immune response. Sunitinib, rather than sorafenib (classical first-line drug for HCC), abolished tumor-induced profound immunotolerance to ignite immunological antitumor immune response by significantly suppressing Treg production and PD-1 expression [12]. Additionally, compared to sorafenib, sunitinib has a partial overlap but a significantly distinct target inhibition spectrum [9]. Chemoimmunotherapy of sunitinib combined with anti-PDL-1 antibody has been reported to increase antitumor immunity, inhibit tumor growth, and achieve optimal tumor control [13]. Sunitinib can not only inhibit tumor growth by blocking the signaling pathways of receptor tyrosine kinase, including c-KIT, FLT-3, RET, etc., but also inhibit angiogenesis by targeting vascular endothelial growth factor receptors (VEGFR) and platelet-derived growth factor receptors (PDGFR) [14]. Despite its efficacy, the severe side effects of sunitinib, such as hypertension, myelosuppression, hypothyroidism, and neutropenia, limit its further clinical application in the treatment of HCC [15,16]. Furthermore, the emergence of drug resistance to small-molecule targeted drugs is also a major challenge that cannot be overlooked in the treatment of HCC [17,18]. Among them, the issue of sunitinib resistance has attracted increasingly widespread attention [19–21]. Therefore, it is urgent to explore novel therapeutic modalities to tackle the problems of serious side effects and drug resistance of Sunitinib. Compared to traditional therapeutic methods of tumors, phototherapy, including photodynamic therapy (PDT) and photothermal therapy (PTT), has great potential applications in oncotherapy due to its high efficiency, low side effects, and no resistance [22–26]. PDT can transform oxygen into toxic reactive oxygen species (ROS), which have the capability to damage lipids, proteins, and DNA, and further kill cancer cells [27–29]. PTT is a promising strategy to rapidly convert localized light into heat in order to kill cancer cells when the temperature exceeds 45 °C [30,31]. However, insufficient oxygen supply (hypoxia) and short half-life (<40 ns) of ROS significantly reduced antitumor efficacy of PDT [32]. The upregulation of the heat shock protein expression and the damage to surrounding tissue severely restrict the clinical application of PTT [24]. Therefore, it is suggested that other therapeutic approaches should be combined with PDT/PTT to address the limitations and enhance the treatment efficacy. In recent years, the application of polymers in the field of medicine and health care has attracted extensive attention. For example, crosslinked polymers with high biocompatibility have attracted much attention in the field of wearable electronics and microfluidic chips for enhancing healthcare status [33–35]. Biodegradable polymers are commonly used for controlled-release drug delivery, tissue engineering, and temporary prosthetic implants.

Herein, we employed an amphiphilic polymer (Ce6-PEG2000-FA) to fabricate a novel nanoparticle by self-assembly method, sunitinib@Ce6-PEG2000-FA NPs (SCF NPs), with synergistic antivascular activity and PDT/PTT [36,37]. Sunitinib can induce apoptosis of vascular endothelial cells, thereby inhibiting tumor angiogenesis. Meanwhile, Ce6-PEG2000-FA can achieve PDT and PTT, killing HCC cells. Furthermore, folate receptors on the surfaces of HCC cells can be targeted by the FA molecules of the Ce6-PEG2000-FA to enhance the active targeting capability of NPs [38]. Accordingly, the resulting SCF NPs not only possess excellent passive (enhanced permeability and retention effect) and proactive (folate receptors) targeting ability towards tumors but also present superior synergistic effects with anti-angiogenic therapy and phototherapy (Scheme 1).

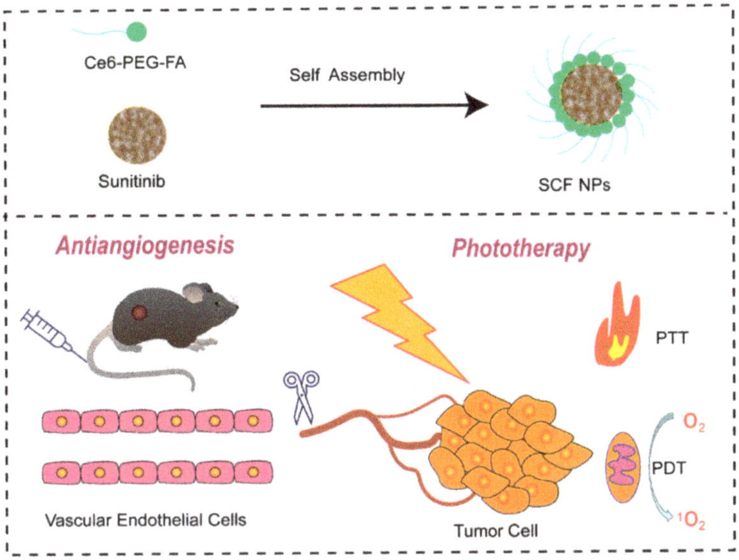

Scheme 1. A schematic illustration depicting the fabrication and application of SCF NPs for antivascular and PDT/PTT therapy at the tumor site is presented.

2. Materials and Methods

2.1. Preparation of SCF NPs

The SCF NPs were prepared by the self-assembly method. Firstly, 2 mg of sunitinib and 10 mg of Ce6-PEG2000-FA were dissolved in 2 mL of THF. Subsequently, the mixed solution was added to 10 mL of ultrapure water under sonication. Then, the THF was removed by rapid stirring for 48 h. Finally, the solution filtered through a 220 nm filter was freeze-dried, and the resulting solid powder was stored in the refrigerator at 4 °C.

Remarks: Poly(ethyleneglycol) (PEG), chlorin e6 (Ce6), nanoparticles (NPs), folic acid (FA), tetrahydrofuran (THF).

2.2. Singlet Oxygen Detection

The $^1O_2$ generation was detected by a singlet oxygen sensor green (SOSG) reagent. Briefly, 5 μM SOSG was dissolved in SCF NPs aqueous solution. The mixed solution was irradiated by a 660 nm laser (150 mW cm^{-2}) for 100 s. The fluorescence was collected every 10 s (excitation wavelength: 404 nm).

2.3. Photothermal Effect

Various concentrations of SCF NPs were irradiated by a 660 nm laser (800 mW cm^{-2}, 10 min). Furthermore, 100 μg/mL SCF NPs was irradiated with different power of 660 nm laser irradiation. Temperature variations were measured by using a FLIR infrared camera.

2.4. Cell Lines and Culture Conditions

Hep-3B cell lines were purchased by the Institute of Biochemistry and Cell Biology, SIBS, CAS (Shanghai, China). Hep-3B Cells were cultured in Dulbecco's modified Eagle's medium (DMEM) with 10% fetal bovine serum at 37 °C atmospheres containing 5% CO_2.

2.5. In Vitro Cytotoxicity Assay

The cytotoxicity of Ce6-PEG2000-FA and SCF NPs was estimated by cell viability assay using the Cell Counting Kit-8 (CCK-8) method. Hep-3B and LO2 cells were incubated in 96-well plates in the presence of SCF NPs for 12 h. The light group was illuminated with a

660 nm laser (600 mW cm^{-2}) for 3 min, followed by incubation for 12 h. Subsequently, CCK-8 solution was added per well, and the cells were incubated for 2 h at 37 °C in the presence of 5% CO_2. Finally, the experimental results were detected with a microplate reader.

2.6. Cellular Uptake

Hep-3B cells were cultured in confocal dishes with incubation for 12 h. Then, 7.6 μg/mL SCF NPs were added into confocal dishes and incubated for 12 h. The cells were washed three times with PBS and fixed with paraformaldehyde for 15 min. DAPI was used to stain the nucleus. The cellular uptake was observed by confocal fluorescence microscope (Olympus FV3000, Tokyo, Japan).

2.7. Hemolysis Assay

Fresh blood from nude mice was incubated with saline containing heparin. The whole blood was washed with saline 3 times to obtain a 4% red blood cell (RBC) suspension. Then, 1.0 mL SCF NPs (10, 25, 50, and 100 μg/mL) and the RBC solution (0.2 mL) were mixed and incubated for 6 h. The mixed solution was photographed after centrifugation (3000 rpm/min). Finally, the absorption of supernatant was recorded at 540 nm by ultraviolet spectrophotometer. Hemolysis rate % = $(A_{SCF\ NPs} - A_{PBS})/(A_{ultrapure\ water} - A_{PBS}) \times 100\%$.

2.8. In Vivo Fluorescence Imaging

200 μL of SCF NPs (100 μg/mL) were intravenously administered to the Hep-3B tumor-bearing nude mice. Photos were photographed at 2, 4, 6, 8, and 12 h by the IVIS Lumina K Series III system (Perkin Elmer, Waltham, MA, USA).

2.9. In Vivo Tumor Therapy

Hep-3B cells were subcutaneously injected into the subcutaneous tissue of male nude mice. When the tumor volumes reached about 75 mm^3, the tumor-bearing nude mice were randomly divided into three groups (control, SCF NPs without laser irradiation, SCF NPs with laser irradiation). After 6 h of tail vein injection of SCF NPs, the tumors of the light group were illuminated by a 660 nm laser for 10 min (600 mW cm^{-2}). All groups were treated every 2 days, and the volumes of the tumors were recorded. After two treatments, one nude mouse was taken from each group for hematoxylin and eosin (H&E) staining of tumor tissue. After treatment for 14 days, the nude mice were sacrificed, and the main organs (heart, liver, spleen, lung, and kidney) were dissected. Histopathological examination of main organs was conducted by H&E staining.

3. Results and Discussion

3.1. Characterization of SCF NPs

As shown in Scheme 1, the water-soluble nanoplatform (named SCF NPs) was obtained by a facile self-assembled approach using amphiphilic polymers (Ce6-PEG2000-FA) (Figure S1) to encapsulate multitarget tyrosine kinase inhibitor (sunitinib) (Figure S2). The detailed synthetic route of SCF NPs was described in the experimental section. The scanning electron microscope (SEM) and transmission electron microscope (TEM) were utilized to evaluate the morphologies of SCF NPs (Figures 1a, S3 and S4), and the results recapitulated that the prepared nanoparticle was uniformly spherical morphology with a diameter of around 58 nm (Figure 1b) and the Zeta potential (Figure 1c) was +25.93 mV, suggesting successful synthesis of the nanoparticles. The suitable size of SCF NPs can facilitate nanoparticles targeting tumor tissue through enhanced permeability and retention (EPR) effect. After incubation with PBS for 14 days, no obvious aggregation or precipitate was observed, indicating superior solubility and good stability of SCF NPs. (Figure S5). In Figure 1d, SCF NPs displayed two obvious absorption peaks at 411 nm and 678 nm, a slight red shift could be observed compared to free Ce6-PEG2000-FA, which proved the successful loading of drugs (Figure 1d). Furthermore, the clear fluorescence peak (Figure 1e) and fluorescence photos of various concentrations (10, 50, 100, 200 µg/mL) (Figure 1f) indicated the potential of SCF NPs for fluorescence imaging applications. To assess the efficiency of PDT in vitro, a typical molecular singlet oxygen sensor green (SOSG) was used as the probe to detect the $^1O_2$ generation in SCF NPs aqueous solution via detecting the change of fluorescence intensity spectrum. Under laser irradiation (150 mW cm^{-2}), the fluorescence intensity increased gradually, which proved that SCF NPs possess remarkable $^1O_2$ generation performance (Figure 1g). Furthermore, the $^1O_2$ generation ability of SCF NPs was further verified by monitoring the changes in the ultraviolet absorption spectrum of 1,3-diphenylisobenzofuran (DPBF) at 418 nm with 660 nm laser irradiation.

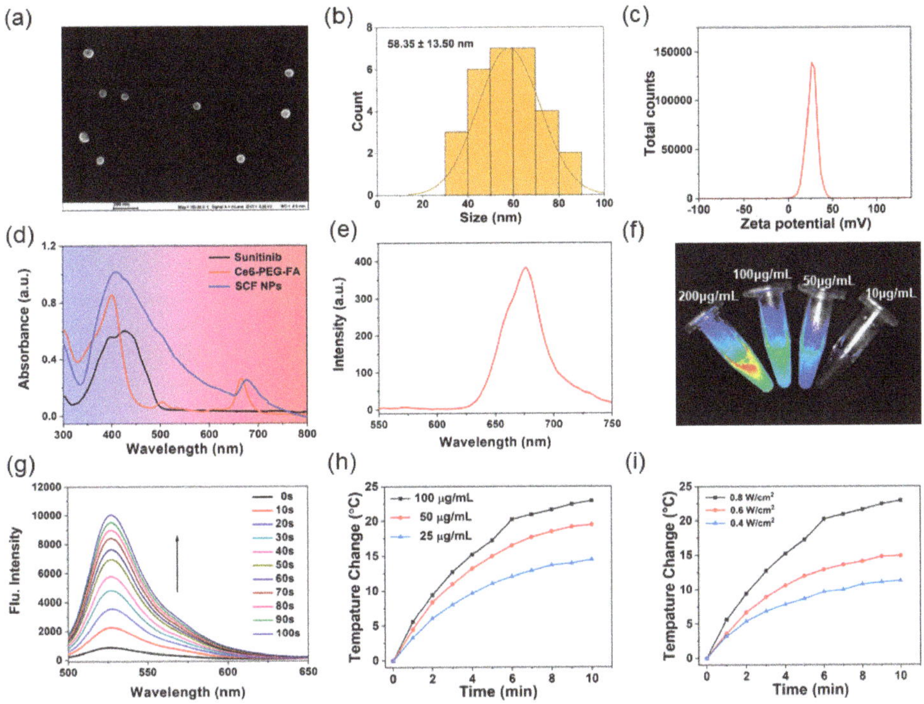

Figure 1. (**a**) SEM image of SCF NPs, scale bar: 200 nm. (**b**) Size distribution of SCF NPs. (**c**) Zeta potential of SCF NPs. (**d**) Ultraviolet absorption spectrum of sunitinib, Ce6-PEG2000-FA, and SCF NPs (Ce6-PEG-FA donated as Ce6-PEG2000-FA). (**e**) The fluorescence emission spectrum of SCF NPs in PBS. (**f**) The fluorescence imaging intensity of various concentration SCF NPs. (**g**) Fluorescence spectra of SOSG mixed with SCF NPs excited by 660 nm laser irradiation. (**h**) Photothermal property of SCF NPs at different concentrations under laser irradiation (660 nm, 800 mW cm^{-2}). (**i**) Photothermal property of SCF NPs (100 µg/mL) with different power densities.

The absorption intensities of DPBF mixed with SCF NPs in PBS (pH 7.4) time-dependently decrease under the 660 nm laser irradiation (Figure S6). Meanwhile, after the conjugation of PEG2000 with Ce6, SCF NPs still exhibited superior $^1O_2$ photogeneration owing to the action of Ce6-PEG2000-FA (Figure S7). These results unambiguously demonstrated that SCF NPs can be employed as an outstanding PDT therapeutic agent for tumor therapy. Considering the existing near-infrared-region (NIR) absorption of SCF NPs, the photothermal effect was further investigated in PBS aqueous solution. The temperature changes of SCF NPs at different concentrations were investigated using a thermal imaging device under illumination conditions (800 mW m^{-2}). The temperature changes of SCF NPs increased gradually with the increasing concentrations (Figure 1h). The temperature of SCF

NPs increased by 14.5 °C even at a low concentration (25 μg/mL), and the photothermal performance may be attributed to the presence of Ce6. The photothermal properties of Ce6 were retained after successful conjugation with PEG2000, and with increasing laser power, a corresponding increase in temperature was observed at a concentration of 100 μg/mL of SCF NPs (Figure 1i). Taken together, these results suggested that SCF NPs could serve as a potential nanoplatform for both PDT and PTT.

3.2. Cellular Experiment of SCF NPs

The intracellular uptake of SCF NPs was observed using a confocal fluorescence microscope with Ce6 as a fluorescent probe. As shown in Figure 2a, the green fluorescence of Ce6 could be observed within the cytoplasm of Hep-3B cells by confocal fluorescence microscopy, suggesting that SCF NPs are efficiently endocytosed by tumor cells. Biosafety assessment is critical before nanocarriers are used for biomedical research. The Hep-3B and human normal hepatocytes (LO2) were employed to evaluate the biocompatibility of Ce6-PEG2000-FA. CCK-8 assays were conducted to detect the cytotoxicity of Ce6-PEG2000-FA. As described in Figure S8, the result of CCK-8 indicated that the Ce6-PEG2000-FA showed superior biosafety. Both cell lines maintained high cell viability (>90%) even at concentrations greater than 200 μg/mL. As the nanocarrier of SCF NPs, Ce6-PEG2000-FA not only have a good encapsulation effect but also show little toxicity to normal cell and tumor cell. Then, the efficacy of combination therapy comprising phototherapy and molecular targeted therapy was evaluated. As indicated in Figure 2b, the cell survival rates of tumor cells treated with SCF NPs with or without laser radiation were concentration dependent. It is worth noting that dark toxicity of SCF NPs is rather high, with 46.5% cell viability at a concentration of 8 μg/mL. This phenomenon may be attributed to the successful loading of sunitinib, effective cellular uptake of nanoparticles, and smooth intracytoplasmic release of sunitinib. Importantly, SCF NPs exhibited strong phototoxicity, with 30.9% cell viability at a concentration of 8 μg/mL. The IC50 values of SCF NPs with or without laser irradiation are 7.8 and 4.3 μg/mL for Hep-3B cells, respectively. Therefore, SCF NPs showed remarkable synergistic therapeutic outcomes based on phototherapy and molecular targeted therapy. Subsequently, we further investigated the intracellular ROS generation of various groups by using DCFH-DA as the fluorescence probe. In Figure 2c, the strongest fluorescence emission could be detected in the group of MCS NPs with laser irradiation. The quantitative analysis of the fluorescence intensity further confirmed this result (Figure 2d). Notably, the group of SCF NPs without light irradiation also displayed a slight fluorescent signal, which may be attributed to sunitinib-induced generation of reactive oxygen species (ROS) [39,40]. In order to verify the antivascular properties of SCF NPs, a blood vessel formation experiment was performed. As displayed in Figure 2e,f, SCF NPs can be successfully taken up by HUVEC cells, and SCF NPs group could effectively fight against angiopoiesis (vascular length, tightness, and intersections significantly decreased) than the control group. Therefore, a synergistic therapeutic strategy combining phototherapy with molecular targeted therapy can effectively eliminate tumor cells in vitro.

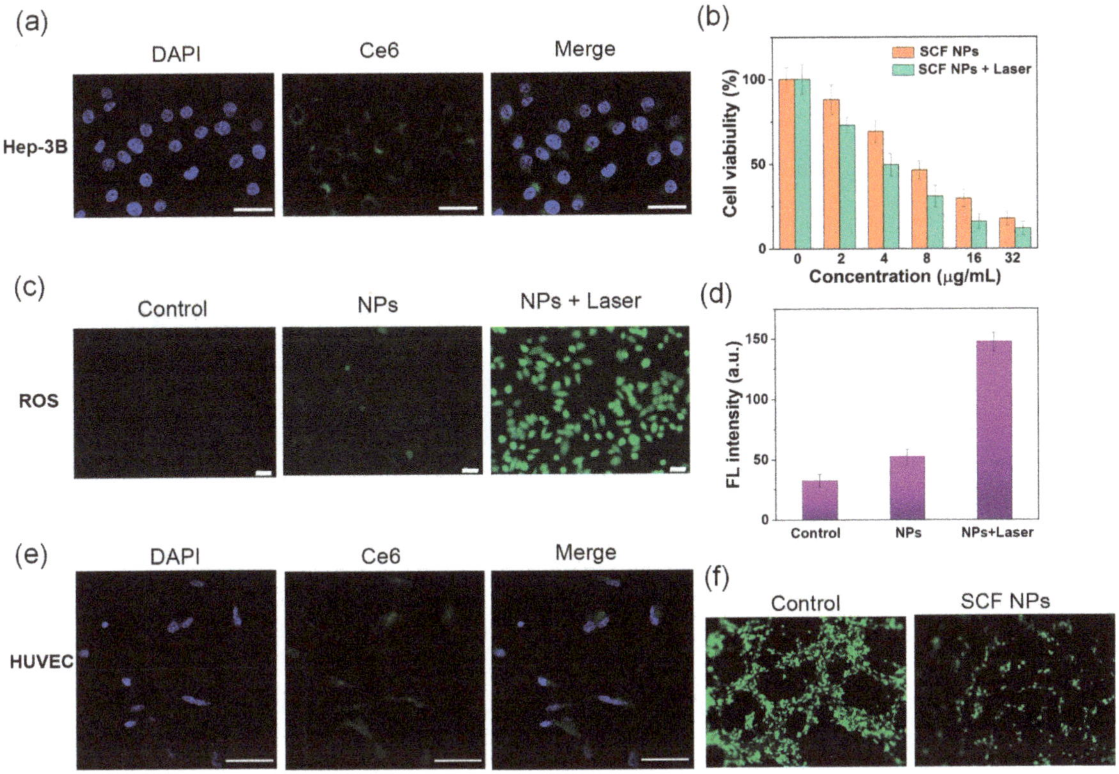

Figure 2. (**a**) Cellular uptake of SCF NPs in Hep-3B cells, scale bar: 30 μm. (**b**) In vitro cytotoxicity of SCF NPs for Hep-3B cell measured by CCK-8. (**c**) DCFH-DA probe staining to detect the generation of $^1O_2$. Green fluorescence represents the generation of $^1O_2$. Scale bar: 50 μm. (**d**) Fluorescence intensity quantitative analysis of DCFH-DA. (**e**) Cellular uptake of SCF NPs in HUVEC cells, scale bar: 30 μm. (**f**) In vitro anti-angiogenesis assay of control group and SCF NPs group.

3.3. Multimodal Imaging and In Vivo Experiments

The imaging performance of nanoparticles is of great significance for early diagnosis, targeted treatment, and intraoperative guidance of tumors. To investigate the in vitro imaging properties of SCF NPs, a hemolysis experiment was performed to assess the biocompatibility of nanoparticles. As illustrated in Figure 3a, the hemolysis rate (2.91%) was less than 5% even at a concentration of 100 μg/mL SCF NPs, demonstrating satisfactory biosafety in vivo. Subsequently, we used real-time fluorescence imaging to assess the targeting performance of SCF NPs in vivo. Figure 3b,c showed the real-time fluorescence imaging and fluorescence intensity change in vitro. Before intravenous injection of SCF NPs, no fluorescence signals at the tumor site could be observed. After intravenous injection of nanoparticles, a gradual increase in fluorescence signal at the tumor site was observed, reaching a maximum after approximately six hours, followed by a gradual decrease. The results of fluorescence imaging experiments confirmed the superior tumor-targeting properties of SCF NPs, which may be attributed to the size-dependent passive targeting properties and the folate-relevant active targeting properties. Owing to the outstanding photothermal peculiarity of SCF NPs, after 6 h intravenous injection of SCF NPs, the real-time photothermal photos at the tumor site can be visually evaluated by the infrared camera. The thermal imaging photos indicated the effective enrichment of SCF NPs at tumor sites. As shown in Figure 4a,b, the temperature of the tumor area in mice injected with SCF NPs can be elevated by more than 17 °C upon 660 nm laser irradiation, providing an effective photothermal treatment. In comparison, the temperature of mice injected with saline only increased slightly by about 3.7 °C. In a word, the multimodal imaging of SCF NPs could provide effective diagnosis and precise intraoperative navigation. The excellent tumor-targeting ability and imaging performance of SCF NPs encourage us to further evaluate therapeutic efficacy in vivo. To validate the antitumor effect of SCF NPs in vivo, tumor volumes were recorded every two days during 14 days of treatment. As shown in Figure 4c,d, compared to the control group, the tumor growth was effectively inhibited but not eliminated entirely for the group treated with SCF NPs. Noteworthy, we surprisingly observed that the SCF NPs group with laser radiation can completely eradicate tumors after two or three treatments without any recrudescence. The results indicate that SCF NPs can remarkably inhibit tumor growth by combining phototherapy and molecular targeted therapy. After two treatments, tumors were removed from nude mice and stained for hematoxylin and eosin (H&E). The histopathology results indicate that SCF NPs with laser radiation could effectively induce cell apoptosis and necrosis of tumor tissues (Figure 4e). To evaluate the toxicology of SCF NPs on major organs (heart, liver, spleen, lung, and kidney), after 14 days of treatment, mice were sacrificed and dissected to recover major organs, which were subsequently fixed in 10% buffered formalin and embedded in paraffin wax for routine histological examination via H&E staining (Figure 5). The experimental results showed no obvious damage, which indicating the prepared SCF NPs have no obvious toxicity. These results indicated that SCF NPs exhibited promising antitumor efficacy through PDT/PTT and molecular targeting therapy.

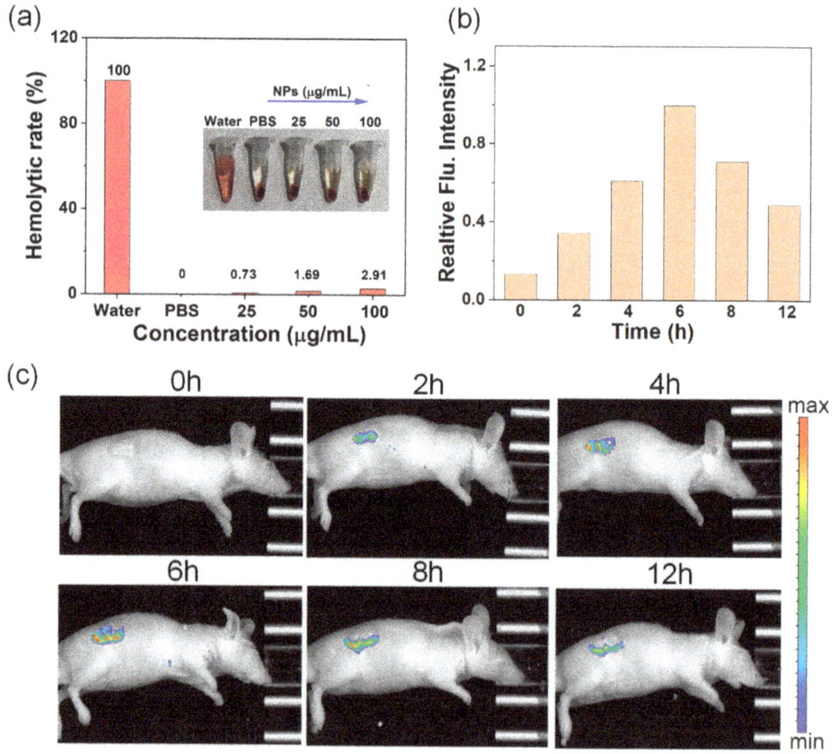

Figure 3. (**a**) Hemolysis experiment with different concentrations (25, 50, 100 μg/mL) of SCF NPs. (**b**) Quantification of the fluorescence signal of Hep-3B tumor-bearing mice before and after intravenous injection with SCF NPs. (**c**) Real-time fluorescence imaging of SCF NPs.

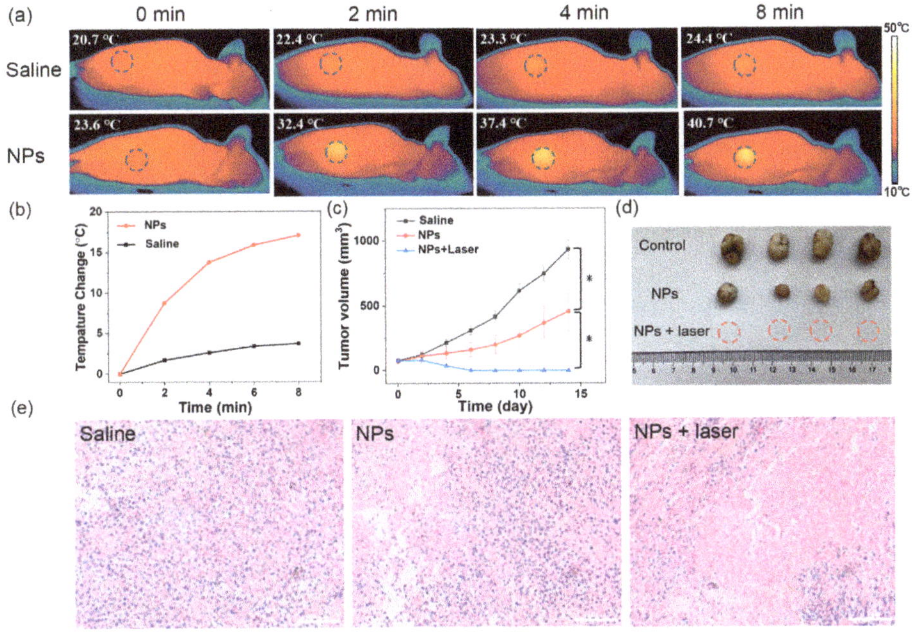

Figure 4. (**a**) IR thermal images of mice bearing Hep-3B tumors injected with saline or SCF NPs, irradiated with 660 nm laser (600 mW cm^{-2}). (**b**) The curves of temperature changes at the tumor site under laser irradiation. (**c**) Tumor volume growth curves of various groups. * $p < 0.05$. (**d**) Images of dissected tumors from different groups (Control, SCF NPs, SCF NPs with laser radiation). (**e**) H&E staining of tumor tissue for each group, scale bars: 50 μm.

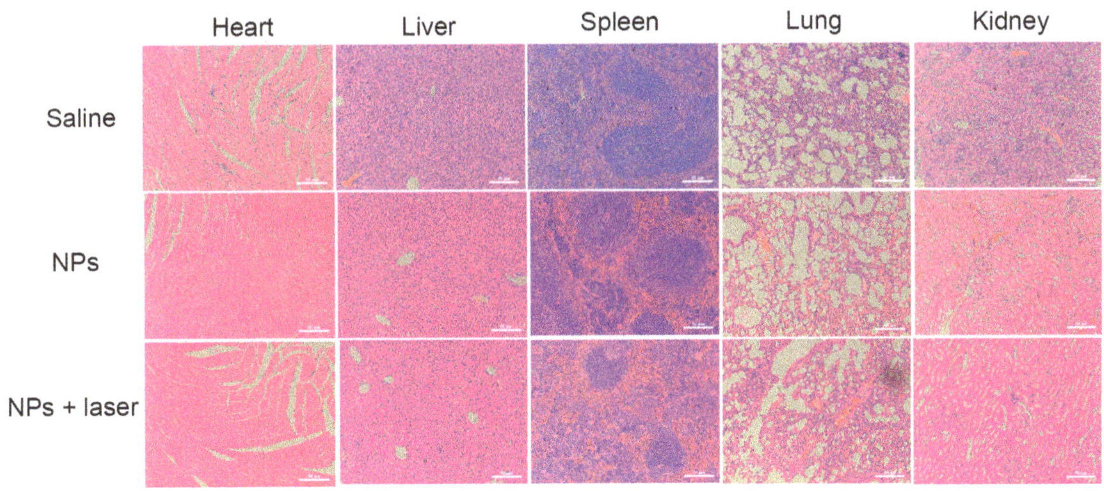

Figure 5. H&E staining for major organ sections (heart, liver, spleen, lung, and kidney) from different groups of mice. Scale bar: 50 μm.

4. Conclusions

In summary, we developed a multifunctional nanoplatform (SCF NPs) by utilizing a novel amphiphilic polymer (Ce6-PEG2000-FA) to deliver a multitarget tyrosine kinase inhibitor (sunitinib). The Ce6-PEG2000-FA exhibited outstanding biocompatibility, and the final prepared nanoparticles possess a diameter of around 58 nm, which can target tumor tissue through the EPR effect. The folic acid group in Ce6-PEG2000-FA can enhance the active targeting performance of nanoparticles. The SCF NPs showed antivascular ability due to the successful loading of sunitinib. The SCF NPs demonstrated superior therapeutic effects based on PDT, PTT, and molecular targeted therapy. Simultaneously, SCF NPs exhibited remarkable photothermal and fluorescence imaging properties, which can be utilized for tumor diagnosis and intraoperative navigation. This novel sunitinib-based nanoplatform for multimodal therapy offers a promising tactic for the treatment of HCC.

Supplementary Materials: The following supporting information can be downloaded at: https://www.mdpi.com/article/10.3390/mi14030613/s1, Figure S1: The chemical formula of Ce6-PEG2000-FA; Figure S2: The chemical formula of sunitinib; Figure S3: SEM images of SCF NPs; Figure S4: TEM image of SCF NPs; Figure S5: The photo of PBS aqueous solution after incubuating SCF NPs for 14 days; Figure S6: The degradation of DPBF with the presence of SCF NPs under 660 nm laser irradiation; Figure S7: The degradation of DPBF with the presence of Ce6-PEG2000-FA under 660 nm laser irradiation; Figure S8: Cytotoxicity assay of Hep-3B and LO2 cell incubuated with different concentrations (20, 40, 60, 80, 100, 200 μg/mL) of Ce6-PEG2000-FA.

Author Contributions: W.X. and M.Y. contributed equally. Conceptualization, M.Y. and Y.Z.; wrote the manuscript and performed the experiments, W.X.; methodology, W.X.; polishing the articles, X.D. and H.P.; data analysis, Y.Y. and J.W.; writing original draft preparation, W.X.; funding acquisition, Y.Z. All authors have read and agreed to the published version of the manuscript.

Funding: The work was funded by the National Natural Science Foundation of China (81872255 and 62041101), the Natural Science Foundation of Jiangsu Province (BK20200710), the Social Development Project of Jiangsu Province (BE2022812), the Leading-edge Technology Programme of Jiangsu Natural Science Foundation (BK20212021).

Data Availability Statement: The data that support the findings of this study are available from the corresponding author upon reasonable request.

Conflicts of Interest: The authors declare no conflict of interest.

References

1. Parkin, D.M.; Bray, F.; Ferlay, J. Pisani, P. Global Cancer Statistics, 2002. *CA Cancer J. Clin.* **2005**, *55*, 74–108. [CrossRef]
2. Chen, J.G.; Zhang, S.W. Liver Cancer Epidemic in CHINA: Past, Present and Future. *Semin. Cancer Biol.* **2011**, *21*, 59–69. [CrossRef] [PubMed]
3. Llovet, J.M.; Zucman-Rossi, J.; Pikarsky, E.; Sangro, B.; Schwartz, M.; Sherman, M.; Gores, G. Hepatocellular Carcinoma. *Nat. Rev. Dis. Prim.* **2016**, *2*, 16018. [CrossRef] [PubMed]
4. Raoul, J.L.; Forner, A.; Bolondi, L.; Cheung, T.T.; Kloeckner, R.; de Baere, T. Updated Use of TACE for Hepatocellular Carcinoma Treatment: How and When to Use It Based on Clinical Evidence. *Cancer Treat. Rev.* **2019**, *72*, 28–36. [CrossRef] [PubMed]
5. McGlynn, K.A.; Petrick, J.L.; El-Serag, H.B. Epidemiology of Hepatocellular Carcinoma. *Hepatology* **2021**, *73* (Suppl. S1), 4–13. [CrossRef] [PubMed]
6. Tung-Ping Poon, R.; Fan, S.T.; Wong, J. Risk Factors, Prevention, and Management of Postoperative Recurrence after Resection of Hepatocellular Carcinoma. *Ann. Surg.* **2000**, *232*, 10–24. [CrossRef]
7. El-Serag, H.B.; Rudolph, K.L. Hepatocellular Carcinoma: Epidemiology and Molecular Carcinogenesis. *Gastroenterology* **2007**, *132*, 2557–2576. [CrossRef]
8. Ayuso, C.; Rimola, J.; Vilana, R.; Burrel, M.; Darnell, A.; García-Criado, Á.; Bianchi, L.; Belmonte, E.; Caparroz, C.; Barrufet, M.; et al. Diagnosis and Staging of Hepatocellular Carcinoma (HCC): Current Guidelines. *Eur. J. Radiol.* **2018**, *101*, 72–81. [CrossRef]
9. Zhu, A.X.; Raymond, E. Early Development of Sunitinib in Hepatocellular Carcinoma. *Expert Rev. Anticancer Ther.* **2009**, *9*, 143–150. [CrossRef]
10. Wu, Y.; Zhang, Y.; Qin, X.; Geng, H.; Zuo, D.; Zhao, Q. PI3K/AKT/mTOR Pathway-Related Long Non-Coding RNAs: Roles and Mechanisms in Hepatocellular Carcinoma. *Pharmacol. Res.* **2020**, *160*, 105195. [CrossRef]
11. Qing, X.; Xu, W.; Zong, J.; Du, X.; Peng, H.; Zhang, Y. Emerging Treatment Modalities for Systemic Therapy in Hepatocellular Carcinoma. *Biomark. Res.* **2021**, *9*, 64. [CrossRef]
12. Qi, X.; Yang, M.; Ma, L.; Sauer, M.; Avella, D.; Kaifi, J.T.; Bryan, J.; Cheng, K.; Staveley-O'Carroll, K.F.; Kimchi, E.T.; et al. Synergizing Sunitinib and Radiofrequency Ablation to Treat Hepatocellular Cancer by Triggering the Antitumor Immune Response. *J. ImmunoTherapy Cancer* **2020**, *8*, e001038. [CrossRef]

13. Li, G.; Liu, D.; Cooper, T.K.; Kimchi, E.T.; Qi, X.; Avella, D.M.; Li, N.; Yang, Q.X.; Kester, M.; Rountree, C.B.; et al. Successful Chemoimmunotherapy Against Hepatocellular Cancer in a Novel Murine Model. *J. Hepatol.* **2017**, *66*, 75–85. [CrossRef] [PubMed]
14. Klug, L.R.; Khosroyani, H.M.; Kent, J.D.; Heinrich, M.C. New Treatment Strategies for Advanced-Stage Gastrointestinal Stromal Tumours. *Nat. Rev. Clin. Oncol.* **2022**, *19*, 328–341. [CrossRef] [PubMed]
15. Zhu, A.X.; Sahani, D.V.; Duda, D.G.; di Tomaso, E.; Ancukiewicz, M.; Catalano, O.A.; Sindhwani, V.; Blaszkowsky, L.S.; Yoon, S.S.; Lahdenranta, J.; et al. Efficacy, Safety, and Potential Biomarkers of Sunitinib Monotherapy in Advanced Hepatocellular Carcinoma: A Phase II Study. *J. Clin. Oncol.* **2009**, *27*, 3027–3035. [CrossRef] [PubMed]
16. Kollmannsberger, C. Sunitinib Side Effects as Surrogate Biomarkers of Efficacy. *Can. Urol. Assoc. J.* **2016**, *10*, S245–S247. [CrossRef]
17. Xia, S.; Pan, Y.; Liang, Y.; Xu, J.; Cai, X. The Microenvironmental and Metabolic Aspects of Sorafenib Resistance in Hepatocellular Carcinoma. *EBioMedicine* **2020**, *51*, 102610. [CrossRef]
18. Xu, W.J.; Ye, C.P.; Qing, X.; Liu, S.L.; Lv, X.Y.; Wang, W.J.; Dong, X.C.; Zhang, Y.W. Multi-Target Tyrosine Kinase Inhibitor Nanoparticle Delivery Systems for Cancer Therapy. *Mater. Today Bio* **2022**, *16*, 100358. [CrossRef]
19. Broxterman, H.J.; Gotink, K.J.; Verheul, H.M. Understanding the Causes of Multidrug Resistance in Cancer: A Comparison of Doxorubicin and Sunitinib. *Drug Resist. Updat.* **2009**, *12*, 114–126. [CrossRef]
20. Ferrari, S.M.; Centanni, M.; Virili, C.; Miccoli, M.; Ferrari, P.; Ruffilli, I.; Ragusa, F.; Antonelli, A.; Fallahi, P. Sunitinib in the Treatment of Thyroid Cancer. *Curr. Med. Chem.* **2019**, *26*, 963–972. [CrossRef]
21. Rausch, M.; Rutz, A.; Allard, P.M.; Delucinge-Vivier, C.; Docquier, M.; Dormond, O.; Wolfender, J.L.; Nowak-Sliwinska, P. Molecular and Functional Analysis of Sunitinib-Resistance Induction in Human Renal Cell Carcinoma Cells. *Int. J. Mol. Sci.* **2021**, *22*, 6467. [CrossRef]
22. Yuan, Y.; Zhang, C.J.; Xu, S.; Liu, B. A Self-Reporting AIE Probe with a Built-in Singlet Oxygen Sensor for Targeted Photodynamic Ablation of Cancer Cells. *Chem. Sci.* **2016**, *7*, 1862–1866. [CrossRef] [PubMed]
23. Chen, H.; Tian, J.; He, W.; Guo, Z. H2O2-Activatable and O2-Evolving Nanoparticles for Highly Efficient and Selective Photodynamic Therapy against Hypoxic Tumor Cells. *J. Am. Chem. Soc.* **2015**, *137*, 1539–1547. [CrossRef] [PubMed]
24. Liang, P.; Huang, X.; Wang, Y.; Chen, D.; Ou, C.; Zhang, Q.; Shao, J.; Huang, W.; Dong, X. Tumor-Microenvironment-Responsive Nanoconjugate for Synergistic Antivascular Activity and Phototherapy. *ACS Nano* **2018**, *12*, 11446–11457. [CrossRef]
25. Xu, W.; Xu, L.; Jia, W.; Mao, X.; Liu, S.; Dong, H.; Zhang, H.; Zhang, Y. Nanomaterials Based on Phase Change Materials for Antibacterial Application. *Biomater. Sci.* **2022**, *10*, 6388–6398. [CrossRef]
26. Xu, W.J.; Qing, X.; Liu, S.L.; Yang, D.L.; Dong, X.C.; Zhang, Y.W. Hollow Mesoporous Manganese Oxides: Application in Cancer Diagnosis and Therapy. *Small* **2022**, *18*, 2106511. [CrossRef]
27. Dolmans, D.E.; Fukumura, D.; Jain, R.K. Photodynamic Therapy for Cancer. *Nat. Rev. Cancer* **2003**, *3*, 380–387. [CrossRef]
28. Chen, D.; Xu, Q.; Wang, W.; Shao, J.; Huang, W.; Dong, X. Type I Photosensitizers Revitalizing Photodynamic Oncotherapy. *Small* **2021**, *17*, e2006742. [CrossRef] [PubMed]
29. Xie, J.; Wang, Y.; Choi, W.; Jangili, P.; Ge, Y.; Xu, Y.; Kang, J.; Liu, L.; Zhang, B.; Xie, Z.; et al. Overcoming Barriers in Photodynamic Therapy Harnessing Nano-Formulation Strategies. *Chem. Soc. Rev.* **2021**, *50*, 9152–9201. [CrossRef]
30. Yang, N.; Cao, C.; Li, H.; Hong, Y.; Cai, Y.; Song, X.; Wang, W.; Mou, X.; Dong, X. Polymer-Based Therapeutic Nanoagents for Photothermal-Enhanced Combination Cancer Therapy. *Small Struct.* **2021**, *2*, 2100110. [CrossRef]
31. Ou, C.; Na, W.; Ge, W.; Huang, H.; Gao, F.; Zhong, L.; Zhao, Y.; Dong, X. Biodegradable Charge-Transfer Complexes for Glutathione Depletion Induced Ferroptosis and NIR-II Photoacoustic Imaging Guided Cancer Photothermal Therapy. *Angew. Chem. Int. Ed. Engl.* **2021**, *60*, 8157–8163. [CrossRef]
32. Zou, J.; Zhu, J.; Yang, Z.; Li, L.; Fan, W.; He, L.; Tang, W.; Deng, L.; Mu, J.; Ma, Y.; et al. A Phototheranostic Strategy to Continuously Deliver Singlet Oxygen in the Dark and Hypoxic Tumor Microenvironment. *Angew. Chem. Int. Ed. Engl.* **2020**, *59*, 8833–8838. [CrossRef]
33. Li, Q.; Liu, Z.; Zheng, S.; Li, W.; Ren, Y.; Li, L.; Yan, F. Three-Dimensional Printable, Highly Conductive Ionic Elastomers for High-Sensitivity Iontronics. *ACS Appl. Mater. Interfaces* **2022**, *14*, 26068–26076. [CrossRef]
34. Ge, G.; Zhang, Y.Z.; Zhang, W.; Yuan, W.; El-Demellawi, J.K.; Zhang, P.; Di Fabrizio, E.; Dong, X.; Alshareef, H.N. Ti(3)C(2)T(x) MXene-Activated Fast Gelation of Stretchable and Self-Healing Hydrogels: A Molecular Approach. *ACS Nano* **2021**, *15*, 2698–2706. [CrossRef] [PubMed]
35. Ge, G.; Mandal, K.; Haghniaz, R.; Li, M.; Xiao, X.; Carlson, L.; Jucaud, V.; Dokmeci, M.R.; Ho, G.W.; Khademhosseini, A. Deep Eutectic Solvents-Based Ionogels with Ultrafast Gelation and High Adhesion in Harsh Environments. *Adv. Funct. Mater.* **2023**, *33*, 2207388. [CrossRef]
36. Ge, W.; Wang, L.; Zhang, J.Y.; Ou, C.J.; Si, W.L.; Wang, W.J.; Zhang, Q.M.; Dong, X.C. Self-Assembled Nanoparticles as Cancer Therapeutic Agents. *Adv. Mater. Interfaces* **2021**, *8*, 2001602. [CrossRef]
37. Zhao, L.P.; Zheng, R.R.; Chen, H.Q.; Liu, L.S.; Zhao, X.Y.; Liu, H.H.; Qiu, X.Z.; Yu, X.Y.; Cheng, H.; Li, S.Y. Self-Delivery Nanomedicine for O(2)-Economized Photodynamic Tumor Therapy. *Nano Lett.* **2020**, *20*, 2062–2071. [CrossRef] [PubMed]
38. Liu, M.C.; Liu, L.; Wang, X.R.; Shuai, W.P.; Hu, Y.; Han, M.; Gao, J.Q. Folate Receptor-Targeted Liposomes Loaded with a Diacid Metabolite of Norcantharidin Enhance Antitumor Potency for H22 Hepatocellular Carcinoma both in vitro and in vivo. *Int. J. Nanomed.* **2016**, *11*, 1395–1412. [CrossRef]

39. Paech, F.; Abegg, V.F.; Duthaler, U.; Terracciano, L.; Bouitbir, J.; Krähenbühl, S. Sunitinib Induces Hepatocyte Mitochondrial Damage and Apoptosis in Mice. *Toxicology* **2018**, *409*, 13–23. [CrossRef]
40. Bouitbir, J.; Alshaikhali, A.; Panajatovic, M.V.; Abegg, V.F.; Paech, F.; Krähenbühl, S. Mitochondrial Oxidative Stress Plays a Critical Role in the Cardiotoxicity of Sunitinib: Running Title: Sunitinib and Oxidative Stress in Hearts. *Toxicology* **2019**, *426*, 152281. [CrossRef]

micromachines

Article

Holographic Microwave Image Classification Using a Convolutional Neural Network

Lulu Wang

Biomedical Device Innovation Center, Shenzhen Technology University, Shenzhen 518118, China; wanglulu@sztu.edu.cn or lwang381@hotmail.com

Abstract: Holographic microwave imaging (HMI) has been proposed for early breast cancer diagnosis. Automatically classifying benign and malignant tumors in microwave images is challenging. Convolutional neural networks (CNN) have demonstrated excellent image classification and tumor detection performance. This study investigates the feasibility of using the CNN architecture to identify and classify HMI images. A modified AlexNet with transfer learning was investigated to automatically identify, classify, and quantify four and five different HMI breast images. Various pre-trained networks, including ResNet18, GoogLeNet, ResNet101, VGG19, ResNet50, DenseNet201, SqueezeNet, Inception v3, AlexNet, and Inception-ResNet-v2, were investigated to evaluate the proposed network. The proposed network achieved high classification accuracy using small training datasets (966 images) and fast training times.

Keywords: microwave imaging; breast cancer; deep learning; AlexNet; transfer learning

Citation: Wang, L. Holographic Microwave Image Classification Using a Convolutional Neural Network. *Micromachines* **2022**, *13*, 2049. https://doi.org/10.3390/mi13122049

Academic Editors: Xiao Xiao and Gang Ge

Received: 11 November 2022
Accepted: 22 November 2022
Published: 23 November 2022

1. Introduction

Breast cancer is the leading cause of female cancer deaths [1]. Previous studies showed that early breast cancer detection methods combined with suitable treatment could improve survival rates significantly [2]. X-ray mammography is the current gold-standard imaging tool for diagnosing breast cancer, but it produces harmful radiation and is unsuitable for dense breasts [3]. Microwave imaging has been proposed as one of the most potential breast imaging tools [4]. Researchers have extensively investigated microwave imaging in many aspects, including measurement of the microwave dielectric properties of breast tissues [5,6], image algorithms [7,8], numerical models [9,10], data acquisition systems [11–13], microwave antennas [14–16], clinical trials [17,18], image enhancement and improvement methods [19–21], and image classification [22–24]. If microwave images contain specific qualitative and quantitative indicators, this may help characterize benign and malignant tumors and predict disease. However, this work is challenging because this interdisciplinary study involves several disciplines, such as microwave science, medical imaging, machine learning, and computer vision.

Over the past two decades, deep learning has attracted increasing attention and has achieved excellent performance in medical image classification and disease detection [25,26]. For example, Chen et al. employed the biclustering mining method in ultrasound images to identify breast lesions with accuracy, sensitivity, and specificity of 96.1, 96.7, and 95.7%, respectively [27]. However, the image datasets were too small to implement generalizations. Le et al. applied a deep neural network to enhance microwave images [28]. Khoshdel et al. investigated the feasibility of using 3D U-Net architecture to improve microwave breast images [29]. Rana et al. investigated machine learning for breast lesion detection using microwave radar imaging [22]. Mojabi et al. applied convolutional neural networks (CNN) to microwave and ultrasound images to classify uncertainty quantification and breast tissue [24]. However, obtaining big microwave image datasets for training networks is challenging.

AlexNet is one of the most popular CNN architectures, and it is exploited on ImageNet datasets (including 50 million images) [30]. Previous studies demonstrated that small datasets (a few hundred) are employed for image classification [31]. However, small datasets are unsuitable for training networks due to easy overfitting. With the help of transfer learning, the training process can be conducted on a personal computer using small datasets [32].

In our previous studies, the holographic microwave imaging (HMI) method was proposed and tested for breast lesion detection [33–35]. This paper investigates the feasibility of using modified AlexNet with transfer learning to identify, classify, and quantify five classes of HMI datasets (fatty, dense, heterogeneously dense, very dense, and very dense breasts containing tumors), thereby solving the highly subjective judgment problem of lesions or abnormal tissues. Experimental validations are conducted on realistic MRI-based breast models to investigate the effectiveness and accuracy of modified AlexNet with transfer learning. In addition, a comparison study of several deep learning networks, including ResNet18, ResNet50, ResNet101, GoogLeNet, Inception v3, AlexNet, and VGG19, was conducted to evaluate the performance of HMI image classification. The research findings not only extend the application of deep learning but also help to understand microwave science from the perspective of deep learning with computer vision. The rest of this paper is organized as follows: Section 2 describes the proposed materials and method. Section 3 presents experimental validations and results. Section 4 concludes the study.

2. Materials and Method

2.1. Convolutional Neural Network

A typical CNN contains an input layer (that receives pixel values), a convolution layer (that extracts image features), a pooling layer (that reduces the pixels to be processed and formulates abstract elements), and an output layer (that maps the extracted features into classification vectors corresponding to the feature categories) that can be described as:

$$\begin{aligned} z^l &= W^l * x^{l-1} + b^l \\ a^l &= \sigma\left(z^l\right) \end{aligned} \tag{1}$$

where l denotes the lth layer and $*$ is a convolution operation. W^l, b^l, and z^l denote the weights matrix, bias matrix, and weighted input of the lth layer. σ is the nonlinear activation function. When $l = 2$, $x^{2-1} = x^1$ is the image matrix whose elements are pixel values. When $l > 2$, x^{l-1} is the feature maps matrix a^{l-1}, which is extracted from the $(l-1)$th layer, i.e., $x^{l-1} = a^{l-1} = \sigma\left(z^{l-1}\right)$. Let L be the output layer and a^L is the final output vector.

Nonlinear activation functions are employed from the second layer to the last layer. The cost function is:

$$E_0^L = -\frac{1}{n}\sum_{i=1}^{n}\sum_{k=1}^{N}\left[t_k^L ln a_k^L + \left(1 - t_k^L\right) ln\left(1 - a_k^L\right)\right] \tag{2}$$

where n is the training number and N is the number of neurons in the output layer corresponding to the N classes. t_k^L is the targeted value corresponding to the kth neuron of the output layer and a_k^L is the actual output value of the kth neuron of the output layer.

The output layer error can be defined as:

$$\delta^L = \frac{\partial E_0^L}{\partial z^L} \tag{3}$$

where $\partial(\cdot)$ denotes the partial derivative operation. $l = \{ L-1,\ L-2,\ \ldots,\ 2\}$, then:

$$\delta^l = W^{l+1}\delta^{l+1}\circ\sigma'\left(z^l\right) \tag{4}$$

where $\circ$ is the Hadamard product. The partial derivative from E_0^L to W^{l+1} and b^l can be calculated as follows:

$$\left.\begin{array}{r}\frac{\partial E_0^L}{\partial W^l} = \frac{\partial E_0^L}{\partial a^l} \circ \frac{\partial a^l}{\partial W^l} = \delta^l \circ x^{l-1} \\ \frac{\partial E_0^L}{\partial b^l} = \frac{\partial E_0^L}{\partial a^l} \circ \frac{\partial a^l}{\partial b^l} = \delta^l\end{array}\right\} \quad (5)$$

the changes can be computed by:

$$\left.\begin{array}{r}\Delta W^l = -\eta \frac{\partial E_0^L}{\partial W^l} \\ \Delta b^l = -\eta \frac{\partial E_0^L}{\partial b^l}\end{array}\right\} \quad (6)$$

where η denotes the learning rate.

The ResNet architecture reduces training errors and network layers [36]. Adding a quick identity link to the primary network unit is the key to the ResNet architecture:

$$H(X) = F(X) + X \quad (7)$$

where $H(X)$ is the ideal image and $F(X)$ is the residual map.

2.2. Datasets

As shown in Table 1, publicly available MRI-derived breast phantoms from 9 human subjects were used to develop realistic breast models by converting pixel values in MRI images to complex-valued permittivity [37,38]. Figure 1 shows a sample (breast 9) of 12 phantoms and the real and imaginary parts of the relative complex-valued permittivity. Figure 2 shows the real and imaginary parts of 12 breast phantoms. The HMI method was applied to generate HMI breast image datasets using the developed, realistic numerical microwave breast models. The numerical model simulated a sphere-shaped inclusion as a tumor (radius of 5 and 10 mm).

Table 1. Characteristics of breast phantoms.

Number	Phantom Class	Quantity	Model	Size
No 1	I: fatty	253	RGB	310 × 355 × 253
No 2	I: fatty	288	RGB	267 × 375 × 288
No 3	II: dense	307	RGB	316 × 352 × 307
No 4	II: dense	270	RGB	300 × 382 × 270
No 5	II: dense	251	RGB	258 × 253 × 251
No 6	III: heterogeneously dense	202	RGB	269 × 332 × 202
No 7	III: heterogeneously dense	248	RGB	258 × 365 × 248
No 8	III: heterogeneously dense	273	RGB	219 × 243 × 273
No 9	IV: very dense	212	RGB	215 × 328 × 212
No 10	V: very dense breast contains two tumors	212	RGB	215 × 328 × 212
No 11	V: very dense breast contains two tumors	212	RGB	215 × 328 × 212
No 12	V: fatty breast contains two tumors	253	RGB	310 × 355 × 253

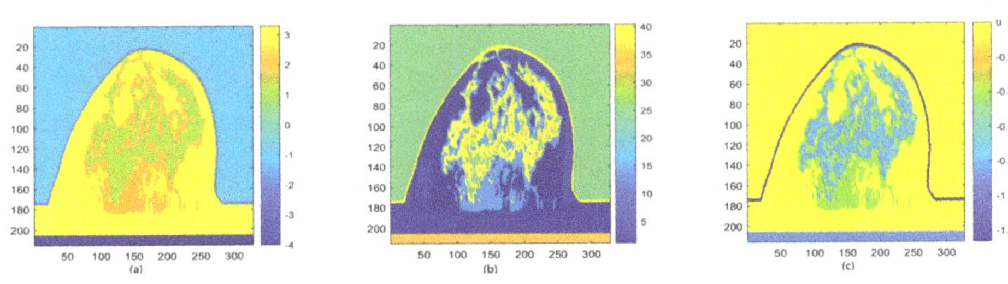

Figure 1. (**a**) example of 12 breast phantoms (breast 9); (**b**) real part of the relative complex-valued permittivity of breast 9; and (**c**) imaginary part of the relative complex-valued permittivity of breast 9.

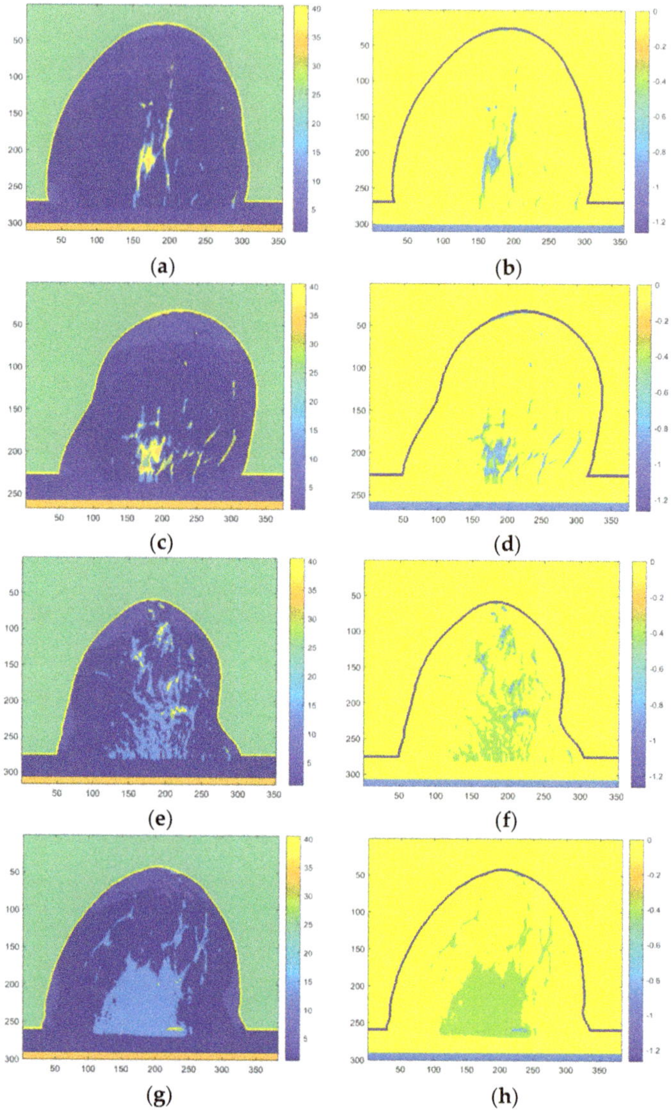

Figure 2. Cont.

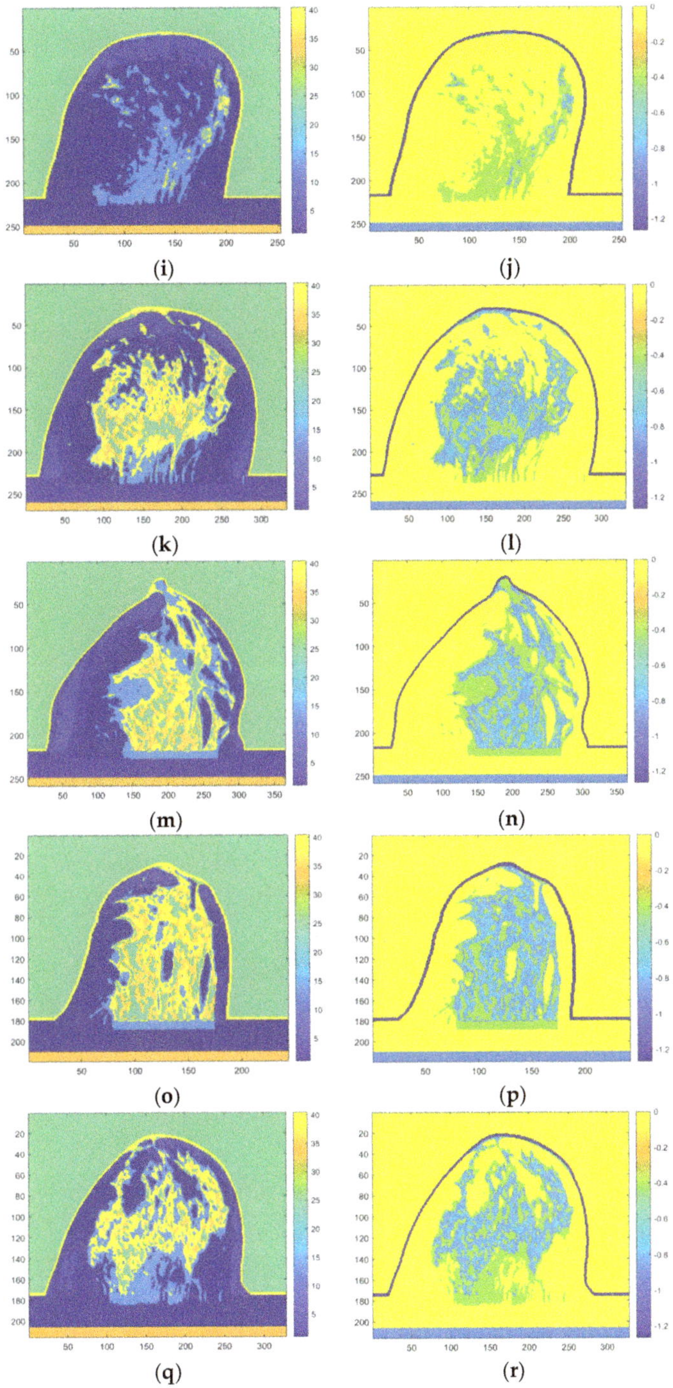

Figure 2. *Cont.*

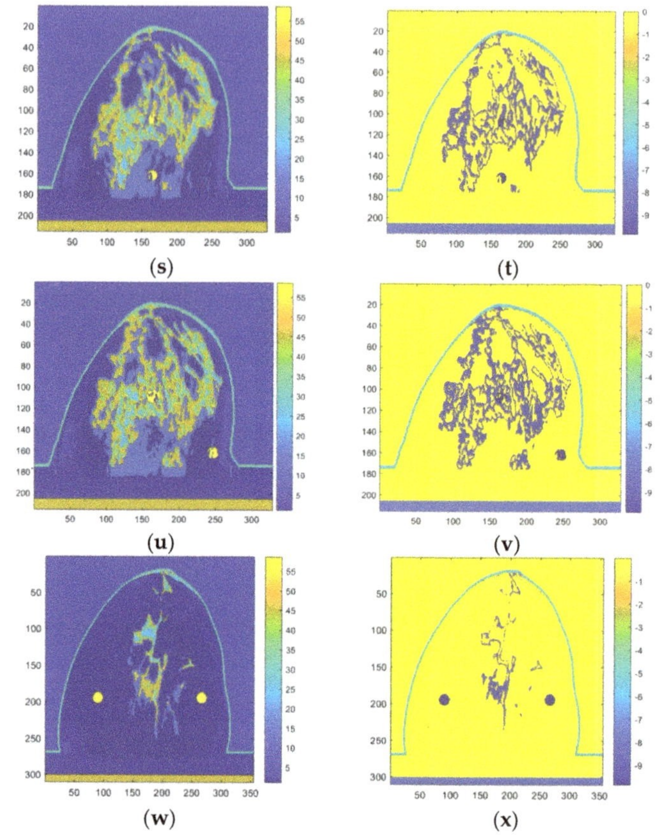

Figure 2. (**a**,**b**) real and imaginary parts of the relative complex-valued permittivity of breast 1; (**c**,**d**) real and imaginary parts of the relative complex-valued permittivity of breast 2; (**e**,**f**) real and imaginary parts of the relative complex-valued permittivity of breast 3; (**g**,**h**) real and imaginary parts of the relative complex-valued permittivity of breast 4; (**i**,**j**) real and imaginary parts of the relative complex-valued permittivity of breast 5; (**k**,**l**) real and imaginary parts of the relative complex-valued permittivity of breast 6; (**m**,**n**) real and imaginary parts of the relative complex-valued permittivity of breast 7; (**o**,**p**) real and imaginary parts of the relative complex-valued permittivity of breast 8; (**q**,**r**) real and imaginary parts of the relative complex-valued permittivity of breast 9; (**s**,**t**) real and imaginary parts of the relative complex-valued permittivity of breast 10; (**u**,**v**) real and imaginary parts of the relative complex-valued permittivity of breast 11 and; (**w**,**x**) real and imaginary parts of the relative complex-valued permittivity of breast 12.

This study used two datasets to train and test the CNN networks (see Table 2). Dataset 1 consists of the real part of HMI breast images, and dataset 2 consists of the imaginary part of HMI breast images. According to [37], the dataset in this study includes five classes of HMI images (12 phantoms), which are fatty, dense, heterogeneously dense, very dense, and breasts containing tumors. Class V was identified based on tumors that existed, and three Class V models were investigated in this study (see Table 1).

Table 2. Training parameters.

Dataset	1	2
Modality	Real part of HMI breast	Imaginary part of HMI breast
Number of phantoms	12	12
Classes of images	5	5
Number of HMI images	1379	1379
Image size	227 × 227 × 3	227 × 227 × 3
Number of training images	966	966
Number of validation images	275	275
Number of test images	138	138
Number of Class I	160	160
Number of Class II	457	457
Number of Class III	444	444
Number of Class IV	108	108
Number of Class V	210	210
Cross-validation group	8-fold	8-fold
Maximum number of epochs	50	50
Minimum batch size	25	25
Validation frequency	30	30
Initial learning rate	0.0003	0.0003

2.3. *Training and Testing Data*

2.3.1. Image Segmentation

An original HMI image contains different types of tissues with different sizes and cannot be applied directly for classification. We applied the image segmentation method to partition each original HMI image into sub-images and created the total of the sub-images. Sub-image properties are 227 × 227 pixels (a RGB image). The segmentation method helps to change the representation to a more meaningful and easier-to-analyze image while changing the scale to fit AlexNet. Image segmentation makes HMI images in each sub-image more uniform, which is suitable for classification and facilitates the final determination of the percentage of each mechanism. In addition, to ensure the authenticity of extracted features from the training dataset, image augmentation techniques such as rotation, height, and width shift were not used to ensure the integrity of the original images.

2.3.2. Image Labeling

Both datasets 1 and 2 were classified into five classes (see Figure 2 and Table 2). The fatty breast (class I) consists of skin, muscle, and fat tissue. Dense breast (Class II) consists of skin, muscle, fat, and dense tissue (which has higher dielectric properties than fatty tissue). Heterogeneously dense breast tissue (Class III) consists of skin, muscle, fat tissue, and heterogeneously dense tissue. A very dense breast (Class IV) consists of skin, muscle, fat, dense tissue (which has higher dielectric properties than fat), and very dense fatty tissues (which have higher dielectric properties than fat and dense tissues). A breast contains tumors (Class V) consisting of skin, muscle, fat, heterogeneously dense tissue, and two tumors.

The created HMI images illustrate the application behavior of the trained network. Therefore, their sub-images were not labeled. Different numbers of sub-images from each class were selected for manual labeling and then used for training and testing the proposed network. Training and testing datasets were utterly independent to ensure the reliability and stability of the proposed method.

For each dataset, 70% of the total images were used to train the proposed network, 20% of the total images were used to validate the network, and 10% of the total images were used to test the network. All breast image datasets were resized to 227 × 227 × 3 pixels. The training image dataset was applied to tune the network parameters using a gradient-based method. The testing image dataset was involved in the testing process to generate predictions. Table 2 shows the parameters used for training the networks.

2.4. Network Architecture

2.4.1. Modified AlexNet

AlexNet is the most popular CNN architecture due to its better performance in image classification. Thus, this study applied a modified AlexNet with transfer learning (see Table 3) to HMI images to improve image classification accuracy. Table 3 shows the structure of modified AlexNet with transfer learning. The first convolution layer of the network takes input datasets and passes them through convolution filters. Thus, the input image is required to be resized to 227 × 227 × 3 pixels, corresponding to the breadth, height, and three-color channels representing the depth of the input image. The last convolutional layer implements the reconstructed image process, aggregating the high-resolution patch-wise representations to produce the output image. The cross-entropy loss function is used to reduce errors. The batch normalization function is performed before each activation function to solve overfitting problems. The ReLU layer provides faster and more efficient training, mapping negatives, and maintaining positive values. The max pooling layer simplifies the output and reduces the resolution by reducing the number of parameters needed to learn. The fully connected layer combines all features to classify the images into four classes. The SoftMax function normalizes the output of the fully connected layer.

Table 3. AlexNet with transfer learning.

Schematic	No.	Name	Type	Activations	Weights & Bias
	1	data	Image input	227 × 227 × 3	
	2	conv1	Convolution	55 × 55 × 96	Weights: 11 × 11 × 3 × 96; bias: 1 × 1 × 96
	3	relu1	ReLu	55 × 55 × 96	
	4	norm1	Cross-channel normalization	55 × 55 × 96	
	5	pool1	Max pooling	27 × 27 × 96	
	6	conv2	Grouped convolution	27 × 27 × 96	
	7	relu2	ReLU	27 × 27 × 256	Weights: 5 × 5 × 48 × 128; bias: 1 × 1 × 128 × 2
	8	norm2	Cross-channel normalization	27 × 27 × 256	
	9	pool2	Max pooling	13 × 13 × 256	
	10	conv3	Convolution	13 × 13 × 384	Weights: 3 × 3 × 25 × 384; bias: 1 × 1 × 384
	11	relu3	ReLU	13 × 13 × 384	
	12	conv4	Grouped convolution	13 × 13 × 384	Weights: 3 × 3 × 192 × 192; bias: 1 × 1 × 192 × 2
	13	relu4	ReLU	13 × 13 × 384	
	14	conv5	Grouped convolution	13 × 13 × 256	Weights: 3 × 3 × 192 × 128; bias: 1 × 1 × 128 × 2
	15	relu5	ReLU	13 × 13 × 256	
	16	pool5	Max pooling	6 × 6 × 256	
	17	fc6	Fully connected	1 × 1 × 4096	Weights: 7029 × 9216; bias: 4096 × 1
	18	relu6	ReLU	1 × 1 × 4096	
	19	drop6	Dropout	1 × 1 × 4096	
	20	fc7	Fully connected	1 × 1 × 4096	Weights: 4096 × 4096; bias: 4096 × 1
	21	relu7	ReLU	1 × 1 × 4096	
	22	drop7	Dropout	1 × 1 × 4096	
	23	fc8	Fully connected	1 × 1 × 4	Weights: 4 × 4096; bias: 4 × 1
	24	softmax	SoftMax	1 × 1 × 4	
	25	output	Classification output		

2.4.2. Transfer Learning

As shown in Table 3, the last three layers of AlexNet were replaced by transfer learning to avoid overfitting. The proposed AlexNet network consists of a pre-trained network and a transferred network. The parameters in the pre-trained network were trained on publicly available ImageNet datasets. Therefore, it could be adapted to extract features from the HMI image dataset. The parameters in the transferred network represent a small part of the proposed AlexNet network. Thus, a small training dataset can meet the requirements of transfer learning.

2.5. Data Analysis and Image Processing

MATLAB version R2020a with the deep learning library tool was used for data analysis and image processing. The proposed network was developed on a laptop (ThinkPad P53) with an Intel i7-8700K CPU (2.60 GHz) and 256 GB of RAM. Stochastic gradient descent with momentum (SGDM) was selected to train the transferred part of AlexNet.

The MATLAB Transfer Learning of Pretrained Network for Classification tool was used to train and test various deep learning networks using dataset 2, including ResNet18, GoogLeNet, ResNet101, VGG19, ResNet50, DenseNet201, SqueezeNet, Inception v3, AlexNet, and Inception-Res-Net-v2.

2.6. Performance Metrics

The overall performance of the proposed architecture depends on the evaluation matrix, which contains True Positives (TP), False Positives (FP), False Negatives (FN), and True Negatives (TN). The AlexNet architecture was evaluated on the testing dataset using four performance metrics, including precision and accuracy. Precision quantifies the exactness of a model and represents the ratio of carcinoma images accurately classified out of the union of predicted same-class images [39].

$$\text{Precision} = \frac{\text{TP}}{\text{TP} + \text{FP}} \tag{8}$$

where TP refers to images correctly classified as breast tumor images and FP represents the typical images mistakenly classified as breast tumor images.

Accuracy evaluates the correctness of a model and is the ratio of the number of images accurately classified out of the total number of testing images.

$$\text{Accuracy} = \frac{\text{TP} + \text{FN}}{\text{TP} + \text{TN} + \text{FP} + \text{FN}} \tag{9}$$

where TN refers to the correctly classified standard images.

3. Results and Discussion

3.1. Results

Figure 3a shows the training progress of the proposed network using dataset 1 and the SGDM method, including classification accuracy and cross-entropy loss for each epoch of training and validation. At 50 epochs, the highest classification accuracy of training and validation was 100 and 100%, respectively, and the lowest cross-entropy loss of training and validation was 0 and 0%, respectively. The training time was 11 min and 13 s for training 966 images from dataset 1.

Figure 3b displays the training progress of modified AlexNet with transfer learning using dataset 2 and the SGDM method. At 50 epochs, the highest classification accuracy of training and validation was 100 and 100%, respectively, and the lowest cross-entropy loss of training and validation was 0 and 0%, respectively. The training time was 10 min and 55 s for training 966 images from dataset 2.

As shown in Figure 4a, the performance of the proposed network was evaluated using the confusion matrix on testing images (from dataset 1). The actual horizontal row and predicted vertical column demonstrate the classification accuracy and sensitivity of the proposed network, respectively. For example, in the first row, 16 images were used to classify Class IV in the testing dataset, and 16 images (100%) were classified accurately. Therefore, the classification accuracy of Classes I, II, III, IV, and V was 100, 100, 100, 91.7, and 67.7%, respectively. In the first column, 16 images were used to predict class I of the testing images (from dataset 1), where 16 images (100%) were classified accurately. The sensitivity of Classes I, II, III, IV, and V was 100, 78.3, 97.7, 100, and 100%, respectively.

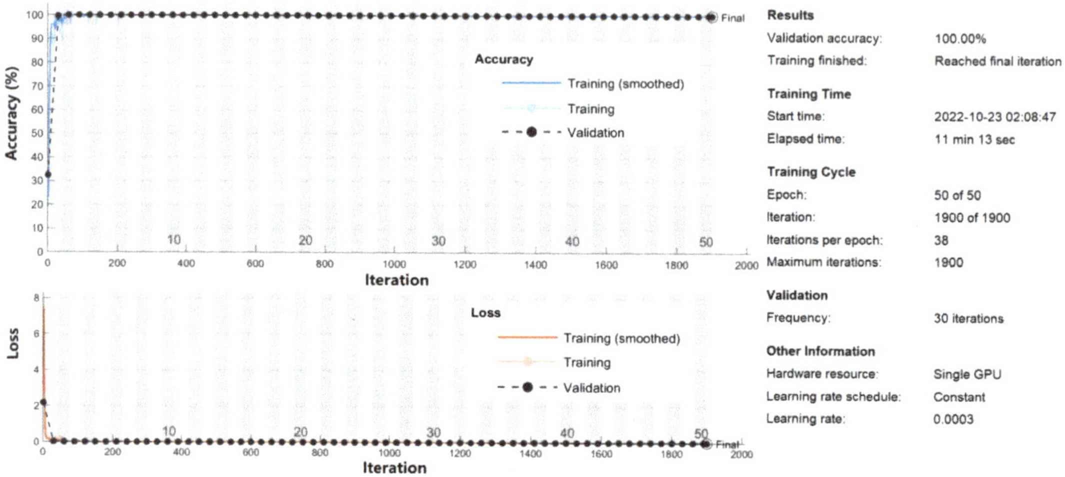

(**a**)

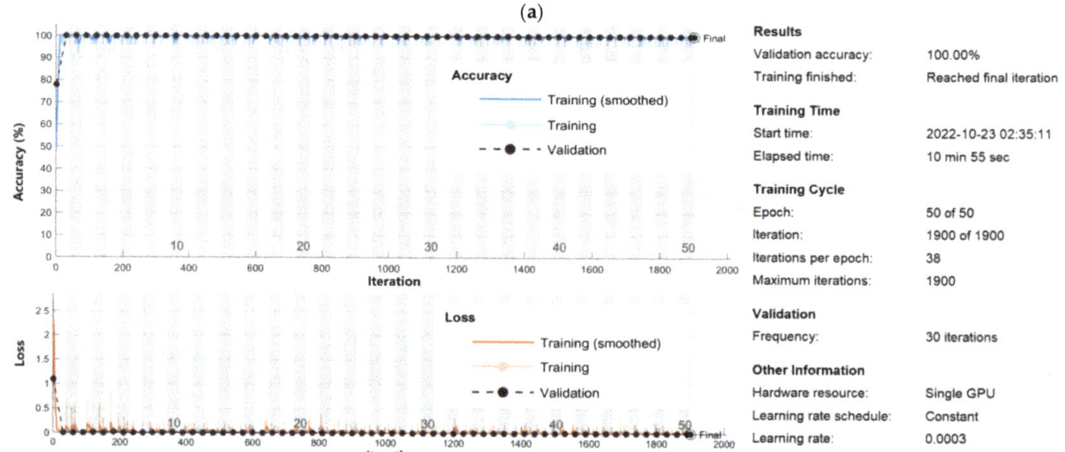

(**b**)

Figure 3. Training progress of the proposed network using (**a**) dataset 1 and (**b**) dataset 2.

Figure 4b shows the performance of modified AlexNet with transfer learning on testing images (from dataset 2). In the first row, 16 images were used to classify Class I in the testing dataset, and 16 images (100%) were classified accurately. The proposed network obtained a classification accuracy of 100, 100, 100, 100, and 100% for Classes I, II, III, IV, and V, respectively. In the first column, 16 images were used to predict Class I in the testing images, where 16 images (100%) were classified accurately. The proposed network obtained a sensitivity of 100, 100, 100, 100, and 100% for Classes I, II, III, IV, and V, respectively.

Figure 5a,b demonstrate the randomly selected 16 examples of training images (from dataset 1) and randomly selected 16 examples of testing images (from dataset 1) using AlexNet with a transfer learning network, respectively.

Figure 6a,b display the randomly selected 16 examples of training images (from dataset 2) and randomly selected 16 examples of testing images (from dataset 2), respectively.

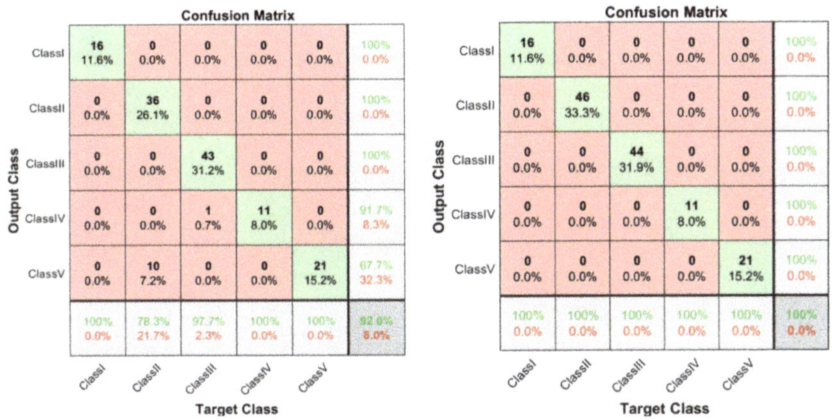

(a) (b)

Figure 4. (**a**) Confusion matrix of testing dataset 1; (**b**) Confusion matrix of testing dataset 2.

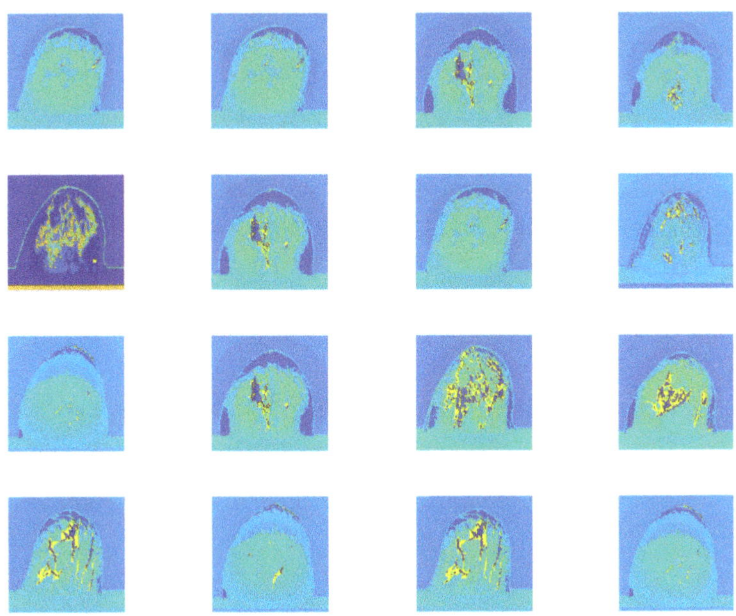

(**a**)

Figure 5. *Cont.*

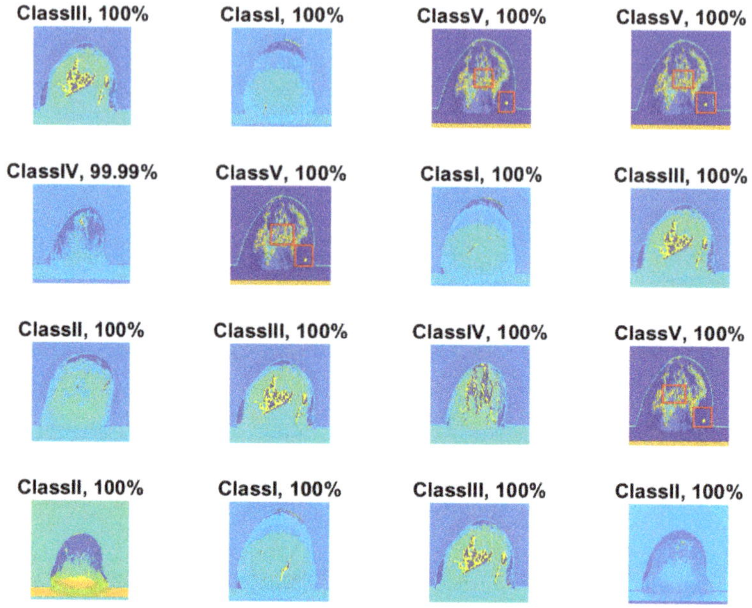

(b)

Figure 5. (**a**) randomly selected 16 examples of training images (from dataset 1), and (**b**) randomly selected 16 examples of testing images (from dataset 1).

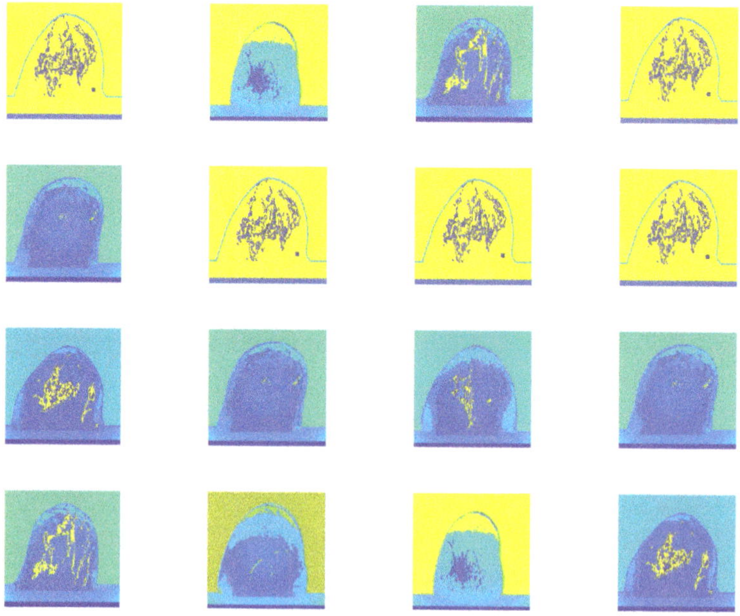

(a)

Figure 6. *Cont.*

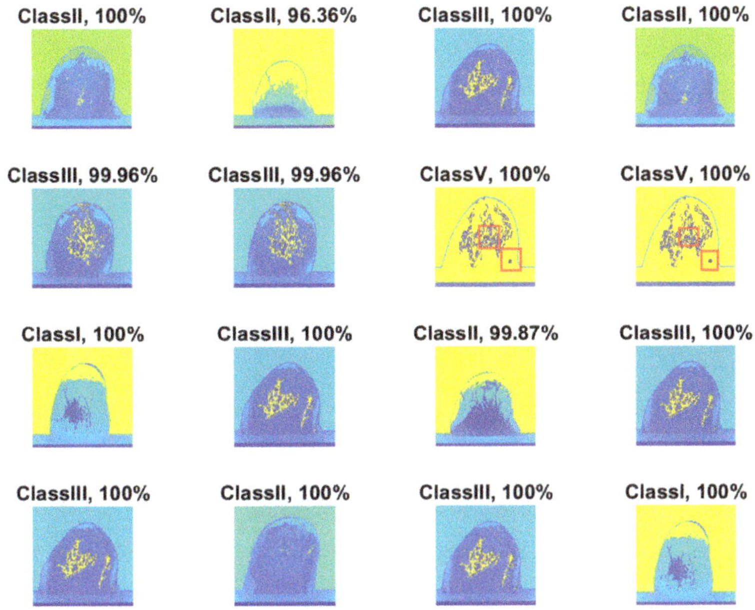

(b)

Figure 6. (**a**) randomly selected 16 examples of training images (from dataset 2); (**b**) randomly selected 16 examples of testing images (from dataset 2).

Table 4 presents the prediction results of dataset 2 using several deep learning networks. MobileNet-v2 obtained the highest accuracy (96.84%), and the training time was 28 min and 38 s. AlexNet used the shortest training time (3 min and 4 s) with relatively low accuracy (79.89%), Inception-ResNet-v2 obtained the lowest accuracy (79.34%) and used a long training time (106 min and 48 s), and DenseNet201 used the longest training time (132 min and 25 s) with relatively high accuracy (96.01%). Modified AlexNet with transfer learning achieved higher classification accuracy than other deep learning networks, which is suitable for classifying HMI images.

Table 4. HMI image classification using different deep learning networks.

Architecture	Accuracy	Training Time	Result
MobileNet-v2	96.84%	28 min 38 s	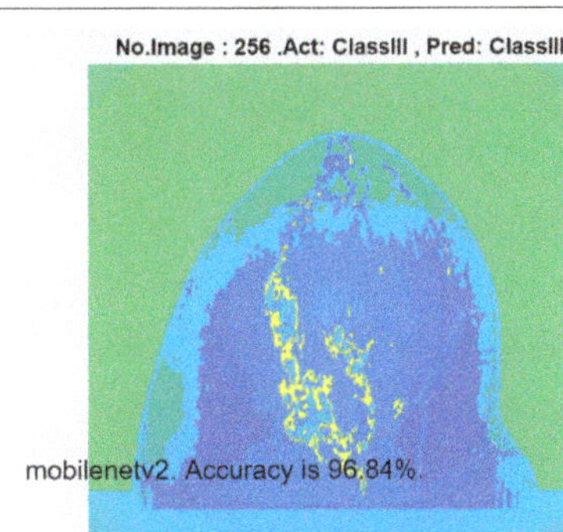

Table 4. *Cont.*

Architecture	Accuracy	Training Time	Result
DenseNet201	96.01%	132 min 25 s	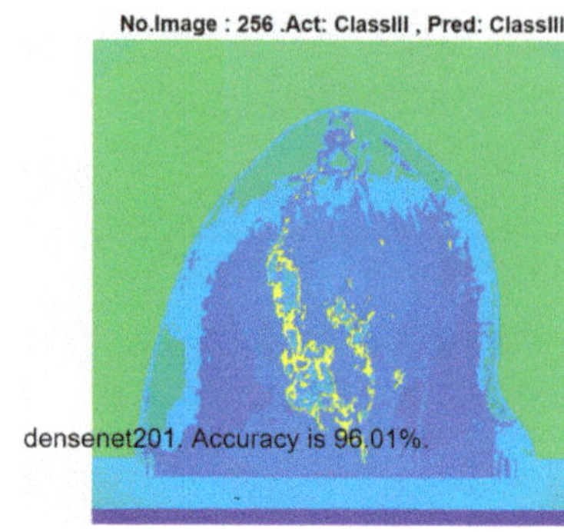
SqueezeNet	92.98%	16 min 3 s	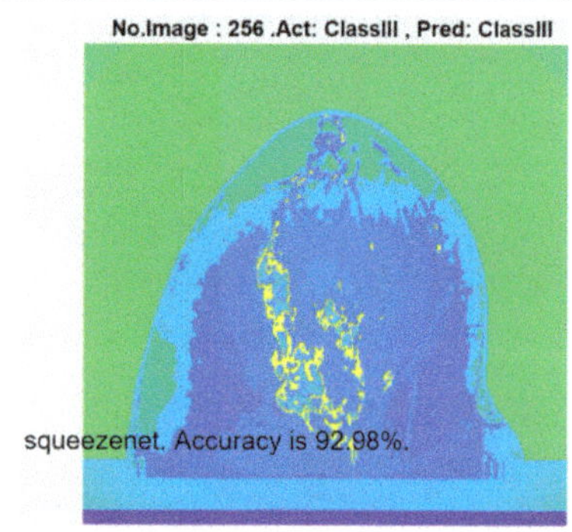
Inception-v3	86.24%	11 mins 30 s	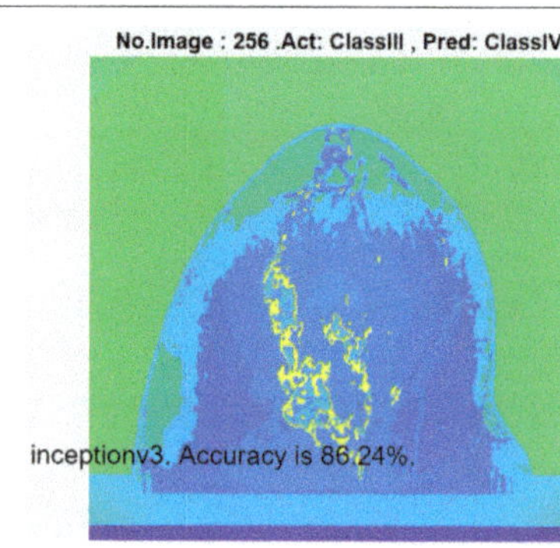

Table 4. *Cont.*

Architecture	Accuracy	Training Time	Result
ResNet101	84.73.%	43 min 5 s	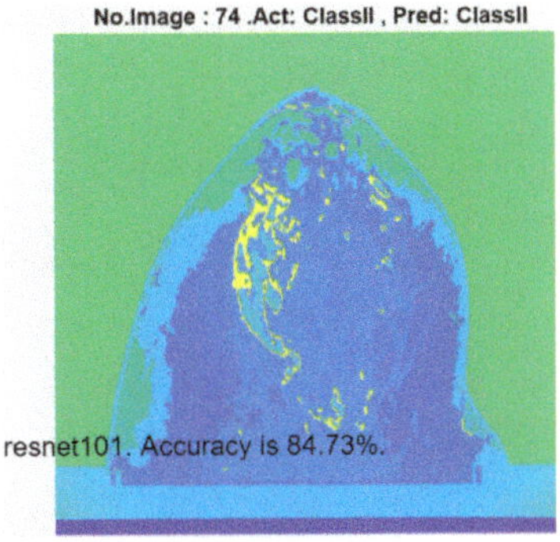
GoogLeNet	81.02%	7 min 48 s	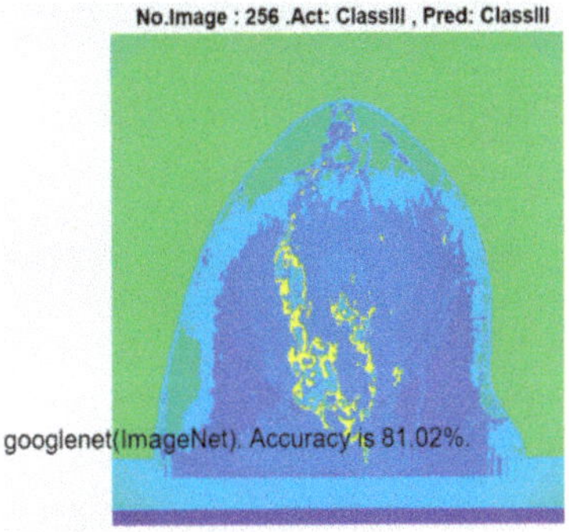
AlexNet	80.33%	5 min 39 s	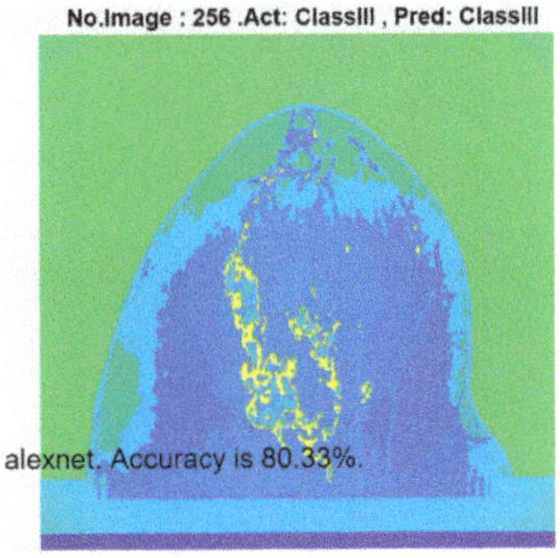

Table 4. *Cont.*

Architecture	Accuracy	Training Time	Result
ResNet50	78.40%	36 min 16 s	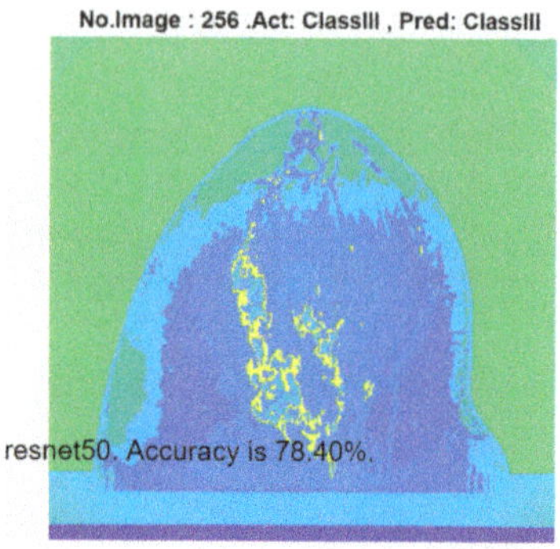
ResNet18	77.30%	11 min 45 s	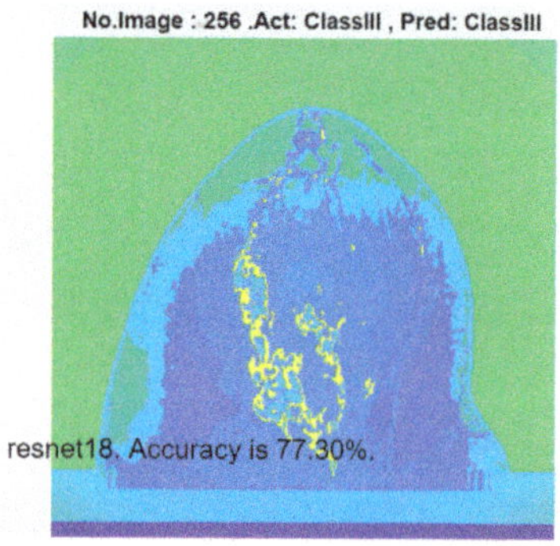
Inception-ResNet-v2	73.18%	106 mins 48 s	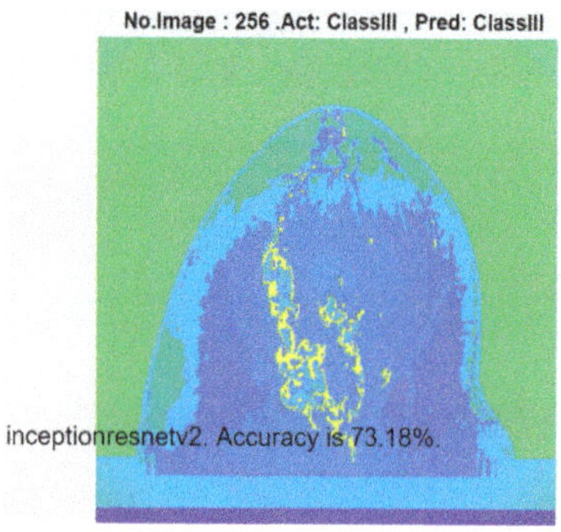

3.2. Discussion

In this study, five classes of breast phantoms were developed using the method presented in [37]. The initial HMI breast images were created using the HMI method detailed in [33]. The initial images were analyzed and processed using the proposed CNN architecture. The proposed architecture offered higher classification accuracy and sensitivity for image dataset 2 (imagery-part HMI images; see Figure 4a) than image dataset 1 (real-part HMI images; see Figure 4b). For image dataset 1, the modified AlexNet with transfer learning offers higher classification accuracy for classes I–III (100%) than classes IV (91.7%) and V (67.7%), and higher sensitivity for classes I (100%), IV (100%), and V (100%) than classes II (78.3%) and III (97.7%). However, no significant difference in classification accuracy and sensitivity was obtained for dataset 2. Figure 4 demonstrates that image datasets affect the performance classification accuracy and sensitivity of the modified AlexNet with transfer learning.

The randomly selected 16 testing examples of image dataset 1 are shown in Figure 5b, and the 16 randomly selected testing examples of image dataset 2 are shown in Figure 6b. Although a classification accuracy of 100% was obtained for examples of image dataset 1 (see Figure 5b), it does not mean that the classification accuracy of dataset 1 is as high as 100%. For example, the classification accuracy rates of 91.7% and 67.7% were obtained for classes IV and V, respectively (see Figure 4a). Although the proposed CNN architecture provides accuracy and sensitivity of 100% to classify dataset 2 (see Figure 4b), the classification accuracy of some testing examples is below 100% (96.36–99.96%; see Figure 6b). This may be caused by MATLAB calculation errors.

Compared with some popular deep learning networks (see Table 4), modified AlexNet with transfer learning has apparent advantages in classification accuracy and training time. For example, modified AlexNet with transfer learning obtained higher accuracy (100% vs. 96.84%) and required shorter training time (10 min 55 s vs. 28 min 38 s) to classify image dataset 2 than MobileNet-v2. The experimental results demonstrated that the modified AlexNet with transfer learning could identify, classify, and quantify HMI images with high accuracy, sensitivity, and reasonable training time. Several factors may affect the test results, including image preprocessing, the number of training images (in percentages), the total number of image datasets, and MATLAB calculation errors.

4. Conclusions

In this study, the CNN architecture was introduced for analyzing HMI images. A modified AlexNet with transfer learning was developed to identify, classify, and quantify five classes of HMI images (fatty, dense, heterogeneously dense, very dense, and very dense breasts containing tumors). Various experimental validations were conducted to validate the performance of the proposed network. Various popular deep learning networks, including AlexNet, were studied to evaluate the proposed network. Results demonstrated that the proposed network could automatically identify and classify HMI images more accurately (100%) than other deep learning networks. In conclusion, the proposed network has the potential to become an effective tool for analyzing HMI images using small training datasets, which offers promising applications in the microwave breast imaging field.

Funding: This research was funded by the International Science and Technology Cooperation Project of the Shenzhen Science and Technology Commission (GJHZ20200731095804014).

Institutional Review Board Statement: Not applicable.

Data Availability Statement: Data and code are available from the corresponding authors upon reasonable request.

Conflicts of Interest: The author declares no conflict of interest.

References

1. Siegel, R.; Miller, K.; Fuchs, H.; Jemal, A. Cancer statistics, 2022. *CA Cancer J. Clin.* **2022**, *72*, 7–33. [CrossRef] [PubMed]
2. Yang, Y.; Yin, X.; Sheng, L.; Xu, S.; Dong, L.; Liu, L. Perioperative chemotherapy more of a benefit for overall survival than adjuvant chemotherapy for operable gastric cancer: An updated meta-analysis. *Sci. Rep.* **2015**, *5*, 12850. [CrossRef] [PubMed]
3. Magna, G.; Casti, P.; Jayaraman, S.V.; Salmeri, M.; Mencattini, A.; Martinelli, E.; Di Natale, C. Identification of mammography anomalies for breast cancer detection by an ensemble of classification models based on artificial immune system. *Knowl. Syst.* **2016**, *101*, 60–70. [CrossRef]
4. Meaney, P.M.; Golnabi, A.H.; Epstein, N.R.; Geimer, S.D.; Fanning, M.W.; Weaver, J.B.; Paulsen, K.D. Integration of microwave tomography with magnetic resonance for improved breast imaging. *Med. Phys.* **2013**, *40*, 103101. [CrossRef]
5. Lazebnik, M.; McCartney, L.; Popovic, D.; Watkins, C.B.; Lindstrom, M.J.; Harter, J.; Sewall, S.; Magliocco, A.; Booske, J.H.; Okoniewski, M.; et al. A large-scale study of the ultrawideband microwave dielectric properties of normal breast tissue obtained from reduction surgeries. *Phys. Med. Biol.* **2007**, *52*, 2637. [CrossRef] [PubMed]

6. Lazebnik, M.; Popovic, D.; McCartney, L.; Watkins, C.B.; Lindstrom, M.J.; Harter, J.; Sewall, S.; Ogilvie, T.; Magliocco, A.; Breslin, T.M.; et al. A large-scale study of the ultrawideband microwave dielectric properties of normal, benign and malignant breast tissues obtained from cancer surgeries. *Phys. Med. Biol.* **2007**, *52*, 6093–6115. [CrossRef]
7. Elahi, M.A.; O'Loughlin, D.; Lavoie, B.R.; Glavin, M.; Jones, E.; Fear, E.C.; O'Halloran, M. Evaluation of Image Reconstruction Algorithms for Confocal Microwave Imaging: Application to Patient Data. *Sensors* **2018**, *18*, 1678. [CrossRef]
8. Moloney, B.M.; O'Loughlin, D.; Abd Elwahab, S.; Kerin, M.J. Breast Cancer Detection–A Synopsis of Conventional Modalities and the Potential Role of Microwave Imaging. *Diagnostics* **2020**, *10*, 103. [CrossRef]
9. Soltani, M.; Rahpeima, R.; Kashkooli, F.M. Breast cancer diagnosis with a microwave thermoacoustic imaging technique—A numerical approach. *Med. Biol. Eng. Comput.* **2019**, *57*, 1497–1513. [CrossRef]
10. Rahpeima, R.; Soltani, M.; Kashkooli, F.M. Numerical Study of Microwave Induced Thermoacoustic Imaging for Initial Detection of Cancer of Breast on Anatomically Realistic Breast Phantom. *Comput. Methods Programs Biomed.* **2020**, *196*, 105606. [CrossRef]
11. Meaney, P.M.; Fanning, M.W.; Li, D.; Poplack, S.P.; Paulsen, K.D. A clinical prototype for active microwave imaging of the breast. *IEEE Trans. Microw. Theory Tech.* **2000**, *48*, 1841–1853.
12. Islam, M.; Mahmud, M.; Islam, M.T.; Kibria, S.; Samsuzzaman, M. A Low Cost and Portable Microwave Imaging System for Breast Tumor Detection Using UWB Directional Antenna array. *Sci. Rep.* **2019**, *9*, 15491. [CrossRef]
13. Adachi, M.; Nakagawa, T.; Fujioka, T.; Mori, M.; Kubota, K.; Oda, G.; Kikkawa, T. Feasibility of Portable Microwave Imaging Device for Breast Cancer Detection. *Diagnostics* **2021**, *12*, 27. [CrossRef] [PubMed]
14. Srinivasan, D.; Gopalakrishnan, M. Breast Cancer Detection Using Adaptable Textile Antenna Design. *J. Med. Syst.* **2019**, *43*, 177. [CrossRef]
15. Misilmani, H.M.E.; Naous, T.; Khatib, S.K.A.; Kabalan, K.Y. A Survey on Antenna Designs for Breast Cancer Detection Using Microwave Imaging. *IEEE Access* **2020**, *8*, 102570–102594. [CrossRef]
16. Sheeba, I.R.; Jayanthy, T. Design and Analysis of a Flexible Softwear Antenna for Tumor Detection in Skin and Breast Model. *Wirel. Pers. Commun.* **2019**, *107*, 887–905. [CrossRef]
17. Meaney, P.M.; Kaufman, P.A.; Muffly, L.S.; Click, M.; Poplack, S.P.; Wells, W.A.; Schwartz, G.N.; di Florio-Alexander, R.M.; Tosteson, T.D.; Li, Z.; et al. Microwave imaging for neoadjuvant chemotherapy monitoring: Initial clinical experience. *Breast Cancer Res.* **2013**, *15*, R35. [CrossRef]
18. Sani, L.; Ghavami, N.; Vispa, A.; Paoli, M.; Raspa, G.; Ghavami, M.; Sacchetti, F.; Vannini, E.; Ercolani, S.; Saracini, A.; et al. Novel microwave apparatus for breast lesions detection: Preliminary clinical results. *Biomed. Signal Process. Control* **2019**, *52*, 257–263. [CrossRef]
19. Williams, T.C.; Fear, E.C.; Westwick, D.T. Tissue sensing adaptive radar for breast cancer detection-investigations of an improved skin-sensing method. *IEEE Trans. Microw. Theory Tech.* **2006**, *54*, 1308–1314. [CrossRef]
20. Abdollahi, N.; Jeffrey, I.; LoVetri, J. Improved Tumor Detection via Quantitative Microwave Breast Imaging Using Eigenfunction-Based Prior. *IEEE Trans. Comput. Imaging* **2020**, *6*, 1194–1202. [CrossRef]
21. Coşğun, S.; Bilgin, E.; Çayören, M. Microwave imaging of breast cancer with factorization method: SPIONs as contrast agent. *Med. Phys.* **2020**, *47*, 3113–3122. [CrossRef] [PubMed]
22. Rana, S.P.; Dey, M.; Tiberi, G.; Sani, L.; Vispa, A.; Raspa, G.; Duranti, M.; Ghavami, M.; Dudley, S. Machine Learning Approaches for Automated Lesion Detection in Microwave Breast Imaging Clinical Data. *Sci. Rep.* **2019**, *9*, 10510. [CrossRef] [PubMed]
23. Edwards, K.; Khoshdel, V.; Asefi, M.; LoVetri, J.; Gilmore, C.; Jeffrey, I. A Machine LearningWorkflow for Tumour Detection in Breasts Using 3D Microwave Imaging. *Electronics* **2021**, *10*, 674. [CrossRef]
24. Mojabi, P.; Khoshdel, V.; Lovetri, J. Tissue-Type ClassificationWith Uncertainty Quantification of Microwave and Ultrasound Breast Imaging: A Deep Learning Approach. *IEEE Access* **2020**, *8*, 182092–182104. [CrossRef]
25. Roslidar, R.; Rahman, A.; Muharar, R.; Syahputra, M.R.; Munadi, K. A review on recent progress in thermal imaging and deep learning approaches for breast cancer detection. *IEEE Access* **2020**, *8*, 116176–116194. [CrossRef]
26. Bakx, N.; Bluemink, H.; Hagelaar, E.; Sangen, M.; Hurkmans, C. Development and evaluation of radiotherapy deep learning dose prediction models for breast cancer. *Phys. Imaging Radiat. Oncol.* **2021**, *17*, 65–70. [CrossRef]
27. Chen, Y.; Ling, L.; Huang, Q. Classification of breast tumors in ultrasound using biclustering mining and neural network. In Proceedings of the 2016 9th International Congress on Image and Signal Processing, BioMedical Engineering and Informatics (CISP-BMEI), Datong, China, 15–17 October 2016; pp. 1788–1791.
28. Li, Y.; Hu, W.; Chen, S.; Zhang, W.; Ligthart, L. Spatial resolution matching of microwave radiometer data with convolutional neural network. *Remote Sens.* **2019**, *11*, 2432. [CrossRef]
29. Khoshdel, V.; Asefi, M.; Ashraf, A.; Lovetri, J. Full 3D microwave breast imaging using a deep-learning technique. *J. Imaging* **2020**, *6*, 80. [CrossRef]
30. Deng, J.; Dong, W.; Socher, R.; Li, L.-J.; Li, K.; Li, F.-F. Imagenet: A Large-Scale Hierarchical Image Database. In Proceedings of the 2009 IEEE Conference on Computer Vision and Pattern Recognition, Miami, FL, USA, 20–25 June 2009; pp. 248–255.
31. Liu, G.; Mao, S.; Kim, J.H. A mature-tomato detection algorithm using machine learning and color analysis. *Sensors* **2019**, *19*, 2023. [CrossRef]
32. Lu, S.; Lu, Z.; Zhang, Y.D. Pathological brain detection based on AlexNet and transfer learning. *J. Comput. Sci.* **2019**, *30*, 41–47. [CrossRef]

33. Wang, L.; Simpkin, R.; Al-Jumaily, A. Holographic microwave imaging array: Experimental investigation of breast tumour detection. In Proceedings of the 2013 IEEE International Workshop on Electromagnetics, Applications and Student Innovation Competition, Hong Kong, China, 1–3 August 2013; pp. 61–64.
34. Wang, L.; Fatemi, M. Compressive Sensing Holographic Microwave Random Array Imaging of Dielectric Inclusion. *IEEE Access* **2018**, *6*, 56477–56487. [CrossRef]
35. Wang, L. Multi-Frequency Holographic Microwave Imaging for Breast Lesion Detection. *IEEE Access* **2019**, *7*, 83984–83993. [CrossRef]
36. He, K.; Zhang, X.; Ren, S.; Sun, J. Deep residual learning for image recognition. In Proceedings of the IEEE Conference on Computer Vision and Pattern Recognition, Las Vegas, NV, USA, 27–30 June 2016; pp. 770–778.
37. Burfeindt, M.J.; Colgan, T.J.; Mays, R.O.; Shea, J.D.; Behdad, N.; Van Veen, B.D.; Hagness, S.C. MRI-Derived 3-D-Printed Breast Phantom for Microwave Breast Imaging Validation. *IEEE Antennas Wirel. Propag. Lett.* **2012**, *11*, 1610–1613. [CrossRef] [PubMed]
38. Italian National Research Council. An Internet Resource for the Calculation of the Dielectric Properties of Body Tissues in the Frequency Range 10 Hz–100 GHz. Available online: http://niremf.ifac.cnr.it/tissprop (accessed on 23 October 2022).
39. Sahlol, A.T.; Yousri, D.; Ewees, A.A.; Al-Qaness, M.; Elaziz, M.A. COVID-19 image classification using deep features and fractional-order marine predators algorithm. *Sci. Rep.* **2020**, *10*, 15364. [CrossRef] [PubMed]

micromachines

MDPI

Article

A Novel Fluidic Platform for Semi-Automated Cell Culture into Multiwell-like Bioreactors

Francesca Maria Orecchio [1], Vito Tommaso [2], Tommaso Santaniello [1,*], Sara Castiglioni [2], Federico Pezzotta [1], Andrea Monti [3], Francesco Butera [3], Jeanette Anne Marie Maier [1,2] and Paolo Milani [1]

1 Interdisciplinary Centre of Excellence for Nanostructured Materials and Interfaces (C.I.Ma.I.Na.), Department of Physics, University of Milan, Via Giovanni Celoria, 16, 20133 Milan, Italy; francesca.orecchio@unimi.it (F.M.O.); federico.pezzotta@unimi.it (F.P.); jeanette.maier@unimi.it (J.A.M.M.); paolo.milani@mi.infn.it (P.M.)

2 Department of Biomedical and Clinical Sciences, University of Milan, Via Giovanni Battista Grassi 74, 20157 Milan, Italy; vito.tommaso@unimi.it (V.T.); sara.castiglioni@unimi.it (S.C.)

3 Dolphin Fluidics S.r.l., Via Leonardo Da Vinci, 40, 20094 Corsico, Italy; a.monti@dolphinfluidics.com (A.M.); f.butera@dolphinfluidics.com (F.B.)

* Correspondence: tommaso.santaniello@unimi.it; Tel.: +39-2-50317437

Abstract: In this work, we developed and characterized a novel fluidic platform that enables long-term in vitro cell culture in a semi-automated fashion. The system is constituted by a control unit provided with a piezoelectric pump, miniaturized valves, and a microfluidic network for management and fine control of reagents' flow, connected to a disposable polymeric culture unit resembling the traditional multiwell-like design. As a proof of principle, Human Umbilical Vein Endothelial Cells (HUVEC) and Human Mesenchymal Stem Cells (hMSC) were seeded and cultured into the cell culture unit. The proliferation rate of HUVEC and the osteogenic differentiation of hMSC were assessed and compared to standard culture in Petri dishes. The results obtained demonstrated that our approach is suitable to perform semi-automated cell culture protocols, minimizing the contribution of human operators and allowing the standardization and reproducibility of the procedures. We believe that the proposed system constitutes a promising solution for the realization of user-friendly automated control systems that will favor the standardization of cell culture processes for cell factories, drug testing, and biomedical research.

Keywords: cells-on-a-chip; microfluidics; fluid automation; micro-bioreactors; smart fluidics

Citation: Orecchio, F.M.; Tommaso, V.; Santaniello, T.; Castiglioni, S.; Pezzotta, F.; Monti, A.; Butera, F.; Maier, J.A.M.; Milani, P. A Novel Fluidic Platform for Semi-Automated Cell Culture into Multiwell-like Bioreactors. *Micromachines* **2022**, *13*, 994. https://doi.org/10.3390/mi13070994

Academic Editor: Xiao Xiao

Received: 31 May 2022
Accepted: 22 June 2022
Published: 24 June 2022

1. Introduction

Traditional in-vitro techniques for cell culture allow the growth and proliferation of cells in an artificial and controlled environment and play a key role in elucidating physiology, disease mechanisms, and in testing new drugs [1–4]. These well-established approaches, which rely on 2D culture of cell lines and primary cells in standard Petri dishes and flasks, are still the most diffused in biology and biotechnology research, as well as in the pharmaceutical industry, cell factories, and other biomedical application areas [5].

Nowadays, cell cultivation practices are not very different from those used on a large scale in the second half of the 20th century [6]. Although these procedures enabled the understanding of a number of phenomena concerning cellular life and activity and proved to be determinants for applications such as drug testing and tissue regeneration, they present several intrinsic limitations. In fact, traditional cell culture systems require numerous and complex manual handling methods, limiting the standardization of the cultivation in terms of cell yield and activity [7]. Moreover, long-term cell maintenance is expensive and time consuming, requiring high quantities of reagents, test samples, and effort. The transition to user-friendly automated devices would favor the standardization of the cultivation process and the reduction of production costs and preparation times for cell culture through the decrease of the number of interventions by the operator [8].

Microfluidics and lab-on-a-chip technologies constitute a strategical solution to favor this transition. Performing cellular assays in microfluidic devices provided the advantages of reduced reagent consumption by replacing traditional Petri dishes and flasks with micro-culture systems, offering a cost-effective route for high-throughput cell production and culture in a controlled environment [9]. The micro-bioreactor is a simple yet archetypal microfluidic device that operates similarly to a standard multi-well plate, but with integrated fluidics circuits [10]. The design of this system is an array of cylindrical culture chambers connected to a network of micro- and/or millimeter-sized channels for the injection and withdrawal of cell media and fluids of interest for a specific cell assay [11]. Soft lithography and polymer micro-fabrication technologies allowed for the rapid prototyping of such devices based on polymeric biocompatible materials, such as polydimethylsiloxane (PDMS), polyethylene glycol diacrylate (PEGDA), hydrogels, and thermoplastics [12–15]. The in-vitro culture in fluidics micro-bioreactors has been reported in the literature, both in static and dynamic conditions, for a variety of cells, including stem cells, immortalized tumor cells (HeLa), cardiomyocytes and skeletal muscle cells of murine myoblasts, CrFk-type fibroblasts, and HN9.10e neuronal cell models [16–19].

At the present time, the operation of such micro-devices is restricted to the use of relatively complex, specific, and bulky external equipment and macro-to-micro interface systems for fluidic control, requiring the constant contribution of human operators to manage fluids flow in a manual and not standardized manner [20–22]. These approaches recently stimulated a paradigmatic question in the lab-on-a-chip community about the effective impact of microfluidic devices in standard cell culture: are we dealing with "lab-on-a-chip" or "chip-in-a-lab" systems? [23]. The usability and readout reliability of micro-bioreactors would massively benefit from the systematic implementation of a suitable integrated control system for precise fluid handling in a remote control framework. This is a key factor to render cell culture protocols automatic, minimizing the operator's contribution, and to drastically reduce the number of connection tubes and inlet and outlet ports and external bulky equipment.

Several works have been reported in the literature concerning automated cell culture on-chip, as automation is widely recognized as a benchmark parameter in modern microfluidic systems for cell biology [24–26]. At the research laboratory level, the minimization and miniaturization of additional peripheral equipment is considered a key requirement for the usability of automated devices in cell culture, along with the need for systems that are ready-to-use in a plug-and-play manner [24,27]. Representative examples of microfluidic platforms that target fulfilling these requirements and that are already on the market are the automation systems produced by LabSmith (uProcess™), PreciGenome (iFlow Touch™), and ElveFlow (Liquid Handling Pack). To our knowledge, these and other similar commercial systems only focus on providing the source for a pressure-driven flow. In order to realize a more complex fluidic system to achieve high-integration and high-parallelization (i.e., increasing system compactness and enabling the precise management of multiple channels), there would be a large number of tubes and interconnections involved, as well as relatively large valves, increasing the overall device dimension, and making the platform setup and usage cluttered and error prone.

Here, we present a novel fluidic platform for semi-automated cell culture into fluidic miniaturized bioreactors, prototyped based on a traditional multiwell-like design. The platform is composed of an electronic system equipped with integrated micro-pumps, valves, and microfluidic circuits for the management and fine control of reagents' flow and a disposable polymeric cell culture unit provided with fluidic channels and culture sites. This system combines the features of standard cell culture techniques (monolayer culture in cylindrical wells) with the possibility to conduct specific culture protocols remotely, simplifying the entire process, minimizing the contribution of human operators, and allowing the systematic culture of complex cell systems. At odds with what is on the market and reported in the literature, our fluidic system presents active valves interlocked on a fluidic backplane with relatively complex interconnections. The combination between the complexity of the fluidic design and the use of embedded micro-valves actuated by shape memory alloys actuators renders the system compact and highly performing in managing multiple fluidic channels suitable for cell culture protocols. Our fluidic platform was tested on two different primary human cell types, Umbilical Vein Endothelial Cells (HUVEC) and Mesenchymal Stem Cells (hMSC). The cells were cultured in a semi-automated fashion and morphology, known to predict cell health, which was analyzed by confocal microscopy. In addition, the proliferation rate of HUVEC and the osteogenic differentiation of hMSC were investigated. All the data obtained were compared to the cells cultured in standard culture dishes.

System Design and Semi-Automated Culture Approach

The system (Figure 1) consists of two main units: (i) a fluidic control unit, for the fine control; transport; and perfusion of reagents, culture media, and cells within the system. The fluid transportation takes place within an integrated polymeric fluidic platform, supplied with a network that includes several precision valves and a piezoelectric pump of micro-channels for the handling of the liquids of interest. (ii) A disposable multi-well like cell culture unit, produced in polydimethyilsiloxane (PDMS) via soft-lithography, connected to the fluidic unit through silicone tubes.

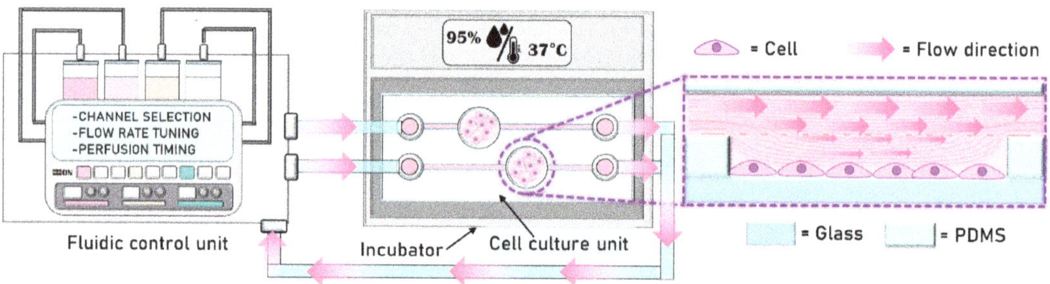

Figure 1. Schematic representation of the microfluidic control platform, including the control module and disposable PDMS unit.

The feasibility of the fluidic platform and PDMS-based device for semi-automated on-chip cell culture was assessed using HUVEC and hMSC as test cellular models. In both cases, the culture protocol operations, such as the change of culture medium and the injection of reagents of interest, were timed and controlled externally via the fluidic control unit, while the fluidic cell culture unit was positioned in an incubator at 37 °C and 5% CO_2. We chose 500 µL/min as the flow rate, providing a laminar flow and low values of shear stress during the assays. With this setup and conditions, we were able to perform a 3-days-long cell culture experiment for HUVEC and a 4-days-long culture for hMSC.

2. Materials and Methods

2.1. Fluidic Control Unit

Implemented by Dolphin Fluidics in cooperation with memetis GmbH, the fluidic control unit is a pressure-driven flow system for fluid transportation and delivery. The source fluids are stored in six chemically inert glass vessels (fluid reservoir bottles) in the back of the fluidic control unit (Figure 2a). These vessels can be pressurized by a single piezoelectric air pump from TTP Ventus Ltd. (Royston, UK), which operates completely silently. A thin PTFE tube (inner diameter 0.8 mm) also reaches through the cap of each vessel, down to its bottom. This way, the internal air pressure will drive fluid from the vessel through the tube when its external end is open.

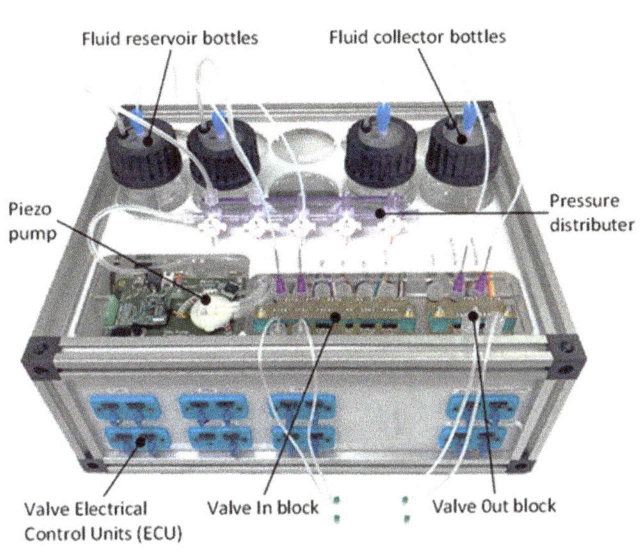

(a)

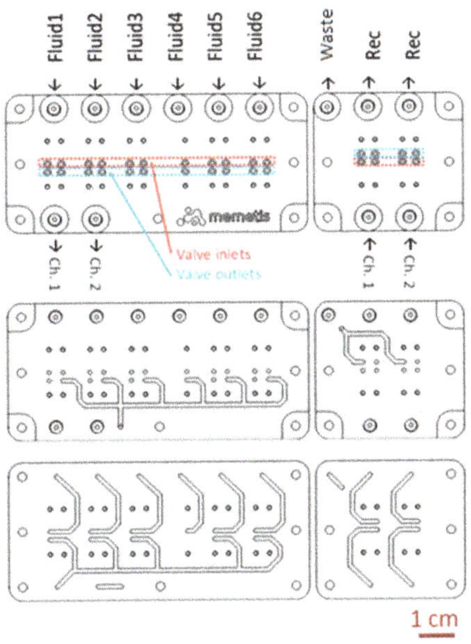

(b)

Figure 2. (**a**) Fluidic control unit for cell culture integrated in a housing; (**b**) schematic top view of fluid distribution circuit (top); cut view of first channel level (middle); cut view of second channel level (bottom). The system is designed to distribute different input fluids to the fluidic cell culture unit and from there to collector bottles. The fluids are transported by pressure-driven flow.

Distribution of the six supply fluids to the inlets of the cell culture platform is managed by a custom 3D-printed biocompatible manifold within the fluidic control unit (Figure 2b), which has two internal levels of flow paths. Eleven normally closed microvalves "Series 09" by Memetis are flange-mounted to the manifold to allow automatic control of fluid pathways ("in-block"). Fluids that leave the fluidic platform are handled by four additional valves ("out-block") on the same manifold and directed to three collector vessels. The schematic diagram of the fluidic circuit is reported into the Supplementary Material File.

The control electronics for the silently operating miniature valves, integrated in the front part of the unit, support automated valve control via USB connection and I2Cprotocol, as well as manual operation using toggle buttons in the front panel. Due to their functional principle, the valves do not only support fully closed/fully open switching states, but can also adopt intermediate states to enable precise flow rate adjustments via closed-loop control. The maximum pressure achievable with this system is 2 bar, while the maximum flow rate value is 80 mL/min, which can be obtained at a pressure of 1 bar. Both pressure and flow rate resolution are +/−2%.

The extreme compactness of the valves with an installation pitch of only 5 mm is enabled by a fatigue-free shape memory alloy (SMA) thin film actuator, allowing for more than 20 million switching cycles. A low-power consumption of 0.15 W holding power (0.3 W switching power) per valve assures that heat introduction into the controlled media is minimal.

2.2. *Disposable Cell Culture Unit Fabrication*

The disposable cell culture unit was produced by means of soft-lithography, using micro-molding of PDMS, a soft polymer, typically employed for prototyping this kind of fluidic system. In fact, PDMS is an elastomeric polymer with suitable properties for biomedical applications, including physiological indifference, excellent resistance to biodegradation, biocompatibility, chemical stability, gas permeability, excellent optical transparency, and can be easily processed using replica molding [28–30].

Suitable master molds (Figure 3a) were fabricated by numerically controlled micro-milling of aluminum slabs to obtain protruding microstructures, resembling the final device design (Figure 3b–d). The disposable unit is designed with cylindrical culture wells with a 15 mm diameter, typical of that of a standard 24 multi-well plate, and is provided with micro-fluidic channels with a submillimeter-sized cross section (800 μm) for fluid transportation into the culture sites. The channels and chamber height were dimensioned to enable low flow rates (100 to 500 μL/min) with associated low values of shear stress affecting the cultured cells during chamber perfusion.

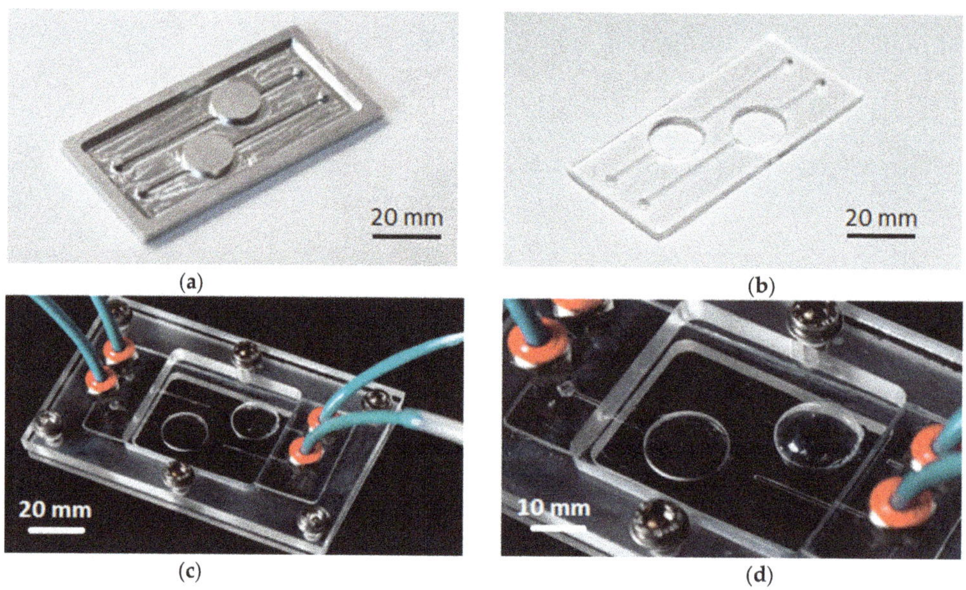

Figure 3. (**a**) Master obtained by micro-milling aluminum components, resembling the final cell culture unit's microstructures; (**b**) cell culture unit, sealed and bonded to a PDMS membrane (upper part) and to a glass coverslip (bottom part); (**c**) final PDMS cell culture unit sandwiched between PC layers. PC were provided with inlet and outlet ports, positioned to match the ends of the PDMS cell culture unit, machined to accommodate standard fittings for the four ports for flowing fluids through the cell culture unit; (**d**) close-up view of the cell culture chambers, sealed on top with the oxygen-permeable PDMS membrane.

PDMS (Sylgard 184 Silicone Elastomer Kit, Daw Corning) was cast over the master to replicate the fluidic network according to the following protocol. The elastomeric part was mixed with a curing agent at a standard ratio of 10:1 w/w. The mixture was stirred for 20 min and placed in a vacuum desiccator to remove air bubbles. The PDMS was then poured into the aluminum master and placed in oven for about 2 h and 15 min at 80 °C. After fabrication, the PDMS platform was bonded to 0.1-mm-thick glass coverslips using oxygen plasma treatment to seal the bottom part of the channels. The treatment was performed inside a low-pressure Zepto Plasma Surface treatment machine, purchased from Diener Electronic GmbH (Ebhause, Germany). The oxygen pressure inside the chamber was fixed at 1.0 mbar, and the surface treatment was applied at a power of 80 W for 1 min and 30 s. Surfaces were bonded together by direct contact, applying pressure manually.

A 0.03-mm-thick PDMS membrane serving as the roof for the culture wells was produced by spin coating. This specific thickness allows for oxygen exchange between the cell culture chambers and the incubation environment [31,32]. The PDMS-membrane was stacked over the cell culture unit via the same optimized oxygen plasma treatment (Figure 3b). The PDMS cell culture unit was then finally sandwiched into a manifold, consisting of two fastened polycarbonate (PC) frames produced by means of CNC milling, in order to ensure fluidic seal of the system (Figure 3c).

2.3. Semi-Automated Cell Culture

The device has two independent cell culture chambers of 1.9 cm^2 surface area and containing 500 μL of medium and cells. Cell cultivation took place directly on glass, since no cell adhesion coatings were added at the bottom of the culture chambers. In order to ensure sterilization and removal of circulating bubbles, we preliminarily fluxed 70% ethanol inside the chip, connected to the machine, and in each channel of the fluidic platform. We then exposed the whole system to UV-c bactericidal action for 30 min, inside the biological hood. Each beaker, inside the platform, was preliminarily. We next proceeded to seed cells inside the chip, linking it to the microfluidic control platform through silicone tubes: The microfluidic control platform enabled us to perform semi-automated cells seeding in a controlled manner at a flow rate of 500 μL/min. These procedures are strictly carried out inside the biological hood, except for the sterilization and bubbles removal, which can take place on a lab bench.

2.3.1. Culture of HUVEC

HUVEC were purchased from Lonza (Basel, Switzerland) and cultured at 37 °C and 5% CO_2 in EBM-2 medium (Lonza), supplemented with 10% fetal bovine serum and EGM-2 SingleQuots Supplements (Lonza). They were used up to passage 4. 10^4 HUVEC were seeded on each chamber of the cell culture unit and on glass coverslips (13 mm in diameter) allocated on a 24-well plate as control. After 3 days, the cells were trypsinized via microfluidic control platform, in a semi-automated manner, collected in the collector bottle, and counted using a cell counter. The experiment was repeated five times in duplicate. At the end of the experiments, some samples were fixed for confocal analysis.

2.3.2. Culture of hMSC

hMSC were purchased from CliniSciences and were cultured at 37 °C and 5% CO_2 in Dulbecco's Modified Eagle's Medium containing 10% fetal bovine serum and 2 mM glutamine (culture medium, CM). All the reagents for hMSC culture were from Sigma-Aldrich. The cells were used between passage 2 and 6. In total, 3.5×10^4 hMSC were seeded on each chamber of the cell culture unit and on glass coverslips as control. To induce osteogenic differentiation, once the cells were confluent, the CM was changed via microfluidic control platform with an osteogenic medium (OM) containing 2×10^{-8} M 1α,25-Dihydroxyvitamin D3, 10 mM β-glycerolphosphate, and 0.05 mM ascorbic acid (Sigma-Aldrich, St. Louis, MO, USA). After 24 h of culture, the cells were trypsinized and collected for Real-Time PCR or fixed for confocal analysis.

2.3.3. Real-Time PCR

Total RNA was extracted by the PureLink RNA Mini kit (Thermo Fisher Scientific, Waltham, MA, USA). Single-stranded cDNA was synthesized from 0.3 μg RNA in a 20 μL final volume using High Capacity cDNA Reverse Transcription Kit, with RNase inhibitor (Thermo Fisher Scientific) according to the manufacturer's instructions. Real-time PCR was performed on 20 ng of cDNA using TaqMan™ Fast Universal PCR Master Mix (Thermo Fisher Scientific, Waltham, MA, USA) and TaqMan Gene Expression Assays (FAM) (Thermo Fisher Scientific, Waltham, MA, USA). The following primers were used: RUNX2 (Hs00231692_m1), Sp7 (Hs01866874_s1), COL1A1 (Hs00164004_m1), and SPP1 (Hs00959010_m1). The housekeeping gene GAPDH (Hs99999905_m1) was used as an internal reference gene. The reactions were performed with CFX96 Real-Time PCR Detection

System (Bio-Rad, Hercules, CA, USA). Relative changes in gene expression were analyzed with the $2^{-\Delta\Delta Ct}$ method.

2.3.4. Confocal Microscopy

HUVEC and hMSC were fixed in phosphate buffered saline containing 4% paraformaldehyde and 2% sucrose pH 7.6 and permeabilized with Triton 0.3%. HUVEC were incubated with anti-cyclophilin (CYP) D and anti-VE-Cadherin overnight at 4 °C, and stained with Alexa Fluor 488 and 647 secondary antibodies respectively (ThermoFisher Scientific, Waltham, MA, USA). hMSC were incubated with anti-CYP D overnight at 4 °C, and stained with Alexa Fluor 488 secondary antibodies (ThermoFisher Scientific, Waltham, MA, USA). Rhodamine-labeled phalloidin was used to visualize the cytoskeleton. 4′,6-Diamidine-2′-phenylindole dihydrochloride (DAPI, Sigma) was used to stain the nuclei. Finally, cells were mounted with ProLong Gold Antifade Mountant (Invitrogen, Carlsbad, CA, USA) and images were acquired using a 40× objective in oil by a SP8 Leica confocal microscope.

2.3.5. Statistical Analysis

Data are expressed as the mean ± standard deviation. The data were analyzed using one-way ANOVA. Statistical significance was defined as p-value ≤ 0.05.

3. Results & Discussion

The validation of the disposable cell culture unit and the fluidic control platform was assessed by culturing two different cell types, HUVEC and hMSC. HUVEC were cultured for 3 days and the proliferation rate and imaging analyses were performed. hMSC were cultured for 3 days in CM till they reached the confluence and for an additional day in the presence of the OM to induce the osteogenic differentiation. Osteogenic marker expression and imaging analyses were then performed. Both cell cultures were performed in a semi-automated fashion, providing fresh medium exchange every 24 h in a semi-automated manner at 500 μL/min flow rate. We selected this flow rate value according to the channels geometry and dimensions in order to keep low shear stresses in the culture chamber during perfusion [33,34]. As shown in Figure 4, HUVEC cultured in the cell culture unit grows as fast as the cells cultured on a standard glass coverslip. Taken together, these results suggest that the design of our cell culture unit provides the proper amount of gas and nutrients exchange between cultured cells and inner/outer environments, allowing them to grow and proliferate according to standard culture on wells. Moreover, the use of precision miniaturized valves and a pulsation-free pump embedded in the control unit is crucial for avoiding harsh shear stress during medium perfusion and to keep a laminar flow throughout the performed assays, avoiding fluidic-mechanical stresses on the cell cultures.

By confocal microscopy, we analyzed HUVEC actin cytoskeleton. Figure 4b demonstrates that the cells cultured on cell culture unit maintain the same polygonal morphology of the control cells seeded on glass coverslips.

In both samples, the actin is organized as a network of well-defined stress fibers spanning the entire cell body. We also analyzed the vascular endothelial (VE)-cadherin, which is an endothelial-specific adhesion molecule located at junctions between endothelial cells. A correct VE-cadherin-mediated adhesion is fundamental not only to control vascular permeability but also to modulate cell proliferation and apoptosis [31]. As shown in Figure 4b, VE-cadherin is well expressed at the cell junctions of both samples.

Then we focused on mitochondria analysis since their morphology reflects the function of the organelles [35,36] and is continuously modified in response to the different functional requirements of the cell. The anti-CYP D antibody highlights a complex mitochondrial network with elongated mitochondria in HUVEC cultured in the cell culture unit and on a traditional coverslip (Figure 4b).

We also tested hMSC as a model to investigate cell differentiation in response to specific stimuli. Initially, the cells were cultured in CM for 3 days in the chip and on a glass coverslip, and then analyzed by confocal microscope. Figure 5a confirms the similar cellular morphology of the two samples with a well-organized actin fiber-network. In both samples, the mitochondria appeared elongated, forming a complex network.

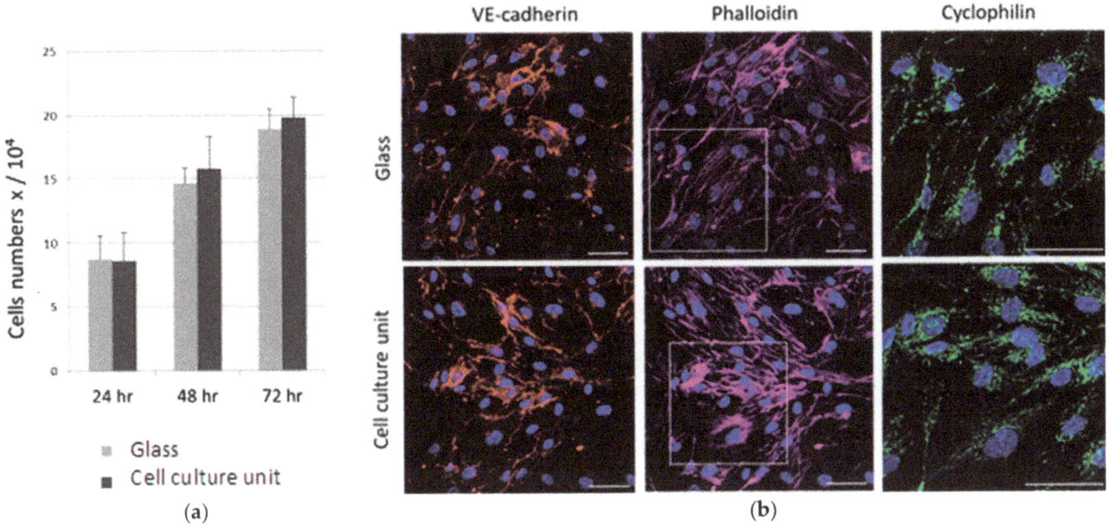

Figure 4. HUVEC were seeded via microfluidic control platform on cell culture unit or on glass coverslips as control. (**a**) The cells were counted using an automated cell counter. Data are the means ± standard deviation of five independent experiments in duplicate. (**b**) HUVEC were cultured on cell culture unit and on glass coverslips for 72 h. The cells were fixed and observed by confocal microscopy after phalloidin, Ve-cadherin, cyclophilin F, and DAPI staining. White square is zoomed-in cyclophilin staining image. Scale bar: 50 µm.

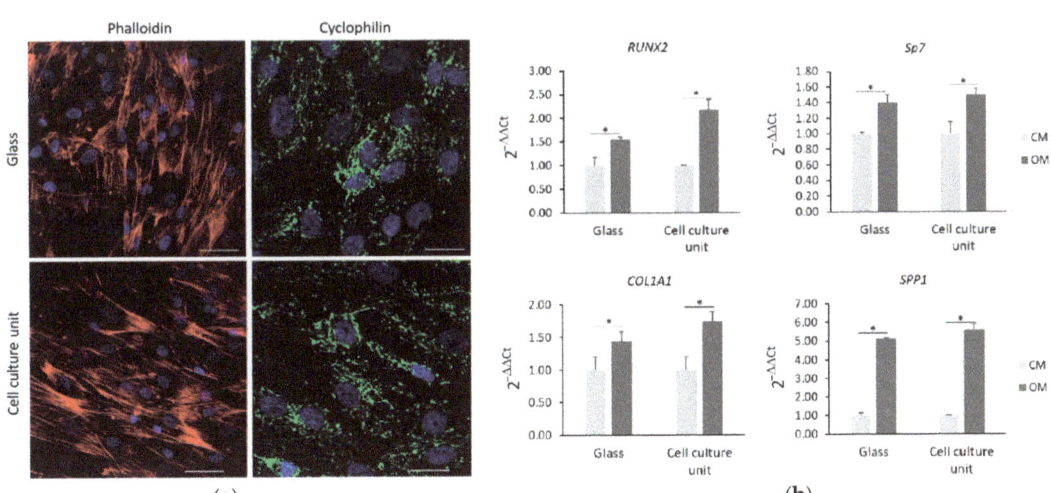

Figure 5. hMSC were cultured on cell culture unit and on glass coverslips. (**a**) After 72 h, cells were observed by confocal microscopy after phalloidin, cyclophilin F, and DAPI staining. Scale bar: 50 µm. (**b**) Confluent hMSC were cultured for 24 h in OM. Real-Time PCR was performed on RNA extracted from hMSC using primers designed on *RUNX2*, *Sp7*, *COL1A1*, and *SPP1* sequence (* $p \leq 0.05$).

Once the cells reached the confluence, the medium was changed with the OM and the hMSC were cultured for an additional 24 h. Then, by real time PCR we analyzed the expression of some osteogenic markers. In particular, we analyzed *RUNX2*, which is the first gene to be expressed during osteogenesis, and *Sp7* and collagen 1A1 (*COL1A1*), which are involved in the early stages of differentiation and osteopontin (*SPP1*), which is one of the most abundant non-collagenous components of the bone extracellular matrix. Figure 5b shows that 24 h of culture in the presence of the osteogenic cocktail induces a similar upregulation of all the osteogenic markers both in the cells cultured on the cell culture unit and in the cells cultured on glass coverslips.

According to these results, the culture system proposed proved to be efficient and time saving, reproducing in a semi-automated framework the traditional culture and analysis protocols in a standard multi-well plate-like disposable culture unit. This system will allow to perform different experiments and assays in parallel, scaling-up for high-throughput cell culture processes in a controlled manner. In fact, the control unit offers the possibility to choose a unique and personal set of media/reagents perfusion time, flow rate, and static/dynamic culture conditions. Moreover, the integrated platform avoids the use of bulky and highly specific equipment and interfaces, while reducing to the minimum the complexity of assembling the units. The optimization of this platform, in view of a full automation of the process, will also allow to culture and co-culture complex biological systems (e.g., primary cells, stem cell lines and organoids) without the request for human operators' contribution. The combination of these features makes this automated culture system a promising solution to implement automation in cultivation processes for cell factories, cells-based precision medicine, and biomedical research.

4. Conclusions

In this study, we developed and evaluated the suitability of a novel fluidic platform for the semi-automated culture of cells into miniaturized devices. The platform comprises two main units: a fluidic control unit for the fine control and perfusion of reagents, and a disposable multi-well like cell culture unit produced in PDMS via soft-lithography. HUVEC and hMSC were seeded via the fluidic platform in a semi-automated fashion, setting the proper timing and flow rate, in order to avoid harsh shear stress on cells during media perfusion and to keep a laminar flow throughout the performed assays. In the fluidic system, the cells are vital and maintain their morphology. The HUVEC proliferation rate and the hMSC osteogenic differentiation potential are comparable to the respective control cells cultured on glasses coverslips

The semi-automated approach for cell culture reported here offers several advantages, such as the reduction of time, reagents, and media consumption, and enables the reproduction of traditional culture and analysis protocols using a standard multi-well plate-like disposable culture unit, massively decreasing the operators' efforts. We plan to develop and optimize a fully automated system to perform remotely different cultures and assays in parallel, scaling-up the system for a high-throughput culture process in a time saving and cost-effective manner. In this way, the entire procedures will be simplified, minimizing the contribution of human operators and allowing the standardization and reproducibility of the in-vitro culture of complex cell systems, such as organoids.

Supplementary Materials: The following supporting information can be downloaded at: https://www.mdpi.com/article/10.3390/mi13070994/s1, Figure S1: Schematic diagram of the fluidic circuit of the control platform.

Author Contributions: Conceptualization: T.S., F.B., J.A.M.M. and P.M.; Methodology, F.M.O., V.T., S.C., F.P., A.M., T.S., F.B., J.A.M.M. and P.M. Software, A.M. and F.B.; Validation, F.M.O., V.T., A.M. and S.C.; Writing—original draft preparation, F.M.O. and S.C.; Writing—review and editing, S.C., T.S., J.A.M.M. and P.M.; Supervision, P.M., J.A.M.M., T.S. and S.C.; Project administration, P.M., J.A.M.M., T.S., F.B. and S.C.; Funding acquisition, P.M., J.A.M.M. and T.S. All authors have read and agreed to the published version of the manuscript.

Funding: This work was supported by Regione Lombardia under the Program "Call Hub Ricerca e Innovazione" (POR-FESR 2014-2020, project ID 1170989—PRINTMED-3D).

Institutional Review Board Statement: Not applicable.

Informed Consent Statement: Not applicable.

Acknowledgments: We thank Francesco Cavaliere and Enea Boria from the Mechanical Workshop of the Department of Physics (University of Milan) for their technical support, and Laura Locatelli from the Department of Biomedical and Clinical Sciences (University of Milan) for confocal microscopy. We also greatly acknowledge Marcel Gültig and Hinnerk Oßmer, Co-Founders & Managing Directors of memetis, for their work on the fluidic platform implementation.

Conflicts of Interest: The authors declare no conflict of interest.

References

1. Mittal, R.; Woo, F.W.; Castro, C.S.; Cohen, M.A.; Karanxha, J.; Mittal, J.; Chhibber, T.; Jhaveri, V.M. Organ-on-chip models: Implications in drug discovery and clinical applications. *J. Cell. Physiol.* **2019**, *234*, 8352–8380. [CrossRef]
2. Rauti, R.; Ess, A.; Le Roi, B.; Kreinin, Y.; Epshtein, M.; Korin, N.; Maoz, B.M. Transforming a well into a chip: A modular 3D-printed microfluidic chip. *APL Bioeng.* **2021**, *5*, 026103. [CrossRef]
3. Jing, L.; Fan, S.; Yao, X.; Zhang, Y. Effects of compound stimulation of fluid shear stress plus ultrasound on stem cell proliferation and osteogenesis. *Regen. Biomater.* **2021**, *8*, rbab066. [CrossRef]
4. Siddique, A.; Pause, I.; Narayan, S.; Kruse, L.; Stark, R.W. Endothelialization of PDMS-based microfluidic devices under high shear stress conditions. *Colloids Surf. B Biointerfaces* **2021**, *197*, 111394. [CrossRef]
5. Jensen, C.; Teng, Y. Is It Time to Start Transitioning from 2D to 3D Cell Culture? *Front. Mol. Biosci.* **2020**, *7*, 33. [CrossRef]
6. Segeritz, C.-P.; Vallier, L. Cell Culture: Growing Cells as Model Systems In Vitro. In *Basic Science Methods for Clinical Researchers*; Elsevier: Amsterdam, The Netherlands, 2017; pp. 151–172. [CrossRef]
7. Kusena, J.W.T.; Shariatzadeh, M.; Studd, A.J.; James, J.R.; Thomas, R.J.; Wilson, S.L. The importance of cell culture parameter standardization: An assessment of the robustness of the 2102Ep reference cell line. *Bioengineered* **2021**, *12*, 341–357. [CrossRef]
8. Mandenius, C.-F. Conceptual Design of Micro-Bioreactors and Organ-on-Chips for Studies of Cell Cultures. *Bioengineering* **2018**, *5*, 56. [CrossRef]
9. Tehranirokh, M.; Kouzani, A.Z.; Francis, P.S.; Kanwar, J.R. Microfluidic devices for cell cultivation and proliferation. *Biomicrofluidics* **2013**, *7*, 051502. [CrossRef]
10. Torino, S.; Corrado, B.; Iodice, M.; Coppola, G. PDMS-Based Microfluidic Devices for Cell Culture. *Inventions* **2018**, *3*, 65. [CrossRef]
11. Low, L.A.; Mummery, C.; Berridge, B.R.; Austin, C.P.; Tagle, D.A. Organs-on-chips: Into the next decade. *Nat. Rev.* **2021**, *20*, 345–361. [CrossRef]
12. Merkel, T.; Bondar, V.; Nagai, K.; Freeman, B.; Pinnau, I. Gas Sorption, Diffusion, and Permeation in Poly(Dimethylsiloxane). *J. Polym. Sci. Part B Polym. Phys.* **2000**, *38*, 415. [CrossRef]
13. Halldorsson, S.; Lucumi, E.; Gómez-Sjöberg, R.; Fleming, R.M.T. Advantages and challenges of microfluidic cell culture in polydimethylsiloxane devices. *Biosens. Bioelectron.* **2015**, *63*, 218–231. [CrossRef]
14. Khot, M.I.; Levenstein, M.A.; de Boer, G.N.; Armstrong, G.; Maisey, T.; Svavarsdottir, H.S.; Andrew, H.; Perry, S.L.; Kapur, N.; Jayne, D.G. Characterising a PDMS based 3D cell culturing microfluidic platform for screening chemotherapeutic drug cytotoxic activity. *Sci. Rep.* **2020**, *10*, 15915. [CrossRef]
15. Santaniello, T.; Yan, Y.; Tocchio, A.; Martello, F.; Gassa, F.; Webb, P.; Zhao, W.; Tamplenizza, M.; Schulte, C.; Liu, Y.; et al. On-chip single cell funneling operated by microfabricated thermo-responsive hydrogel layers. *J. Micromech. Microeng.* **2015**, *25*, 075004. [CrossRef]
16. Mark, D.; Haeberle, S.; Roth, G.; Von Stetten, F.; Zengerle, R. Microfluidic Lab-on-a-Chip Platforms: Requirements, Characteristics and Applications. In *Microfluidics Based Microsystems*; Springer: Dordrecht, The Netherlands, 2013; pp. 305–376. [CrossRef]
17. Govoni, M.; Muscari, C.; Guarnieri, C.; Giordano, E. Mechanostimulation Protocols for Cardiac Tissue Engineering. *BioMed Res. Int.* **2013**, *2013*, 918640. [CrossRef]
18. Ochs, J.; Biermann, F.; Piotrowski, T.; Erkens, F.; Nießing, B.; Herbst, L.; König, N.; Schmitt, R.H. Fully Automated Cultivation of Adipose-Derived Stem Cells in the StemCellDiscovery—A Robotic Laboratory for Small-Scale, High-Throughput Cell Production Including Deep Learning-Based Confluence Estimation. *Processes* **2021**, *9*, 575. [CrossRef]
19. Tandon, N.; Marolt, D.; Cimetta, E.; Vunjak-Novakovic, G. Bioreactor Engineering of Stem Cell Environments. *Biotechnol. Adv.* **2013**, *31*, 1020–1031. [CrossRef]
20. Yourek, G.; McCormick, S.M.; Mao, J.J.; Reilly, G.C. Shear stress induces osteogenic differentiation of human mesenchymal stem cells. *Regen. Med.* **2010**, *5*, 713–724. [CrossRef]
21. van Duinen, V.; Trietsch, S.J.; Joore, J.; Vulto, P.; Hankemeier, T. Microfluidic 3D cell culture: From tools to tissue models. *Curr. Opin. Biotechnol.* **2015**, *35*, 118–126. [CrossRef]

22. Suh, Y.J.; Pandey, M.; Segall, J.E.; Wu, M. Tumor spheroid invasion in epidermal growth factor gradients revealed by a 3D microfluidic device. *Phys. Biol.* **2022**, *19*, 036002. [CrossRef]
23. Mohammed, M.I.; Haswell, S.; Gibson, I. Lab-on-a-chip or Chip-in-a-lab: Challenges of Commercialization Lost in Translation. *Procedia Technol.* **2015**, *20*, 54–59. [CrossRef]
24. Coluccio, M.L.; Perozziello, G.; Malara, N.; Parrotta, E.; Zhang, P.; Gentile, F.; Limongi, T.; Raja, P.M.; Cuda, G.; Candeloro, P.; et al. Microfluidic platforms for cell cultures and investigations. *Microelectron. Eng.* **2018**, *208*, 14–28. [CrossRef]
25. Jaccard, N.; Macown, R.J.; Super, A.; Griffin, L.D.; Veraitch, F.S.; Szita, N. Automated and online characterization of adherent cell culture growth in a microfabricated bioreactor. *J. Lab. Autom.* **2014**, *19*, 437–443. [CrossRef]
26. Perozziello, G.; Møllenbach, J.; Laursen, S.; Di Fabrizio, E.; Gernaey, K.; Krühne, U. Lab on a chip automates in vitro cell culturing. *Microelectron. Eng.* **2012**, *98*, 655–658. [CrossRef]
27. Ortseifen, V.; Viefhues, M.; Wobbe, L.; Grünberger, A. Microfluidics for Biotechnology: Bridging Gaps to Foster Microfluidic Applications. *Front. Bioeng. Biotechnol.* **2020**, *8*, 589074. [CrossRef]
28. van Poll, M.L.; Zhou, F.; Ramstedt, M.; Hu, L.; Huck, W.T.S. A Self-Assembly Approach to Chemical Micropatterning of Poly(dimethylsiloxane). *Angew. Chem. Int. Ed.* **2007**, *46*, 6634–6637. [CrossRef]
29. Miranda, I.; Souza, A.; Sousa, P.; Ribeiro, J.; Castanheira, E.M.S.; Lima, R.; Minas, G. Properties and Applications of PDMS for Biomedical Engineering: A Review. *J. Funct. Biomater.* **2022**, *13*, 2. [CrossRef]
30. Shakeri, A.; Khan, S.; Didar, T.F. Conventional and emerging strategies for the fabrication and functionalization of PDMS-based microfluidic devices. *Lab Chip* **2021**, *21*, 3053–3075. [CrossRef]
31. Khan, O.F.; Sefton, M.V. Endothelial cell behaviour within a microfluidic mimic of the flow channels of a modular tissue engineered construct. *Biomed. Microdevices* **2011**, *13*, 69–87. [CrossRef]
32. Kim, M.-C.; Lam, R.H.W.; Thorsen, T.; Asada, H.H. Mathematical analysis of oxygen transfer through polydimethylsiloxane membrane between double layers of cell culture channel and gas chamber in microfluidic oxygenator. *Microfluid. Nanofluid.* **2013**, *15*, 285–296. [CrossRef]
33. Vestweber DChau, L.T.; Rolfe, B.E.; Cooper-White, J.J. A microdevice for the creation of patent, three-dimensional endothelial cell-based microcirculatory networks. *Biomicrofluidics* **2011**, *5*, 034115. [CrossRef]
34. Cimetta, E.; Figallo, E.; Cannizzaro, C.; Elvassore, N.; Vunjak-Novakovic, G. Micro-bioreactor arrays for controlling cellular environments: Design principles for human embryonic stem cell applications. *Methods* **2009**, *47*, 81–89. [CrossRef] [PubMed]
35. Vestweber, D. VE-Cadherin. *Arter. Thromb. Vasc. Biol.* **2008**, *28*, 223–232. [CrossRef] [PubMed]
36. Westermann, B. Bioenergetic role of mitochondrial fusion and fission. *Biochim. Biophys. Acta* **2012**, *1817*, 1833–1838. [CrossRef]

micromachines

MDPI

Article

Accurate and Automatic Extraction of Cell Self-Rotation Speed in an ODEP Field Using an Area Change Algorithm

Haiyang Wu [1], Dan Dang [2,*], Xieliu Yang [1], Junhai Wang [1], Ruolong Qi [1,*], Wenguang Yang [3] and Wenfeng Liang [1,*]

1 School of Mechanical Engineering, Shenyang Jianzhu University, Shenyang 110168, China; haiywu@foxmail.com (H.W.); yang.xieliu@sjzu.edu.cn (X.Y.); jhwang@sjzu.edu.cn (J.W.)
2 School of Science, Shenyang Jianzhu University, Shenyang 110168, China
3 School of Electromechanical and Automotive Engineering, Yantai University, Yantai 264005, China; yangwenguang@ytu.edu.cn
* Correspondence: dangdan@sjzu.edu.cn (D.D.); qiruolong@foxmail.com (R.Q.); liangwf@sjzu.edu.cn (W.L.)

Abstract: Cells are complex biological units that can sense physicochemical stimuli from their surroundings and respond positively to them through characterization of the cell behavior. Thus, understanding the motions of cells is important for investigating their intrinsic properties and reflecting their various states. Computer-vision-based methods for elucidating cell behavior offer a novel approach to accurately extract cell motions. Here, we propose an algorithm based on area change to automatically extract the self-rotation of cells in an optically induced dielectrophoresis field. To obtain a clear and complete outline of the cell structure, dark corner removal and contrast stretching techniques are used in the pre-processing stage. The self-rotation speed is calculated by determining the frequency of the cell area changes in all of the captured images. The algorithm is suitable for calculating in-plane and out-of-plane rotations, while addressing the problem of identical images at different rotation angles when dealing with rotations of spherical and flat cells. In addition, the algorithm can be used to determine the motion trajectory of cells. The experimental results show that the algorithm can efficiently and accurately calculate cell rotation speeds of up to ~155 rpm. Potential applications of the proposed algorithm include cell morphology extraction, cell classification, and characterization of the cell mechanical properties. The algorithm can be very helpful for those who are interested in using computer vision and artificial-intelligence-based ideology in single-cell studies, drug treatment, and other bio-related fields.

Keywords: ODEP; self-rotation; area change

Citation: Wu, H.; Dang, D.; Yang, X.; Wang, J.; Qi, R.; Yang, W.; Liang, W. Accurate and Automatic Extraction of Cell Self-Rotation Speed in an ODEP Field Using an Area Change Algorithm. *Micromachines* **2022**, *13*, 818. https://doi.org/10.3390/mi13060818

Academic Editor: Xiao Xiao

Received: 26 April 2022
Accepted: 22 May 2022
Published: 24 May 2022

1. Introduction

With the intensive research into micro and nano mechanics and the rapid advancement of microscopy, cellular level research has become a hot topic in many fields. As the most basic unit of life in higher organisms, cells can not only grow, develop, proliferate, age, and die independently in an adapted environment, but they can also organically form tissues and organs with different morphologies and functions in order to maintain the functions necessary for life, thereby forming a complete living organism [1]. Cells have specific tissue and organ functions, and are therefore regarded as single units to examine and analyze the multidimensional characteristics of cells during proliferation, differentiation, movement, and other physiological activities. This is significant for revealing the complex and diverse changes in living bodies, as well as their distribution patterns in time and space [2,3]. The multidimensional information of cells includes cellular morphological information, electrophysiological information, and mechanical properties, among others [4–10]. The mechanical properties of cells play a pivotal role in characterizing different cellular states and functions. Cells live in a complex mechanical and chemical environment where changes in the physical mechanics of the cell activate cellular signaling channels and induce reorganization of the cytoskeleton [11]. At different levels of the cellular hierarchy, changes in the cellular organization can be characterized by changes in the specific mechanical properties. Clearly, cells of different structures vary greatly in several aspects, including biological activities, such as cell differentiation, growth, and adhesion, and the pathogenic mechanisms of life, such as oxidative stress, viral attack, and parasites of life [12,13]. Thus, quantifying the mechanical properties of living cells provides information about the actual conditions of the cells. To understand how cells respond to their surroundings, it is necessary to characterize their mechanical properties. Researchers at the University of Texas discovered, for the first time, that changes in the mechanical properties of cells may be responsible for tumorigenesis [14]. Changes in the mechanical properties of cells will eventually lead to uncontrolled cell division, and such uncontrolled cells have a low mortality rate and give rise to the growth of malignant tumors. Plenty of research on cancer has shifted focus to the biochemical perspective in order to address the various interdependent factors associated with biochemical carcinogenesis. Meanwhile, the mechanical properties of cells affect their ability to be deformed, to be moved, and to perceive external stimuli in their microenvironment. Research has shown that changes in the mechanical properties of cells define the direction for the investigation of many cellular diseases [15]. New techniques for the precise capture and manipulation of cells at a microscale have also made it possible to characterize the mechanical properties of cells.

Several techniques are currently available for cell manipulation and are being used in different fields, such as optical, ultrasound, magnetic, and electric fields [16–22]. Optical tweezers, which use lasers to generate forces to trap cells, require high-intensity light. This method has several limitations. Specifically, the cells may easily become damaged; the applied range is limited by the area in which the optical tweezer system can operate, which impedes the manipulation of cells in large quantities. Meanwhile, acoustic forces generated using ultrasound can be used to trap or lift cells. Nowadays, it is commonplace to manipulate cells using electrical forces, such as dielectrophoresis (DEP) [23–25]. A new technique for manipulating tiny particles based on DEP forces, also known as optically induced dielectrophoresis (ODEP), was first introduced by Ming et al. [26]. ODEP can realize label-free and non-invasive manipulation of cells [27,28] and microparticles [29,30] through optically projected patterns serving as virtual electrodes. Hence, fabricated metal electrodes for generating a non-uniform electric field in DEP are not required [31]. In addition, ODEP can manipulate individual cells or particles, which is not easy for DEP [32]. ODEP combines the advantages of optical tweezers and DEP for the manipulation, sorting, organization, and pattern formation of micro and nano particles, offering a powerful technique for optical manipulation. Subsequently, by combining microfluidic, electrophoretic, and microscopic imaging techniques in a rotating chamber, L. Huang et al. generated a rotating electric field by applying an appropriate alternating current (AC) signal to the electrodes to drive controlled three-dimensional (3D) rotation of a single cell, enabling 3D cell imaging [33]. The area-specific membrane capacitance and cytoplasmic conductivity of different cell types were determined through tests. The 3D cell imaging of cancer cells and normal leukocytes demonstrated that there were subtle differences in the geometric parameters of the two cell types. Some cells with specific intrinsic dielectric properties can also self-rotate in a linearly polarized (i.e., non-rotating) AC field. For example, the self-rotation of Melan-A pigment cells can be induced through a specific optical electrode pattern that generates ODEP forces and a bandwidth of AC bias frequencies [34]. The self-rotation phenomenon is used to obtain the behavioral information of pigment cells so as to elucidate their physical properties, which in turn allows for the separation of pigment cells from non-pigment ones. Our group proposed a method that can extract the mechanical properties of cells by manipulating Raji cell translation and self-rotation through DEP/ODEP-based microfluidics [32,35]. An advantage of this method is that it allows individual cells to rotate in a known and controlled direction in a rotation-free electric field, without any requirement on the threshold of the electric field. This method opens up further possibilities for the label-free identification of cancer and related clinical applications. This method achieves rotational control of the cells, but relies on rotation speed data to obtain the final measurements of the physical properties. Usually, the rotation speed of cells is measured visually by manually timing their rotation under a video microscope using a stopwatch [33,35]. However, manual measurement can be tedious and prone to large errors. Therefore, an automated computer-vision-based method is required to identify the morphology of the cell accurately and quickly, and to measure its rotation speed. A number of cell rotation speed extraction methods have been proposed by various research groups. For example, a computer-based real-time machine vision algorithm was first proposed by G. De Gasperis et al. for measuring cell rotation and analyzing rotational spectra [36]. However, this method has the limitation that it can only measure in-plane rotation when only a single cell is present in the image. A multi-cell visual tracking algorithm capable of achieving an automatic process for determining cell information has been reported [37]. The algorithm uses the Sobel operator to extract horizontal and vertical gradients in order to identify cells in a computationally cumbersome manner. It is difficult for this proposed algorithm to obtain out-of-plane rotating cell speeds [37]. Our group proposed an automated algorithm based on image matching that can accurately determine the rotation speed of cancer cells in an ODEP-based microfluidic chip [27,38–42]. An optical-flow-based method was proposed in order to obtain 2D kinematic field data from the image sequences. Then, the data were back-propagated onto a 3D sphere model to calculate the rotation axis and rotation speed of the cell [43]. Both methods are effective at measuring the self-rotation speed of cells in the ODEP field, but the algorithms are too complex and computationally intensive. To improve the automation and efficiency of the cell self-rotation speed extraction, we recently developed a new area change algorithm to overcome the limitations of the existing methods. Specifically, this algorithm provides a simplified process for cell self-rotation speed extraction compared with previous algorithms. In addition, this algorithm employs three key techniques to achieve the following goals: (1) It reduces the influence of experimental conditions on the measurement results by removing vignettes and improving the contrast of cells, and it can extract a complete cell morphological profile. This demonstrates the good adaptability of the algorithm. (2) After selecting cells of interest in the first frame, the algorithm calculates the area of the cells, regardless of whether they are rotating in-plane or out-of-plane and without interference from other cells within the image. This eliminates the need to manually observe the video, and thereby reduces the reliance on the manual intervention for results and ensures higher stability. (3) The algorithm denoises the area change curve using a convolution method, allowing for more accurate extraction of the cell rotation period and more accurate measurements of the cell rotation speed.

In this paper, we proved that the self-rotation of cells can be characterized automatically, efficiently, and accurately through captured videos of multiple cells experiencing different motions using a new area change algorithm. We presented the key steps involved in the proposed algorithm in detail, including outline extraction and speed calculation based on the microscopic images of cells. We first calculated the self-rotation speed of Raji cells under a given AC bias parameter to demonstrate the specific calculation process. Then, this process was repeated to calculate the self-rotation speed of Raji cells under several other AC bias parameters.

2. Theory

In general, ODEP makes use of the light-sensitive properties of the low-conductivity optical semiconductor material to generate an electric field perpendicular to the chip between the upper and bottom electrodes of the chip. The structure of the electrodes is determined by the incident optical pattern. In addition, the optical semiconductor material has an extremely high resistance in the dark region, and most of the applied voltage falls on the hydrogenated amorphous silicon (a-Si:H) layer. The optical semiconductor material generates a large number of optically induced carriers in the lighted region, thus sharply increasing the local conductivity. When a voltage is applied, different partial pressures are generated in the bright and dark regions, so that an inhomogeneous electric field is created between the upper and lower pole plates of the chip. If suspended particles are in close proximity to this inhomogeneous electric field, the interaction between the two particles and the electrically polarized dipole moment of the liquid solution generates a force, which is called the DEP function and is defined as ODEP. Generally, when the polarization capacity of the particles is greater than that of the solution, the particles are induced to move to a higher electric field gradient by generating a positive DEP force. On the contrary, a negative dielectrophoretic force is induced to move to a lower electric field gradient. This means that DEP forces can be either positive or negative. In the case of an inhomogeneous electric field, the DEP force on a spherical particle can be expressed as [44].

$$\left\langle \vec{F}_{DEP} \right\rangle = 2\pi R^3 \varepsilon_m \mathrm{Re}[\mathrm{K}(\omega)] \nabla \left| \vec{E}rms \right| \tag{1}$$

where E_{rms} is the square root mean of the electric field strength; R is the cell radius; ε_m denotes the dielectric constant of the liquid medium; ω is the angular frequency with the expression $\omega = 2\pi f$, where f is the frequency of the voltage applied across the liquid medium; and Re $[K(\omega)]$ is the real part of the Clausius–Mossotti (CM) factor that determines the direction of the DEP force. For cells, the expression for Re $[K(\omega)]$ is usually written as

$$\mathrm{Re}[\mathrm{K}(\omega)] = \mathrm{Re}\left[-\frac{\omega^2(\tau_1\tau_2 - \tau_c\tau_2{}') + j\omega(\tau_2{}' - \tau_1 - \tau_2) - 1}{\omega^2(\tau_c\tau_2{}' + \tau_1\tau_2) - j\omega(\tau_2{}' + 2\tau_1 + \tau_2) - 2}\right] \tag{2}$$

where $\tau_c = \varepsilon_c/\sigma_c$, $\tau_1 = \varepsilon_m/\sigma_m$, $\tau_2 = C_{mem}/\sigma_c$, and $\tau_2{}' = C_{mem}/\sigma_m$; the subscripts c, mem, and m denote the cytoplasm, cell membrane, and liquid medium, respectively; C_{mem} denotes the capacitance value of the cell membrane; and ε and σ denote the dielectric constant and conductivity, respectively. When the electric field frequency approaches zero, the size of the real part of the CM factor is mainly affected by the conductivity of the solution and the particles. On the contrary, when the electric field frequency is very high, the real part of the CM factor is only affected by the dielectric constant. If the positive and negative values of the CM factor are known at different frequencies, the particles can be manipulated by modulating specific frequencies to produce positive and negative DEP forces.

In terms of cell manipulation, previous studies have not only reported the translational motion of cells in ODEP fields, but also demonstrated that certain types of cells self-rotate in linearly polarized AC fields along an axis perpendicular to the electric field lines [27,42]. Turcu et al. proposed that a theoretical analysis can explain the fact that certain cells with specific dielectric properties can self-rotate in a non-rotating electric field [45]. Based on previous studies, our group attempted to describe the rotational behavior of colored cells with different contents of intrinsic melanin in a linearly polarized AC electric field [35]. We also experimentally confirmed the rotation phenomenon by implanting foreign particles into the cells. Prior to inoculation, these cells did not initially rotate in the presence of an externally applied non-rotating AC electric field. In general, the rotation equation is defined as

$$T = \frac{9}{4} V \varepsilon_1 E_0^2 \frac{\varepsilon_r - \sigma_r}{(\varepsilon_r + 2)(\sigma_r + 2)} \left(\frac{\mathrm{X} + \mathrm{X}_0}{(\mathrm{X} + \mathrm{X}_0)^2 + 1} + \frac{\mathrm{X} - \mathrm{X}_0}{(\mathrm{X} - \mathrm{X}_0)^2 + 1} \right) \tag{3}$$

where V is the volume of cells. The other symbols are as follows:

$$\begin{cases} \varepsilon_r = \varepsilon_p/\varepsilon_m \\ \sigma_r = \sigma_p/\sigma_m \\ \omega_0\tau = \mathrm{X}_0 \\ \omega\tau = \mathrm{X} \\ \tau = (\varepsilon_p + 2\varepsilon_m)/(\sigma_p + 2\sigma_m) \end{cases} \tag{4}$$

where ω_0 is the angular frequency of the cellular motion. From the above equation, we can clearly conclude that the electrophysiological information and mechanical properties of the cell can be extracted from its behavioral information, as the self-rotation varies with frequency.

3. Materials and Methods

3.1. Experimental Setup and Working Principles

An ODEP-based microfluidic platform was designed as the experimental setup in this study. Its structure is shown schematically in Figure 1. The system has been described in detail in our previous work [42]. We used a CCD camera to record the movement of cells in the ODEP chip as an image injected through a charge-coupled projector. The ODEP chip consists of four layers from top to bottom. The top layer is a transparent film of indium tin oxide (ITO) glass; the middle part is a custom-designed microchannel with a hydrogenated amorphous silicon (a-Si:H) layer, which is connected to the top layer by a fabricated tape; and the bottom part is an ITO glass substrate with a hydrogenated amorphous silicon (a-Si:H) film deposited on top.

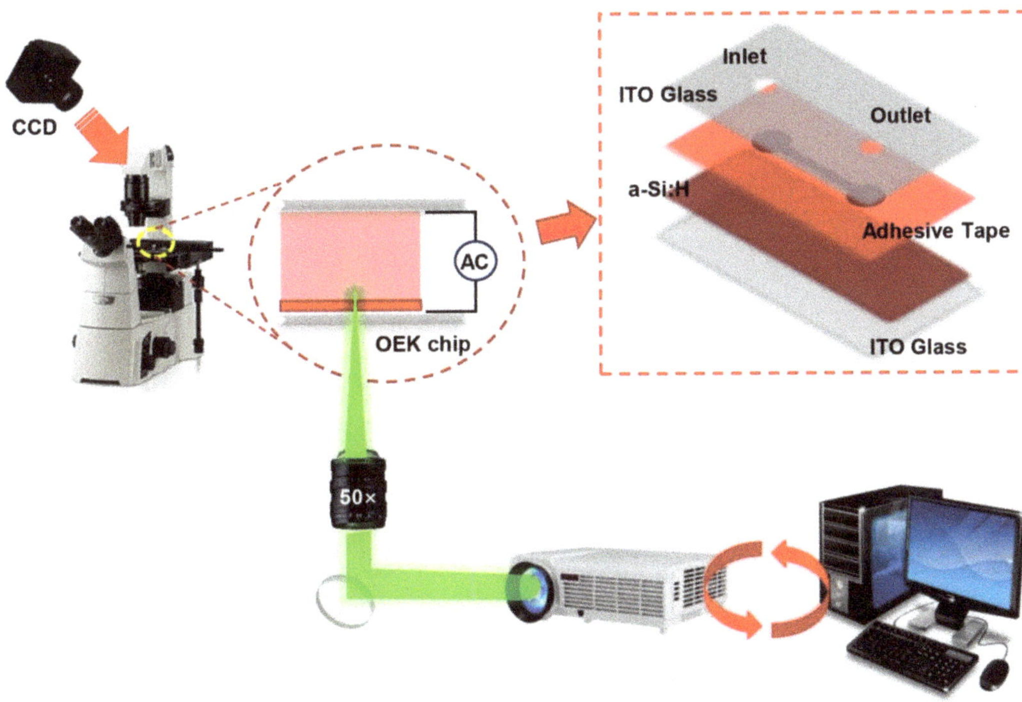

Figure 1. Schematic illustration of the ODEP-based microfluidic platform used as the experimental setup. The inset is an exploded view of the ODEP chip.

When the designed optical pattern was projected onto the a-Si:H surface, the electron–hole pairs were excited and enhanced by the migration of electrons from the valence band to the conduction band of the a-Si:H layer, resulting in a large number of optically induced carriers and a significant increase in local conductivity. When a voltage was applied, different partial voltages were generated in the light and dark regions, which in turn created an inhomogeneous electric field between the upper and lower pole plates of the chip. The electric field at the end of the liquid chamber then increased sharply above the locally illuminated a-Si:H region, as most of the applied voltage was shifted sharply into the liquid chamber. As a result, an inhomogeneous non-rotating electric field was generated in the liquid chamber.

3.2. Cell Preparation

The cells used in the experiments were Raji cells bought from the Cell Bank of the Chinese Academy of Sciences (Shanghai, China). They were suspended in the cell culture medium supplied by the Roswell Park Memorial Institute (RPMI1640, Thermo Fisher Science, Bridgewater, NJ, USA), which contained 10% (v/v) fetal bovine serum, 1% penicillin (v/v) (100 U/ml), and 1% streptomycin (v/v) (100 µg/ml), and were incubated at 37 °C in a humidified atmosphere of 5% CO_2. The diameter of the Raji cells was about 12 µm.

Before the experiment, 1 mL of the Raji cell suspension was centrifuged at 1000 rpm for 5 min at 4 °C and then the supernatant was removed. The collected Raji cells were resuspended in 1 mL of RPMI-1640 culture medium and centrifuged again with the same parameters to remove the remaining culture medium. The remaining Raji cells were then resuspended in 1 mL of isotonic solution for further experiments. The isotonic solution was prepared from 8.5% (ω/v) sucrose, 0.3% (ω/v) glucose, and 0.5% (ω/v) bovine serum albumin (BSA) in deionized water. The purpose of using BSA was to reduce the affinity between the cells and the a-Si:H substrate. The conductivity of the isotonic solution was measured using a conductivity meter (Cond3110, VWR International, RADNOR, PA, USA) at 1.5×10^{-2} S/m. The cell concentration of the cell suspension was kept constant at 1×10^5 cells/mL.

3.3. Self-Rotation Speed Extraction

Generally speaking, the projected area between each frame captured by the CCD camera varied as the cells rotated in an ODEP field. Therefore, a curve of cell area changes could be obtained by measuring the area of the rotating cells in all of the frames. As the images captured by the camera contained multiple cycles of cell rotations, the cell area change curve varied periodically. Moreover, the time required to perform a complete rotation could be obtained by selecting the trough points of adjacent cycles in the cell area change curve, i.e., the frames with the smallest area points in the adjacent rotation cycles. Therefore, the cell area change curve is a reliable method for obtaining the rotation speed of the cell.

Considering the characteristics of the cells, the imaging conditions, the low contrast between cells and the image background, the unclear cell edges, and the presence of impurities such as regionally brighter light sources and broken incomplete cells in the image background, a fluorescent microscope was used to capture the experimental images under poor lighting conditions. Through image processing, a complete and smooth cell outline was efficiently segmented without destroying the integrity of the characteristic points inside the cells. In addition, the self-rotation speed of the self-rotating cells in the cell physical field was extracted efficiently, automatically, and accurately by extracting the morphological characteristics of the cell from the 2D cell microscopic images. This algorithm contains three important steps for achieving more accurate and faster speed extraction. First, in order to extract the cell outline completely, the original cell image was vignetted to reduce errors in cell image segmentation, thus improving the accuracy of area calculation. Second, the rotating cells were selected, and the areas of those cells in each frame of the image were extracted and used to plot a cell area change curve. Third, the cell area change curve was denoised to obtain a periodically changing curve. The specific steps are described in Figure 2.

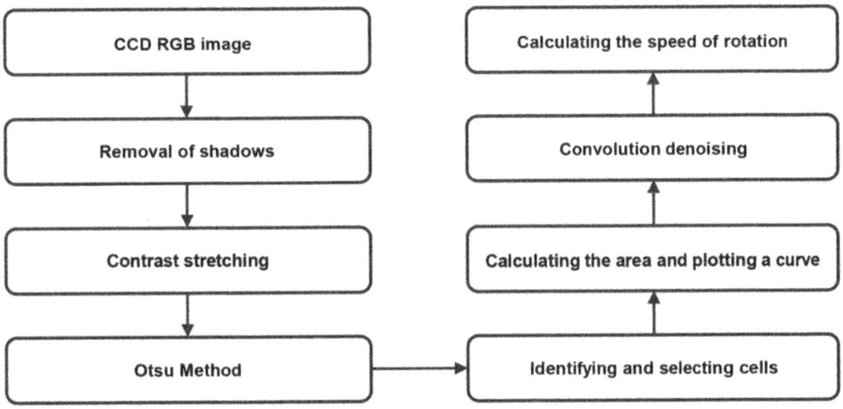

Figure 2. Flow chart of the algorithm for extracting cell speed based on area change.

3.3.1. Removing shadows

As a result of the presence of a light source in the center of the image, the corners of the image were less bright or saturated than the middle part of the image. The dark shadows in the corners of the image needed to be removed, so as to minimize the noise in the image after binarization of the final segmentation, as shown in Figure 3.

Figure 3. Binarization of cell images without removing the dark corners.

The first method we used was shadow correction based on entropy minimization [46]. Minimizing the entropy of the whole image has also been proven to be an effective method for shadow correction. However, the optimization process could cause the ordinary entropy to fall into a local optimum. Therefore, according to the concept of logarithmic entropy, if the histogram of an original normal image has a single-peaked distribution, then for an image with dark corners, there must be another distribution of low luminosity in the histogram. The purpose of correcting dark corners is to make the distribution of low luminosity closer to the original normal luminosity. Ordinary entropy calculations will not decrease until there is a partial overlap between the two histograms. Before this, the entropy is always increasing, while the logarithmic entropy at least remains constant until there is no overlap. This, in turn, offers a better way to obtain a global optimum. The luminance is first mapped logarithmically and the mapping equation is as follows

$$i(L) = (N - 1)\log(1 + L)/\log 256 \tag{5}$$

where L is the input image of cells, i is the grey-scale value, and N is the pixel value.

The first method has a slightly slower calculation speed, and typically takes 15,000 ms to process a 1392 × 1040 color map. Therefore, we used gamma correction as the second method [47]. As the gamma curve of the image was edited for non-linear tone editing of the image, the dark and bright parts of the image signal were detected and the ratio of the two was increased to improve the image contrast. Then, gamma correction enhanced the storage accuracy of the dark luminance, i.e., it removed the dark parts by increasing the luminance of the dark corners.

$$f(I_{output}) = I_{input}{}^{\gamma} \tag{6}$$

The value of γ determines how the grey scale is mapped between the input image and the output image, i.e., whether the contrast in the lower or higher grey-scale areas is enhanced. If $\gamma > 1$, this indicates enhanced contrast in the higher grey-scale areas of the image; if $\gamma < 1$, this indicates enhanced contrast in the lower grey-scale areas of the image. This results in a brightness-enhanced image.

3.3.2. Enhancing contrast

As a result of the low contrast between the cells and the background and the presence of brighter noise within the image, a segmented linear method was used to highlight the targets or grey-scale intervals of interest and suppress grey-scale areas not of interest. The grey-scale interval [a, b] was extended, and the grey-scale intervals [0, a] and [b, M_f] were compressed (as shown in Figure 4). By adjusting the position of the fold inflection points and controlling the slope of the segmented straight lines, the grey-scale intervals of interest could be extended. Ultimately, the parameters were adjusted to enhance the contrast between the cells and the background and to reduce the contrast between the light source and the background, thereby allowing for clearer observation of the cell outline.

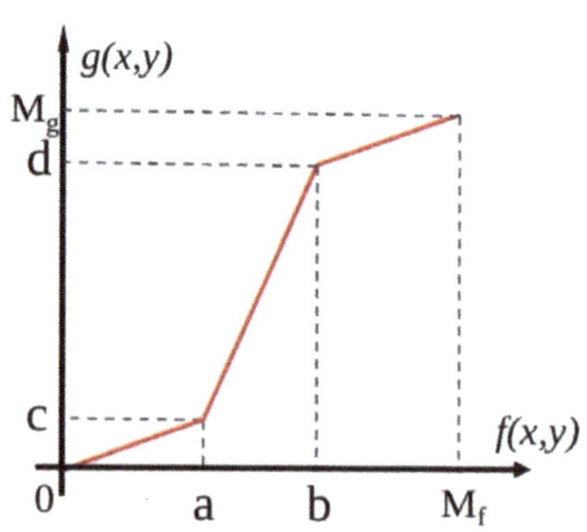

Figure 4. Grey-scale stretching curve.

As a result of the generally low concentration of grey values in the pixels, the whole image can be very dark and not quite as contrasted. With grey-scale transformation, as shown in Equation (7), we stretched the grey values to a specified interval to enhance the contrast significantly.

$$I(x,y) = \frac{I(x,y) - I_{\min}}{I_{\max} - I_{\min}}(MAX - MIN) + MIN \tag{7}$$

where I_{max} and I_{min} are the minimum and maximum grey values of the original image, respectively. *MAX* and *MIN* are the minimum and maximum grey values of the grey space, respectively.

3.3.3. Binarization

After linear stretching of the image contrast, the morphological characteristics of the cells could be distinguished more clearly. At this point, the Otsu method was used for binarization [48]. The image was segmented into two parts—the background and the foreground—according to the grey-scale characteristics of the image. As variance is a measure of the uniformity of the grey-scale distribution, the greater the interclass variance between the background and foreground, the greater the difference between the two parts of the image. When part of the foreground was mis-segmented into the background or part of the background was mis-segmented into the foreground, the interclass variance between the two parts became smaller. Therefore, the segmentation that resulted in the largest interclass variance had the lowest probability of misclassification [49]. The Otsu method is less affected by image brightness and contrast, which can further reduce the effect of low contrast in dark parts on the segmentation results.

3.3.4. Post-Processing

As observed from the binarized image, the cell outline was not smooth and complete, and there were still some impurities around the cells. Therefore, the image needed to be post-processed. Using a morphological approach, objects with small areas in the binary image were removed to reduce the excess noise. In addition, an expansion operation was performed to smooth out the cell outline. Finally, a complete and smooth cell outline was obtained.

3.3.5. Extraction of Cell Behavior

After the previous morphological manipulation, narrow junctions in the binary image were disconnected, small protrusions were eliminated, and cell contours were smoothly extracted. All adjacent white pixels in the binary image were labelled based on the traversal of the image, and pixels with the same marker formed a connected domain. White pixels in different connected domains were labelled differently so that each connected domain in the image could be extracted. Finally, all cells in the resulting binary image were identified and labelled using an external rectangle to frame the connected domains and a shape center to determine the location of the connected domains. A cell of interest was selected and extracted separately. Its area was calculated using a morphological method. When the cell was rotated in-plane, the smallest outer rectangular box of the cell was determined and the cell area within a quarter of the box was extracted. The cell rotation period was determined by the change in cell area within this section.

This experimental image had 159 frames. The above operation was performed on each frame to obtain an area change curve of this cell projected onto the 2D plane as it self-rotated in the physical field. From the area change curve, it could be tentatively determined that the area of the cell varied periodically, i.e., the speed of its self-rotation could be determined by the area change. To obtain a clearer period, the area change curve was smoothed and denoised to filter out the excess peaks using a convolution method.

Finally, the self-rotation speed was extracted through the processed area change curve. When the cells were rotated out-of-plane along a fixed axis, the area of the cell image varied at different moments during one rotation cycle. Then, the trough of a cell area change cycle was labeled where the cell area reached a minimum value. When the cell area shifted to the trough point of the next cycle, the cell would rotate for a complete week. The final rotation speed was calculated using Equation (8).

$$n = \frac{60 f_{\text{fps}}}{X_i - X_{i-1}} \tag{8}$$

where f_{fps} denotes the frame rate of the video, f_{fps} =15 fps; X_i and X_{i-1} denote the frame numbers of the wave valley points in two adjacent cycles of area change; and n is the average self-rotation speed (in rpm).

In addition, for extracting the displacement information of the cells, the frame difference method was used to detect the moving targets in each frame of the image. Tracking of the moving targets was achieved using a time-series-based prediction model and the displacement trajectory of the cell was plotted.

4. Results and Discussion

4.1. Self-Rotation Speed of Raji Cells under a Given AC Bias Parameter

In our experiment, the bias voltage was set to 10 V_{PP} and Raji cells with a diameter of approximately 12 μm were observed at a 70 kHz AC frequency. A total of 159 cell images were acquired using a CCD camera, and the last 59 frames of five cycles of stabilization were used to calculate the average rotation speed. The original image and the image with dark corners removed are presented in Figure 5. Figure 6 shows the effect of a contrast-enhanced image. The image was binarized using the Otsu method, as shown in Figure 7. Figure 8 shows the way that the cell was identified and labeled, and Figure 9 shows a selected cell of interest. Figure 10 shows the area change curve obtained based on the area calculated for all of the frames. The data selected from frames 101 to 159 were convolved and denoised to obtain a clear area cycle change curve with an average self-rotation speed of 86 rpm over five cycles, as shown in Figure 11. The motion trajectory of the cell of interest is shown in Figure 12. The method for detecting the cell area when the cell was in-plane rotation is shown in Figure 13.

Figure 5. (**a**) Original cell image taken by CCD and (**b**) cell image with dark corners removed.

Figure 6. Grey-scale image after contrast stretching.

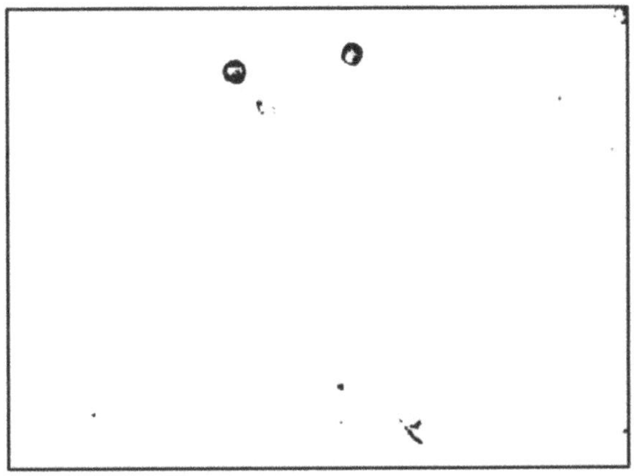

Figure 7. Image binarized using the OTSU method.

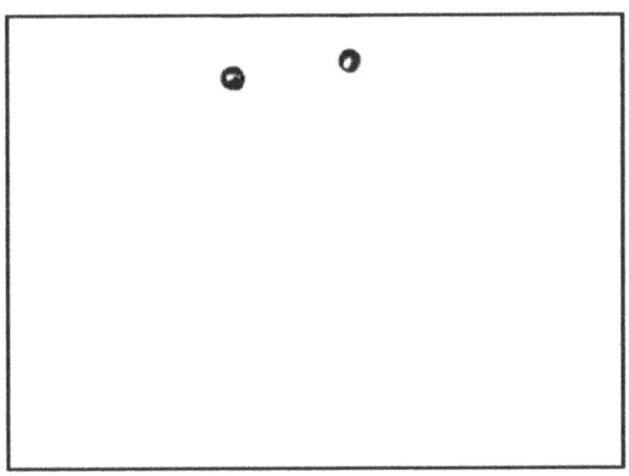

Figure 8. Complete cell outline after the image was denoised using the morphological method.

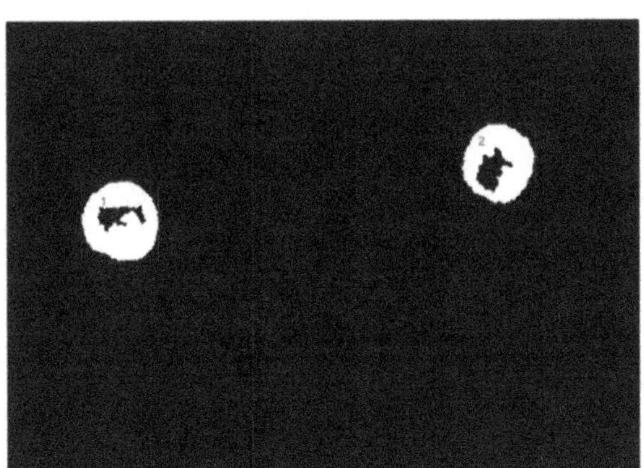

Figure 9. Cell identification and labelling.

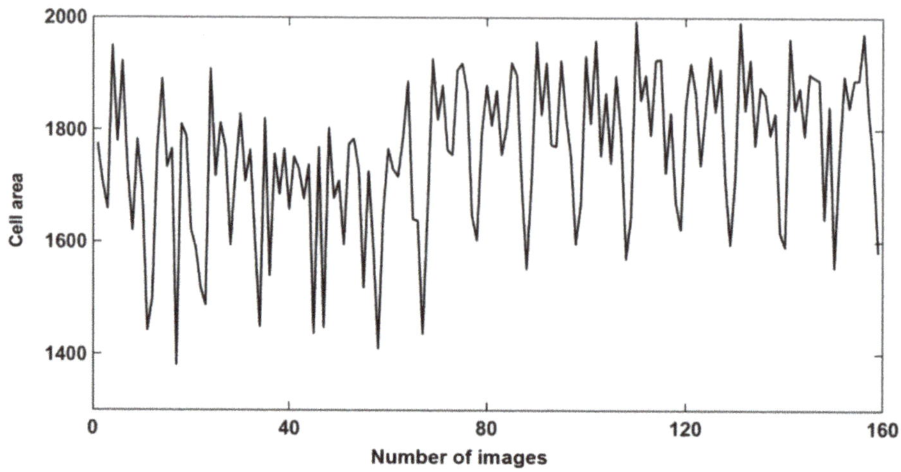

Figure 10. Area change curve based on the area of the cell of interest calculated in 159 images.

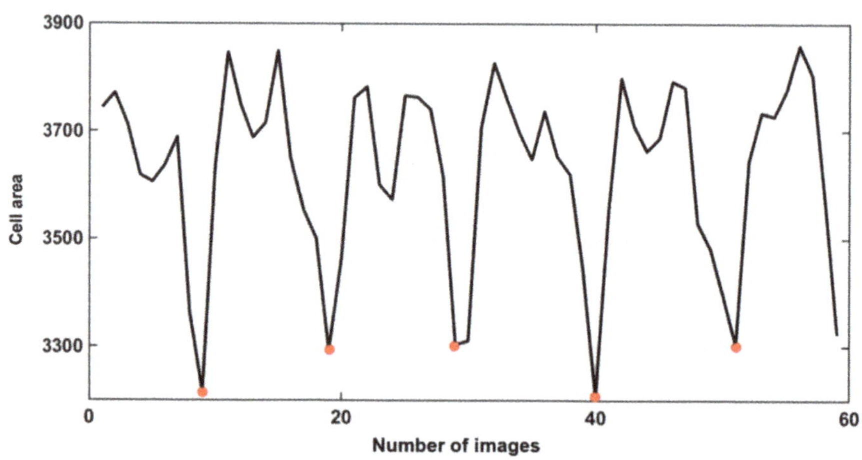

Figure 11. A smooth area change curve obtained based on data selected from the last 59 frames, which were then convolved and denoised.

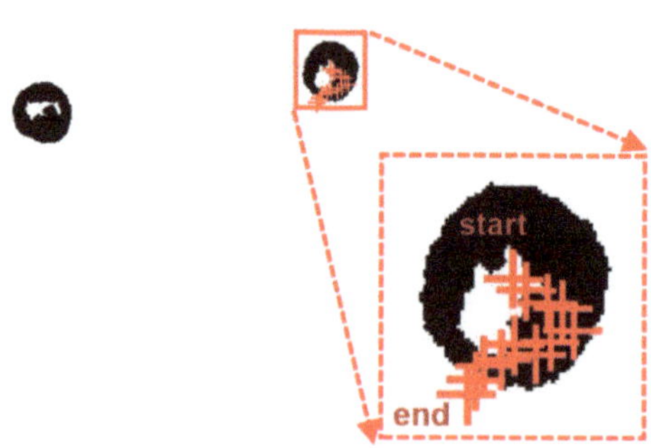

Figure 12. Motion trajectory of the cell of interest.

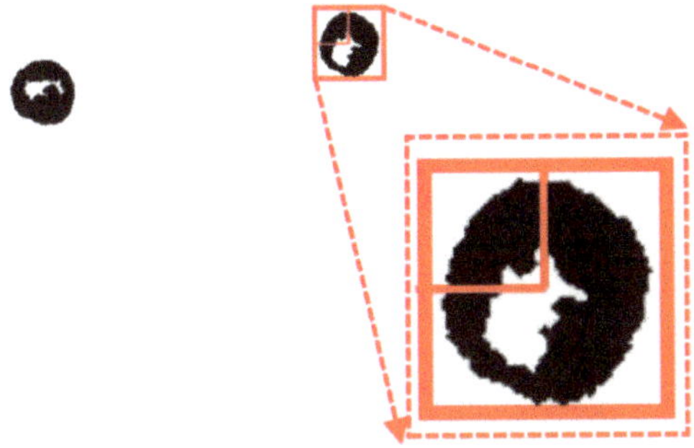

Figure 13. A method for calculating the cell area when the plane is rotated.

4.2. Self-Rotation Speed of Raji cells under Different AC Bias Parameters

By varying the applied AC frequency and bias potential, it was possible to characterize the change in the self-rotation speed of Raji cells under different AC bias parameters. Figure 14 shows the self-rotation speeds of Raji cells as a function of different AC frequencies at a voltage maintained at 10 Vpp, using only one Raji cell for the measurements. The Raji cell was capable of self-rotation motion at a frequency scope ranging from 20 kHz to 180 kHz. The extracted self-rotation speed of the Raji cell proved an increase from 20 kHz to 65 kHz and then the self-rotation speed reached a maximum value of 129.4 rpm at 65 kHz. Furthermore, the characterization of the self-rotation speed of the Raji cells with respect to the AC bias potential was also investigated, as shown in Figure 15. In this figure, the AC frequency was maintained at 60 kHz. The results demonstrated a quadratic relationship between the self-rotation speed of the Raji cells and AC bias potential.

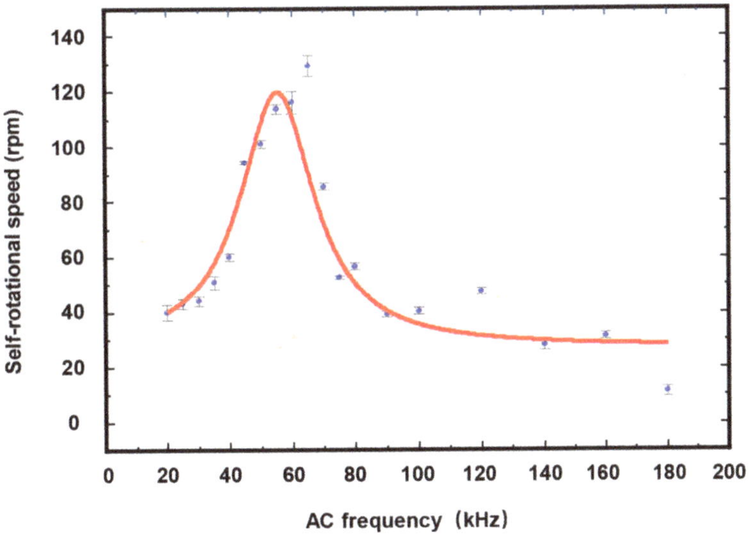

Figure 14. Self-rotational speeds of Raji cells with different AC frequencies at an external AC bias potential of 10 V_{pp}. The measurement points in this figure were obtained for one Raji cell only. At each measurement point, the period of stable continuous cell rotation was selected for measurement, and the number of cell rotation periods was {3, 3, 5, 6, 4, 5, 5, 5, 5, 3, 5, 6, 5, 5, 3, 5, 2, 2, 2, 2, 2, 3, 2}. In the figure, each data point represents a mean ± maximum deviation.

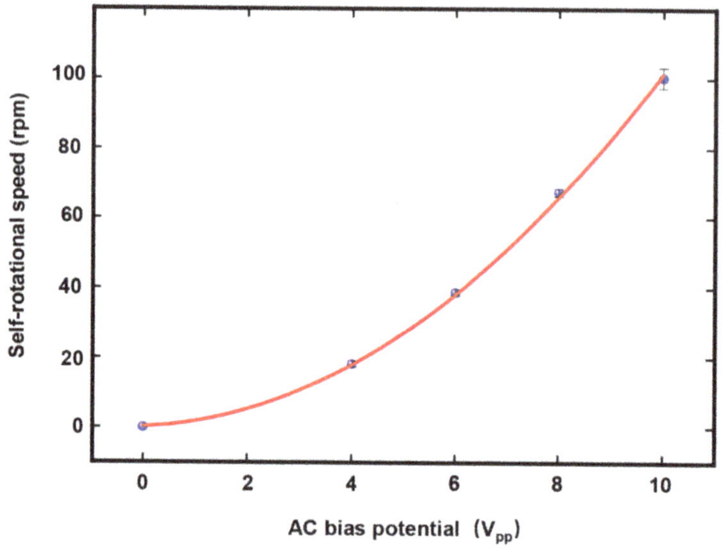

Figure 15. Self-rotational speeds of Raji cells with different AC bias potentials at an external AC frequency of 60 kHz. The measurement points in this figure were obtained for one Raji cell only. At each measurement point, the period of stable continuous rotation of the cell was selected for measurement, and the number of rotation periods for the applied AC bias potential of {4, 6, 8, 10} VPP was {1, 2, 3, 5}, respectively. In the figure, each data point represents a mean ± maximum deviation.

4.3. Discussion

The algorithm provides an effective method to calculate the self-rotation speed of cells in an ODEP field and to depict the motion trajectory of the cells. It can extract the rotation speed of cells quickly and accurately under different experimental conditions. All of the steps, except the selection of cells of interest, can be done automatically, eliminating interference with the experimental results through manual intervention. For cells with a homogeneous structure, the algorithm can calculate their rotation speed based on changes in their internal structure. In the case of in-plane rotation, the algorithm can also offer a highly adaptable and stable method for calculation. In the future, the algorithm can potentially provide some new ideas for the fast calculation of other rotational targets and rotational methods. The algorithm can also be combined with various microfluidic rotation platforms to expand its scope of application. For example, it can be used with a number of structure-based, acoustic-based, or optical-based microfluidic platforms for cell manipulation, which involve rotation and translation to extract the physiological properties of cells from their behavioral information [5–56]. Capturing cells of interest when they are rotating with high-speed translation will be a challenge that can, for example, be addressed by increasing the frame rate of cameras. In addition, multi-target rotation speed detection should be a priority area of research in future studies.

5. Conclusions

In this study, an area change algorithm was presented for the accurate extraction of self-rotation speeds and translational motions of cancer cells suspended in an ODEP chip, i.e., an AI-powered optical−electrical−microfluidic platform. The algorithm works well with microscopic images of cells under poor lighting conditions and with low contrast. The algorithm enables the automatic identification of cells and the extraction of their self-rotation speed. With a pre-processing stage that removes dark corners for increased contrast, the algorithm can clearly extract the cell outline, proving to be effective in processing cell micrographs under poor lighting conditions and with low contrast. A convolution method is used to process the area change curve, from which the cell rotation period can be determined clearly to help measure the self-rotation speed of the cells more accurately. The algorithm can process spherical and flat cells in the same image rotating at different angles by extracting the outline of the internal cell structure. In addition, it has a strong capability for processing cells rotating in-plane and out-of-plane. Furthermore, the algorithm can obtain the motion trajectory of cells of interest to allow for a more comprehensive analysis of the morphological and behavioral information of cells. Potential applications of this algorithm include separating different types of cancer cells in different populations, distinguishing normal cells from cancer cells, and characterizing the morphological and electrical properties of cells.

Author Contributions: Conceptualization, W.L.; methodology, H.W.; validation, X.Y. and J.W.; writing—original draft preparation, H.W.; writing—review and editing, R.Q., D.D., W.Y. and W.L. All authors have read and agreed to the published version of the manuscript.

Funding: This research was funded by the National Natural Science Foundation of China (grant number 61973224, 61925307, and 61803323) and the Young Top Talent of Xingliao Program (grant number XLYC2007186).

Acknowledgments: This work was supported by the National Natural Science Foundation of China (project nos. 61973224, 61925307, and 61803323) and the Young Top Talent of Xingliao Program (project no. XLYC2007186).

Conflicts of Interest: The authors declare no conflict of interest.

References

1. Kanehisa, M. Toward understanding the origin and evolution of cellular organisms. *Protein Sci.* **2019**, *28*, 1947–1951. [CrossRef] [PubMed]
2. Cohen, M.C.; Cohen, S. Cytokine Function: A Study in Biologic Diversity. *Am. J. Clin. Pathol.* **1996**, *105*, 589–598. [CrossRef] [PubMed]
3. De la Fuente, I.M.; López, J.I. Cell motility and cancer. *Cancers* **2020**, *12*, 2177. [CrossRef] [PubMed]
4. Ilyas, A.; Asghar, W.; Ahmed, S.; Lotan, Y.; Hsieh, J.-T.; Kim, Y.-T.; Iqbal, S.M. Electrophysiological analysis of biopsy samples using elasticity as an inherent cell marker for cancer detection. *Anal. Methods* **2014**, *6*, 7166–7174. [CrossRef]
5. Cadwell, C.R.; Scala, F.; Li, S.; Livrizzi, G.; Shen, S.; Sandberg, R.; Jiang, X.; Tolias, A.S. Multimodal profiling of single-cell morphology, electrophysiology, and gene expression using Patch-seq. *Nat. Protoc.* **2017**, *12*, 2531–2553. [CrossRef] [PubMed]
6. Mitchison, J.M.; Swann, M.M. The mechanical properties of the cell surface. *J. Exp. Biol.* **1954**, *31*, 443–460. [CrossRef]
7. Wu, P.H.; Aroush, D.R.B.; Asnacios, A.; Chen, W.C.; Dokukin, M.E.; Doss, B.L.; Durand, P.; Ekpenyong, A.; Guck, J.; Guz, N.V.; et al. Comparative study of cell mechanics methods. *Nat. Methods* **2018**, *15*, 491. [CrossRef]
8. Basoli, F.; Giannitelli, S.M.; Gori, M.; Mozetic, P.; Bonfanti, A.; Trombetta, M.; Rainer, A. Biomechanical characterization at the cell scale: Present and prospects. *Front. Physiol.* **2018**, *9*, 1449. [CrossRef]
9. Montell, D.J. Morphogenetic cell movements: Diversity from modular mechanical properties. *Science* **2008**, *322*, 1502–1505. [CrossRef]
10. Ford, J. Red blood cell morphology. *Int. J. Lab. Hematol.* **2013**, *35*, 351–357. [CrossRef]
11. Bukoreshtliev, N.V.; Haase, K.; Pelling, A.E. Mechanical cues in cellular signalling and communication. *Cell Tissue Res.* **2012**, *352*, 77–94. [CrossRef] [PubMed]
12. Butcher, D.T.; Alliston, T.; Weaver, V.M. A tense situation: Forcing tumour progression. *Nat. Rev. Cancer* **2009**, *9*, 108–122. [CrossRef] [PubMed]
13. Zhao, X.; Jin, L.; Shi, H.; Tong, W.; Gorin, D.; Kotelevtsev, Y.; Mao, Z. Recent advances of designing dynamic surfaces to regulate cell adhesion. *Colloids Interface Sci. Commun.* **2020**, *35*, 100249. [CrossRef]
14. Katira, P.; Zaman, M.H.; Bonnecaze, R.T. How changes in cell mechanical properties induce cancerous behavior. *Phys. Rev. Lett.* **2012**, *108*, 028103. [CrossRef] [PubMed]
15. Darling, E.M.; Di Carlo, D. High-Throughput assessment of cellular mechanical properties. *Annu. Rev. Biomed. Eng.* **2015**, *17*, 35–62. [CrossRef] [PubMed]
16. Laurell, T.; Petersson, F.; Nilsson, A. Chip integrated strategies for acoustic separation and manipulation of cells and particles. *Chem. Soc. Rev.* **2007**, *36*, 492–506. [CrossRef] [PubMed]
17. Zhang, Y.; Zhao, J.; Yu, H.; Li, P.; Liang, W.; Liu, Z.; Lee, G.-B.; Liu, L.; Li, W.J.; Wang, Z. Detection and isolation of free cancer cells from ascites and peritoneal lavages using optically induced electrokinetics (OEK). *Sci. Adv.* **2020**, *6*, eaba9628. [CrossRef]

18. Ozkan, M.; Wang, M.; Ozkan, C.; Flynn, R.; Esener, S. Optical manipulation of objects and biological cells in microfluidic devices. *Biomed. Microdevices* **2003**, *5*, 61–67. [CrossRef]
19. Stevenson, D.J.; Gunn-Moore, F.; Dholakia, K. Light forces the pace: Optical manipulation for biophotonics. *J. Biomed. Opt.* **2010**, *15*, 041503. [CrossRef]
20. Guo, F.; Mao, Z.; Chen, Y.; Xie, Z.; Lata, J.P.; Li, P.; Ren, L.; Liu, J.; Yang, J.; Dao, M.; et al. Three-dimensional manipulation of single cells using surface acoustic waves. *Proc. Natl. Acad. Sci. USA* **2016**, *113*, 1522–1527. [CrossRef]
21. Ahmed, D.; Ozcelik, A.; Bojanala, N.; Nama, N.; Upadhyay, A.; Chen, Y.; Hanna-Rose, W.; Huang, T.J. Rotational manipulation of single cells and organisms using acoustic waves. *Nat. Commun.* **2016**, *7*, 11085. [CrossRef] [PubMed]
22. Yi, C.; Li, C.-W.; Ji, S.; Yang, M. Microfluidics technology for manipulation and analysis of biological cells. *Anal. Chim. Acta* **2006**, *560*, 1–23. [CrossRef]
23. Takahashi, Y.; Miyata, S. Continuous ES/Feeder cell-sorting device using dielectrophoresis and controlled fluid flow. *Micromachines* **2020**, *11*, 734. [CrossRef] [PubMed]
24. Zhang, Z.; Zheng, T.; Zhu, R. Characterization of single-cell biophysical properties and cell type classification using dielectrophoresis model reduction method. *Sens. Actuators B Chem.* **2019**, *304*, 127326. [CrossRef]
25. Nie, X.; Luo, Y.; Shen, P.; Han, C.; Yu, D.; Xing, X. High-throughput dielectrophoretic cell sorting assisted by cell sliding on scalable electrode tracks made of conducting-PDMS. *Sens. Actuators B Chem.* **2020**, *327*, 128873. [CrossRef]
26. Chiou, P.Y.; Chang, Z.; Wu, M.C. A novel optoelectronic tweezer using light induced dielectrophoresis. In Proceedings of the 2003 IEEE/LEOS International Conference on Optical MEMS, Waikoloa, HI, USA, 18–21 August 2003; pp. 8–9.
27. Liang, W.; Zhang, K.; Yang, X.; Liu, L.; Yu, H.; Zhang, W. Distinctive translational and self-rotational motion of lymphoma cells in an optically induced non-rotational alternating current electric field. *Biomicrofluidics* **2015**, *9*, 014121. [CrossRef]
28. Liang, W.; Liu, L.; Zhang, H.; Wang, Y.; Li, W.J. Optoelectrokinetics-based microfluidic platform for bioapplications: A review of recent advances. *Biomicrofluidics* **2019**, *13*, 051502. [CrossRef]
29. Liang, W.; Wang, S.; Dong, Z.; Lee, G.-B.; Li, W.J. Optical Spectrum and electric field waveform dependent optically-induced dielectrophoretic (ODEP) micro-manipulation. *Micromachines* **2012**, *3*, 492–508. [CrossRef]
30. Liang, W.; Liu, L.; Wang, J.; Yang, X.; Wang, Y.; Li, W.J.; Yang, W. A Review on Optoelectrokinetics-Based Manipulation and Fabrication of Micro/Nanomaterials. *Micromachines* **2020**, *11*, 78. [CrossRef]
31. Liang, W.; Liu, N.; Dong, Z.; Liu, L.; Mai, J.; Lee, G.-B.; Li, W.J. Simultaneous separation and concentration of micro- and nano-particles by optically induced electrokinetics. *Sens. Actuators A Phys.* **2013**, *193*, 103–111. [CrossRef]
32. Chau, L.-H.; Liang, W.; Cheung, F.W.K.; Liu, W.K.; Li, W.J.; Chen, S.-C.; Lee, G.-B. Self-Rotation of cells in an irrotational AC E-Field in an Opto-Electrokinetics Chip. *PLoS ONE* **2013**, *8*, e51577. [CrossRef] [PubMed]
33. Huang, L.; Liang, F.; Feng, Y. A microfluidic chip for single-cell 3D rotation enabling self-adaptive spatial localization. *J. Appl. Phys.* **2019**, *126*, 234702. [CrossRef]
34. Chau, L.H.; Ouyang, M.; Liang, W.; Lee, G.B.; Li, W.J.; Liu, W.K. Inducing self-rotation of Melan-a cells by ODEP. In Proceedings of the 2012 7th IEEE International Conference on Nano/Micro Engineered and Molecular Systems (NEMS), Kyoto, Japan, 5–8 March 2012; pp. 195–199.
35. Ouyang, M.; Cheung, W.K.; Liang, W.; Mai, J.D.; Liu, W.K.; Li, W.J. Inducing self-rotation of cells with natural and artificial melanin in a linearly polarized alternating current electric field. *Biomicrofluidics* **2013**, *7*, 054112. [CrossRef] [PubMed]
36. De Gasperis, G.; Wang, X.; Yang, J.; Becker, F.F.; Gascoyne, P.R.C. Automated electrorotation: Dielectric characterization of living cells by real-time motion estimation. *Meas. Sci. Technol.* **1998**, *9*, 518–529. [CrossRef]
37. Liu, N.; Lin, Y.; Peng, Y.; Xin, L.; Yue, T.; Liu, Y.; Ru, C.; Xie, S.; Dong, L.; Pu, H.; et al. Automated Parallel Electrical Characterization of Cells Using Optically-Induced Dielectrophoresis. *IEEE Trans. Autom. Sci. Eng.* **2020**, *17*, 1084–1092. [CrossRef]
38. Yang, X.; Niu, X.; Liu, Z.; Zhao, Y.; Zhang, G.; Liang, W.; Li, W.J. Accurate Extraction of the Self-Rotational Speed for Cells in an Electrokinetics Force Field by an Image Matching Algorithm. *Micromachines* **2017**, *8*, 282. [CrossRef]
39. Liang, W.; Wang, Y.; Zhang, H.; Liu, L. Characterization of the self-rotational motion of stored red blood cells by using optically-induced electrokinetics. *Opt. Lett.* **2016**, *41*, 2763. [CrossRef]
40. Liang, W.; Zhao, Y.; Liu, L.; Wang, Y.; Li, W.J.; Lee, G.-B. Determination of Cell Membrane Capacitance and Conductance via Optically Induced Electrokinetics. *Biophys. J.* **2017**, *113*, 1531–1539. [CrossRef]
41. Liang, W.; Yang, X.; Wang, J.; Wang, Y.; Zhang, H.; Yang, W.; Liu, L. Label-free characterization of different kinds of cells using optoelectrokinetic-based microfluidics. *Opt. Lett.* **2020**, *45*, 2454–2457. [CrossRef]
42. Liang, W.; Zhao, Y.; Liu, L.; Wang, Y.; Dong, Z.; Li, W.J.; Lee, G.B.; Xiao, X.; Zhang, W. Rapid and label-free separation of Burkitt's lymphoma cells from red blood cells by optically-induced electrokinetics. *PLoS ONE* **2014**, *9*, e90827. [CrossRef]
43. Zhao, Y.; Jia, D.; Sha, X.; Zhang, G.; Li, W.J. Determination of the three-dimensional rate of cancer cell rotation in an optically-induced electrokinetics chip using an optical flow algorithm. *Micromachines* **2018**, *9*, 118. [CrossRef] [PubMed]
44. Jones, T.B. Basic theory of dielectrophoresis and electrorotation. *IEEE Comput. Graph. Appl.* **2003**, *22*, 33–42. [CrossRef] [PubMed]
45. Turcu, I. Electric field induced rotation of spheres. *J. Phys. A Math. Gen.* **1987**, *20*, 3301. [CrossRef]
46. Lopez-Fuentes, L.; Oliver, G.; Massanet, S. Revisiting Image Vignetting Correction by Constrained Minimization of Log-Intensity Entropy. In *International Work-Conference on Artificial Neural Networks*; Springer: Cham, Switzerland, 2015; pp. 450–463. [CrossRef]
47. Shi, Y.; Yang, J.; Wu, R. Reducing illumination based on nonlinear gamma correction. In Proceedings of the 2007 IEEE International Conference on Image Processing, San Antonio, TX, USA, 16 September–19 October 2007; Volume 1, pp. I-529–I-532.

48. Otsu, N. A threshold selection method from gray-level histograms. *IEEE Trans. Syst. Man Cybern.* **1979**, *9*, 62–66. [CrossRef]
49. Xu, X.; Xu, S.; Jin, L.; Song, E. Characteristic analysis of Otsu threshold and its applications. *Pattern Recognit. Lett.* **2011**, *32*, 956–961. [CrossRef]
50. Heijmans, H.J. Connected morphological operators for binary images. *Comput. Vis. Image Underst.* **1999**, *73*, 99–120. [CrossRef]
51. Huang, L.; Wang, W. 3D Electro-Rotation of single cells. *Synth. Lect. Biomed. Eng.* **2019**, *14*, 119p. [CrossRef]
52. Tang, T.; Hosokawa, Y.; Hayakawa, T.; Tanaka, Y.; Li, W.; Li, M.; Yalikun, Y. Rotation of biological cells: Fundamentals and applications. *Engineering* **2021**. [CrossRef]
53. Läubli, N.; Shamsudhin, N.; Ahmed, D.; Nelson, B.J. Controlled three-dimensional rotation of single cells using acoustic waves. *Procedia CIRP* **2017**, *65*, 93–98. [CrossRef]
54. Tang, Q.; Liang, F.; Huang, L.; Zhao, P.; Wang, W. On-chip simultaneous rotation of large-scale cells by acoustically oscillating bubble array. *Biomed. Microdevices* **2020**, *22*, 13. [CrossRef]
55. Liang, W.; Yang, X.; Wang, J.; Wang, Y.; Yang, W.; Liu, L. Determination of dielectric properties of cells using AC Electrokinetic-based microfluidic platform: A review of recent advances. *Micromachines* **2020**, *11*, 513. [CrossRef] [PubMed]
56. Løvmo, M.K.; Pressl, B.; Thalhammer, G.; Ritsch-Marte, M. Controlled orientation and sustained rotation of biological samples in a sono-optical microfluidic device. *Lab Chip* **2021**, *21*, 1563–1578. [CrossRef] [PubMed]

MDPI
St. Alban-Anlage 66
4052 Basel
Switzerland
www.mdpi.com

Micromachines Editorial Office
E-mail: micromachines@mdpi.com
www.mdpi.com/journal/micromachines

www.ingramcontent.com/pod-product-compliance
Lightning Source LLC
Chambersburg PA
CBHW041942260926
51850CB00018B/419
* 9 7 8 3 7 2 5 8 1 1 0 0 7 *